COMBINATORIAL CHEMISTRY AND MOLECULAR DIVERSITY IN DRUG DISCOVERY

COMBINATORIAL CHEMISTRY AND MOLECULAR DIVERSITY IN DRUG DISCOVERY

Edited By

ERIC M. GORDON
Versicor, Inc.
Fremont, California

and

JAMES F. KERWIN, JR.
Abbott Laboratories
Abbott Park, Illinois

 **WILEY-LISS**

A JOHN WILEY & SONS, INC., PUBLICATION
New York · Chichester · Weinheim · Brisbane · Singapore · Toronto

This book is printed on acid-free paper. ⊚

Copyright © 1998 by Wiley-Liss, Inc. All rights reserved.

Published simultaneously in Canada.

While the authors, editors and publisher believe that drug selection and dosage and the specification and usage of equipment and devices, as set forth in this book, are in accord with current recommendations and practice at the time of publication, they accept no legal responsibility for any errors or omissions, and make no warranty, expressed or implied, with respect to material contained herein. In view of ongoing research, equipment modifications, changes in governmental regulations and the constant flow of information relating to drug therapy, drug reactions, and the use of equipment and devices, the reader is urged to review and evaluate the information provided in the package insert or instructions for each drug, piece of equipment, or device for, among other things, any changes in instructions or indication of dosage or usage and for added warnings and precautions.

Library of Congress Cataloging-in-Publication Data:
Combinatorial chemistry and molecular diversity in drug discovery/edited by Eric M. Gordon and James F. Kerwin., Jr.
    p. cm.
    "A Wiley-Liss publication."
    Includes bibliographical references and index.
    ISBN 0-471-15518-7 (cloth : alk. paper)
    1. Combinatorial chemistry. 2. Drugs--Design. I. Gordon, Eric M. II. Kerwin, James F.
RS419.C656   1988
615'.19--dc21
                                     97-38036

Printed in the United States of America.

10 9 8 7 6 5 4 3 2 1

CONTENTS

FOREWORD: COMBINATORIAL CHEMISTRY AT A CROSSROADS

Walter H. Moos
MitoKor, San Diego, California

The present monograph is the most comprehensive overview yet published in combinatorial chemistry and molecular diversity (CCMD). Moreover, the authors of the individual chapters make up a veritable "Who's Who" in the field. The sections and chapters address all of the fundamentals of CCMD, with a bias toward those topics of highest interest to academic or industrial bioorganic, medicinal, organic, and pharmaceutical chemists, as well as to "drug hunters" in healthcare research settings, such as biotechnology and pharmaceutical companies. Thus, this book should admirably serve anyone who wishes to review or study the state of the art in combinatorial diversity and related biomedical arenas.

In an increasingly cost-conscious healthcare environment, the role of CCMD is clear: Help to make pharmaceutical research (and development) better, faster, and cheaper! With timeframes now exceeding 15 years from the start to the finish of the pharmaceutical R&D process, anything that helps will no doubt find a legitimate place in the industry (as well as in some departments in academia) (Figure 1). Though "faster" and "cheaper" may well pertain in many situations, "better" is perhaps the wrong label for a technology that, at its heart, must play a leading though sometimes supportive role alongside a series of already strong technology bases. Further, with chemistry research accounting for no more than about 10% of the total cost of pharmaceutical R&D, there are practical limits to the financial impact that CCMD can have on the aggregate expenses of the industry (Figure 2).

In summarizing the field, one needs to consider a variety of issues that range from history to philosophy to empiricism. Though history is often rewritten by the leading practitioners of a field, and CCMD may be no different over time,

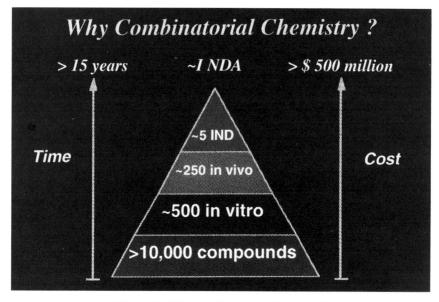

Figure 1. Why combinatorial chemistry?

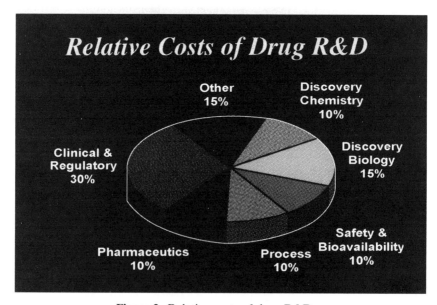

Figure 2. Relative costs of drug R&D.

a few basic origins seem clear: (1) Nature beat us to it. (2) Peptide chemists were next. (3) Many others then jumped on the bandwagon. (4) Truly viable pharmaceutical approaches have been evident only in the last couple of years.

Combinatorial chemistry and its intrinsic game of building blocks encompass many complementary technologies, all of which have matured significantly since the early days (Figure 3). Thus, the nuts and bolts of CCMD include, among other considerations, library strategies, tactics, and design principles; solid-phase methods; encoding and spatial separation paradigms; scaffold concepts; and quality assurance and control issues. The evolution of the field from peptide and other natural biooligomer and biopolymer libraries to today's wide array of heterocyclic libraries, plus extensions into unnatural polymers and materials science, bode well for CCMD continuing to play an important role in science beyond the present decade and into the next millennium.

Of course, the evolution of the field is far from complete. For example, until recently, there has been far too little attention given to the analytical chemistry side of the combinatorial story. One result? Often, groups have chased leads that were either impossible to find, or not the structure expected. It is interesting to speculate whether the field will move back heavily to solution-phase methods after the solid-phase aspects are more fully worked out (Figure 4). Perhaps this is to be expected of the entrepreneurs of the field,

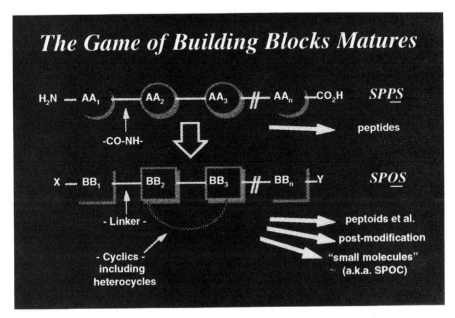

Figure 3. The game of building blocks matures.

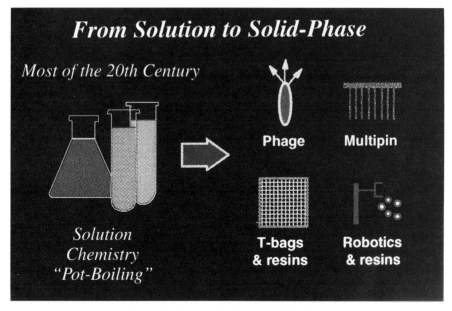

Figure 4. From solution to solid phase.

who are always looking for the next uncharted path forward. From the earliest days of deconvolutions, to various strategies for identifying leads, including encoding with taggants, the face of drug screening and lead identification has changed immeasurably and forever (Figures 5 and 6).

The informatics side of the equation appears to be on a trajectory that will allow better use of computational tools, but more needs to be done. Structure-based design of libraries is being employed increasingly, and facile combinatorial database and on-line interfaces are improving. Still, the question of "how high is up" remains more of a philosophical argument than one based on widely understood facts. There is still considerable bias as to which set of specifics is "better," for example, a library of 100 compounds versus a library of 1,000,000 compounds, when in truth each problem may require its own unique solution (Figure 7).

While drug discovery-related applications of CCMD have been mined progressively in recent years, drug development extrapolations are still in their infancy. Pharmaco- and toxicokinetics are immediately the most amenable to combinatorial techniques, and are beginning to show results.

There are some who maintain that CCMD is a tool of the 1990s only because other requisite tools, such as high-throughput screening, had to evolve first (Figure 8). There are others who point to the key role that automation and microprocessors play in most modern technology platforms. These people are all correct to some extent, but may be missing the point. Certainly all of

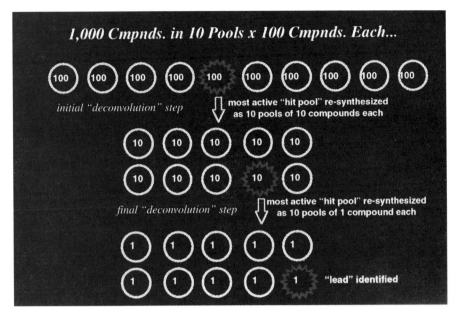

Figure 5. Deconvolution: 1000 compounds in 10 pools times 100 compounds each.

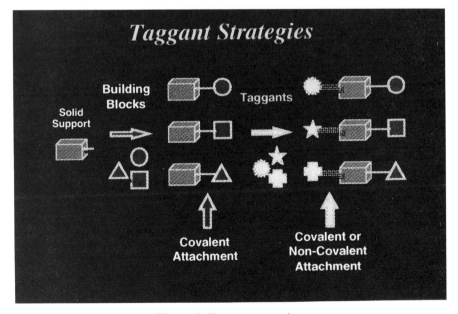

Figure 6. Taggant strategies.

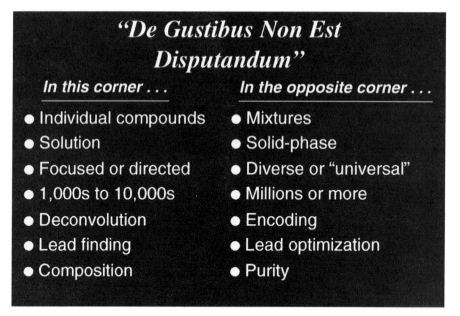

Figure 7. *De gustibus non est disputandum* (roughly: matters of taste can't be argued).

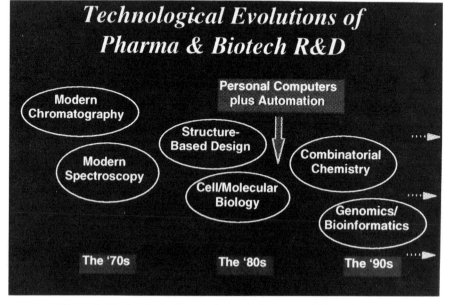

Figure 8. Technological evolutions of pharma and biotech R&D.

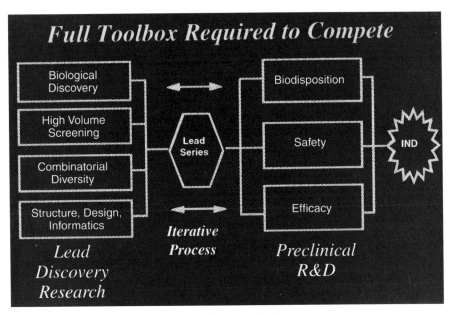

Figure 9. Full toolbox required to compete.

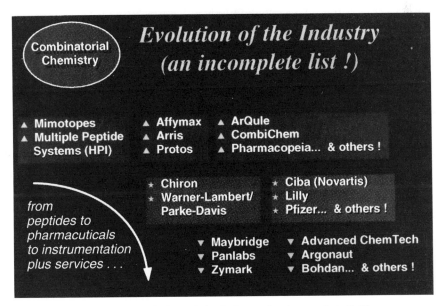

Figure 10. Evolution of the industry (an incomplete list!).

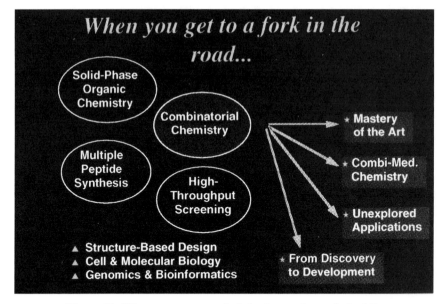

Figure 11. When you get to a fork in the road . . . (take it!).

these advances must cooperate to make the most of any one of the CCMD approaches outlined in this book. Nonetheless, at its most fundamental level, CCMD can and will stand on its own for years to come, even with the need for a complete toolbox to compete most effectively (Figure 9).

To date, CCMD has spawned an industry that has raised well over one billion dollars in capital (Figure 10). When major financial stakes exist, intellectual property positions and resulting patent battles are soon to follow, and a few cases have already been tested.

Alongside solid-phase and other chemistries, multiple-peptide and compound syntheses, high-throughput screening, and with the support of structure-based design, cell and molecular biology, and genomics and bioinformatics, CCMD may have reached a proverbial crossroads. Hopefully, the field will proceed in many new directions in the years to come (Figure 11). There is still much to do in mastering the art and in marrying the new tools with the old, and there are undoubtedly several areas as yet unanticipated or at least unexplored.

All of these topics, and more, are discussed in the chapters that follow. Enjoy!

PREFACE

The field of drug discovery stands on the threshold of a new era in which more new synthetic molecules will be prepared in the next few years, than have ever existed. Preparation of such numbers of small, organic molecules, relies on the new techniques, tools, and strategies of combinatorial chemistry. The "libraries" of compounds which emerge from combinatorial chemical syntheses are used to search for novel drug leads or to optimize the activity and properties of a known bioactive hit. Although the first successes with small molecule chemical library approaches are barely a few years old, most research groups now believe that the widespread use of combinatorial and related synthetic chemistry technologies will soon revolutionize drug discovery.

Less than five years ago almost all organic synthesis performed in the service of drug discovery was carried out by the "hand crafted" serial synthesis and testing of individual molecular entities. This status quo had evolved from many years of steady, significant, incremental improvements in synthetic organic chemistry and the analytical tools of structural characterization. Over many years, synthetic chemistry in pharmaceutical drug discovery had grown to mirror in strategy and execution the practice and philosophy of academic chemistry, specifically in recreating and practicing the highly sophisticated art of "natural product total synthesis". The late R. B. Woodward, the most dominant organic chemist of the 20th century, had an overpowering influence on creative organic chemical thinking in the latter half of the century, and the depth of acceptance of the movement he symbolized inevitably retarded initiatives that might have ultimately lead researchers to stepping outside his broadly accepted paradigm of synthesis. Most synthetic chemists who joined the American pharmaceutical industry after the late fifties had been trained to some degree in the "Woodwardian style" of synthetic chemistry and continued to practice it in industry. However, in many cases the synthetic targets of drug research were not actually related to natural materials.

By the early 1990's unprecedented pressure was being put on the world-wide drug industry to both produce more cost-effective medicines and also increase productivity. Studies showed that the typical cost of producing a new registered molecule in the major pharmaceutical companies was in the range of $7–9,000 per entity. In addition to pharmacoeconomics, a confluence of several other factors helped to set the stage for the widespread adoption, development and acceptance of combinatorial chemistry and chemical library approaches. Progress and refinements in high throughput automated screening, in which thousands to millions of compounds could be routinely evaluated by robotic techniques, exacerbated a thirst for new molecular entities to feed batteries of primary screens directed towards novel drug targets. Proceeding in chronological parallel with improvements in screening capacity, successful methods began to be developed for creating and evaluating molecular diversity based on polymeric molecules. In fact, the primary tools and technologies necessary to create peptide and oligonucleotide libraries had been available for some years, but the impetus and conceptual insights required to enable the process did not emerge until the middle 1980's.

Early efforts in polymeric molecule library production were viewed with intellectual curiosity by the traditional drug discovery community, but were generally considered less relevant to pharmaceutical discovery because the libraries were confined to peptide and oligonucleotide structures which had historically not led to drug entities. Having recently emerged from frustrating, prolonged, world-wide efforts to turn peptidomimetic leads for renin inhibitors and later, HIV protease inhibitors, into drugs; the industry was particularly wary of peptide-based leads. Interplay of the aforementioned forces along with a widespread feeling of change developing within an increasingly unsettled pharmaceutical industry, led researchers in the early 1990's to seriously ponder the future genesis of drug leads. New lead structures for drug discovery had historically emerged by screening of microbial fermentation broths, plant extracts, compound collections, and more recently by mechanism and structure based design approaches, especially when the biological target was an enzyme. Combinatorial organic synthesis, which might emulate the progress achieved with peptide libraries, began to be considered as a long range solution to cutting the time and high costs associated with serial organic synthesis of drug leads. Enthusiasm began to mount around the thought that if one could create large numbers of non-polymeric small molecules simultaneously, great leverage would be in hand for discovering new and useful molecules on an accelerated time-frame. Not far into the present decade, the idea crystallized in the minds of a number of researchers, that if sound synthetic strategies could be devised, the necessary chemical tools might already exist to efficiently create large collections of molecules within which might be found the pharmacophores of the future.

This book attempts to capture in one volume, a significant portion of the myriad of research areas into which the combinatorial chemistry field has recently spread. Because of the immense breadth of today's research on combinatorial chemistry, it has rapidly become difficult to speak with real authority except in those areas where ones own research is performed. Especially for that reason we thought it best to invite a broad collection of expert authors who each would contribute in their special area of expertise. Their collective efforts form the basis for this volume which describes as broadly as has yet been published, the young but exploding field of combinatorial chemistry.

The speed with which powerful new technology spreads in the late twentieth century is impressive to behold. The rate of progress in applying and improving these techniques is also equally staggering. There have been few historical examples of fundamental changes in the way we think about chemistry, synthesis and drug discovery, that have spread with the rapidity and impact of combinatorial chemistry. We hope that this book will be a reference volume for those already working in, or coming into the field. On the other hand, after watching this field over the past six years, we also recognize that six years hence the contents of this volume may be more of historical interest than at its cutting edge.

The material has been organized into six parts: (1) an introduction to combinatorial chemistry and molecular diversity, and sections on (2) small molecule libraries, (3) automation, analytical, and computational methods, (4) biological diversity, (5) screening, and (6) combinatorial drug screening and development. The chapters are supplemented with figures, tables, and references to related books and articles. This volume should be of interest to medicinal chemists, pharmacologists, molecular biologists, biochemists, enzymologists, and research managers in industry, academia, and government who are involved in drug discovery research.

Every project of this size is dependent on many people besides the scientific authors and editors. Since the inception of the idea for this volume, Ms. Colette Bean, Editor of John Wiley & Sons has been instrumental in moving the work forward at every stage of the project. We are indebted to Ms. Kristin Cooke and Mrs. Camille Pecoul Carter our managing editors who ably coordinated the book through the production process of copyediting, typesetting, proofreading and printing. The constant efforts of Ms. Terri Nebozuk are gratefully acknowledged. Finally, although many skilled people contributed to create this volume, the inevitable faults and omissions which are contained herein are solely those of the Editors.

Eric M. Gordon   James F. Kerwin, Jr.
Versicor, Inc.   *Abbott Laboratories*
Fremont, California   *Abbott Park, Illinois*

CONTRIBUTORS

Valery V. Antonenko, Affymax Research Institute, 3410 Central Expressway, Santa Clara, CA 95051

John J. Baldwin, Pharmacopeia, Inc., 101 College Road East, Princeton, NJ 08540

Judd Berman, Glaxo Wellcome Research Institute, Five Moore Drive, Research Triangle Park, NC 27709

Bruce A. Beutel, Pharmaceutical Products Division, Abbott Laboratories, Bldg, AP10, Dept. 4CP, 100 Abbott Park Road, Abbott Park, IL 60064-3500

Rene A. Braeckman, Chiron Corporation, 4560 Horton Street, Emeryville, CA 94608

Robert D. Brown, Pharmaceutical Research Division, Abbott Laboratories, Abbott Park, IL 60064-3500

Mark G. Bures, Pharmaceutical Research Division, Abbott Laboratories, Abbott Park, IL 60064-3500

David A. Campbell, Affymax Research Institute, 3410 Central Expressway, Santa Clara, CA 95051

John C. Chabala, Pharmacopeia, Inc., 101 College Road East, Princeton, NJ 08540

Daniel Chelsky, Pharmacopeia, Inc., 101 College Road East, Princeton, NJ 08540

Anthony W. Czarnik, IRIORI, 11149 North Torrey Pines Road, La Jolla, CA 92037

George Detre, Affymax Research Institute, 3410 Central Expressway, Santa Clara, CA 95051

Jonathan A. Ellman, Department of Chemistry, University of California at Berkeley, Berkeley, CA 94720

William L. Fitch, Affymax Research Institute, 3410 Central Expressway, Santa Clara, CA 95051

Jacqueline A. Gibbons, Chiron Corporation, 4560 Horton Street, Emeryville, CA 94608

Dane A. Goff, Chiron Corporation, 4560 Horton Street, Emeryville, CA 94608

Mikhail F. Gordeev, VERSICOR, Inc., 34790 Ardentech Court, Fremont, CA 94555

Eric M. Gordon, VERSICOR, Inc., 34790 Ardentech Court, Fremont, CA 94555

J. Russell Grove, Affymax Research Institute, 3410 Central Expressway, Santa Clara, CA 95051

Steven E. Hall, Sphinx Pharmaceuticals, A Division of Eli Lilly & Co., 4615 University Drive, Durham, NC 27707

Ronald H. Hoess, Dupont–Merck Pharmaceutical Co., Experimental Station E328/B33, Wilmington, DE 19880-0328

Christopher P. Holmes, Affymax Research Institute, 4001 Miranda Avenue, Palo Alto, CA 94304

Russell J. Howard, Affymax Research Institute, 3410 Central Expressway, Santa Clara, CA 95051

Jeffery W. Jacobs, VERSICOR, Inc., 34790 Ardentech Court, Fremont, CA 94555

Stephen W. Kaldor, Lilly Research Laboratories, A Division of Eli Lilly & Co., Lilly Corporate Center, Indianapolis, IN 46285

James F. Kerwin, Pharmaceutical Products Division, Abbott Laboratories, Abbott Park, IL 60064-3500

Chaitan Khosla, Departments of Chemistry and Chemical Engineering, Stanford University, Stanford, CA 94305-5025

Nicolay Kulikov, Affymax Research Institute, 3410 Central Expressway, Santa Clara, CA 95051

Kit Lam, Arizona Cancer Center, Tucson, AZ 85737

Michael Lebl, Trega Biosciences, San Diego, CA 92121

Gary C. Look, Affymax Research Institute, 3410 Central Expressway, Santa Clara, CA 95051

Derek Maclean, Affymax Research Institute, 3410 Central Expressway, Santa Clara, CA 95051

Eric J. Martin, Chiron Corporation, 4560 Horton Street, Emeryville, CA 94608

Yvonne C. Martin, Pharmaceutical Products Division, Abbott Laboratories, Abbott Park, IL 60064-3500

Houng-Yau Mei, BioOrganic Chemistry Section, Department of Chemistry, Parke–Davis Pharmaceutical Research, Division of Warner–Lambert Company, Ann Arbor, MI 48105

Walter H. Moos, MitoKor, 11494 Sorento Valley Road, San Diego, CA 92121

Edmund J. Moran, Ontogen Corp., 2325 Camino Vida Roble, Carlsbad, CA 92009

Marc Navre, Affymax Research Institute, 3410 Central Expressway, Santa Clara, CA 95051

Michael Needels, Affymax Research Institute, 4001 Miranda Avenue, Palo Alto, CA 94304

Cindy Nguyen, Affymax Research Institute, 3410 Central Expressway, Santa Clara, CA 95051

Zhi-Jie Ni, VERSICOR, Inc., 34790 Ardentech Court, Fremont, CA 94555

Dinesh V. Patel, VERSICOR, Inc., 34790 Ardentech Court, Fremont, CA 94555

Michael R. Pavia, Millenium Pharmaceuticals Inc., Canbridge, MA 02139

Lutz S. Richter, Chiron Corporation, 4560 Horton Street, Emeryville, CA 94608

John Schullek, Affymax Research Institute, 4001 Miranda Avenue, Palo Alto, CA 94304

Bruce Seligmann, SIDDCO, Tucson, AZ 85718

Lihong Shi, Affymax Research Institute, 3410 Central Expressway, Santa Clara, CA 95051

Miles G. Siegel, Lilly Research Laboratories, A Division of Eli Lilly & Co., Lilly Corporate Center, Indianapolis, IN 46285

Nolan H. Sigal, Pharmacopeia, Inc., 101 College Road East, Princeton, NJ 08540

Michael J. Sofia, Transcell Technologies, Inc., 8 Cedar Brook Drive, Cranbury, NJ 08512

Kerry L. Spear, Chiron Corporation, 4560 Horton Street, Emeryville, CA 94608

Jeffrey Sugarman, Affymax Research Institute, 4001 Miranda Avenue, Palo Alto, CA 94304

Arathi Sundaram, Affymax Research Institute, 3410 Central Expressway, Santa Clara, CA 95051

Steven A. Sundberg, Affymax Research Institute, 3410 Central Expressway, Santa Clara, CA 95051

Eric W. Taylor, Chiron Corporation, 4560 Horton Street, Emeryville, CA 94608

Alex A. Virgilio, Department of Chemistry, University of California at Berkeley, Berkeley, CA 94720

David Weininger, Daylight Chemical Information Systems, Inc., Santa Fe, NM 87501

Zhengyu Yuan, VERSICOR, Inc., 34790 Ardentech Court, Fremont, CA 94555

Ronald N. Zuckerman, Chiron Corporation, 4560 Horton Street, Emeryville, CA 94608

PART I

COMBINATORIAL CHEMISTRY AND MOLECULAR DIVERSITY: AN INTRODUCTION

1

HISTORICAL OVERVIEW OF THE DEVELOPING FIELD OF MOLECULAR DIVERSITY

JOHN C. CHABALA

Pharmacopeia, Inc., Princeton, New Jersey

Molecular diversity is the differences in physical properties that exist among different molecules. These properties can be expressed as differences in shape or size, polarity and charge, lipophilicity, polarizability, or flexibility. Since these properties are inherent to the individual molecules, molecular diversity does not apply to the bulk properties of collections of molecules* or to supermolecular materials. The existence of living systems and all their variations in form and action arise from the molecular diversity of their constituent biomolecules. Biological materials were among the first sources of molecular diversity that were exploited by humans, primarily for their medicinal properties. A wide variety of valuable medicinal plants and microorganisms have been identified, and powerful active principles have been isolated from these and other biological sources, materials such as morphine to relieve pain, quinine to prevent malaria, penicillin to treat bacterial infection, and cortisone to reduce inflammation. Although the molecular diversity of natural products was the first source of useful drugs, the growth of the chemical industry in the late nineteenth and early twentieth centuries provided a source of unnatural,

---

* This does not imply that useful variations in bulk properties cannot be generated by combinatorial approaches. See Xiang X-D, Sun X, Briceño G, Lou Y, Wang K-A, Chang H, Wallace–Freedman WG, Chen S-W, Schultz PG (1995): A combinatorial approach to materials discovery. *Science* 268:1738–1740.

Combinatorial Chemistry and Molecular Diversity in Drug Discovery, Edited by
Eric M. Gordon and James F. Kerwin, Jr.
ISBN 0-471-15518-7   Copyright © 1998 by Wiley-Liss, Inc.

synthetic compounds that were evaluated for utility in treating disease. From these pioneering efforts arose drugs such as antipyrene and aspirin to reduce fever and pain, chloroform and ether to induce anesthesia, and salversan to treat syphilis.

Discovery of the anesthetic properties of chloroform and the antisyphilitic properties of salversan represent, respectively, early successes in screening and optimization of drugs. The paradigm set by these approaches has been refined over the course of the twentieth century, but remains largely unchanged in its fundamental principles. Modern drug discovery now focuses on specific biomolecular targets that are believed to be important in disease, and systematically tests large numbers of compounds from both natural and synthetic sources in an attempt to identify a lead compound that favorably alters the function of the biomolecular target. Such lead compounds rarely possess a set of properties that makes them useful as drugs, and they must be systematically optimized by varying their functional groups in order to improve potency, selectivity, oral bioavailability, toxicity, and so on. This approach has had enormous success, having identified important drugs, including Prozac, Tagamet, Capoten, and Claritin.

The competition to discover new drugs has always been great, but the pressure to identify, optimize, develop, and market novel drugs more rapidly and cost effectively than in the past has recently intensified. Cost containment by governmental agencies and managed care organizations has reduced profit margins, particularly for drugs that have little therapeutic advantages or generic equivalents. Yet the productivity of pharmaceutical research and development, ranging between 12 and 30 new drugs each year, has not changed between 1976 and 1994, despite a 10-fold increase in spending.[1] The ability to identify novel drugs is presently not limited by the availability of novel biological targets. Molecular biological approaches have identified a plethora of potentially useful targets, and advances in genomics are exponentially expanding the number of such targets. The key limiting factor in identifying new drug candidates is the availability of diverse collections of chemical compounds.

Although a wide variety of sources of natural products has been accessed and directed searches for new sources in unusual ecological niches and in ethnobotanical materials have been undertaken, the rate of discovery of truly novel natural product drugs has actually decreased. While many pharmaceutical companies have assembled sample collections of synthetic compounds, these collections largely arose from the optimization efforts on previous programs and consequently contain a limited diversity of structural types. The largest of these collections hold a few hundred thousand compounds which can be completely evaluated in modern high-throughput assays within a few months. The promise of rational drug design has not been fully realized, since this approach is limited by the availability of soluble or crystalline targets, the need for structures of target–effector complexes, and the insufficient precision of current computational models.

Thus need for additional sources of molecular diversity is particularly acute. The new and rapidly developing field of molecular diversity has capitalized on advances in synthetic chemistry, solid-phase synthesis, microlithography, molecular biology, and automation and robotics to enable the preparation of large numbers of highly diverse molecules. Although the application of modern molecular diversity generation has been largely confined to the identification of new pharmaceuticals, this approach can be fruitfully applied to any endeavor that seeks to identify molecules with specific molecular properties. Other areas that will likely benefit from the availability of increased molecular diversity are agrochemicals, flavors and fragrances, separation science, material science, and catalysis.

MOLECULAR DIVERSITY FROM BIOLOGICAL SOURCES

Among the first sources of increased biomolecular diversity was the manipulation of microorganisms to alter the fermentation yields of a natural product or its analogs. These techniques, which evolved from the 1940s, first varied the fermentation conditions and later employed chemical or radiological mutagenesis of the producing microorganism. Supplying unnatural compounds in controlled growth media sometimes provided unusual products, but the results were largely unpredictable. DNA synthesis and molecular biological techniques of cloning and expression, developed during the late 1970s and early 1980s, provided powerful new tools for increasing diversity of biogenic materials. As the biosynthetic pathways of complex natural products were elucidated it became possible to manipulate the biosynthetic enzymes to produce novel substances. In 1987 Malpartida et al.[2] demonstrated that polyketide biosynthetic genes could be homology cloned from among different *Streptomyces* species and suggested that these synthetic genes could yield novel natural products by recombination or mutagenesis. This prediction has been proven correct, and recently several microbiological systems have been described that permit the generation of novel "unnatural natural products" by the combinatorial cloning of biosynthetic enzymes into microorganisms stripped of their natural biosynthetic machinery. Both condensed[3] and macrocyclic[4] products have been created, and such approaches promise the ability to access whole new collections of molecules, which not only increase molecular diversity but also can form the substrates for subsequent combinatorial chemical elaboration.

The desire to more rapidly define the epitope selectivity of antigens provided a strong impetus for generating large collections of peptides, and the first such approach utilized display of peptides on the surface of phage particles. In 1985, Smith[5] reported the cloning, expression, and detection of a foreign peptide on the surface of a bacteriophage, and in 1988 Parmley and Smith[6] suggested that a library of randomly created peptide sequences could form an "epitope library" for mapping antibody specificities. The creation of such peptide libraries containing from about 10^7 to more than 10^8 members, and their successful screening for binding to proteins[7-9] was independently reported by three groups in 1990. Large phage display libraries have since been successfully prepared and screened against a wide variety of important biological targets. Phage display has also been employed to create libraries of proteins and protein domains and has had a revolutionary impact on antibody engineering, an area that has recently been reviewed.[10,11] In 1992, Cull et al.[12] described the peptide-on-plasmid technique that permits the intracellular expression, lysis and isolation for testing, and clonal expansion of peptides on the C-terminus of a fusion protein. Unlike phage display, the peptide-on-plasmid approach allows the preparation of peptide libraries with free carboxy termini. Recently, Mattheakis et al.[13] have reported an *in vitro* transcription and translation system based on *E. coli* wherein decapeptide libraries of 10^{12} members are created as RNA–peptide–polysome complexes that can be screened and the structures of active peptides deduced by conversion of the bound RNA to cDNA, which is subsequently amplified by PCR. These advances in the creation of peptides, active proteins, and mutagenized libraries of proteins will open new avenues for analyzing the factors governing binding and catalysis and will permit the generation of new proteins with specific properties.

Libraries of peptides have been a major focus of biogenic diversity, but libraries of oligonucleotides have also been prepared.[14] Using an approach that self-selects and amplifies active members, researchers have identified a variety of oligonucleotides that bind to peptide ligands[15] and enzymes.[16,17] This area is the subject of a recent review.[18]

MOLECULAR DIVERSITY FROM CHEMICAL SYNTHESIS

Although biological systems can generate enormously large collections of compounds, such collections utilize a small set of reaction types and synthons and consequently access only a small fraction of potential molecular diversity. The highly efficient reactions utilized by biological systems are exquisitely sensitive to the structures of the reactants, and small variations in structure frequently lead to greatly diminished yields. Chemical synthesis can employ a far broader set of starting materials and reactions than biological systems, and consequently can in principle create far greater diversity of structures. Oligomeric biomolecules offer the advantage of ease of detection and structure determination, in part because biologically generated reagents can be employed and in part because there is a long history of microtechniques for their analysis. Although easily prepared, biooligomers generally possess poor oral bioavailability, membrane penetration, and metabolic stability, which limits their utility as drugs. In spite of advances in separation and detection techniques, the isolation and structure determination of nonoligomeric natural products such as polyketides remains a tedious and time-consuming process.

All of the 10 largest selling drugs in 1994 were small organic molecules, i.e., compounds with molecular weights of <700. Eight of these were heterocycles and one, Mevacor, was a natural product. Because of the greater likelihood of identifying small molecules with oral bioavailability, central nervous system penetration, or appropriate duration of action, there is a great incentive to develop methods to increase the molecular diversity of sets of such molecules. Two key challenges must be met to harness the full potential of the molecular diversity of synthetic organic compounds. The first is to develop a variety of reactions that reliably produce useful yields of products from the broadest set of starting materials. The second challenge is to efficiently identify the synthetic products, especially in large collections of molecules, sometimes referred to as the "address problem." Progress in addressing these issues began with the chemical synthesis of biooligomers, but within the past several years collections of $>10^4$ small organic molecules have been prepared, and several strategies have been successfully developed for identifying individual members of these libraries.

The synthetic chemical generation of molecular diversity is conceptually and experimentally rooted in peptide synthesis. The chemical synthesis of peptides on polymeric supports, solid-phase synthesis, first reported by Merrifield[19] in 1963, demonstrated the advantages of increased yield and facile isolation by this technique. Over the subsequent 30 years vast improvements have been made in the efficiency of peptide synthesis by employing optimized protecting groups, coupling reagents, solid-phase linking groups, and polymeric supports. The extension of solid-phase synthesis to the preparation of small organic molecules has occurred only within the past decade, in part because the pressures to rapidly prepare large numbers of such molecules have only recently greatly increased, and in part because the synthesis of small organic compounds required the optimization of novel varied organic reactions on solid phase. In addition, the solvents and reaction conditions employed in general organic synthesis required resins with swelling properties different from the resins typically employed in biooligomer synthesis. A major advance in resin support was made by Bayer and Rapp,[20] who described Tentagel, a polyethylene glycol grafted polystyrene resin that has proven to have wide utility in general organic solid-phase synthesis. The great variety of linkers developed originally for peptide synthesis has facilitated the construction of heterocycles and other small organic compounds on solid phase, and several improvements have been described, including versatile o-nitrobenzyl photolabile linkers[21] and linkers that leave hydrogen, rather than heteroatoms, on the cleavage product.[22] Improved analytical techniques employing NMR of ^{13}C-enriched materials,[23] as well as direct ^{1}H NMR using magic-angle techniques[24] have greatly assisted in the optimization of reactions on solid phase. Single resin bead FTIR has also been employed to follow reactions.[25] As more research groups focus on solid-phase synthesis, more useful techniques for conveniently following the course of reactions will appear, as evidenced, for example, by reports of qualitative and quantitative colorimetric reagents developed to assist in following reactions on solid phase.[26]

Parallel Synthesis

Using traditional manual organic synthesis a chemist can usually prepare only about 25 to 50 compounds per year. The simultaneous synthesis of multiple products in discrete reaction vessels, often with the aid of robotics or other automation, is known as parallel synthesis. Parallel synthesis has increased chemical productivity, and hence accessible molecular diversity, 100-fold or more. The first molecules prepared using parallel synthesis were peptides, whose rapid synthesis was motivated by the same goal as phage display: the desire to rapidly identify continuous peptide epitopes. Geysen and coworkers in 1984 reported the simultaneous synthesis of 96 peptides on pins in a format compatible with microtiter plate assays[27] originally developed for serological analysis. As with phage display, this technique was soon expanded beyond epitope mapping to explore the specificities of peptide receptors.[28,29] This approach has subsequently been expanded to include synthesis on a wide variety of solid supports. In 1985 Houghten[30] reported the synthesis of peptides on resin beads held in porous polypropylene bags, the so-called tea-bag method. This is a particularly flexible technique that has been utilized to prepare over 150 peptides[31] simultaneously in multimilligram quantities. The development of commercial multiple peptide and oligonucleotide synthesis machines and pipetting robots[32,33] has facilitated the rapid parallel synthesis of hundreds to thousands of molecules.

Among the first reports of small organic parallel synthesis was the preparation of benzodiazepines and hydantoins reported in 1993 by DeWitt et al.,[34] who used a proprietary semiautomated synthesis device wherein reactions are performed on resin beads contained within fritted glass reactors. A library of benzodiazepines prepared on polymeric pins was reported by Bunin and Ellman in 1994.[35] During the past two years an exponentially increasing number of publications have reported general organic reactions performed on solid phase,[36] typically on resin beads. These reactions include such valuable operations as reductive amination[37,38]; Heck,[39] Stille,[40,41] and Suzuki[42,43] coupling; enolate alkylation[42]; hydride reductions[44]; cycloaddition reactions[45–47]; and carbohydrate synthesis,[48] to list but a few. The growing interest in solid-phase synthesis in academic and industrial laboratories portends an ever-increasing growth of the synthetic repertoire on solid phase and a consequent explosive growth in molecular diversity. Multiple peptide synthesis machines have been modified for use in general organic synthesis, and several groups in Germany and the United States are constructing machines to perform parallel synthesis.

Combinatorial Synthesis

The combinatorial synthesis approach to creating molecular diversity is fundamentally different from parallel synthesis in that combinatorial synthesis performs multiple reactions in one reaction vessel and can create all possible products from a set of reactants by performing many fewer discrete reaction steps. The simplest manifestation of this approach is to allow several reagents to react in solution at the same time to form all possible products. This approach is usually unproductive unless the reagents are few and their reactivities are carefully matched, because the most rapidly reacting reagent combinations form the bulk of the product, and the separation and isolation of products quickly becomes limiting. However, this approach can be employed in those cases where reaction rates are approximately equal, as in the preparation of a library of over 97,000 members by reaction of mixtures of amines with 9,9-dimethylxanthene-2,4,5,7-tetracarboxylic acid tetrachloride as reported by Carell et al.[49] The identification of individual active compounds from this library required the repetitive resynthesis and retesting of the most active smaller subsets of the library.[50]

Combinatorial synthesis is generally performed on solid phase, either as a two-dimensional array or on resin particles, because solid-phase synthesis somewhat simplifies the problem of differential reactivity and product identification. While the simultaneous coupling of multiple peptides has also been reported,[51] the different rates of coupling of amino acid pairs nevertheless makes it difficult to achieve equimolar amounts of products. The first example of combinatorial synthesis in two-dimensional arrays was an extension of synthesis on pins known as the "mimetope strategy" reported by Geysen et al. in 1986.[45,52] In this approach, one or more amino acids in a peptide are held constant while others are coupled as mixtures, permitting the generation of library pools of $>10^8$ peptides. The most active pool determines the best amino acid at a given variable position, and repetitive resynthesis and testing of smaller pools permits the elucidation of active sequences. A particularly elegant form of combinatorial synthesis is light-directed spatially addressable parallel chemical synthesis reported by Fodor et al. in 1991.[53] Photolithography combined with Merrified chemistry creates up to 65,000 individual reaction centers on a solid surface. Peptides are constructed at each center by selected photodeprotection and use of opaque masks, followed by peptide coupling. A single peptide is created at each site whose sequence is known from its coordinate location.

A revolutionary advance in combinatorial synthesis is the portion-mixing approach of resin beads initially described by Furka and coworkers in 1988,[54-56] a technique also known as split synthesis[57] or divide, couple, and recombine.[58] This approach involves dividing a portion of resin into a number of lots, allowing each lot to react with a different synthon, and then mixing all the lots. The mixed resin lots are then redivided and each new lot that contains all the products of the first step is allowed to react with a different synthon to produce all possible products. In its simplest form, the total number of compounds prepared is the product of the number of lots in which each step is divided. For example, if the procedure were performed five times with division into lots of 10, 8, 12, 9, and 13 each, a total of $10 \times 8 \times 12 \times 9 \times 13 = 112,320$ compounds would be produced in only 52 separate reactions. Houghten and coworkers have extensively utilized split synthesis and have prepared libraries of mixtures of up to 50 million different peptides.[59] Active sequences can be determined by a deconvolution approach similar to that employed with combinatorial libraries prepared on pins. However, this approach has been refined by generating sublibraries wherein the amino acid at a given position is held constant in a sublibrary others are varied. Simultaneous biological assessment of all pools indicates amino acids preferred at a given position and expedites the identification of the most potent sequences.[60] The split-synthesis process has been automated using a robotic synthesizer.[61]

Although split synthesis can rapidly generate large libraries of compounds, the efficient identification of the active members of these libraries still presents a significant challenge. Split synthesis has generally employed resin beads of <140-μm diameter that bind <400 pmol of an organic compound. If the compound is a natural peptide or oligonucleotide its structure can be determined directly using the amounts available from a single resin bead. Structure elucidation is much more difficult with other types of compounds, although advances in mass spectrometry can assist in the identification of peptides[62] and unnatural oligomers such as oligo-N-alkylglycines (peptoids),[63,64] or of mixtures of small numbers of compounds.[65] However, large libraries of non-oligomeric small molecules present structure elucidation problems for which no direct solutions currently exist. To address this problem, several approaches have been developed that utilize encoding molecules whose structures represent the structure of the library member. Encoding combinatorial peptide libraries with DNA codes was first proposed by Brenner and Lerner in 1992[66] and was put into practice by Needles et al.[67] the following year. Peptides themselves have also been used as encoding molecules. In these cases the encoding molecules are cosynthesized with the ligand at each step of split synthesis, and the ligand structure is determined by sequencing the encoding oligomer.[68,69] In 1993, Still et al.[70] reported an alternative approach that utilized a set of small molecule tags that are employed as a binary coding system for split synthesis. The tags contain electrophoric groups that are sensitively analyzed by gas chromatography employing electron capture detection. Still and coworkers have utilized this technique to identify several peptidosteroid receptors demonstrating selectivity towards target peptides.[71] Using a recently described improvement whereby the tags are attached via carbene insertions directly into the bead matrix,[72] Baldwin et al.[73] prepared an encoded library of 1179 benzopyrans, and Burbaum et al.[74] identified nanomolar small molecule inhibitors of carbonic anhydrase. The latter workers demonstrated an iterative encoded library synthesis approach that further optimized these inhibitors for isozyme selectivity.

FUTURE DIRECTIONS

Although molecular diversity has had a long history, only within the past decade has the focus shifted to the preparation of a wide variety of organic compounds. The development of novel solid-phase organic reactions is increasing exponentially, and new techniques continue to be developed to prepare and identify organic compounds prepared in parallel and combinatorial synthesis. The manipulation of the biosynthetic machinery of microorganisms promises to be an increasing source of novel structures, and the marriage of combinatorial chemical synthesis with these novel unnatural products will expand accessible molecular diversity yet further. The advent of machines to perform general organic synthesis will accelerate the discovery and optimization of novel drugs, as will the application of encoding techniques. Recently, the use of radio-frequency transponders has been suggested as a means of tracking particles used for combinatorial synthesis, and research directed toward preparing and testing large numbers of small quantities of organic compounds on microfabricated surfaces has been initiated. Combinatorial approaches are likely to be extended beyond pharmaceutical applications to separation science, flavors and fragrances, and agrochemicals. The process of increasing molecular diversity may well become in a sense autocatalytic as combinatorial approaches are applied to the rapid identification of catalysts of the very reactions used to prepare libraries of compounds. The application of molecular biology and genomics that has dominated drug discovery will no longer find chemistry the limiting commodity, and much more effort will be expended to develop novel ultrahigh-throughput screening techniques to efficiently and rapidly evaluate the wealth of organic compounds that will become available.

REFERENCES

1. Pharmaceutical Research and Manufacturers of America (January 1995): *New Drug Approvals in 1994.*

2. Malpartida F, Hallam SE, Kieser HM, Motamedi H, Hutchinson CR, Butler MJ, Sugden DA, Warren M, McKillop C, Bailey CR, et al. (1987): Homology between Streptomyces genes coding for synthesis of different polyketides used to clone antibiotic biosynthetic genes. *Nature* 325:818–821.

3. McDaniel R, Ebert–Khosla S, Hopwood DA, Khosla C (1993): Engineered biosynthesis of novel polyketides. *Science* 262:1546–1550.

4. Kao CM, Katz L, Khosla C (1994): Engineered biosynthesis of a complete macrolactone in a heterologous host. *Science* 265:509–512.

5. Smith GP (1985): Filamentous fusion phage: novel expression vectors that display cloned antigens on the virion surface. *Science* 228:1315–1317.

6. Parmley SF, Smith GP (1988): Antibody-selectable filamentous fd phage vectors: affinity purification of target genes. *Gene* 73:305–318.

7. Scott JK, Smith GP (1990): Searching for peptide ligands with an epitope library. *Science* 249:386–390.

8. Cwirla S, Peters EA, Barrett RW, Dower WJ (1990): Peptides on phage: a vast library of peptides for identifying ligands. *Proc Natl Acad Sci USA* 87:6378–6382.

9. Devlin JJ, Panganiban LC, Devlin PE (1990): Random peptide libraries: a source of specific protein binding molecules. *Science* 249:404–406.

10. Hoogenboom HR, Marks JD, Griffiths AD, Winter G (1992): Building antibodies from their genes. *Immunol Rev* 130:41–68.

11. Soderlind E, Simonsson AC, Borrebaeck CAK (1992): Phage display technology in antibody engineering of phagemid vectors and in vitro maturation systems. *Immunol Rev* 130:109–124.

12. Cull MG, Miller JF, Schatz PJ (1992): Screening for receptor ligands using large libraries of peptides linked to the C terminus of the lac repressor. *Proc Natl Acad Sci USA* 89:1865–1869.

13. Mattheakis LC, Bhatt R, Dower WJ (1994): An in vitro polysome display system for identifying ligands from very large peptide libraries. *Proc Natl Acad Sci USA* 91:9022–9026.

14. Irvine D, Tuerk C, Gold L (1991): SELEXION: systematic evolution of ligands by exponential enrichment with integrated optimization by nonlinear analysis. *J Mol Biol* 222:739–761.

15. Nieuwlandt D, Wecker M, Gold L (1995): In vitro selection of RNA ligands to substance P. *Biochemistry* 34:5651–5659.

16. Allen P, Worland S, Gold L (1995): Isolation of high-affinity RNA ligands to HIV-1 integrase from a random pool. *Virology* 209:327–336.

17. Schneider DJ, Feigon J, Hostomsky Z, Gold L (1995): High-affinity ssDNA inhibitors of the reverse transcriptase of type 1 human immunodeficiency virus. *Biochemistry* 34:9599–9610.

18. Gold L (1995): Oligonucleotides as research, diagnostic, and therapeutic agents. *J Biol Chem* 270:13581–13584.

19. Merrifield RB (1963): Solid phase peptide synthesis, I: the synthesis of a tetrapeptide. *J Am Chem Soc* 85:2149–2154.

20. Bayer E, Rapp W (1992): Polystyrene-immobilized PEG chains: dynamics and application in peptide synthesis, immunology, and chromatography. In Harris JM, ed. *Poly(Ethylene Glycol) Chemistry.* New York: Plenum, pp 325–345.

21. Holmes CP, Jones DG (1995): Reagents for combinatorial synthesis: development of a new o-nitrobenzyl photolabile linker for solid phase synthesis. *J Org Chem* 60:2318–2319.

22. Plunkett MJ, Ellman JA (1995): A silicon-based linker for traceless solid-phase synthesis. *J Org Chem* 60:6006–6007.

23. Look GC, Holmes CP, Chinn JP, Gallop MA (1994): Methods for combinatorial organic synthesis: the use of fast ^{13}C NMR analysis for gel phase reaction monitoring. *J Org Chem* 59:7588–7590.

24. Fitch WL, Detre G, Holmes CP, Shoolery JN, Keifer PA (1994): High-resolution ^{1}H NMR in solid-phase organic synthesis. *J Org Chem* 59:7955–7956.

25. Yan B, Kumaravel G, Anjaria H, Wu A, Petter RC, Jewell CF Jr, Wareing JR

(1995): Infrared spectrum of a single resin bead for real-time monitoring of solid-phase reactions. *J Org Chem* 60:5736–5738.

26. Chu SS, Reich SH (1995): NPIT: a new reagent for quantitatively monitoring reactions of amines in combinatorial synthesis. *Bioorg Med Chem Lett* 5:1053–1058.

27. Geysen HM, Meloen RH, Barteling SJ (1984): Use of peptide synthesis to probe viral antigens for epitopes to a resolution of a single amino acid. *Proc Natl Acad Sci USA* 81:3998–4002.

28. Wang JX, Bray AM, DiPasquale AJ, Maeji NJ, Geysen HM (1993): Application of the multipin peptide synthesis technique for peptide receptor binding studies: substance P as a model system. *Bioorg Med Chem Lett* 3:447–450.

29. Spellmeyer DC, Brown S, Stauber GB, Geysen HM, Valerio R (1993): Endothelin receptor ligands: replacement net approach to SAR determination of potent hexapeptides. *Bioorg Med Chem Lett* 3:519–524.

30. Houghten RN (1985): General method for the rapid solid-phase synthesis of large numbers of peptides: specificity of antigen-antibody interactions at the level of individual amino acids. *Proc Natl Acad Sci USA* 82:5131–5135.

31. Beck–Sickinger AG, Durr H, Jung G (1991): Semi-automated T-bag peptide synthesis using 8-fluorenylmethoxycarbonyl strategy and benzotriazol-1-yltetramethyluronium tetrafluroborate activation. *Pept Res* 4:88–94.

32. Schnorrenberg G, Gerhardt H (1989): Fully automatic simultaneous multiple peptide synthesis in micromolar scale — rapid synthesis of series of peptides for screening in biological assays. *Tetrahedron* 45:7759–7764.

33. Gausepohl H, Boulin C, Kraft M, Frank RW (1992): Automated multiple peptide synthesis. *Pept Res* 5:315–320.

34. DeWitt SH, Kiely JK, Stankovic CJ, Schroeder MC, Cody DMR, Pavia MR (1993): "Diversomers": an approach to nonpeptide, nonoligomeric chemical diversity. *Proc Natl Acad Sci USA* 90:6909–6913.

35. Bunin BA, Ellman JA (1994): The combinatorial synthesis and chemical and biological evaluation of a 1,4-benzodiazepine library. *Proc Natl Acad Sci USA* 91:4708–4712.

36. Henderson I, Baldwin JJ (1996): Recent advances in the generation of small molecule combinatorial libraries: encoded split-synthesis and solid-phase synthetic methodology. *Med Res Rev* 16:391–405.

37. Gordon DW, Steele J (1995): Reductive alkylation on a solid phase: synthesis of a piperazinedione combinatorial library. *Bioorg Med Chem Lett* 5:47–50.

38. Stanková M, Issakova O, Sepetov NF, Krchnák V, Lam KS, Lebl M (1994): Application of one-bead one-structure approach to identification of nonpeptidic ligands. *Drug Dev Res* 33:146–156.

39. Yu K-L, Deshpande MS, Vyas DM (1994): Heck reaction in solid-phase synthesis. *Tetrahedron Lett* 35:8919–8922.

40. Deshpande MS (1994): Formation of carbon–carbon bond on solid support: application of the Stille reaction. *Tetrahedron Lett* 35:5613–5614.

41. Plunkett MJ, Ellman JA (1995): Solid-phase synthesis of structurally diverse 1,4-benzodiazepine derivatives using the Stille coupling reaction. *J Am Chem Soc* 117:3306–3307.

42. Backes BJ, Ellman JA (1994): Carbon–carbon bond-forming methods on solid

support: utilization of Kenner's "safety-catch" linker. *J Am Chem Soc* 116:11171–11172.

43. Frenette R, Friesen RW (1994): Biaryl synthesis via Suzuki coupling on a solid support. *Tetrahedron Lett* 35:9177–9180.

44. Kurth MJ, Ahlberg Randall LA, Chen C, Melander C, Miller RB (1994): Library-based lead compound discovery: antioxidants by an analogous synthesis/deconvolutive strategy. *J Org Chem* 59:5862–5864.

45. Murphy MM, Schullek JR, Gordon EM, Gallop MA (1995): Combinatorial organic synthesis of highly functionalized pyrrolidines: identification of a potent angiotensin converting enzyme inhibitor from a mercaptoacyl proline library. *J Am Chem Soc* 117:7029–7030.

46. Beebe X, Schore NE, Kurth MJ (1995): Polymer-supported synthesis of cyclic ethers: electrophilic cyclization of isoxazolines. *J Org Chem* 60:4196–4203.

47. Beebe X, Chiappari CL, Olmstead MM, Kurth MJ, Schore NE (1995): Polymer-supported synthesis of cyclic ethers: electrophilic cyclization of tetrahydrofuroisoxazolines. *J Org Chem* 60:4202–4212.

48. Danishefsky SJ, McClure KF, Randolph JT, Ruggeri RB (1993): A strategy for the solid-phase synthesis of oligosaccharides. *Science* 260:1307–1309.

49. Carell T, Wintner EA, Bashir–Hashemi A, Rebek J Jr (1994): A novel procedure for the synthesis of libraries containing small organic molecules. *Angew Chem Int Ed Engl* 33:2059–2061.

50. Carell T, Wintner EA, Rebek J Jr (1994): A solution-phase screening procedure for the isolation of active compounds from a library of molecules. *Angew Chem Int Ed Engl* 33:2061–2064.

51. Geysen HM, Rodda SJ, Mason TJ (1986): A priori delineation of a peptide which mimics a discontinuous antigenic determinant. *Mol Immunol* 23:709–715.

52. Geysen HM, Rodda SJ, Mason TJ, Tribbick J, Schoofs PG (1987): Strategies for epitope analysis using peptide synthesis. *J Immunol Methods* 102:259–274.

53. Fodor SPA, Read JL, Pirrung MC, Stryer L, Lu AT, Solas D (1991): Light-directed, spatially addressable parallel chemical synthesis. *Science* 251:767–773.

54. Furka Á, Sebestyén F, Asgedom M, Dibó G (1988): Cornucopia of peptides by synthesis. *Abstr 14th Int Congr Biochem, Prague, Czechoslovakia* 5:47.

55. Furka Á, Sebestyén F, Asgedom M, Dibó G (1988): More peptides by less labor. *Abstr 10th Int Symp Med Chem., Budapest, Hungary* 288.

56. Furka Á, Sebestyén F, Asgedom M, Dibó G (1991): General method for rapid synthesis of multicomponent peptide mixtures. *Int J Pept Protein Res* 37:487–493.

57. Lam KS, Salmon SE, Hersh EM, Hruby VJ, Kazmierski WM, Knapp RJ (1991): A new type of peptide library for identifying ligand-binding activity. *Nature* 354:82–84.

58. Houghten RA, Pinilla C, Blondelle SE, Appel JR, Dooley CT, Cuervo, JH (1991): Generation and use of synthetic peptide combinatorial libraries for basic research and drug discovery. *Nature* 354:84–86.

59. Houghten RA, Appel JR, Blondelle SE, Cuervo JH, Dooley CT, Pinilla C (1992): The use of synthetic peptide combinatorial libraries for the identification of bioactive peptides. *BioTechniques* 13:412–421.

60. Pinilla C, Appel JR, Blanc P, Houghten RA (1992): Rapid identification of high-

affinity peptide ligands using positional scanning synthetic peptide combinatorial libraries. *BioTechniques* 13:901–905.

61. Zuckermann RN, Kerr JM, Siani MA, Banville SC, Santi DV (1992): Identification of highest affinity ligands by affinity selection from equimolar peptide mixtures generated by robotic synthesis. *Proc Natl Acad Sci USA* 89:4505–4509.

62. Brummel CL, Lee INW, Zhou Y, Benkovic SJ, Winograd N (1994): A mass spectrometric solution to the address problem of combinatorial libraries. *Science* 264:399–402.

63. Simon RJ, Martin EJ, Miller SM, Zuckermann RN, Blaney JM, Moos WH (1994): Using peptoid libraries [oligo N-substituted glycines] for drug discovery. *Techniques Protein Chem* 5:533.

64. Zambias RA, Boulton DA, Griffin PR (1994): Microchemical structural determination of a peptoid covalently bound to a polymeric bead by matrix-assisted laser desorption ionization time-of-flight mass spectrometry. *Tetrahedron Lett* 35:4283–4286.

65. Carell T, Wintner EA, Sutherland AJ, Rebek J, Jr, Dunayevskiy YM, Vouros P (1995): New promise in combinatorial chemistry: synthesis, characterization, and screening of small-molecule libraries in solution. *Chem Biol* 2:171–183.

66. Brenner S, Lerner RA (1992): Encoded combinatorial chemistry. *Proc Natl Acad Sci USA* 89:5381–5383.

67. Needels MC, Jones DG, Tate EH, Heinkel GL, Kochersperger LM, Dowe WJ, Barrett RW, Gallop MA (1993): Generation and screening of an oligonucleotide-encoded synthetic peptide library. *Proc Natl Acad Sci USA* 90:10700–10704.

68. Kerr JM, Banville SC, Zuckermann RN (1993): Encoded combinatorial peptide libraries containing non-natural amino acids. *J Am Chem Soc* 115:2529–2531.

69. Nikolaiev V, Stierandova A, Krchnak V, Seligmann B, Lam KS, Salmon SE, Lebl M (1993): Peptide-encoding for structure determination of nonsequenceable polymers within libraries synthesized and tested on solid-phase supports. *Pept Res* 6:161–170.

70. Ohlmeyer MHJ, Swanson RN, Dillard LW, Reader JC, Asouline G, Kobayashi R, Wigler M, Still WC (1993): Complex synthetic chemical libraries indexed with molecular tags. *Proc Natl Acad Sci USA* 90:10922–10926.

71. Boyce R, Li G, Nestler P, Suenaga T, Still WC (1994): Peptidosteroid receptors for opioid peptides: sequence-selective binding using a synthetic receptor library. *J Am Chem Soc* 116:7955–7956.

72. Nestler HP, Bartlett PA, Still WC (1994): A general method for molecular tagging of encoded combinatorial libraries. *J Org Chem* 59:4723–4724.

73. Baldwin JJ, Burbaum JJ, Henderson I, Ohlmeyer MHJ (1995): Synthesis of a small molecule library encoded with molecular tags. *J Am Chem Soc* 117:5588–5589.

74. Burbaum JJ, Ohlmeyer MHJ, Reader JC, Henderson I, Dillard LW, Li G, Randle TL, Sigal NH, Chelsky D, Baldwin JJ (1995): A paradigm for drug discovery employing encoded combinatorial libraries. *Proc Natl Acad Sci USA* 92:6027–6031.

2

STRATEGIES IN THE DESIGN AND SYNTHESIS OF CHEMICAL LIBRARIES

ERIC M. GORDON

Versicor Inc., South San Francisco, California

In the past few years there has been an intense interest in the generation of small molecule libraries through the techniques of combinatorial chemistry, the rapid synthesis of large numbers of diverse, nonpolymeric, organic molecules. In this chapter we analyze problems posed by the implementation of combinatorial chemistry as the technology specifically relates to drug discovery. We identify productive strategies and define some of the pivotal points in the creation of small molecule libraries. Combinatorial chemistry is a type of synthetic strategy that leads to large chemical libraries. Chemical libraries are intentionally created collections of differing molecules that may be screened for sets of preselected criteria. In the case of drug discovery, such criteria usually involve some form of molecular recognition and/or biological activity. The combinatorial process proceeds by the systematic interconnection of a set, or sets, of chemical building blocks. Building blocks (BBs) are small, reactive molecules that may be intercombined with each other or with members of other BB families (amines, carboxylic acids, aldehydes, etc.) in varied ways to eventually produce a combinatorial library containing many different products. Thus by employing a building block collection (vide infra), and systematically assembling these blocks in many combinations it is possible to create chemical libraries as populations of molecules.

Combinatorial chemistry may be envisioned as a collection of strategies,

Combinatorial Chemistry and Molecular Diversity in Drug Discovery, Edited by
Eric M. Gordon and James F. Kerwin, Jr.
ISBN 0-471-15518-7   Copyright © 1998 by Wiley-Liss, Inc.

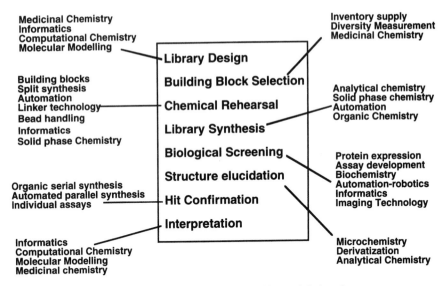

Figure 2.1. Critical technologies in the combinatorial chemistry process.

technologies, synthesis techniques, instruments, screening methods, and chemical and computational analysis tools, which may be appropriately configured to produce chemical libraries and extract relevant information contained therein (Figure 2.1). With perhaps the exception of a significant engineering/instrumental component in combinatorial technologies, the disciplines that compose the work have not really changed from traditional drug discovery research, but the degree of cross-disciplinary interaction has dramatically increased. Hence, in contrast to more traditional approaches, combinatorial chemistry involves a high degree of multidisciplinary integration.

ISSUES IN LIBRARY DESIGN

Strategic Considerations

Synthetic chemical library methods made their first appearance in the development of peptide libraries. The ready availability of a large and structurally diverse range of amino acid building blocks, a highly refined, generic coupling chemistry, and the fact that small peptides are biologically and pharmaceutically key molecules focused early efforts on peptide chemistry as a useful vehicle for exploring the power and procedural issues attached to combinatorial ligand discovery. Figure 2.2 illustrates the conceptual jump that had to be made in the early 1990s. The top of Figure 2.2 represents the interconnection of various building blocks by a single repetitive coupling chemistry, as to form peptides or oligonucleotides. The challenge faced was to use similar concepts

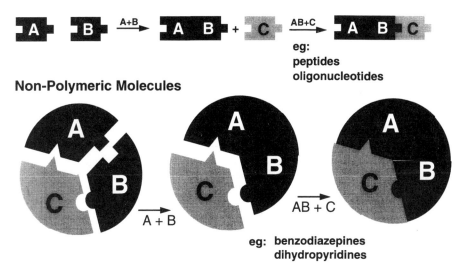

Figure 2.2. Combinatorial chemistry.

and techniques of this early type of "beads on a string" approach, but to apply them to producing libraries of nonpolymeric small molecules as illustrated at the bottom of Figure 2.2.

A primary objective of producing small molecule libraries by combinatorial chemistry is to provide collections of compounds suitable for drug discovery screening or drug development optimization. When complete, the combinatorial drug discovery exercise should have created a stable population of low molecular weight entities, free of reactive and toxicity-causing functionality. In the early 1990s, the central unanswered question might have been Can we routinely make libraries of nonpolymeric small molecules? By the mid 1990s, this question had been answered to the satisfaction of most workers. A new key question has emerged for drug researchers: How are small molecule combinatorial technologies to be best utilized for drug discovery?

A striking aspect of enabling combinatorial chemistry approaches compared to more traditional (serial) drug discovery schemes is the apparent discontinuity in strategies and tactics that must be brought to bear. Though the principles underlying chemical reactions are invariant, the strategy and tactics of combinatorial organic synthesis as it relates to drug discovery diverges markedly from those of synthesis done serially. Whereas synthesis done traditionally in medicinal chemistry serves the goal (usually) of producing a single product of previously specified structure, the goal of combinatorial chemistry is to create searchable *populations* of molecules.

To move beyond polymeric diversity into the realm of nonpolymeric small molecule libraries, we need to meet at least two main requirements: first, it

Combinatorial Synthesis of Small Molecules

Expand Scope of Solid Phase Organic Chemistry

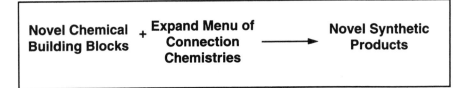

Novel Chemical Building Blocks + **Expand Menu of Connection Chemistries** ⟶ **Novel Synthetic Products**

Known Pharmacophores
"Pharmacophores of the Future"

Figure 2.3. Generalizing combinatorial organic synthesis.

is necessary to expand the scope of applicable solid-phase interconnection chemistries; and second a large and diverse supply of chemical building blocks needs to be available. The interconnection of these building blocks by solid-phase synthetic organic chemistry to yield novel synthetic libraries that contain member molecules of pharmaceutical interest is discussed below (Figure 2.3). Until recently, most solid-phase synthesis was directed toward peptide and oligonucleotide preparation. In the recent past, widespread efforts in many laboratories have shown that most organic transformations can be adapted to solid support. Some of the types of organic transformations performed in our laboratories at Affymax during the period 1991–1995 are listed in Figure 2.4. This represents but a small fraction of the scope of organic chemistry

Types of reactions

- Cycloadditions; [3+2]; [2+2]; [4+2]
- Condensations
- Photocleavages
- Oxidations / Reductions
- Michael Reactions
- Nucleophilic displacements
- Reductive aminations

- Alkylations - N, S, O, C
- Organometallic
- Mitsunobu Reactions
- Hydrolysis
- Phosphorylations
- Acylations - N, O

Figure 2.4. Combinatorial organic synthesis: solid-phase chemistry.

adapted to solid phase. In extrapolating the rate of progress visible in the recent chemical literature, it seems likely that a large body of solid-phase chemistries will soon be available to the combinatorial chemist. Chapter 16 of this book addresses this issue in detail.

Building Blocks

An essential starting point for the generation of molecular diversity is an assortment of small, reactive molecules, which may be considered chemical building blocks. The universe of structural diversity accessible through assembly of even a small set of building block elements is potentially large, and unleashing the power inherent in the building block approach is crucial to the success of the combinatorial method. The building block argument is easily illustrated, and its implications are profound. Theoretically, the number N of different individual compounds contained in a library prepared by an ideal combinatorial synthesis is determined by two factors: the number of blocks available for each step, b; and the number of synthetic steps in the reaction scheme, x. If an equal number of building blocks are used in each reaction step, then $N = b^x$. If the number of building blocks for each step varies (e.g., b, c, d in a 3-step synthesis), then $N = bcd$. Exploitation of a basis set of (for example) 100 interchangeable building blocks permits the theoretical synthesis of 100 million tetrameric or 10 billion pentameric chemical entities (see Figure 2.5).

Building blocks are the fuel that drives the combinatorial chemistry process. The judicious selection of building blocks for combinatorial library synthesis

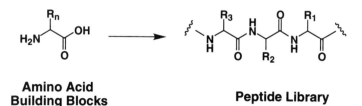

Amino Acid Building Blocks   **Peptide Library**

Basis Set of 100 Building Blocks
Tripeptide : 100^3 = 1 million cpds
Tetrapeptide : 100^4 = 100 million
Pentapeptide : 100^5 = 10 billion

Basis Set of 1000 Building Blocks
Tripeptide: 1000^3 = 1 billion cpds
Tetrapeptide : 1000^4 = 1 trillion
Pentapeptide : 1000^5 = 1 quadrillion

Figure 2.5. Creating chemical diversity from a set of building blocks.

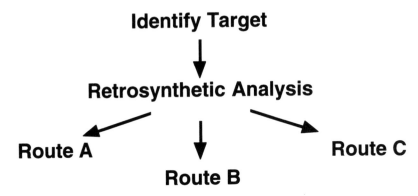

Figure 2.6. Retrocombinatorial analysis.

is a crucial decision that directly impacts the potential for successful library preparation as well as the scope of molecular diversity that is ultimately achieved. The acquisition of BBs is a resource-intensive activity. Once formed, these collections should be fully leveraged by their integration into many varied synthetic schemes and into the production of various libraries, which collectively form a useful portfolio of molecular diversity.

Building block strategies vary markedly depending on whether the intended library is for random screening or for lead development. Building blocks for focused libraries are clearly not randomly selected, and must be derived by a retrosynthetic analysis of the target molecule around which the library is to be exploded (Figure 2.6). Imposed on an analysis of various retrosynthetic disconnection's possibilities, two questions must be asked: Is the chemistry in the forward direction viable, and are the requisite building blocks available or accessible? Focused library preparation puts a great deal of importance on answering both of these questions. A simple example of the retrocombinatorial approach is illustrated in Figure 2.7. Straightforward dissection of a tripeptidyl phosphonate leads to the conclusion that protected BB sets of aminophospho-nates, α-hydroxy acids, and amino acids are necessary to prosecute the synthesis in the forward direction. While many BBs of these types may be acquired, frequently specific building blocks will not be commercially available (i.e., protected aminophosphonates). In this case, such building blocks must be synthesized serially or in parallel.

Building blocks come in chemical families, such as amines, carboxylic acids, and aromatic aldehydes. A critical mass of individual blocks in each family

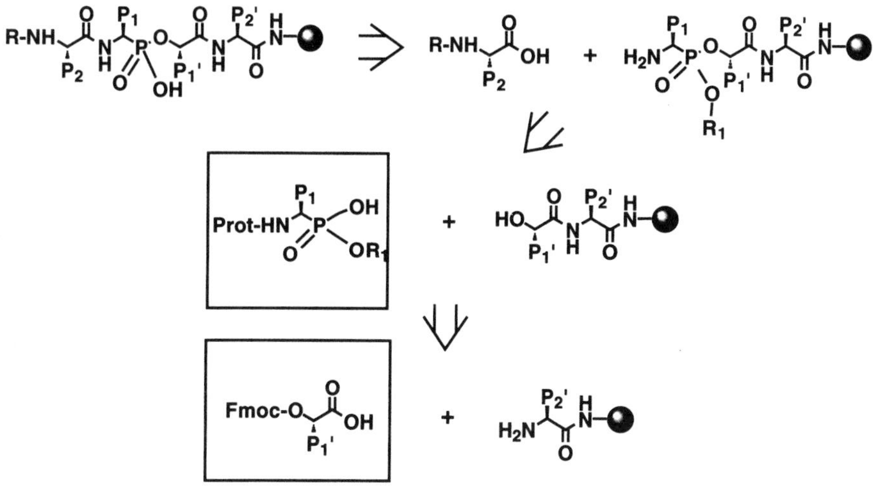

Figure 2.7. Retrocombinatorial analysis of peptidyl phosphonates.

is necessary, as well as a useful number of building block families themselves. Building block acquisition is labor and time intensive and once acquired, BBs should be leveraged to greatest effect. Selection of various chemistries that rely on similar sets of BBs can be an efficient approach. Since the exercise of combinatorial chemistry is justified primarily by time savings in molecule preparation, it is desirable to acquire BBs commercially. Serial preparation of BBs effectively transfers the labor of serial synthesis and partially defeats the purpose of a library approach. Hence, serial BB synthesis is most justifiable in focused library preparation where BBs are not otherwise available. Many factors (besides availability) may come into play in the selection of BBs; including computational analysis for diversity quantitation, representation of various charge types and functional groups, accessing a spread of physical chemical properties, and utilization of conformational space.

A strong argument may also be made to include BBs of historical medicinal chemistry importance. The author believes that there is a crucial role for medicinal chemical knowledge as a bias for library construction. In one sense we may view the many years of serial drug discovery as a sifting for productive starting points. These starting points may indeed still be valuable (benzodiazepines, dihydropyridines, etc.), but overindulgence in pursuit of known scaffolds to the exclusion of new exploration will insure that the "pharmacophores of the future" will not appear in the library. Broad-based success should come from a balanced molecular diversity portfolio containing both conservative structures (known pharmacophores from historical medicinal chemistry) and intentionally created novel types of molecules. The resulting archived portfolio of molecular diversity will thus be biased by a medicinal chemical historical perspective, and be structurally broad enough to identify the "pharmacophores of the future."

The drug discovery path is long, and it is clear that the most time will be saved and most value created, if when library members are identified by screening, they already possess desirable pharmaceutical properties. In other words, the products of diversity creation in a good library should already "look like drugs." Some believe that a population of molecules for screening is best when there are large numbers of library members (quantity). An alternative point of view has been that relatively small, but high-quality (i.e., knowledge-based) libraries will give rise to optimal results. Insufficient data at this time precludes us from assessing what may be termed the quantity versus quality argument. This issue will be answered in the future, and the answers should instruct us greatly in the rules of molecular recognition.

Library Design and Synthesis

Strategy in the design and synthesis of chemical libraries often run counter to the highly refined synthetic strategies of the "total synthesis of natural products" approach, whose thinking and philosophy has dominated organic synthesis for much of the past 50 years. Using combinatorial chemistry to address the creation of molecules for drug discovery, our concern is now directed to the preparation of a group rather than an individual molecule, and the consequences of that shift demand recognition of a set of different strategic considerations. Since library synthesis by combinatorial chemistry is best thought of as an iterative procedure, the interpretation of one library may be viewed as a prelude to designing the next one.

An essential element of the combinatorial discovery process is that one must be able to extract the information made available by library screening. Put another way, creating large quantities of molecular diversity for ligand discovery is insufficient unless there is a format at hand to capture the information, which in this case is the chemical structures of active compounds. Hence, attention must be directed to the assay methodologies employed in screening combinatorial libraries and to the interrelationship between these methods, the diversity display formats, and the mechanisms for determining the structure of selected molecules that emerge from library evaluation.

An underlying thesis mentioned previously is that combinatorial technologies may be thought of as a collection of strategies, technologies, chemistries, screens, instrumentation, and information management tools that will be blended together case by case to address specific drug discovery problems. In our view it is unlikely that a single protocol for implementing combinatorial chemistry will be applicable in all situations. To evolve effective strategies for library design, it is useful to consider the projected role of chemical libraries in drug discovery. Two key, divergent embodiments of the usual combinatorial discovery objectives may be specified: (1) *random screening,* where the task is to identify a novel lead compound in the absence of any structural or mechanistic information about the biological target macromolecule; and (2) *directed screening* or *chemical analoging,* where the objective is to evaluate closely related structural analogs of a lead molecule, establish SAR, and optimize biological potency and other pharmaceutically relevant properties. These objectives raise different issues in terms of both the synthetic combinatorial strategies and assay parameters that must be applied.

Broad screening relies on evaluating large numbers of very diverse molecules. This can be addressed through preparation of large libraries or a portfolio of many varied small libraries. Finding significant hits for a new target is akin to searching for a rare event. Screening procedures have to be able to detect bioactivity in an ocean of inactivity. In contrast, focused libraries or analog libraries are created to "explode" around already identified leads or starting points. In the analog library case, many or all of the library members may be active, and the challenge is often to discriminate among them. Analog libraries and broad screening libraries can require totally different strategies, in terms of library size, bead size (loading), encoding, etc. (see Figure 2.8).

For the purposes of primary lead libraries for screening, we may ask the question, What events need to happen, and what limits should be imposed in the combinatorial process, in order for libraries to contain molecules of interest for drug discovery? As illustrated in Figure 2.9, by making some simple assumptions about the molecular weight of individual building blocks (i.e., average BB = molecular weight 150), we see that, in principle, syntheses of only a few steps (2–4 steps) are necessary to access complex molecules that might be of interest from the drug discovery perspective. If the desired molecular weight range for final library members is 300–750, then only 2–5 BBs need be connected in a synthetic sequence. This means that an upper limit of only 1–4 reactions is necessary to reach this practical goal. This is a very desirable

Broad Screening	Chemical Analoging
Large library size	Moderate library size
Broad structural diversity	Narrow structural diversity
No specific structural goal	Specific structural role
Many building blocks	Specific retro building blocks
Undefined order of combination	Specific order of combination

Figure 2.8. Combinatorial chemistry: two divergent themes.

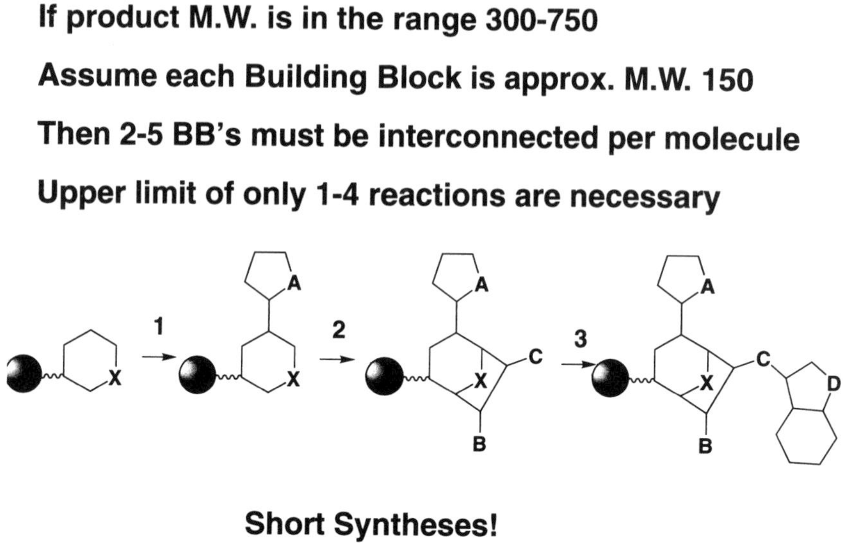

If product M.W. is in the range 300-750

Assume each Building Block is approx. M.W. 150

Then 2-5 BB's must be interconnected per molecule

Upper limit of only 1-4 reactions are necessary

Short Syntheses!

Figure 2.9. What needs to happen in combinatorial chemistry?

outcome, because it informs us that one need only plan and perform short syntheses to yield useful results. Hence, the number of synthetic steps adapted to solid-phase chemistry is relatively small and may even be carried out under automated control. The challenge in library design and synthesis is thus to reduce this simple idea to practice.

Since it takes only a few synthetic steps to prepare libraries of interest, it follows that we must be judicious in the selection of the chemical processes and the individual building blocks brought to the process, for it is these that will ultimately determine the constitution of the library members. The use of diversity measurements may improve the library overall quality. The interesting question, What is molecular diversity and how do you measure it?, has come under recent consideration. Diversity measuring tools are still in their infancy, but the use of these to aid in the selection of BBs and to judge the merits of virtual libraries is rapidly growing.

In practicing combinatorial chemistry for drug discovery, we have identified a pattern that generally follows the steps shown in Figure 2.10. Aspects of these issues are discussed in the balance of the chapter. Following the building block selection described above, the process enters into what we term the "rehearsal" period. This is generally the longest and most resource-intensive part of the library process. This is the time in which chemistry is adapted to solid support and many different building blocks are "auditioned" using the chemistry that library construction will exploit. Since in split–pool methods it will usually be impractical to chemically characterize each member of the library, the rehearsal period is the time for ensuring a high degree of quality

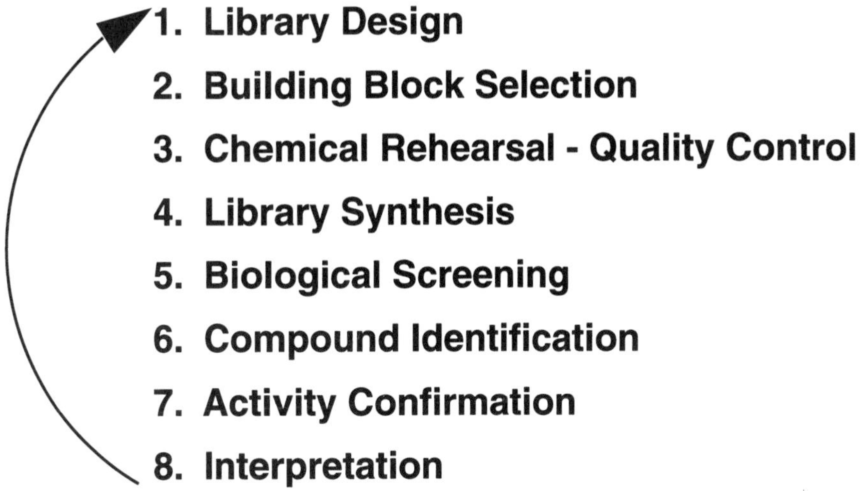

1. **Library Design**

2. **Building Block Selection**

3. **Chemical Rehearsal - Quality Control**

4. **Library Synthesis**

5. **Biological Screening**

6. **Compound Identification**

7. **Activity Confirmation**

8. **Interpretation**

Figure 2.10. The combinatorial chemistry process.

control in the final library. The goal is to enter into the library building process with as high a degree of confidence in the chemistry and BBs as possible. BBs that when tested individually (best in parallel) fail to perform as desired in the rehearsal are excluded from the final library. In this way the library becomes enriched in those BBs that are broadly suitable to the library chemistry. The rehearsal process illustrates the relationship between split-pool synthesis and parallel synthesis. Rehearsing is an exercise in parallel synthesis. A successful rehearsal provides sufficient information to prepare both a high-quality parallel array and a split–pool library of many members.

Library synthesis itself is relatively short once the background activities are complete. Screening of libraries is considered in depth in Chapters 22 and 23. Hits in screening need of course to be associated with their corresponding structural information. Hits in parallel arrays are identified by their spatial address in the array. Split–pool libraries must be deconvoluted, which often reduces to a problem in parallel synthesis arrays. In the case of encoded libraries (vide infra), the tag must be read and interpreted. In all cases it is wise to independently confirm, by separate synthesis and testing, structures that are presumed to possess interesting biological activity (activity confirmation).

The final step in the process; interpretation is one in which huge amounts of library screening data must be digested. An upward arrow signifies that the combinatorial chemistry process is an iterative one, in which the data of interpretation is used to formulate the design of a new library. In the course of this process, information necessary to generate a pharmacophoric model will be collected. Computational tools that can effectively capture this information will be important in the iteration of the library cycle.

Analysis of Synthetic Strategies

Selection criteria for desirable synthetic routes to library preparation differ from tradition in that efficiency and reliability of the reactions must be considered across a broad range of substrate possibilities rather than just one substrate. Fortunately, in many cases the chemical literature already contains rich sources of just this sort of information, which has been compiling for over 100 years! This valuable resource should be fully mined before extensive work is devoted to new chemistries.

The investment made in adapting a certain chemistry to solid support should be fully exploited by using each solid supported intermediate in a number of different ways. Synthetic schemes should be intentionally designed to lead to new opportunities, instead of dead-ending. The result is a "synthetic tree" wherein a short sequence of reactions form the tree trunk; numerous branches, which capitalize on the chemical sequence of the trunk, sprout off as branches of the main theme. A high level of efficiency in the library preparation effort can be achieved in this way. A related term that has come into frequent use is the notion of creating "libraries from libraries." This strategy is limited by the need to stay below a threshold molecular weight of approximately 750. As mentioned above, an important criterion is whether the products of diversity generation "look like drugs," and that their gross properties (molecular weight, reactivity, etc.) are appropriate. Also valuable is identification of reactions that "cascade" into other reactions. Multicomponent chemical reactions can also serve key roles for specific types of library synthesis strategies (i.e., in which there is only one handle to introduce diversity).

In the next few sections we explore issues related to the technology and the chemistry of combinatorial chemistry (Figure 2.11).

Formats for Molecular Diversity

Selection of a format in which to create a chemical library is an important decision. Of the three main formats currently available (spatially addressable, split-pool, and encoded libraries), each has its own strengths and is suited to particular types of drug discovery problems. The advantages of parallel synthesis are that discrete compounds are produced, the output is adaptable to almost any screening format, there is opportunity to fully characterize products, time-consuming deconvolution is unnecessary, and collection of a full SAR is possible. The limitations of this process are that it produces relatively small libraries (<1000) and it is labor and reagent intensive. Nevertheless, this method is the preferred format for small libraries, and is useful for BB rehearsal for larger split–pool libraries and as a tool for resynthesis of active pools as discretes.

Split–pool synthesis offers the opportunity to prepare many compounds in very few steps. It is probably the simplest route to large libraries, and makes efficient use of reagents. In practice, split–pool synthesis complements parallel

Technology of Combinatorial Chemistry

•**Formats - Spacial , Encoded, Non-encoded**
•**Instrumentation issues**
•**Information management**
•**Screening Approaches**

Chemistry of Combinatorial Chemistry

• **Strategy**
• **Solid Phase Organic Synthesis**
• **Solid Supports & Linkers**
• **Analytical Methods**

Figure 2.11. Components of combinatorial chemistry.

synthesis. The limitations of the split–pool approach are that mixtures must be assayed. The mixture size rapidly restricts the effective library size. When successful split–pool libraries generate hits, these can require substantial work in their deconvolution. Resynthesis can be time-consuming, and usually limited SAR information is obtained directly from one of these libraries. For larger libraries, this type of format has mostly been supplanted by encoding technologies.

Encoded technologies have now been reduced to practice using various chemical and more recently nonchemical tags (Figure 2.12). This approach permits very large libraries to be constructed, and thus is especially useful for primary screening. In most cases discrete compounds are assayed, thus escaping from the problems associated with screening pools. Although the encoding approach is very powerful, it has restrictions. The use of an orthogonal chemical encoding synthesis complicates the ligand-forming chemistry. In most cases, a single bead assay format is required and the derived SAR depends on the number of beads actually decoded, which takes time. Finally, many library equivalents must be assayed to ensure an adequate sampling. These techniques are all discussed in greater depth throughout the book. From a strategic combinatorial library generation point of view, we believe that the size and purpose of the intended library should direct the format to be used (Figure 2.13). Thus, small libraries are best constructed in parallel synthesis arrays, while large libraries for primary screening best exploit the encoding and split–pool approaches.

Adaptation of "Split Synthesis" in which "tags" are used to record chemical history of individual beads

Sequencable Molecular Tags   **Oligonucleotides**
                                            Peptides

Non-Sequencable Molecular Tags   **Haloaromatics**
                                            Fluorophores

                                            Radiofrequency
Non- Molecular Tags   **Encodable**
                                            Transponders

Figure 2.12. Encoded combinatorial synthesis.

Technology of Combinatorial Chemistry

Application of combinatorial chemical processes to real drug discovery problems requires a technology base that differs substantially from traditional methods. Synthesis on solid supports is probably the most significant difference from traditional solution synthesis of molecules. Success in small molecule combinatorial organic synthesis with various commercially available supports

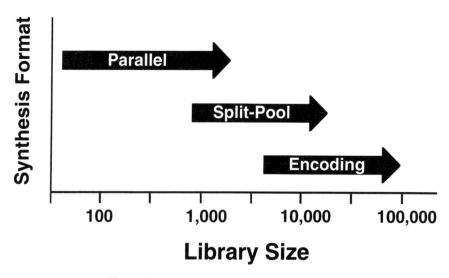

Figure 2.13. Synthesis formats vs. library size.

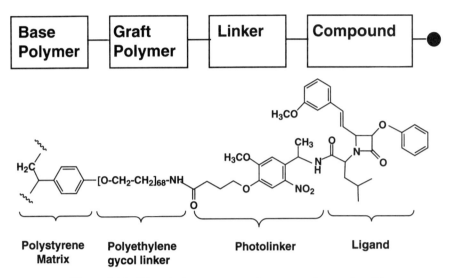

Figure 2.14. What is the structure of compounds on a bead?

has mostly come from empirical studies. Outside of the field of polymer chemistry, only the area of peptide synthesis has historically dealt in substantial detail with issues attached to solvation, mixing, yields, analysis, and so on, of synthesis on solid phase. From a traditional organic chemistry perspective, a logical question arises: What is the structure of a small molecule tethered to a bead? Figure 2.14 shows a monocyclic beta lactam, which extends from the linker arms of a tentagel bead. The beta lactam is appended to a photochemically labile linker, which itself is attached to the solid support by connection to a polyethylene glycol polymer, grafted on top of the base polymer (polystyrene matrix). In principle, the putative ligand may also sometimes bear protecting groups, which may or may not be removed before cleavage from bead. The overall chemical behavior and properties of small ligands tethered to large solid supports continues to be largely an empirical area for combinatorial chemistry, but it is a direction that we can expect to receive significant attention in the future.

In some ways the list of protecting groups that have been increasingly available for many years in organic synthesis is reminiscent of what is now required with linker technology for solid-phase synthesis. At least one function of the usual linker in solid-phase organic synthesis is to tether the growing ligand to the solid support. The linker may also function as a protecting group for a ligand functional group. When cleaved, linkers used in very early solid-phase syntheses left a "signature" on the liberated small molecule. Hence, all members of a library might have a hydroxyl group in the originally tether position. Clearly, it is desirable to have a collection of linkers (i.e., a linker library) to maximize the chemistry that may be orthogonally peformed on the tethered system and to permit manipulation of the final structure of library members. Hence, as in typical organic synthesis where a menu of various protecting groups is desirable, a collection of linkers that are cleaved under different conditions is desirable for solid-phase organic synthesis. A number of mildly cleavable, orthogonal, protecting groups have recently been developed specifically for solid-phase synthesis.

The evaluation of synthetic success is an issue in library preparation. The usual measures of evaluating results in organic synthesis may lose meaning in combinatorial synthesis. The classical notions of such fundamental concepts as purity/homogeneity, yield, exact product structure, relative and absolute stereochemical control, and specific physical properties are less relevant when applied to a broad population of molecules (of course, they may become quite relevant as individuals emerge from a selection process).

The highly refined protocols and methods for individual molecule characterization that have evolved over many years are not entirely applicable to library member characterization, nor are they always practical or necessary. A major dilemma of combinatorial synthesis is the difficulty of confirming the degree to which the expected chemistry has proceeded on the entire population of substrate molecules. The individual members of libraries prepared by parallel synthesis may still be characterized by standard methods, but such characterization soon becomes the rate-limiting step in molecule production.

With current analytical tools, part of the cost of making large numbers of compounds will be our inability to examine each individual to the extent that was formerly considered appropriate. In fact, this cost in forgoing characterization of every individual library member escalates with the creation of large libraries. Even if it were practical, we would still be unable to individually review each library member for composition and purity. It is possible that the luxury of overcharacterization traditionally routinely employed for every new synthetic compound may be fully accorded only those compounds of special biological interest. This growing problem will likely be addressed in the near future as molecule characterization and purity measurements become significant obstacles to research progress. A new consensus will need to be reached within the chemical research community on what constitutes appropriate characterization of chemical libraries.

Automation and Information Management

One question that arose early on in applying the concepts of polymeric (peptidyl) molecular diversity to make nonpolymeric libraries could be stated as follows: Large libraries of peptides can be synthesized because the product is composed of many connected monomeric units; small molecules have only a few monomeric units. Therefore, can you really make large libraries of small molecules? In fact, large libraries of small molecules are well within reach of current technology. The challenge for the combinatorial chemist is to favorably arrange the chemistry for this to happen. Consider the case shown in Figure 2.15, where 2 steps of solid-phase combinatorial chemistry driven by 3 families of commercially readily available materials could theoretically give rise to 1.5

Huge numbers of small molecules can be prepared

Challenge is to arrange the chemistry

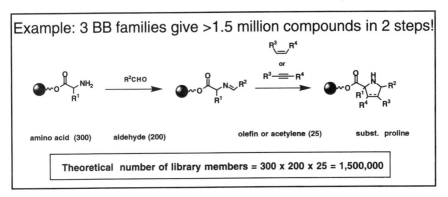

Figure 2.15. Isn't molecular diversity with small molecules hard to achieve?

million different complex proline derivatives. This demonstrates the theoretical viability of creating small molecule diversity.

Not only does Figure 2.15 make clear that the production of large numbers of small molecules are possible by combinatorial methods, it also confirms that instrumental and technological methods will limit library production and size. Though the making of 1.5 million proline derivatives may be theoretically possible (the solid-phase chemistry has been shown to work), it is likely not a task that people will be directly engaged in, day in and day out. As we have capitalized on the significant research opportunities inherent in Figure 2.15, the historical distance between organic chemistry and synthetic instrumentation has largely disappeared, and the past 4 years have seen many novel attempts at creating automated instrumentation for combinatorial chemistry.

Working under the premise that large libraries of compounds will be prepared in a few steps by solid-phase techniques, researchers and engineers have attempted to design synthesizers that are flexible with respect to the chemistries that they can perform. Another consideration in the selection of synthetic schemes for preparing combinatorial libraries is schemes that lend themselves to automated control. The requirements for such instruments may be summarized as follows: access to many reaction vessels; capability to deliver many different building blocks at each chemical step; capability to deliver different building block families at different chemical steps; flexible control of simple reaction conditions (i.e., temperature -20 to $100°C$); and a simple, versatile, and inexpensive machine. The last few points are more than perfunctory; they are real considerations. If combinatorial chemistry is to become fully integrated with the tools of drug discovery, it must be a readily available technique, and not one run by a highly trained, small group of specialists. It is likely that in the next few years most chemist benches will be equipped

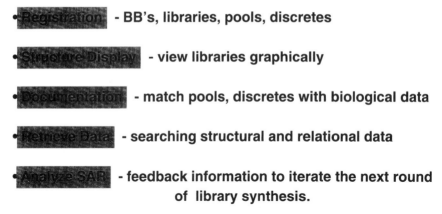

- •Registration - **BB's, libraries, pools, discretes**

- •Structure Display - **view libraries graphically**

- •Documentation - **match pools, discretes with biological data**

- •Retrieve Data - **searching structural and relational data**

- •Analyze SAR - **feedback information to iterate the next round of library synthesis.**

Figure 2.16. Issues in information management.

with dedicated, versatile synthesizers. For this to occur these instruments must be cost-effective and simple to use.

The advent of effective, low-cost automated synthesis instrumentation will certainly reduce the costs of per molecule synthesis and increase productivity, but it will also allow chemists to spend more time engaged in problem-solving research. Automation will be ideal for repetitive tasks such as weighing and dispensing of building blocks, resins, and reagents and pipetting, washing, deprotections, cleavages, characterization, and compound/library registration. Reliable parallel synthesis machines will facilitate the rehearsal of new connection chemistries, the resynthesis of discretes from active pools, and the pursuit of lead optimization with focused libraries.

The rise in use of combinatorial techniques has created a critical need for integrated information-management systems. This need will surely be acerbated with the growing capabilities of automated instrumentation. Some of these information management issues are shown in Figure 2.16.

IMPLICATIONS FOR SCREENING

A discussion of screening techniques and strategies is outside of the purview of this analysis, and is thoroughly discussed in Chapters 22 and 23. However, some features of combinatorial technologies draw molecule making and biological evaluation much closer than has previously been the case. For this reason, the ways and means of combinatorial chemistry have near term implications for the innovation and adoption of future screening techniques. A historical view of molecular diversity suggests that one of the key factors that created a fertile atmosphere for the creation and extension of combinatorial technologies was the success of high throughput screening. The impact of automation on screening engendered a revolution in the process of evaluating biological activity. The growing thirst for molecules to feed automated screening capabilities was a factor in formulating the question, Where will all these new molecules come from? To some extent combinatorial chemistry was a response to increased screening abilities. In the past, a traditional synthetic chemist might make 50 compounds a year for biological testing. With the tools described in this book, it is likely that the number will rise to >10,000 per year. This profound success in molecule making will overwhelm traditional 96-well microtiter screening systems and has already raised the question of how screening methodologies will keep up with the supply of new compounds.

In a sense, the new technology has shifted the equilibrium between molecule making and evaluation. The screening methods in place with the arrival of combinatorial technologies were simply the status quo (96-microtiter systems), and had not been created to intentionally complement the new molecule-making technologies. It seems very probable that in the near future novel screening techniques will emerge that are based on exploiting the maximum from combinatorial library formats in which the molecules are created.

It is useful to think of combinatorial technologies as a type of system where the disparate parts are actually interconnected by various linkages. In factual terms, we recognize that a certain amount of new material is required to achieve a certain concentration for biological testing. To obtain that quantity of material implies a certain loading for beads, a certain cleavage efficiency, which in turn may dictate the size of bead needed to be used (loading, Figure 2.17). Bead size directly impacts the choices available in instrumentation, and the extent of redundancy in library equivalents that will be necessary to obtain a reliable sampling. This total amount of material, adjusted for yields and surplus equivalents, implies what quantities of building blocks must be brought to the library building process. Hence, there is a remarkable interplay of factors throughout the overall library building and evaluation process. This interplay draws molecule making and evaluation closer together in a collaborative sense than has usually been the case. In practice, for speed and coordination, a good argument can be made to have molecule making and molecule evaluation physically proximate to each other.

CONCLUSIONS AND FUTURE DIRECTIONS

Analysis of the processes by which small molecule combinatorial chemistry is performed has led to identification of strategic principles that are useful to the preparation of small molecule libraries. Successful implementation of a combinatorial library approach (including both library synthesis and evaluation) has required a highly interdisciplinary research environment involving chemistry, biology, engineering, instrumentation, informatics, and other fields.

Regardless of whether the objective is a broad discovery search for a suitable starting point or analoging a known lead, a key aspect in the successful

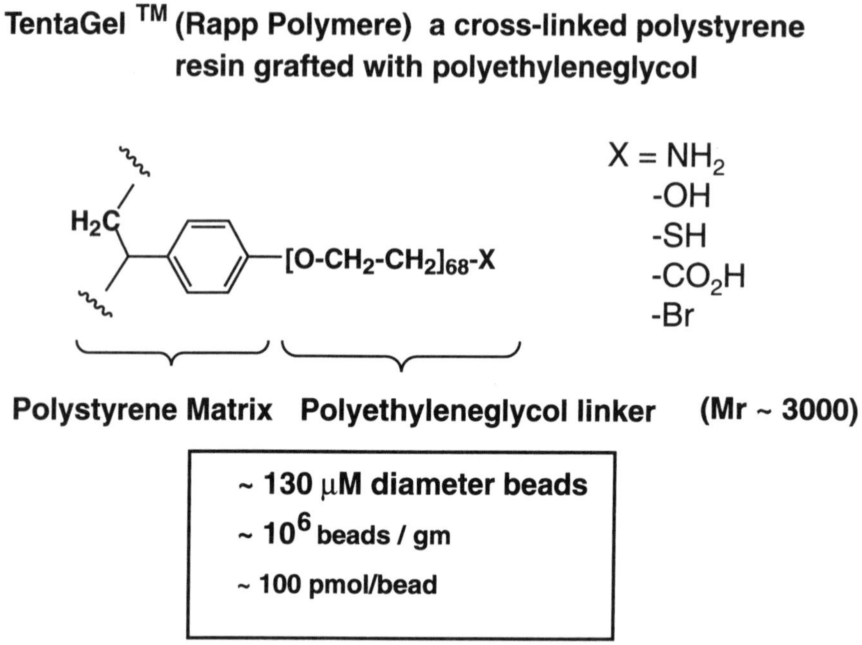

TentaGel ™ (Rapp Polymere) a cross-linked polystyrene resin grafted with polyethyleneglycol

H_2C

—[O-CH$_2$-CH$_2$]$_{68}$-X

X = NH$_2$
-OH
-SH
-CO$_2$H
-Br

Polystyrene Matrix   Polyethyleneglycol linker   (Mr ~ 3000)

~ 130 µM diameter beads
~ 10^6 beads / gm
~ 100 pmol/bead

Figure 2.17. How much material is on a bead?

application of combinatorial technologies to drug discovery is the closely linked integration of synthesis and screening. The creation and evaluation of molecular diversity are two sides of the same coin. In all likelihood, command of a collection of combinatorial tools will be required for general success, rather than adherence to a single rigid protocol. We can expect a continuation in the growing sophistication of synthetic chemistry adapted to solid-phase supports. New linkers (including "traceless" linkers) will continue to be developed to permit greater latitude in tethering various functionality to bead, and in effecting their controlled release. This will directly lead to increased flexibility in the synthetic approaches that are feasible on solid support.

The physical manipulation of building blocks, beads, reagents, and reaction vessels are found to be rate limiting in actual library production. Given the repetitive nature of many of the manipulations required for library construction, an ongoing priority will be to address the issue of automating as many aspects of the generation/evaluation process. The historically distant relationship between instrumental engineering/automation and organic synthesis has substantially narrowed. Recently adaptation of solid-phase combinatorial chemistries to run under automated instrumental control will continue to evolve to include more sophisticated and flexible chemistries.

In the coming years, cloning and sequencing of the human and microbial genomes promises that an unprecedented abundance of new proteins will become available as potential drug targets. Gaining even more prominence than it now assumes will be the issue of discriminating among a myriad of receptors and enzymes to identify valid targets for drug discovery. The rapid ability to access potent and specific ligands for these targets will guide the process of untangling the physiological relevance of endogenous biochemical pathways. Combinatorial methods will be called on to provide molecules quickly and cheaply to drive target validation. In this manner, the identification of leads will benefit from a significant, but hidden benefit that emerges from combinatorial screening: Hits derived from chemical libraries should be readily amenable to combinatorial analoging. New screening technologies will be developed to capitalize on the ready availability of screenable molecules and to cope with higher throughput. The physical size of libraries, the need for preparative amounts of precious building blocks, and the amounts of receptors/enzymes required for extensive screening will continue to drive the field toward minaturization.

The power of combinatorial technologies in generating huge numbers of compounds suggests that in a lead discovery mode, less preconceived bias need be brought to the process of making molecules. In other words, let the numbers do the talking. However, it is important to recognize the profoundly valuable collection of information that has been painstakingly amassed in medicinal chemistry over many decades of single molecule research. Not enough data yet allows us to rigorously compare hit rates for various types of libraries and targets, but anecdotal trends appear to greatly favor those libraries constructed with historical drug discovery principles in mind.

Due to the time and effort required in serial approaches, each target molecule must be selected with great care. Because of the relative ease in creating libraries, little risk is incurred or effort expended in allowing a wide variety of building blocks to participate in diversity generation. Since there is less upfront investment in any individual combinatorially created molecule, the combinatorial chemist can afford to take more risks. We can think in terms of a portfolio of libraries that might be routinely applied to the initiation of a drug discovery search. Combinatorial chemistry will be integrated with the other main tools of drug discovery, including structure- and mechanism-based design, molecular modeling, and computational chemistry, to form a more powerful discovery paradigm. Library design will become an important drug design technique, as researchers learn to bias library synthesis to increase the likelihood of finding high-affinity ligands. As more knowledge of workable strategies for combinatorial synthesis are understood, it is expected that structural and computational input and other rational design information will have an increasing impact on combinatorial drug discovery approaches.

A related, but still immature issue in combinatorial approaches to drug discovery revolves around the idea of quantitation of diversity. An understanding of the concept of measuring molecular diversity could assist in designing libraries to contain maximal structural diversity. This notion arose many years ago in deciding which few representative, highly diverse compounds to select out of large database collections when setting up groups of preliminary screening samples. The huge numbers involved in combinatorial techniques intensify this issue. A number of interesting approaches to the diversity quantitation problem can be expected to emerge.

Another area where considerable effort must be applied is in the registry of libraries and individual library members. Library compounds may not be registered and documented for testing in the same ways as serially produced compounds historically have been, but exactly what changes are necessary remain to be determined. Vast numbers of compounds have been and are being created; keeping track of these and their corresponding biological activities will require innovative database management techniques. Additionally, nomenclature needs to be developed by which one can simply express the constitution, scope, and nature of chemical libraries. Legal issues, including the patenting and documentating of libraries and their component members, will need to be pioneered.

The huge numbers of molecules that will be made and screened will demand a new generation of database management tools. The most effective data analysis strategies will make use of both positive and negative screening information to help construct hypothetical active sites/binding sites of potential drug targets. This information-intensive task is well suited to the information flow from combinatorial chemistry. Analysis of library screening results in turn will drive the next round of library synthesis.

Though the field of combinatorial chemistry is chronologically a new enterprise, the evolution of thought in this fertile area continues to outrace experimental practice. The growth of combinatorial chemistry marks a new level of empowerment for the chemist. Large numbers of molecules that in the recent past might have taken a year or a career to synthesize will now be done in a single experiment. The availability of such numbers of molecules and their ability to be screened for particular criteria allows researchers to expand the scope of questions they may ask and direct efforts to the task of turning leads into drugs.

3

SOLID-PHASE PEPTIDE SYNTHESIS, LEAD GENERATION, AND OPTIMIZATION

BRUCE SELIGMANN
SIDDCO, Tucson, Arizona

MICHAEL LEBL
Trega Biosciences, San Diego, California

KIT S. LAM
Arizona Cancer Center, Tucson, Arizona

Drug discovery has been undergoing revolutionary changes for the last 15 years. The first change was in the shift from whole animal testing of compounds, to mechanistic based in vitro screening. This was based on new knowledge about molecular mechanisms brought about by advances in biochemistry. Some companies began to adapt in vitro mechanistic assays for "high throughput" screening, on the order of 10,000 to 20,000 compounds per year. At the same time, medicinal chemists began to recognize the advances being made in the understanding of the structure of enzymes and proteins, as well as to use x-ray to define the structure of the small molecules they had synthesized which exhibited desired biological activity. This approach enabled chemists to understand how structure might relate to activity in order to design even more active molecules. Computational chemistry came of age in the pharmaceutical industry, focused initially on methods of obtaining structure activity relationships (SAR) between series of active molecules, and then

Combinatorial Chemistry and Molecular Diversity in Drug Discovery, Edited by
Eric M. Gordon and James F. Kerwin, Jr.
ISBN 0-471-15518-7   Copyright © 1998 by Wiley-Liss, Inc.

evolving into de novo rational design efforts using structural information about the target molecule. The explosion of molecular biology went hand-in-hand with this evolution of rational design and mechanistic approaches by providing the means to obtain sufficient quantities of target proteins to both enable high volume assay as well as structural studies on otherwise inaccessible target molecules.

The extraordinary growth of combinatorial chemistry beginning some five years ago has led to yet another revolution, focused directly on the approach used by the medicinal chemist to generate lead compounds and to optimize their activity. However, the impact of combinatorial chemistry does not end with the successful discovery of therapeutic agents. It is now entering another phase even before its impact on lead generation and optimization has been fully developed or implemented by most companies, and even though the solid phase chemistry for non-peptide synthesis remains largely undeveloped. This new role of combinatorial chemistry is being spawned by the rush of genomics information which will produce an incredible number of new potential targets each year, more than overwhelming the capacity of the pharmaceutical industry. What is crucial is no longer the sequencing of the human and other genomes, but instead the identification of targets and their rapid evaluation to enable the industry to identify those with the greatest potential for impact on disease. To-date the use of transgenic experiments, and to a lesser extent, the use of antibodies, have served as the mainstay for determining the role of new, or even somewhat older target proteins in disease. However, combinatorial chemistry in the form of antibody display libraries, peptide display libraries, and synthetic peptide combinatorial libraries offer state-of-the-art tools to the pharmaceutical industry which can be exploited to identify antagonists/inhibitors and agonists of new target molecules which can in turn be used in vivo to validate the potential of exploiting a new target to address specific diseases. With time, non-peptide combinatorial chemistry may similarly be used to address this need, but currently such libraries are limited by the state of synthetic chemistry knowledge. While the potential for target validation is just now beginning to be exploited, results reviewed here clearly indicate that screening synthetic peptide libraries can produce leads for a large variety of targets of therapeutic interest, provide extensive databases of SAR to the medicinal chemist, and when libraries are designed based on the initial leads or other information, provide a valuable tool to rapidly optimize the activity of molecules. The envelope of peptide chemistry has been pushed to enable the reliable, high-purity synthesis of large numbers of peptides containing available proteinogenic and nonproteinogenic amino acids, amines, and carboxylic acids as linear, cyclic, and branched structures in order to produce the greatest diversity among library components and between libraries.

This chapter focuses on synthetic peptide/peptidomimetic combinatorial library synthesis, lead generation, and optimization. Numerous reviews[1-6] have discussed the various formats used to synthesize combinatorial libraries: the synthesis of medium to large numbers of compounds by fixed array combinatorial synthesis, as described by Geysen[7]; the iterative synthesis, assay, resynthesis, and reassay of mixtures described in 1991 by Houghten[8]; the split-and-mix, one-compound–one-bead approach to synthesizing large numbers of single compounds described the same year by Lam et al.[9] (1991); and the current automated parallel synthesis of large numbers of compounds and rational limited deconvolution of relatively small mixtures.[10] Types of assays by which combinatorial libraries have been screened have also been described and reviewed extensively, with the conclusion that combinatorial libraries are amenable to screening with every high throughput assay currently known.

This overview describes specific novel peptide chemistry and the assay results that reveal the potential brought to medicinal chemistry through peptide combinatorial library synthesis and assay. The review of results is divided according to type of target, rather than type of combinatorial approach. The reader is thus encouraged to focus on what has been achieved as it applies to drug discovery regardless of the specific combinatorial chemistry synthetic/screening format. Combinatorial chemistry may reverse the dogma that has developed over the years regarding the therapeutic utility of peptides. Even the authors are divided over whether the ability to synthesize and assay large numbers of peptide and peptidomimetic analogs will produce useful therapeutic agents with acceptable pharmacological properties for a large number of targets.

COMBINATORIAL APPROACHES TO LIBRARY DESIGN, SYNTHESIS, AND SCREENING

There is no universally accepted definition of combinatorial chemistry. For the purpose of this chapter it is defined to encompass the synthesis of compounds from sets of subunit and chemical reactions used in one or more reaction steps. This definition includes both solution and solid-phase synthetic reactions, the synthesis of compounds through several sequential reaction steps in which the same or different sets of subunits and chemical reactions are used, as well as the reaction of multiple subunits in one reaction step to form multicomponent compounds. The compatibility of each building block with each chemical reaction, and the product formed must be known and validated for the process to give useful libraries of compounds. While a distinction has been made in the past by some investigators between oligomer compounds and other multicomponent compounds, there is no difference between a linear or a branching synthetic scheme. Thus, just as in polymer chemistry where there may or may not be cross-linking and branching, so in oligomer chemistry where the subunits or monomers making up the final compound are different rather than the same (thus distinguishing the compounds from polymers of the same subunits) the oligomeric compounds may be linear or branched.

There are several basic strategies to the design, synthesis, and screening of combinatorial libraries. To simplify this review the different strategies are explained below, and the terminology that is used throughout this review is established. Multiple synthesis covers any method in which discrete compounds are synthesized simultaneously to create a library of isolated compounds whose identity is known from the synthesis scheme. Iterative synthesis/ screening involves the synthesis of compounds in such a manner that a mixture results that is not directly resolvable to determine the identity of discrete active compounds, but that instead is resolvable to determine the identity of a specific residue(s) in any mixture that shows activity when assayed. A new set of mixtures is then synthesized based on this information and assayed, and the identity of the next specific residue(s) determined. The iterative process is continued until the identity of a complete, active molecule is determined. This process has several characteristics. First, the pool size diminishes as the iterations proceed, ending at the last step with the synthesis and testing of individual compounds. Second, the relationship of the last residues to be defined depends on the selection of the residues in the initial iterative steps, so there is an algorithmic relationship between the first and last components of the compound to be identified. Methods of deconvolution fall within the iteration definition and will be considered as a form of iterative synthesis/ screening.

Positional scanning is a method of synthesizing several mixtures of compounds such that in each specific residues can be defined. The whole set of mixtures enables the investigator to identify every active residue of a virtual compound without actually having synthesized and tested that compound as a pure entity, only presumably as a component of each mixture. This process does not involve any iterative synthesis and assay steps. Furthermore, unlike the iterative process, there is no dependence of the identification of any one residue on the identity of any other residue within the active molecule. The actual synthesis of the derived active compound(s) must be performed separately, ideally as a multiple synthesis effort.

One-compound–one-bead postassay identification design (PAID) approaches (sequencable/encoded[5,9]) covers a spectrum of approaches that share a synthesis scheme that assures that each resin particle contains only one structural compound. This is efficiently and practically achieved by applying the split-and-mix method of synthesis introduced by Furka[11] and independently by others[8,9] to assure that every amino acid couples in the intended equimolar ratio. This approach has the unique requirement for investigators to maintain the relationship between assay result and the bead on which the compound was synthesized so that they can utilize information on the solid-phase particle to identify the active compound or components of the active compound. It is principally in the method of identification that variations of this approach have been introduced. In some instances the compound itself can be directly sequenced or otherwise identified by analytical methods. In other cases the synthesis is carried out in such a manner that a coding molecule is also synthesized on the solid-phase particle, and it is the coding molecule that is used to identify the active compound. The furthest afield approach introduced to date is the use of freely mixable capsules containing both resin and a microchip which can be read by radio-frequency methods. The basis for split-and-mix libraries lies in the freely mixable character of the support, whether it is a single resin bead or a collection of resin beads in some capsule or a tube of gel. While the one-compound–one-bead postassay identification approach to library synthesis absolutely depends on use of the split-and-mix synthesis process, iterative approaches can also use the split-and-mix method of synthesis.

The concept of the iterative synthesis of mixtures, positional scanning, and efficient screening in solution of one-compound–one-bead libraries raises the issue of synthesizing and testing mixtures of compounds rather than pure, highly characterized compounds. Scientists are accustomed to testing known compounds individually so as to minimize the possibility that some unknown factor or interfering substance may influence the observations being made. In screening large numbers of compounds, and particularly in combinatorial chemistry for the synthesis of large numbers of compounds, there is frequently a great advantage to synthesizing and assaying mixtures or pools of compounds. The immediate question is what risk does this entail, and what pool sizes can be used. The absolute answer must be judged on a case-by-case, experimental basis. However, as seen from all the examples given below, there does not seem to be an adverse effect of employing mixtures of compounds. Direct examination showed that pooling of hundreds of compounds did not mask positive compounds.

Analogs of the RGD peptide sequence were synthesized, pooled, and screened for binding to the gpIIb/IIIa platelet receptor.[12] The library synthesized was of the form $YGRGY_1X_2X_3$, where Y_1 consisted of the randomization of S, D, R, H, E, while 19 L-amino acids were randomized in the X_2 position (omitting isoleucine) and 20 L-amino acids were randomized in the X_3 position, creating a library of 1900 species. The library compounds were cleaved from the resin, pooled according to the residue incorporated at Y_1, and assayed. The YGRGD X_2X_3 pool inhibited the binding of soluble gpIIb/IIIa receptor to fibrinogen-coated plates greater than any of the other pools, exhibiting similar activity assayed as a pool of 360 compounds as did control peptide, RGDS. The identification of YGRGD X_2X_3 was predictable from previously described SAR, and the authors did not define X_2X_3; the focus of this study was to examine the effect of mixtures on assay results. In this study, assaying compounds as mixtures (of 360 closely related compounds) did not negatively influence the binding data and the ability to detect an active compound. The examples cited below all suggest that screening mixtures of compounds is a powerful approach for generating important information about the structure of active compounds, and that to avoid the use of pooling, when otherwise appropriate, is to unnecessarily lessen the amount of information that can be obtained for a particular system.

COMBINATORIAL SYNTHESIS OF PEPTIDES

Choice of Solid Support

Beaded polymer used for library synthesis has to fulfill certain criteria depending on the synthetic and screening strategy. For libraries based on the one-bead–one-compound approach, the size and substitution homogeneity is very important: To be able to evaluate the biological signal created by peptide released from a single bead one must be sure that the amount of compound released from each bead is uniform. Also important is the resin resistance to the formation of clusters (resin stickiness): Clusters will prevent the statistical free redistribution of resin beads and thus substantially lower the number of structures created. The ability of the resin to swell in both organic and aqueous media is especially important when target binding to the beads is used as the criteria for positive bead identification.

Polyacrylamide beads fulfill most of the above-mentioned criteria and was actually the resin used for evaluating technology feasibility.[9] However, polyacrylamide has been replaced by TentaGel, polyoxyethylene-grafted polystyrene,[13] which has become the resin of choice for both peptide and organic solid-phase library synthesis. The PEG-PS resin[14] of Millipore (Perspective Biosystems today has similar composition and properties, but differs in the placement of the chemically reactive group (amino group) in relation to the polystyrene matrix. TentaGel has the functionalizable group at the end of the polyoxyethylene chains, far from the hydrophobic polystyrene chain, a feature that is especially important for compound display on the bead surface for the bead-binding assay. PEG-PS has the functional group next to the polystyrene chain such that the polyoxyethylene chain does not serve as a linker connecting the synthetic compound with the polymer, but rather as modifier of polymer properties.

Alternative carriers to classical resin beads have been tested. Polyacrylamide-grafted polypropylene pins were used for the synthesis of the first library.[15] This type of support was shown to be very useful in multiple-peptide synthesis.[16] It was adapted to larger scale by the application of crowns attached to the pins,[17] and is also useful for nonpeptide synthesis.[18]

Paper is a good support for multiple (SPOT) synthesis of thousands of peptides[19] or for the synthesis of libraries.[19,20] Synthesis on paper is performed by spotting the solution of protected amino acids onto the functionalized cellulose paper in the presence of activating reagent. In this case, the reaction vessel is the carrier itself, liquid manipulation during the synthesis (usual shaking in the case of solid-phase synthesis) is eliminated, and the reaction is driven by the diffusion of liquid in the carrier. This principle of internal volume synthesis was tested using polymeric carriers on a multiple synthesizer utilizing centrifugation for liquid elimination[21] and was found comparable, if not better, than the classical arrangement of solid-phase peptide synthesis.[22] Cotton, the purest form of cellulose, was found to be a convenient solid-phase support, especially for multiple synthesis[23] or library generation.[24]

One specific feature of membrane, sheet, or thread-like carriers is their divisibility. This feature can be used for the synthesis of libraries with a nonstatistical or forced distribution of library members.[25] The synthesis of this library starts with n pieces of the carrier, which are coupled with n different building blocks (1st randomization). Each of the n pieces is then divided into m parts and these smaller parts are distributed into m reaction vessels in which m reactions are performed (second randomization). The process can be repeated as many times as permitted by the handleability of polymer particles. The result of this process is a library of $(n)(m) \cdots = X$ compounds on X polymeric particles, where no compound is missing and none is represented more than once.

Synthetic Issues

Coupling in peptide library synthesis is performed in the usual way—in a battery of bubblers, closed plastic vials, polypropylene syringes,[26] or in tea bags[27]—and standard coupling reagents are used.[28–30] Detailed descriptions of library synthesis procedures have been published.[31,32] Difficult couplings can be forced to completion by the application of more reactive coupling reagents, such as TFFA.[33] However, due to the ease of handling and simplicity in deprotection performed in multiple vessels in parallel, the Fmoc protecting group is favored over the Boc group.

The techniques used to monitor coupling reactions required adaptation. In split-and-mix synthesis,[8,9,11] beads within each reaction mixture contain different peptide sequences, and therefore coupling kinetics for each bead may be different. Since we are interested in knowing not whether the coupling is complete on average, but rather whether coupling is complete on each bead in the sample, it is advisable to follow the reaction at the level of individual beads. This can be achieved by nondestructive methods such as bromphenol blue monitoring.[34] An example of monitoring individual beads is shown in Figure 3.1. Coupling was complete for the majority of beads (clear beads), but coupling on several beads is not complete (colored beads). The classical ninhydrin test[35] did not reveal any problems at this coupling stage.

Equimolarity of incorporation of amino acids into the growing peptides of a library can be achieved in several ways. The most reliable is split synthesis,[8,9,11] which was designed for the purpose of generating equimolar peptide mixtures.[11] Equimolarity, however, can be achieved only in the cases when the number of particles used for the synthesis is substantially larger than the number of synthesized peptides (1,000,000 compounds cannot be synthesized on 100,000 beads). Split synthesis results in a collection of polymeric particles, each containing individual sequences, and was the basis for the development of the one-bead–one-compound library technique.[9] The split method, however, is inconvenient in the case of iterative libraries containing fixed positions inside of the peptide sequence. In this case the synthesis of a library with 20 amino acids in one fixed position and 20 amino acids in positions requiring equimolar

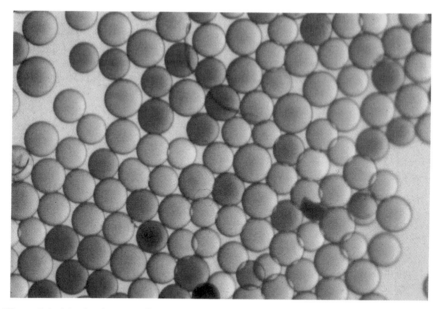

Figure 3.1. Monitoring coupling reactions on individual beads: Use of bromphenol blue to monitor the coupling reaction at the level of individual beads. At the time this photo was taken coupling was complete for the majority of resin beads, indicated by the lack of remaining color, but coupling can be seen to be incomplete of several (colored) beads.

incorporation of amino acids would require the use of 400 reaction vessels. An alternative is the use of a mixture of amino acids in which the ratio is adjusted according to the reactivity of the particular protected building block;[15,36-38] or double or triple coupling of subequimolar (0.8 molar) amount of equimolar mixtures of amino acids.[39-41] With either approach, the library would contain an equimolar representation of each peptide, though each bead would have a collection of peptides synthesized onto it (again with equimolar representation).

Synthesis of libraries can be performed manually or by automation.[42-44] The distribution of the resin is achieved by volume distribution of either a homogeneous (nonsedimenting) suspension of beads in isopycnic solution[42] (Zuckermann et al., 1992b) or a stirred suspension[43] (Saneii et al., 1993). The third design achieves distribution through sedimentation of the suspension in a symmetrical distribution vessel (Figure 3.2).[44]

Libraries immobilized on glass slides have been synthesized by the application of photolithographical techniques.[45,46] The identity of positively reacting molecules is known from the position on the glass surface (see Chapter 4). The solid-phase synthetic principle is also used in the case of liquid-phase combinatorial synthesis.[47] Synthesis is performed on soluble polymer (functionalized polyoxyethylene), which is then precipitated for removal of excess reagents. Biological screening can be performed with compounds directly attached to the soluble polymer.

Figure 3.2. Redistribution of resin: The redistribution of resin into multiple vessels through the uniform settling of resin suspended in liquid media. The resin is thoroughly mixed and then allowed to settle. The design of the settling apparatus direct equal portions of the resin into the reaction vessels for the next step of synthesis.

Structure Determination by Nonencoded Means

Structure determination of the active component of a mixture is not an issue in the techniques utilizing iterative synthesis or positional scanning, the technique in which a certain feature of the library molecules is kept constant and the rest of the molecule structure is randomized. Determining the activity of variously defined mixtures defines the synthesis of the next generation library (or individual compound array in the case of positional scanning),[48,49] and the active structure is defined by a synthetic algorithm. Part of the material used in each step of the synthesis of an iterative library can be saved and used for the synthesis of subsequent library generations. This principle, called recursive deconvolution,[10] is attractive for laboratories evaluating one type of combinatorial library. It is not practical when hundreds of different libraries containing up to 100 building blocks in each randomized position are used, since one library composed of 4 consecutively connected building blocks, each position being randomized by 100 different blocks, would require storage of 1,010,100 resin aliquots.

A technique allowing the rapid determination of the active compound from a mixture is based on an orthogonal mixture principle. In this case two (or more) sets of mixtures are generated in such a way that individual compounds are present in two (or more) different mixtures in identical concentration. The biological activity of two orthogonal mixtures defines the active compound common for both mixtures.[50] The disadvantage of this method is the complexity of the synthesis of the orthogonal libraries (Figure 3.3), but robotic synthesis of the library addresses this problem. Conceptually similar was a library design utilizing enriched and depleted mixtures of amino acids in the construction of library mixtures.[51]

Library Formats

Challenging syntheses are the norm in combinatorial peptide chemistry, not the exception. For example, some biological targets require a free carboxy terminus (or both N- and C-termini free) to be displayed from amino to carboxy terminus. This is technically problematic due to significant problems with racemization in every step. This problem was solved by variously synthesizing a cyclic peptide containing a cleavable linker in its cyclic structure such that cleavage of the intramolecular linker exposed the free terminus of the molecule (Figure 3.4).[52-54]

Screening one-bead–one-compound libraries for activity in solution requires multiple release of an equimolar quantity of compound from individual beads (vide infra).[32,35] This can be achieved by attaching the compound to the bead via a linker, allowing release in two independent steps,[56,57] or by attaching the compound onto a mixture of linkers cleavable under different conditions. The first approach has the advantage of not decreasing the amount of released compounds, since the amount of compound on each bead is multiplied by the

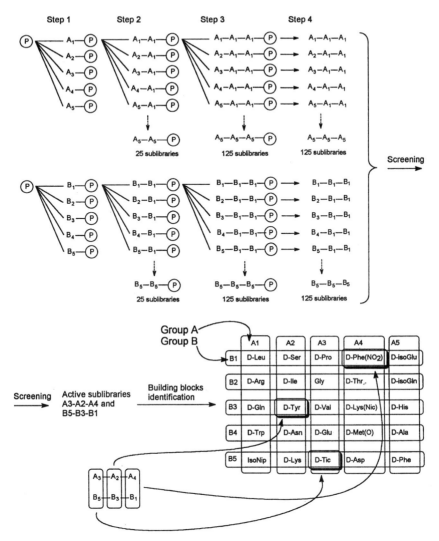

Figure 3.3. Orthogonol Deconvolution of Mixtures: Synthesis of two sets of libraries using orthogonol overlapping building block subsets permits the direct identification of an active compound from a the assay of both sets without resynthesis. Only one compound is common between the active sets. In this example mixtures contained 125 compounds each. Each set contained 125 sublibraries (as depicted for set A and Set B) which were assayed (total of 250 assays) in order to screen and identify the active compound from among the total 15,625 compounds.

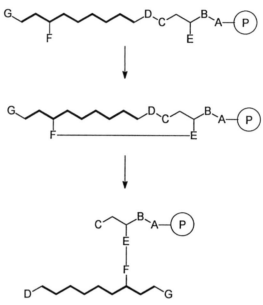

Figure 3.4. Synthetic method for the solid phase synthesis and display of a free carboxy terminus of peptides remaining bound to solid phase particle during assay. The approach depends upon the synthesis of a peptide on a branched linker containing a protected arm, followed by deprotection of the linker arm an cyclization back onto the linker, and then cleavage of the original linker arm on which the linear form of the peptide had been synthesized. The peptide remains bound to the resin (indicated by the "P") through the cyclizing peptide-linker bond.

branching of the linker to accommodate each step of release (the release mechanism is illustrated in Figure 3.5). An alternative method of tiered release is the kinetic release realized by either timed exposure of a relatively stable (benzhydrylamine) attachment to mild cleaving conditions (vapors of trifloro-acetic acid),[58] or exposure of photolytically cleavable attachment to light.[59] Various approaches to stepwise release of the compounds from polymeric carrier have been reviewed.[60]

The principle of a one-bead–one-compound library can be combined with the multiple-defined positional scanning concept[61] to address the problem of generation and screening incomplete libraries. A library of hexapeptides composed from 20 amino acids would contain 64,000 compounds. However, it is probable that not all six amino acids in any one peptide with significant affinity for the biological target are essential for binding. Some of the amino acids in the sequence can be easily replaced without significant loss of activity, whereas several critical amino acids cannot usually be replaced by a substitute. The task is therefore to identify the critical amino acids, or the motif required for binding. If we define the motif as an arrangement of three amino acids, not necessarily contiguous, we can create 20 different positional arrangements

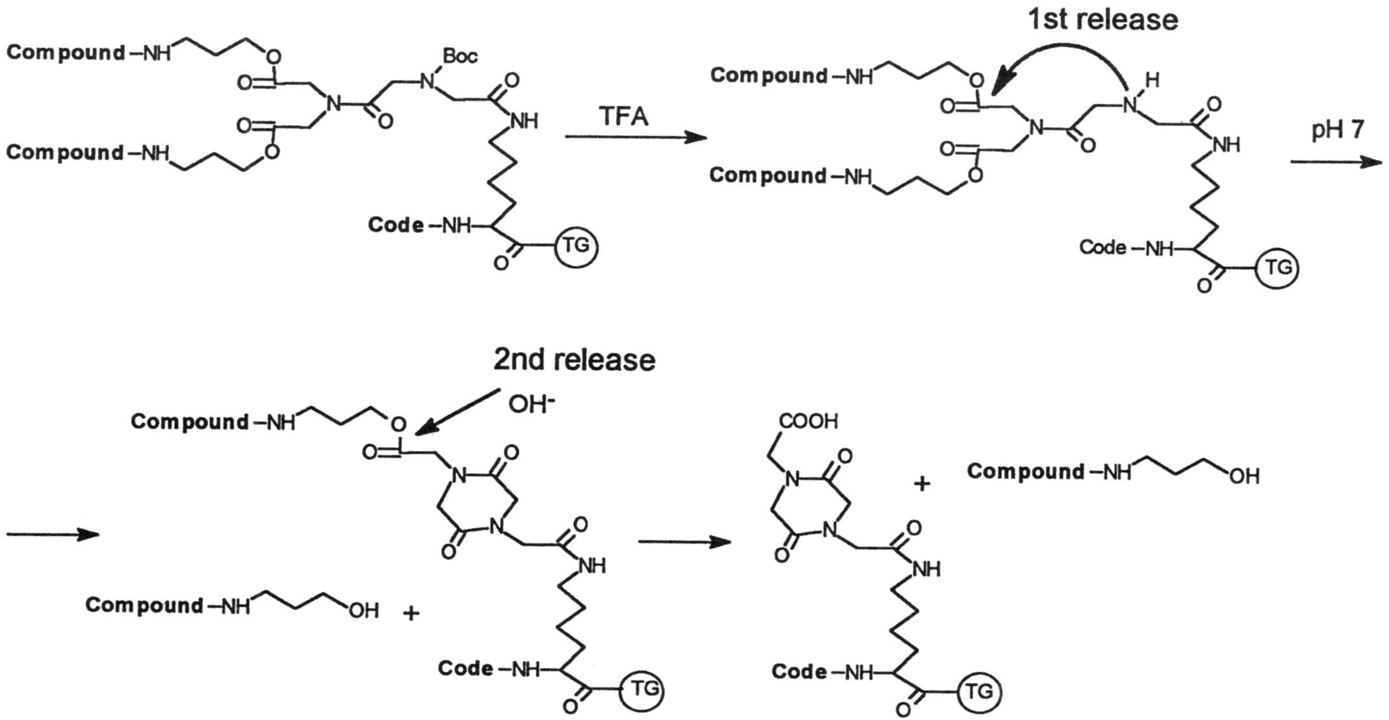

Figure 3.5. Chemistry and method of release of identical copies of a compound from a doubly cleavable branched linker. Branching of the linker enables two equivalents of peptide and one equivalent of coding molecule to be synthesized on the resin (indicated by the "TG") for every equivalent of resin loading capacity. The linker used permits one copy of peptide to be released by acid, removed and tested, and then the second copy of peptide can be released at neutral pH, removed and tested, and finally the identity of peptide can be determined by the coding molecule left on the resin bead. The released peptides have the identical chemical memory (hydroxyl rather than a c-terminal carboxyl group) of the linker attachment.

51

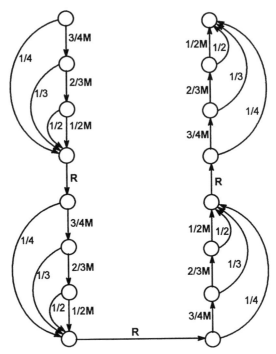

Figure 3.6. Method of synthesis of a "library of libraries": Each solid phase support contain a mixture of compounds representing a library of compounds ranging in length from 3 to 15 residues to evaluate all 3-residue pharmacophores. While it is not necessary for the 3-residue pharmacophore to be composed of contiguous amino acid residues, in peptides greater than six residues in length pharmacophores containing residues separated by more than three residue uni positions (e.g. the pharmacophore of amino acid 1, 5, and 9 of a nonapeptide) will not be represented.

of those three amino acids in the framework of the hexapeptide. In the case of 20 amino acids, each of these positional arrangements contain 8000 individual compositional motifs. Therefore, the complete library of tripeptide motifs in a hexapeptide framework, a "library of libraries," is composed of 160,000 species of motifs. This is a substantial reduction in complexity compared to the 64,000,000 individual compounds in a hexapeptide library.

The synthesis of a library of libraries requires splitting the resin into several aliquots—in the case of the above-mentioned libraries of hexapeptides, into up to six aliquots. In three aliquots a mixture of amino acids is coupled and in three aliquots the position is randomized between the 20 different amino acids during parallel coupling as in the standard one-bead–one-compound synthetic method. In total, 252 couplings are performed in the synthesis of this library (for the scheme see reference 61).

The scheme of synthesis of an alternative library with variable length is depicted in Figure 3.6. At the beginning of synthesis, and after each randomization step, one-quarter of the resin is separated and the mixture of amino acids is coupled to the remaining part. After this coupling, one-third of the resin is separated and the remainder undergoes coupling with the mixture of amino acids. The next coupling is performed with half of the resin from the previous coupling. All portions of the resin are then combined and a randomization is performed. Synthesis of a library of libraries with a three-amino acid motif, by this method, consists of three randomization steps and four stages of multiple couplings of amino acid mixtures. As a result, each solid-phase particle of the library is subjected to three mandatory randomization steps and as many as 12 acylations with the mixture of amino acids. This library, containing peptides of lengths from 3 to 15 residues, consists of 256 positional motif sublibraries. Among sublibraries of up to hexapeptides, all positional motifs are presented. However, because this synthetic scheme does not allow more then three successive acylations with the amino acid mixture, motifs in which pharmacophore positions are separated by more than three adjacent structural unit positions are not represented.

A significant increase in the diversity of peptide libraries can be achieved by using not only alpha but also other amino acids in the construction of the peptidic chain.[60] A peptide backbone can serve as the scaffold onto which a variety of building blocks can be attached via coupling to trifunctional amino acids (aminoglycine, diaminopropionic acid, diaminobutyric acid, ornithine, lysine, iminodiacetic acid, aspartic acid, glutamic acid, serine, hydroxyproline, cysteine, etc.).[54,62,63] Iminodiacetic acid is a convenient structural unit allowing construction of peptide-like libraries.[64] The scheme for its application to library construction is given in Figure 3.7.

Attachment of carboxylic acids onto the free amino groups was the basis for the construction of an alpha, beta, gamma library in which both the backbone and the side-chain arrangement was randomized.[65] On the border of peptidic and nonpeptidic libraries are the peptoids, or NSGs (N-substituted glycine peptides).[66,67] Libraries of peptoids were constructed either by coupling individual N-substituted glycines or by a submonomer approach, in which bromoacetic acid was coupled to the amino acid group of the growing peptoid chain, and primary amines were used to displace bromine in nucleophilic substitution, forming thus another N-substituted glycine in the chain[68] (see Chapter 6).

The linear arrangement of building blocks in peptide-like fashion was the basis for several library designs.[2,6,69,70] Because these designs produce nonpeptidic structures, they are not discussed here in detail.

Cyclization is a generally accepted method for decreasing the conformational flexibility of peptides. This restriction is expected to provide more potent ligands, as in the case of disulfide cyclic libraries containing ligand for the IIbIIIa receptor.[55] Cyclization can change the preference of the biological receptor for ligands, as shown in the study of streptavidin binders from cyclic libraries of various sizes.[71] A cyclic peptide composed of lysines and glutamic acid was used as a template onto which various carboxylic acids were

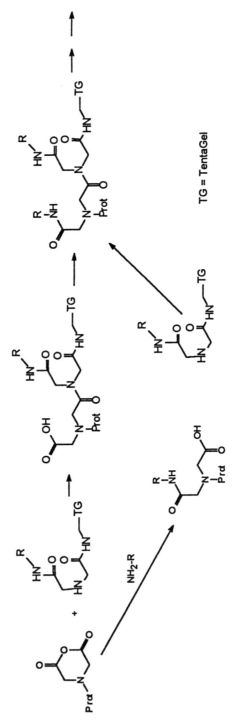

Figure 3.7. Introduction of side chain diversity: Use of a trifunctional amino acid, such as the depicted iminodiacetic acid, as a scaffold on which to synthesize diverse side chains while extending the peptide backbone the desired number of residues. Solid phase is represented by the "TG".

TG = TentaGel

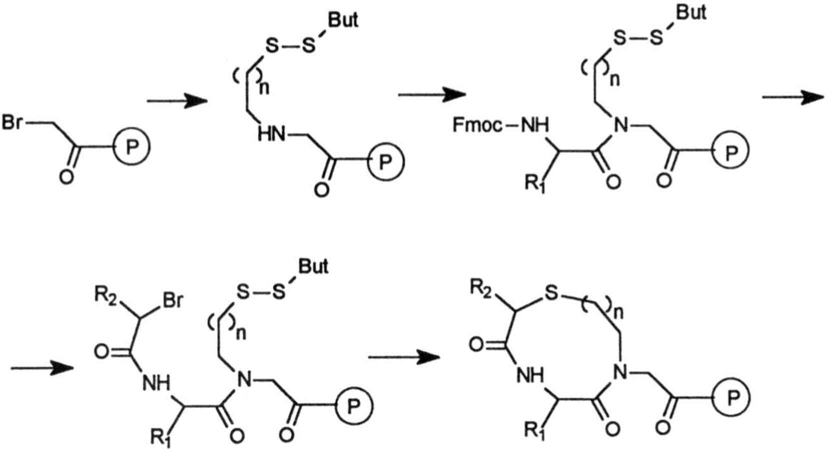

Figure 3.8. Synthesis of a cyclic-turn mimic on solid phase (indicated by "P"): The low molecular weight mimic displays side chains within the same spatial arrangement as the tertiary-turn structure of proteins.

attached.[72] Cyclic peptide libraries were studied by Spatola et al., and an optimal strategy for their synthesis was devised.[73] Ellman et al., synthesized a β-turn mimetic library with the same goal (Figure 3.8).[74]

Conformational restriction has also been achieved by constructing libraries with a bias toward α-helical conformation. One such library was synthesized by randomizing four positions in the sequence of the amphipathic helix (YKLLKKLLKKLKKLLKKL) on either the hydrophobic or hydrophilic sides of the molecule, from which peptides with increased antimicrobial activity were identified.[75]

The combination of peptide structure with nonpeptidic elements has been used for library design. Potent and specific zinc endopeptidase inhibitors were identified in a library of peptides with the amino terminus modified by the Z-Phe(PO$_2$CH$_2$) peptidomimetic group.[76] A similar approach was used in a Pfizer study of the endothelin antagonist developed by Fugisawa.[77] N-terminal substitution was kept intact and all amino acids were randomized by an array of natural and unnatural α- and non-α-amino acids. Promising leads were generated from a library constructed using acylation of the peptide chain by an array of three building blocks containing amino and carboxyl functionality and capped by a set of carboxylic acids.[78] The peptide chain in this case was used as the biasing element targeting the binding pocket of the Src SH3 domain. Vinylogous sulfonyl peptide synthesis was developed,[79] and libraries were used for the studies of synthetic receptors.[80] Boc protected vinylogous sulfonyl chlorides coupling in dichloromethane catalyzed by dimethylamino-pyridine monitored by bromophenol blue (for proper base[DBU] excess during the reaction) was used in the steps of library synthesis. Monitoring of the

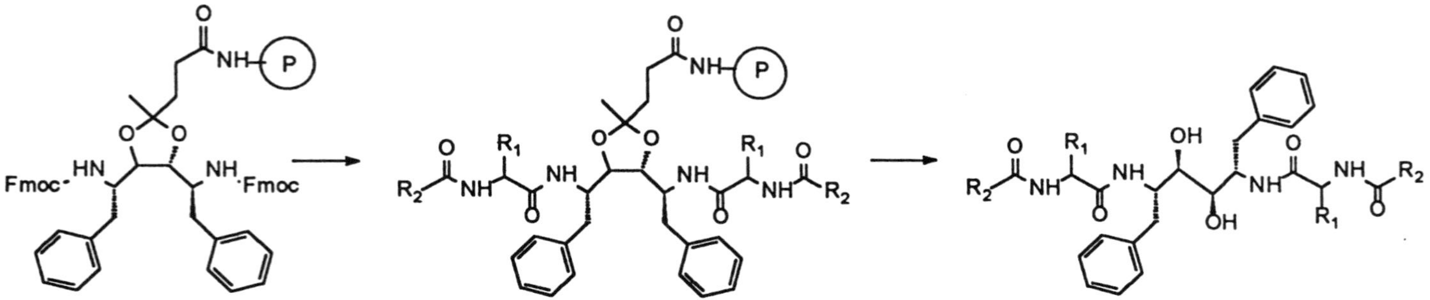

Figure 3.9. Incorporation of a target-specific (e.g. HIV protease) pharmacophore (diol motif) as the basic scaffold of a peptide library synthesized on solid phase (indicated by "P") followed by cleavage and testing of the free peptides.

reaction was found to be critical. Using an excess of the base destroyed the sulfonyl chloride, and even an excess of sulfonyl chloride did not result in complete amino group modification. Libraries of synthetic receptors were generated by the combinatorial synthesis of peptides on scaffold molecules such as a macrocyclic tetramine[81] or steroid molecule.[82]

Palladium-mediated macrocyclization was used in the synthesis of cyclic libraries. The carboxy terminal lysine side chain was acylated by acrylic acid, and the amino terminal group of the linear peptide was acylated by iodobenzoic acid. Pd(O)-mediated cyclization provided clean product in high yield.[83] The synthetic scheme of C2 symmetric inhibitors of HIV protease (Figure 3.9) serves as an example of solid-phase synthesis utilizing a specific feature of the target molecule (in this case the diol structure) to simultaneously protect the functional group critical to the function of the constructed molecules and at the same time use this protection as the attachment to the solid support.[84]

LEAD GENERATION AND OPTIMIZATION

Drug discovery is composed of two major elements: lead generation and lead optimization. The research reviewed in this section demonstrates that combinatorial peptide chemistry is a powerful tool for generating leads using generic libraries designed on general rather than target-specific principles. Combinatorial peptide chemistry is also a powerful tool for the optimization of leads through the synthesis and screening of optimization libraries designed based on a known lead or SAR. It is particularly powerful when combined with traditional medicinal chemistry and molecular modeling, though published examples of this are just now beginning to appear. Table 3.1 lists the targets for which leads have been identified and/or optimized from synthetic combinatorial peptide libraries. This list covers a variety of classes of targets. In some cases, investigators have pursued these targets as a means of validating a new method or the design of a combinatorial library. In other cases, information has been sought about the target and its chemical selectivity. In still other studies, the objective has been to identify or optimize a lead that can be used to modify the activity mediated by a specific target. Details of how each class of target has been pursued are discussed in the following sections.

Identification of Antibody Binding Domains and Putative Peptide Antigens

The identification of peptides that bind specifically to antibodies can be used to define the immunizing epitope within the native antigen. In addition, compounds can be identified that bind and elicit the same function as the native antigen, but that are structurally different, constituting a mimeotope—a mimic of the native antigen epitope. In either case there is the potential for identifying useful vaccines from synthetic combinatorial peptide libraries. There is also the potential to identify peptides that bind a portion of the eCDR region but

TABLE 3.1.

Target Molecule	Scientific Accomplishment	Method	References[a]
Multiple Synthesis			
Anti-VPI mAb	ID epitope, new method	Pin syn/solution assay, biased library[b]	7
Anti-Hemaglutin mAb	ID epitope, new method	Tea bag syn/solution assay, biased library	27
Anti-PLP mAb	ID epitope, new method	Spot syn/screen on polypropylene, biased library	85
Anti-CMV human sera	Map epitope, new method	Spot syn/screen on cellulose, biased library	86
Anti-opiate mAb	ID epitope, new method	Photochem syn & screen on glass, biased library	87
Anti-gp120 mAb	ID epitope	Microplate syn., biased library	88
Anti-gp160 mAb	ID epitope/vaccine	Pin syn/solution assay, biased library	89
Anti-TGFα mAb	Epitope ID	Spot syn/assay	90
Streptavidin	Novel binder	Spot syn/assay, surface presentation	91
DNA Binding	Novel binder	Spot syn/assay, double coupling	39
Vancomycin	Novel binder	Affinity capillary electrophoresis selection	92
c-AMP dep./c-Abl kinases	Kemtide analog	Autophosphorylation of library cpds, LCMS detection, biased library	93
Acetylcholinesterase	Inhibitor SAR	Syn & spot on TLC plate, novel detection, biased library	94
Chymase/ chymotrypsin	Inhibitor SAR, optimize specificity/activity	Pins, biased library	95
NK1 receptor	Antagonist SAR	Pin syn/solution assay, biased library	96,97
NK3 receptor	Novel antagonist, SAR, optimize activity	Microplate syn, novel information rich library	98
Split and Mix, One-Compound–One-Bead, Postassay Identification Format Libraries			
Anti-β-endorphin mAb	ID epitope	New method/on bead	9
Anti-β-endorphin mAb	ID epitope	Double release	55

TABLE 3.1. (*Continued*)

Target Molecule	Scientific Accomplishment	Method	References[a]
Split and Mix, One-Compound–One-Bead, Postassay Identification Format Libraries			
Anti-β-endorphin mAb	ID epitope	On-bead/radiolabel	99
Anti-β-endorphin mAb	ID epitope	On-bead/coded	100
Anti-cMYC mAb	ID etipope	On-bead/coded	101
Anti-gp120 mAb	ID epitope	On-bead/MS sequence	102, 103
Anti-measles virus mAb	ID epitope/vaccine	On-bead	104
B-cell surface idiotype	ID epitope	On-bead	105, 106
MHC Class I (HLA-A2 * B7)	ID binding motif	On-bead	107
Streptavidin/avidin	Novel/specificity	On-bead	9, 108, 109
Neural neurophysin	Oxytocin binding, methods validation	On-bead, magnetic bead selection, biased library	110
Src, Pl-3-kinase SH3	Novel/putative substrates from database search & ligand/SH3 NMR structure determined	On-bead	111 112
c-AMP dep. kinase	Specificity of substrates	On-bead phosphorylation/biased library	113
Src kinase	Novel/putative substrates	On-bead phosphorylation	114, 115
Hepatitis A 3C endoproteinase	Specificity of substrates	On-bead/novel sequencing method, biased library	116
Stromeylisin/collagenase	Novel/specificity of substrates	On-bead/novel fluorescence detection method	117 118, 119
Endopeptidase subtilisin Carlsberg	Novel/substrate pH specificity	On-bead/novel direct fluorescence detection method, biased library	120
Edopeptidase subtilisin Carlsberg	Novel inhibitors	On-bead/novel indirect fluorescence detection method	121
Protein tyrosine phosphatase	Inhibitors, optimize specificity/activity	Solution, rf chip coding, biased library	122
Trypanosomal gPG-kinase	Trypanosom-specific inhibitors	On-bead, fluorescent & Magnetic bead selection	123

TABLE 3.1. (*Continued*)

Target Molecule	Scientific Accomplishment	Method	References[a]
Split and Mix, One-Compound–One-Bead, Postassay Identification Format Libraries			
Leukocyte elastase	Inhibitors, optimize specificity, activity	On-bead, biased library	124
Thrombin	Inhibitor SAR, novel lead/optimize activity	On-bead, biased library	125
Factor Xa	Novel inhibitor lead, SAR/optimize activity	On-bead (lead from generic library, optimize with biased library)	126
Elastase, subtilisin, trypsin, etc.	Inhibitor SAR	Solution/novel affinity column-sequencing method, biased cyclic scaffold library	127
gpllb/llla receptor	Antagonist	On-bead & double release	55
C5a receptor	Agonist–antagonist switch, SAR	Double release, biased library	128
G-protein coupled receptors	Novel agonists & antagonists, SAR	Double release, novel melanocyte functional assay, biased library	129 130
Antimicrobial	Growth inhibitors	Single release, agarose plaque forming assay	131
Indigo carmine	Small molecule	On-bead, direct color detection	132
Iterative Synthesis and Screening Library Methods			
Anti-peptide mAb	ID epitope	New method/solution	8
Anti-peptide mAb	ID epitope	Solution/novel pooling	51
Anti-β-endorphin mAb	ID epitope	Solution	133
Anti-β-endorphin mAb	ID epitope	On-bead	10
Anti-β-endorphin mAb	ID epitope	Liquid syn/screening	47
Anti-TGFa mAb	ID epitope	Solution, spot, double couple	39
Anti-gp120 mAb	ID epitope	Robotics/affinity selection	134
Anti-LPS mAb	Discover novel epitope	NMR friendly scaffold library	135

TABLE 3.1. (*Continued*)

Target Molecule	Scientific Accomplishment	Method	References[a]
Iterative Synthesis and Screening Library Methods			
Anti-lysozyme mAb	Discontinuous epitope presentation	Rigid scaffold library	136
cAMP/cGMP kinases	Novel/substrate specificity	Solution	20
HIV-1 protease	Novel inhibitor/SAR, optimize activity	Solution, biased statin library	137
Trypsin	Novel inhibitor/SAR, optimize activity	Solution, lead generation/ optimization	24, 138
Zinc endopeptidases	Optimize inhibitor selectivity/activity	Solution, biased library	116
Endothelin receptor	Antagonist SAR, optimize activity	Solution, biased substance P library	139 140
Endothelin receptor	Antagonist SAR	Solution, biased antagonist FR-139317 library	77
μ-Opiate receptor	Novel agonists, SAR, optimize analgesic activity & selectivity	Solution	141–3
Antimicrobial	Growth inhibitors	Solution	144 145
Positional Scanning Format Library Methods			
Anti-peptide mAb	ID epitope	New method	146, 147
ACE and other enzymes	Substrate	Solution/novel seq. method	148
Chymotrypsin	Novel inhibitors	Tea bag, cyclic template library	72
IL-6 receptor	Novel antagonists, SAR	Solution, branched "polydentate" library	149
MHC class 1	Novel ligands binding MHC-1, &/or sensitize target cells for cytolysis by T-cell	Solution	150 151
RBC lysis	Inhibition of melittin mediated lysis	Solution	152

[a] Unless noted, libraries used were generic in design. Biased libraries include overlapping, deletion, truncation, or biased amino acid selection based on sequence of native antigen or ligand.

that do not elicit an antibody response, particularly when the native antigen binds through discontinuous epitopes or a conformational epitope.[106] In addition, due to inherent tight binding and a wide variety of commercially available reagents and assays, antibody binding has been used as a method for validating new approaches in the application of synthetic combinatorial chemistry.

Multiple-Synthesis Libraries. In the first study published describing the multiple synthesis of peptides on fixed solid supports in a 96-well microplate array, the target used to validate the approach and demonstrate utility was an antibody.[7] The pins, which had been dipped in reaction solutions to synthesize a library of known compounds, were dipped in assay medium containing a monoclonal antibody raised against the immunologically important coat protein (VPI) of foot-and-mouth disease virus to identify peptides that bound the antibody. The authors first synthesized 108 different overlapping hexapeptides required to cover the complete 213-amino acid sequence of VPI. From the results they identified GDLQVL as the epitope recognized by the antibody. They then synthesized a library of 120 different peptides in which all possible single-residue replacements were made using the 20 proteinogenic amino acids. This identified the two leucine residues as critical for antibody binding.

Another early study defining antibody–peptide recognition was reported by Houghten.[27] He described the multiple synthesis of peptides by segregating resin in individual permeable containers ("tea bags"), then simultaneously coupling amino acids to the segregated resin of a number of such tea bags suspended in a variety of reaction vessels. Each tea bag was tracked through the synthetic scheme so that at the end of the process there was a library of compounds synthesized simultaneously in a one-compound–one-bag format, ready for screening after cleavage from the resin. Using this method 238 different 13-mer peptides were synthesized, reflecting single amino acid variations of a sequence from hemagglutinin protein (residues 98–110, YPYDVPDYASLRS), which has been used to raise a monoclonal antibody. The individual peptides cleaved from the bags of resin were preadsorbed to microtiter plate wells to constitute an ELISA screening assay, and the antibody binding to these peptides was monitored. Houghten demonstrated that with this approach the residues critical for binding could be identified as, D_{101}, D_{104}, and A_{106}, validating the tea-bag multiple-synthesis concept and utility.

Antibody binding was used to validate a multiple-synthesis approach in which peptides are synthesized as spots on commercially available peptide synthesis polypropylene membranes (Millipore, Bedford, MA) and then assayed either while attached to the membranes by an ELISA-type assay or following release from the membranes.[85] A related synthesis method had been described using cellulose membranes.[41,86] The membranes were mounted on a 96-well synthesis device. The library consisted of all possible octamers (1–8, 2–9, . . . , 269–276) derived from bovine myelin proteolipid protein (PLP, 276 amino acid peptide), which had been used as the antigen for producing the target molecule anti-PLP antibody. Synthesis was complete in 16 h. The polypropylene sheet was then incubated with the anti-PLP antibody, and bound antibody was detected with protein A conjugated to alkaline phosphatase via a calorimetric enzyme assay. The dominant epitopes mapped by this approach corresponded to peptides spanning residues 38–46 and residues 195–205. The authors thus not only validated a novel method for multiple synthesis, but also identified the binding epitopes for this anti-PLP antibody.

Frank et al. used serum from humans that was immunopositive for a 58-amino acid sequence from human cytomegalovirus 36/40 K protein (anti-CMV) as a test system.[86] They synthesized a library of 49 overlapping decapeptides on a sheet of cellulose paper, spotting the reaction mixtures as separate points of a grid to form an array of compounds. With the peptides still bound, the cellulose paper was incubated with the CMV positive serum. Positive spots were detected by identifying those to which antibody from the serum bound (using an anti-human β-galactosidase-conjugated second antibody). In this way the authors could map the binding epitope of individual human sera. This method was also used to characterize the epitope recognized by an anti-TGFa antibody, Tab2, by synthesizing a hexapeptide library spotted on cellulose of 1728 mixtures in the format XOOOOX, were X represents the positions where residues were randomized, defined by the spot location, and O represents positions where one of six amino acid mixtures was incorporated (A,P,G; D,E; H,K,R; N,Q,S,T; F,Y,W; I,L,V,M).[90]

A similar library was generated by phage display. The synthetic combinatorial library identified sequences that were native to TGFa (SHFNDC, K_d = 200 nM, and VSHFND, K_d = 80 nM) as well as a large number of other sequences with comparable affinity. The phage library identified the native HFND sequences (ASHFND, K_d = 60 nM; HFNDYL, K_d = 900 nM; and HFNDCI, K_d = 8000 nM), but, unlike the synthetic library, not the complete native sequence VSHFND or any novel sequences.

Stigler et al. characterized the binding epitope of the Fab fragment of an antibody (3D6) raised against the transmembrane protein gp41 of HIV-1 using a library of overlapping peptides, a random library, and molecular modeling.[153] The overlapping peptide library identified CSKLICTTAVPW, which was used to synthesize a secondary library on cellulose sheets to generate a library of all possible L and D conformers (494 in all). This identified critical residues and this information, together with low-resolution X-ray of the peptide bound to the Fab fragment, enabled the authors to construct a model of the bound conformation of the peptide and interactive residues. This study provides a strategy for constructing models of protein–peptide complexes using peptide library results to generate empirical experimental data to augment computer modeling and low resolution crystallographic approaches.

A useful method for determining antibody binding epitopes was demonstrated using photochemistry and light-directed synthesis on a sheet of glass to create an array of overlapping epitopes against which to screen an antibody.[87] The epitope of an antibody, C 32.39 anti-dynorphin-β-antibody, raised against the opioid sequence YGGFLRRQFKVVT was identified by synthesizing a library of 1024 compounds consisting of the single amino acids as well as all di-, tri-, (etc.) peptides to cover all possible deletions (single, double, etc.) and truncations of, while not scrambling, the original sequence. This represents a very different strategy than simply the synthesis of overlapping sequences within the immunizing peptide. Fluorescently labeled antibody bound to the individual compounds displayed as an array on the glass surface was monitored, with the location on the glass giving the identity of each compound. Deletion sequences containing RQFKVV (four in all) exhibited high fluorescence. From the truncation results, loss of T, especially T plus V, from the C-terminus resulted in a loss of activity, while binding increased with truncation of the N-terminus to the second R, again identifying the RQFKVV(T) sequence as the epitope recognized by antibody. H-YGGFLRRQFKVVT-OH bound with an IC_{50} of 0.58 nM, while Ac-RQFKVVT-OH bound with an IC_{50} of 1.7 nM, and Ac-RQFKVVT-NH$_2$ bound with an IC_{50} of 0.49 nM.

Numerous other studies have demonstrated the utility of synthetic peptide libraries in the mapping of antibody binding domains and the production of potent antigens. Pinilla et al. recently demonstrated that calcium independent antigens could be identified by screening a library against a calcium-dependent monoclonal antibody; the anti-FLAG antibody M1, which binds DYKDDDDK-NH$_2$ in the presence of calcium and is used for protein purification.[49] The IC_{50} of DYKDDDDK-NH$_2$ determined with and without calcium was 14 and 274 nM, respectively, while a peptide identified from a hexapeptide library, DYKAKE-NH$_2$, exhibited an IC_{50} of 38 versus 3 nM in the presence and absence of calcium, respectively, and one identified from a decapeptide library, DYKEKELEDD-NH$_2$, exhibited an IC_{50} of 19 versus 2 nM with and without calcium, respectively.

Experiments with an anti-gp120 antibody identified an equipotent sequence containing nonnatural residues from a highly biased decamer library based on the native sequence Ac-RAFHTTGRII-NH$_2$.[88] The library, Ac-RA(X)HT-TG(X)I(X)-NH$_2$, produced an equipotent submillimolar affinity compound, Ac-Ra(L-napthylalanine)HTTG(R)I(L-norvaline)-NH$_2$, and related napthyla-lanine containing analogs due to the pooling method used in screening.

The multipin scanning approach of combinatorial peptide synthesis[154] was used to identify the minimal and optimal length of an endogenous sequence from HIV-1 gp160, which serves as the recognition domain for binding to the MHC class I D^d complex of murine cytotoxic T lymphocytes (CTL) to elicit a gp160 specific CTL response (lysis of peptide-binding, D^d-expressing target cells).[89] The approach consisted of synthesizing an array of either N-terminal and C-terminal truncations of the 15-mer sequence from the native protein (residues 315–329). The minimal sequence required for recognition based on this experiment consisted of an 8-mer, PGRAFVTI, residues 320–327. In the second set of experiments the C-terminus was either extended or truncated, and for each C-terminus (331, 330, 329, 328), a set of six different N-terminally truncated peptides (residues 318 to 323) was synthesized to create, for example, six 331 C-terminal peptides with an N-terminus of residue 318, 317, 316, 315, 314, or 313, respectively. Assaying this set of compounds determined that truncation beyond residue 320 resulted in loss of activity regardless of the C-terminal extension, confirming the minimum sequence, and identifying the most active peptides as the 9-mers PGRAFVTIG and GPGRAFVTI or the 10-mer RGPGRAFVTI (residues 318–327).

A corroborating study using a conventional approach also identified the 10-mer (residues 318–327) as more active than the native 15-mer.[155,156] When the 10-mer peptide was incorporated into VAC recombinants and used to immunize mice, CTL lines obtained from the immunized mice exhibited specific priming by synthetic peptide comparable to that induced by wild-type, full-length gp160, demonstrating that peptides identified by synthetic combinatorial chemistry efforts can be used effectively as the basis for vaccine production.

Split-and-Mix, One-Compound–One-Bead, Postassay Identification Format Libraries. Antibody binding was used to validate the use of the split-and-mix approach to synthesizing and screening libraries in a one-compound–one-bead, postassay, directly sequenceable identification format.[9] A library of peptides was synthesized on beads using the split-and-mix strategy[11] to assure that each amino acid coupled regardless of its inherent reactivity and to assure that there would be only one species of compound on each resin particle (Figure 3.10). With the peptides still attached to the beads, the library was incubated with mouse monoclonal anti-β-endorphin. This is an antibody that recognizes a sequence in the native ligand (binding affinity, 17 nM). Antibody binding to beads was detected using a secondary antibody conjugated to an enzyme reporter (e.g., alkaline phosphatase), the positive beads were collected, and the peptide on each was sequenced to determine the identity of the individual compounds. The ligands identified from peptide libraries in which 19 of the proteinogenic amino acids were coupled at each position included YGGF(X) motifs, among them YGGFQ (15 nM affinity). The authors demonstrated that a positive bead could be treated to remove the target molecule and reassayed to confirm binding or to assess specificity. Furthermore, the bulk of the on-bead library remained after screening (minus any positive beads) and could be reused in additional screens.

A significant problem with the use of streptavidin to identify the bound biotinylated antibody is that streptavidin itself binds to many peptides, and hence can bind directly to beads (examples given below). To avoid this, two independent methods can be used to recognize the bound biotinylated antibody: streptavidin and an anti-mouse antibody. (Lam et al, 1995). Beads were first labeled with the streptavidin/alkaline phosphatase conjugate, colorized (turquoise) with the substrate BICP, and selected. The selected beads were incubated with anti-mouse antibody conjugated with alkaline phosphatase, and colorized with BCIP plus NBT, turning positive beads dark purple.

Split and Mix, One-Bead-One-Compound Synthesis

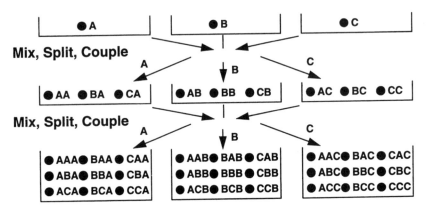

One-Bead-One-Compound Screening

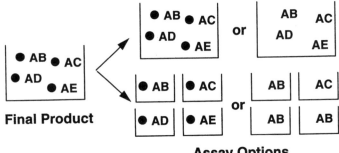

Figure 3.10. Split and mix synthesis and screening to produce one-bead-one-compound libraries for assay as mixtures of compounds or single compounds attached to the solid phase (indicated by solid circles) or as mixtures or single compounds released into solution from the solid phase. Release is carried out in a relational manner (to the solid phase bead of origin), such that once activity is detected of release compound it is possible to directly determine compound identity from chemical material remaining attached to the solid phase bead.

Those beads remaining turquoise (not binding the anti-mouse antibody) were sequenced, and it was shown that the peptides were streptavidin binders. Those that turned dark purple proved to be YGG motif peptides, true anti-β-endorphin binders.

In a follow-up study the approach was further modified to permit the controlled double release and assay of the library peptides in solution.[55] A pH-sensitive linker was employed that was stable in aqueous solution at acidic pH (4.5) but underwent a ring closure and cleavage to release compound (one-third of the total) from each bead at pH 7 in the form of

XXXXX-NH-$(CH_2)_3$-OH.[56] Released in randomly generated pools of 500 beads per well, the mixtures were assayed in an ELISA-type assay to identify active pools. The beads corresponding to these active pools were then recovered and redistributed, 1–2 per well, and subjected to basic pH (0.2% NaOH) to cleave another third of the compound attached through an ester linkage, again releasing molecules with the structure XXXXX-NH-$(CH_2)_3$-OH. After assay and identification of the positive wells, the corresponding beads were recovered and sequenced to determine identity (using the remaining third of compound which had been synthesized onto a protected linker). The identified compounds were synthesized and their activity was confirmed. This follow-up study validated the double-release assay through the identification of the spiked control YGGFL (30 nM) as well as YGGFG (200 nM), YGVFG (1000 nM), and YGAFG (700 nM) from the assay of 500,000 library beads. The double-release assay is a powerful technique, now widely used by many laboratories.

In a related study the on-bead assay was conducted by incubating beads with radiolabeled target molecule, immobilizing the beads in a thin layer of agarose, identifying positive beads by autoradiography, and manually selecting beads for sequencing.[99] The target used was again an anti-β-endorphin monoclonal antibody. From a pentapeptide library the authors identified YGSFE, YLWFQ, and YGAFE, clearly related to the known YGGFL sequence, and in the case of YGAFE, nearly identical to a peptide found using the double-release assay described above. Like Lam et al.,[9] the authors demonstrated that the recovered beads could be treated to dissociate the target and reassayed several times.

An on-bead method using a sequenceable coding molecule, rather than relying on sequencing of the active compound itself, was validated using the same antibody.[100] The assay was carried out using fluorescently labeled antibody and a fluorescence activated cell sorter to select the (positive) fluorescent beads. A heptapeptide library was synthesized on 10-mm macroporous beads, which permitted the target protein to permeate the bead and reach compound synthesized inside. The method was first validated using the known epitope RQFKVVT ($K_d = 0.5$ nM) synthesized onto beads. Two percent of the library beads stained above background and the bound fluorescent antibody could be completed by excess RQFKVVT, demonstrating specificity of binding. Peptides were identified from the library that, when synthesized with the threonine linker (T), ranged in binding affinity from a K_d of 0.3 to 1400 nM. The sequence of the most potent ($K_d = 0.29$ nM) was TFRQFKV(T), a close analog of the native epitope RQFKVV(T) ($K_d = 0.51$ nM).

An alternative coding method was validated using a monoclonal anti-c-MYC antibody as the target protein.[101] In this case the coding molecules consisted of a binary digital code arrangement of electrophores that could be separated and detected by electron capture gas chromatography (see Chapter 14). In this initial work the codes were attached to a fraction of the test compound, but the coding molecules can also be synthesized directly onto the polystyrene bead. In digital coding the position and identity of each residue is represented by a three-"residue" bit code, where no residue is considered one bit of information, just as would be a specific coding residue. The library synthesized was a heptapeptide library of the format H_2N-XXXXXXEEDLGGGG-bead, containing D, E, I, K, L, Q, and S randomized at each position. It was known that the anti-c-MYC antibody 9E10 bound EQKLISEEDL, hence the selection of amino acids for the library. Using the approach described above (mixing library beads with the anti-c-MYC antibody and detecting those beads to which antibody bound using a secondary alkaline phosphatase conjugated antibody) the authors identified and sequenced the codes for 12 beads. Upon resynthesis and testing in solution the three most potent ($IC_{50} \sim 1$ mM) were EQKLIS(EEDL), LQKLIS(EEDL), and QQKLIS(EEDL), all closely related to the known epitope, validating this coding methodology.

Screening using the anti-gp120 antibody was used to validate a laser desorption mass spectroscopic method of sequencing to determine the identity of compounds on positive beads.[102,103] A synthetic hexapeptide library was synthesized in such a manner that at the end of each synthetic step a portion of the sequences were capped to prevent their complete synthesis in the remaining scheme. Thus, each bead contained the complete test peptide as well as capped truncations. The library beads were incubated with the anti-gp120 antibody, and positive beads were identified using a secondary antibody conjugated to alkaline phosphatase. Each positive bead was selected and its peptide contents cleaved and subjected to laser desorption mass spectroscopy. The spectrum contained the molecular ions for the complete compound plus all the possible capped analogs. The differences in molecular weight could be used to identify the composition of the parent peptide. This method can be applied to any molecular library, not just peptide libraries.

From the first study a family of peptides were identified, which provided a consensus binding epitope of (F/P)**GRAF**(Q/X)(F/X). This corresponded well with the peptide from the HIV gp120 protein used to produce the antibody originally, RIQRGP**GRAF**VTIGK ($IC_{50} = 11$ nM), compared to an $IC_{50} = 50$ nM for PGRAFQF. In the second study a pentapeptide library was synthesized, with and without acetylation, and screened, resulting in the identification of a XGRAF motif and an Ac-PG(R/X)AF(X/R)F motif. Ac-PGRAFQF exhibited an IC_{50} of 525 nm. The authors also screened against streptavidin (see below), identifying the HPQ motif with and without acetylation.

Another example of identifying a vaccine from a synthetic combinatorial library has been published.[104] Steward et al. synthesized an octamer library, screened on-bead with a monoclonal antibody raised against the F protein of measles virus (anti-MVF antibody F7-21). Six peptide mimetopes were identified, and when used to immunize mice, one (NIIRTKKQ), which cross-reacted with the antibody, inhibited measles virus plaque formation and acted as a vaccine to protect mice against fatal encephalitis induced by infection with measles.

Iterative Synthesis and Screening Library Methods. Hexapeptides were identified that bound a monoclonal antibody that recognized a 13-residue peptide (Ac-YPYDVPDYASLRS-NH$_2$) using an ELISA-type assay with the library peptides free in solution to identify antibody binding epitopes.[8] A total of 324 different pools were synthesized with the format Ac-O$_1$O$_2$XXXX-NH$_2$ in which the first two positions of each were defined for each pool. Screening identified one pool containing Ac-DVXXXX-NH$_2$. Subsequently, 20 pools were synthesized with the format Ac-DVO$_3$XXX-NH$_2$ and screened to define O$_3$; the process iterated until each position had been defined. In the process of defining the last position a set of individual peptides rather than mixtures was synthesized and assayed, having the same O$_{1-5}$ residues, but variable O$_6$ residues. Among them was the native sequence Ac-DVPDYA-NH$_2$ (IC$_{50}$ = 30 nM), which was also the most active. Corroborating studies using traditional approaches confirm this to be the antigenic determinant of the 13-residue peptide.[157,158] Using this approach the same group defined the binding epitope (STTS) of a monoclonal antibody raised against a surface antigen of hepatitis B, identifying the peptides Ac-STTSLI-NH$_2$ and Ac-STTSLM-NH$_2$ (IC$_{50}$ ~ 1.8 and 2.6 mM, respectively).[146] Similar results were found using a positional scanning library approach (see below).

The success of the iterative approach, together with a novel pooling strategy, was also demonstrated using both an antiserum to a tetrapeptide as well as a monoclonal antibody to a 28-amino acid peptide.[51] The antiserum used was raised against the peptide FMRF amide, but because the authors wished to exclude methionine, cysteine, tryptophane, threonine, and isoleucine from their library (reducing the number of amino acids used to 15) the ligand bound to microplates forming the basis of their antibody capture ELISA assay was FLRF. This strategy was somewhat unusual. The authors divided their amino acid mixtures into three groups, a(L,A,V,F,Y), b(G,S,P,D,E), and g(K,R,H,N,Q), with the individual amino acids present at different molar ratios designed to compensate for differences in reactivity. One strategy would have been to synthesize the library as 81 pools representing all possible positions within the tetramer and each of the three groups of amino acids. The authors instead synthesized a pool of compounds containing equal proportions of all 15 amino acids at each position (a 1:1:1 mixture of amino acid groups a:b:g), measured the activity of this pool (after the compounds were cleaved from the resin), and then compared this to at first four pools, which consisted of

AA1	AA2	AA3	AA4	
	Activity/Interpretation			
1:1:1 a:b:g	1:1:1 a:b:g	1:1:1 a:b:g	1:1:1 a:b:g	null control
2:1 b:g	1:1:1 a:b:g	1:1:1 a:b:g	1:1:1 a:b:g	<;a
1:1:1 a:b:g	2:1 b:g	1:1:1 a:b:g	1:1:1 a:b:g	<;a
1:1:1 a:b:g	1:1:1 a:b:g	2:1 b:g	1:1:1 a:b:g	=;g
1:1:1 a:b:g	1:1:1 a:b:g	1:1:1 a:b:g	2:1 b:g	<;a

Proportionate loss of activity was used to identify critical residues. The first round of screening results indicated a group residues were required at positions 1, 2, and 4 for activity, and g group residues were necessary at position 3. The next round was designed to identify triplets of amino acids from among these pools, using as the null control a, a, g, a, and four libraries:

AA1	AA2	AA3	AA4	
	Activity/Interpretation			
a	a	g	a	null control
1:1:3 LAV	a	g	a	<;F/Y
a	1:1:3 LAV	g	a	=;A/L
a	a	1:1:3 KRH	a	=;K/R
a	a	g	1:1:3 LAV	<;F/Y

The last round used the defined dipeptide mixtures as null control and the following four pools:

AA1	AA2	AA3	AA4	
	Activity/Interpretation			
F/Y	A/L	K/R	F/Y	null control
F	A/L	K/R	F/Y	=;F
F/Y	A	K/R	F/Y	<;L
F/Y	A/L	K	F/Y	<;R
F/Y	A/L	K/R	F	=;F

Thus with three rounds of iteration, synthesizing 12 pools of peptides representing 50,625 peptides, the authors had defined the peptide FLRF.

Having shown that the approach worked, the authors then applied the method to the case of the monoclonal antibody, raised against the 28-amino acid peptide Ac-RTPALG**PQAGID**TNEIAPLEPDAPPDAC amide, and for which the epitope had not been defined. The difference was that a hexapeptide format was pursued, so that there were 6 pools in each round of iteration (18 pools synthesized, representing 16,777,216 peptides), but otherwise the strategy was identical. The result was the definition of RQVGHD amide. The authors used this to identify the consensus region of the 28-mer to be PQAGID (bolded above), and confirmed this by synthesizing seven peptides and assessing their relative activity (shown in brackets) compared to the "native" peptide PQAGID [2.8]: RQVGHD [1.2], QVGHD [>750], RQVGH [128], EQVGHD [14], PQVGHD [0.08], RQAGHD [4], and RQVGID [17]. The loss of activity in the N- and C-terminal deletion analogs demonstrated specificity, and the ~35-fold improvement in activity of PQVGHD over the native peptide PQAGID confirmed specificity and demonstrates the power of the approach for optimization.

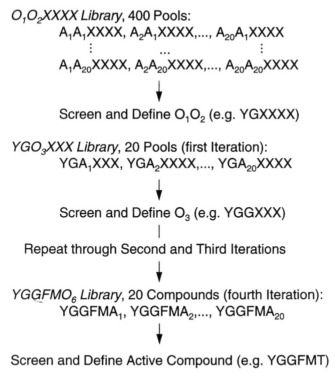

O_1O_2XXXX Library, 400 Pools:
$$A_1A_1XXXX, A_2A_1XXXX,..., A_{20}A_1XXXX$$
$$\vdots \qquad \qquad ... \qquad \qquad \vdots$$
$$A_1A_{20}XXXX, A_2A_{20}XXXX,..., A_{20}A_{20}XXXX$$

Screen and Define O_1O_2 (e.g. YGXXXX)

YGO_3XXX Library, 20 Pools (first Iteration):
$$YGA_1XXX, YGA_2XXXX,..., YGA_{20}XXXX$$

Screen and Define O_3 (e.g. YGGXXX)

Repeat through Second and Third Iterations

$YGGFMO_6$ Library, 20 Compounds (fourth Iteration):
$$YGGFMA_1, YGGFMA_2,..., YGGFMA_{20}$$

Screen and Define Active Compound (e.g. YGGFMT)

Figure 3.11. Iterative Deconvolution of Mixtures: A library composed of 6-mers requires five steps of synthesis and assay of the resulting mixtures, with compounds free in solution. The results of such an approach may be biased by the synthetic sequence, since the first two residues (O_1O_2) defined in the first set of mixtures have considerably more impact on binding than the residue (O_6) identified in the last set of mixtures.

In another study, peptides that bind anti-β-endorphin (3E7) were elucidated using iterative synthesis and screening of a series of hexamer libraries (Figure 3.11). Peptides were cleaved from resin as mixtures and assayed free in solution without separation in an ELISA-type format.[133] The antigen peptide (β-endorphin) was adsorbed onto the surface of microplate wells, the peptide mixture plus antibody was added, incubated, and washed, and then bound antibody was detected. In the first stage 400 libraries were synthesized, each containing specified amino acids (O_1O_2) in the first two positions, with the remainder (X) synthesized by splitting and mixing and cleaved as a complete mixture as regards the X positions (format O_1O_2XXXX). Once the first two positions (A_2A_2) were defined, 20 libraries were synthesized, with the third position (O_3) identified in each (format $A_1A_2O_3XXX$). This process continued until the complete peptide was defined, YGGFMT-NH$_2$ (IC$_{50}$ = 3.2 nM), a peptide identical to the first six residues of β-endorphin.

The most recent variation on the iterative approach is single-step deconvo-

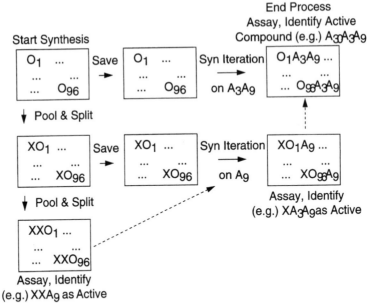

Figure 3.12. Recursive Deconvolution of Mixtures: Compounds composed of 3-mers require three assays and two additional steps of synthesis performed on starting material split and saved from each step of the original synthesis, with the advantage that results are not biased by synthetic sequence. In each case the last building block is added to mixtures in discrete wells of an array, such that the identity of each building block is known from the spatial location in the array of the active mixture. Working backwards (recursively) identifies the active compounds.

lution or recursive deconvolution where automation is utilized to simplify the iterative process (Figure 3.12). In recursive deconvolution the library is split after each synthetic step into samples that are saved for subsequent synthesis and samples that are pooled and split in a many-compounds–one-well format (rather than the one-compound–one-bead format) for the next round of synthesis.[10] Thus, for instance, at the end of the synthesis of a tripeptide library there would be multiple wells containing pools of compounds with the identity of the last amino acid unknown. Screening would identify the last amino acid, and then the investigators would take the previous saved plate of dipeptide mixtures, in which the second residue was known because of well location, synthesize onto these mixtures the identified third residue, and then screen to identify the second residue. Once identified, the investigators would go to the first round of saved single amino acid plates, synthesize the second and third residues onto each, and identify the active tripeptide. The authors synthesized a pentapeptide library, but with only four amino acids (G, L, F, Y) used at each position to create a library of 1024 members, including a peptide to bind the anti-β-endorphin antibody. This approach eliminates the need for tags, dependence on resin bead size to provide sufficient assay material, and some of the synthesis steps required in the iterative approach. The authors screened the library compounds in the anti-β-endorphin antibody ELISA assay with compounds attached to the beads to assess whether compounds in the mixture prevented the antibody from binding to the [Leu$_5$]enkephalin–BSA-coated surface. The authors identified the known peptide NH$_2$-YGGFL as well as two other peptides, NH$_2$-YGGFF and NH$_2$-YGGLL, validating their recursive deconvolution method.

An important issue in all the approaches where a mixture of amino acids is reacted in the same reaction vessel is the fact that some will couple at a faster rate than others, producing an unequal representation within the pool. The rates of reaction are somewhat inherent in the amino acid itself, but are also dependent on the sequence to which the amino acid is being coupled. Many investigators have tried to manage this problem by changing the ratio of amino acids within the mixture inversely according to their general reactivity, using a larger number of equivalents of the slower reacting amino acids to create a "kinetically adjusted" acylating mixture. However, another method has also been described. It had long been accepted in peptide chemistry that for some difficult couplings a process of double coupling was necessary to completely react all available sites on the solid phase. Double coupling means that after the first reaction is complete, the same reaction is repeated again to force coupling to all available reactive sites.

This approach was modified by Kramer et al. to assure that in a mixture there would be essentially an equimolar representation of all amino acids, regardless of the rate at which they reacted with the solid phase.[39] They used a less than saturating mixture in the first coupling reaction, for example, 0.8 equivalents (relative to the number of reactive sites in the solid phase) of an equimolar mixture of amino acids. Thus in the first step the slowest reacting amino acid would be coupled to the solid support with close to the same efficiency as the fastest reacting amino acids. The result was an equimolar representation occupying 80% of the available coupling sites on the solid support at the end of the first reaction. The remaining reactive sites were coupled in the second reaction of the double-coupling protocol, so the fastest reacting amino acids coupled to a <20% excess over the slowest.

The authors validated this approach through the synthesis of libraries on solid-phase resin and on a cellulose sheet, using a variety of targets, among them the epitope definition of an anti-TGFa monoclonal antibody, Tab2. The iterative library design was XXO$_1$O$_2$XX, where in this case the double coupling of a mixture of 19 amino acids was used at the X positions, and O$_1$ and O$_2$ were defined, creating pools of peptides in solution following cleavage from the solid support. A standard competitive ELISA was performed using TGFa bound to microplates and identifying those wells to which horseradish peroxidase labeled anti-TGFa did not bind in the presence of the particular peptide pool. The known epitope was HFND. From the library (containing 47,045,881 compounds in 361 pools) the authors identified XXHFXX, XXFNXX, and

XXNDXX, clearly peptide pools related to, and containing, the known epitope. When these pools were synthesized as spots in a 19×19, O_1 by O_2 array on a sheet of cellulose paper, the same families and conservative substitutions were identified: FN, HF, HY, ND, YF, YN, YY. Using two additional 19×19-array cellulose-bound libraries the authors identified the amino acids permitted on either side of the O_1O_2 motif, synthesizing libraries of the structure X O_1HF O_2X, X O_1FN O_2X, and X O_1ND O_2X.

Positional Scanning Format Library Methods. A decapeptide positional scanning library approach was evaluated using the same anti-Ac-YPYDVPDYASLRS-NH$_2$ antibody.[147] Ten different positional libraries were synthesized, each containing 20 pools with a defined amino acid (O) at the specific position being investigated by that library (O_1XXXXXXXXX, XO_2XXXXXXXX, etc.). From this study 15 peptides were identified with comparable or better activity than the antigenic 13-mer (IC_{50} = 6 nM), and all but one contained a consensus DYA sequence at positions 8, 9, and 10.

A derivative positional scanning method published by Wong et al. was also validated using antibody binding.[159,160] The authors synthesized three libraries of the format Ac-O_1O_2XXXX-NH$_2$, Ac-XXO_3O_4XX-NH$_2$, and Ac-XXXXO_5O_6-NH$_2$, using only 10 "physicochemical representative" proteinogenic amino acids so that each library consisted of 100 dipeptide defined pools of 10,000 compounds each. Screened against a monoclonal antibody against the pili of *Pseudomonas aeruginosa,* the authors identified the best dipeptides, EQ, FI, and PK, corresponding to the known epitope DEQFIPK. They synthesized EQFIPK, and determined an IC_{50} of 1.3 mM, validating the approach. However, when the 8 peptides containing single-amino acid substitutions of the residues representing amino acid classes (D for E; N for Q; W, Y for F; V,M for I; H,R for K) were synthesized and tested, only four exhibited activity similar to that of the lead sequence (IC_{50} = 3.4–0.9 mM). This meant that while the representative algorithm was effective in enriching the active hit rate upon secondary synthesis, if the wrong representative residues had been selected no lead in the case of H or R for K or a less informative lead in the case of N for Q, or Y for F would have been discovered. Therefore, the authors propose that this approach is best used where it is necessary to simplify the synthetic process, such as synthesis and screening libraries of longer peptides (heptamer, octamer, etc.), trading off the risk of missing leads with the ability to actually synthesize and screen a complete library.

Using this approach Pinilla et al, defined the binding epitope (STTS) of a monoclonal antibody raised against a surface antigen of hepatitis B, identifying the peptide consensus Ac-S(S/T)T(P/S)(A/M)(H/M)-NH$_2$, from which the three most active peptides synthesized were Ac-STTSMM-NH$_2$ (IC_{50} = 0.39 mM), Ac-SSTSMM-NH$_2$ (IC_{50} = 0.87 mM), and Ac-STTSAM-NH$_2$ (IC_{50} = 2.8 mM).[146] These results were essentially the same as those obtained by the iterative synthesis library described above.

Liquid-Phase Combinatorial Libraries. As combinatorial chemistry is extended from peptides to nonpeptides, synthesis issues become the major limitation. A method termed liquid-phase synthesis was demonstrated in which the compounds are synthesized onto a chemical phase (polyethylene glycol, MeO-PEG 5000) that can be conveniently crystallized or precipitated by solvent changes. Liquid-phase synthesis permits filtration purification and affords all the advantages of solid-phase synthesis when it comes to the separation of product and reactants, yet permits the reactions to be carried out in solution.[47] This method was validated using anti-β-endorphin as the target and the recursive deconvolution strategy described above. From a pentapeptide library containing G, L, F, and Y, the authors identified YGGFL-MeO-PEG, YGGFF-MeO-PEG, YGGFY-MeO-PEG, and YGGFG-MeO-PEG using the same ELISA assay as described above, though in this case the pools of library compounds were in solution, though not cleaved from the PEG.

Validation of Synthesis, Assay, and Compound Identification Methods. The first description of a fully automated, 96-well, robotic, solid-phase library synthesizer was validated using an anti-gp 120 antibody.[134] This system has been a model for many of the systems now in use and was based on the use of filtration microtiter plates, enabling the removal of solutions without loss of the solid-phase support beads and the addition of solutions to the wells using a single- or multiple-head pipetting device. Individual compounds can be synthesized in each well or, by transferring the beads from wells into a mixing device and then redistributing them among the reaction wells, a library could be synthesized using either the split-and-mix one-compound–one-bead approach or any of the iterative, deconvolution, or positional scanning approaches.

In the example 12 individual peptides were synthesized and tested to validate the quality of synthesis. A library of format RAX$_1$HTTGRIX$_2$ (19 amino acids randomized at each X position) was synthesized and assayed in pools of peptides released from the solid-phase supports using an ELISA assay. After addition of the X$_1$ amino acids the 19 pools (containing 19 peptides differing in X$_2$) were kept separate so that the residue(s) in this position would be defined by the initial assay. The anti-gp120 antibody used had been raised against recombinant gp120, and the binding epitope mapped to a 10-mer sequence (**RAFHTTGRII**). The residues in boldface type had been shown to be important using alanine scans. Screening identified X$_1$ as F = W = Y > H > A = N. Each mixture was separated by affinity chromatography using the anti-gp120 antibody as the affinity support, and the retained fractions were sequenced to identify X$_2$, with the finding I > L > P. It would have also been possible to resynthesize the 19 peptides corresponding to each X$_1$ pool in order to identify X$_2$, but affinity selection as demonstrated in this work can be a powerful approach in certain circumstances.

Scaffold Library. The use of a scaffold library to produce ligands with a known structure was validated by identifying peptides that bind to an anti-lipopolysaccharide (LPS) monoclonal antibody.[135] A library was synthesized into a highly structured peptide that binds zinc in its three-dimensional conformation, PYKPECFKSFSQK(X/S)(X/D)L(S/V)(X/K)HQ(X/R)THTG, presenting the randomized (X) residues to the target molecule on a defined backbone. Because of the peptides zinc binding, the conformation of the peptide mixtures and individual peptides could be confirmed by Co(II)-complex fluorescence analysis, circular dichroism measurements, and transfer NOE NMR. A positional scanning approach was employed, synthesizing five libraries of the format $O_1X_2X_3X_4X_5$, $X_1O_2X_3X_4X_5$, $X_1X_2O_3X_4X_5$, $X_1X_2X_3O_4X_5$, $X_1X_2X_3X_4O_5$, ~26 pools each (17 proteinogenic, 9 nonproteinogenic amino acids), where the O positions were defined according to the pool, and X represented mixtures. Substituting hydrophobic residues at X_1 caused a major change in conformation, and thus only 14 of the amino acids were used at this position. Screening with the anti-LPS antibody produced $H_1F_2V_3Q_4H_5$ as the best binder, and the backbone conformation of these residues was confirmed by physical studies. The authors propose this as a first step toward synthesis of a peptidomimetic and represents an approach that can be generalized for any target molecule.

In another example of the use of a template library, FmocK-P-G-BocK-A-AlocK-P-G-BocK-A was synthesized, cleaved from the resin, and cyclized to serve as a backbone to which amino acids could be attached selectively to the orthogonally protected lysine side chains.[136] A single amino acid was attached to the two Boc-protected lysines, while eight different amino acids were attached, in separate sets of reactions, to each of the other orthogonally protected lysine residue side chains. The limited library that resulted was screened against an anti-lysozyme antibody (HyHEC-5) which recognized a defined discontinuous epitope of lysozyme. Library screening identified a tyrosine that was not present in the native epitope. A similar concept was employed to display four amino acids from a Kemp's triacid backbone.[57]

Identification of Protein and DNA Binding Peptides

The binding to streptavidin has been used to validate new synthetic combinatorial methods. In general, either the direct binding of (alkaline phosphatase-conjugated) streptavidin/avidin to beads is monitored or biotin is bound to microtiter plates, test peptides are added free in solution, and the inhibition of streptavidin/avidin binding is monitored. While not of therapeutic interest, these studies are of scientific note, since they explore the binding of peptides to a protein that binds a nonpeptide (biotin) with very high affinity. Both L- and D-amino acid-containing peptides are bound by streptavidin and avidin. Other more biologically relevant protein–protein interacting systems have also been studied.

Multiple-Synthesis Libraries. Binding to streptavidin was used to validate a surface display array-based library method called Pilot.[91] The effect of the surface on the binding properties of the compounds assayed while remaining attached to the solid phase on which they were synthesized is an issue. Cass et al. attempted to address the surface presentation issue by exploring the binding of ^{125}-I-streptavidin to the sequence YGHPQGG synthesized onto various surfaces. They derivatized polyethylene sheets with a functionalized linker (diamino-substituted polyethylene glycol, Jeffamine ED-600) to which they covalently bonded carboxymethyl dextran (Pharmacia T500) to produce a surface coating that was amino functionalized by coupling tBOC-NH-CH$_2$)$_3$ NH$_2$ followed by cleavage of the protecting tBOC. Arrays of compounds were synthesized onto these thin-film sheets in defined positions to create reusable assay plates.

The strategy used for array design is interesting. A standard mixture of 16 selected amino acids was used (W, amino acids Y,F,D,E,K,R,Nle,V,G, DAla,A,S,H,Q,P,DNap) together with limited positions in which a mixture of two defined amino acids was incorporated. Thus, for instance, an 8×8 array hexapeptide library was generated by creating all possible two-amino acid combinations of the amino acids used in the W mixture in positions 1 and 2 [(Y/F, D/E, K/R, Nle/V, G/DAla, A/S, H/Q, P/DNap) by (Y/F, D/E, K/R, Nle/V, G/DAla, A/S, H/Q, P/DNap)], while the W mixture was used in positions 3, 4, 5, and 6. Probing with streptavidin produced VY. Generating the subarray to define positions 3 and 4 (same format) produced VYGF and VYHP. Generating the final sub-array to define positions 5 and 6 produced VYGFRQ and VYHPQF. Speculating that these two might overlap the authors combined the two motifs to produce HPQVYGFRQ, which was a strong binder compared to the individual peptides.

The identification of DNA binding peptides from a library was demonstrated by Kramer et al.[39] As described above, they synthesized a XXO$_1$O$_2$XX library on a cellulose membrane sheet in the form of pools in a 19×19, O$_1$ by O$_2$, array using the double coupling of 0.8 equimolar (less than saturating) mixtures of amino acids. The cellulose sheet was incubated with a ^{32}P-labeled, double-stranded DNA 15-mer, and analyzed by a Phosphoimager (Molecular Dynamics, Sunnyvale, CA). This led to the identification of a KK, KR, RK, RR consensus for the O$_1$O$_2$ amino acid pair. Considering the crucial role of protein recognition of DNA sequences for transcription, and the current attempts to influence disease states through the selective control of transcription, these results demonstrate the potential for applying peptide combinatorial libraries to these targets.

Split-and-Mix, One-Compound–One-Bead, Postassay Identification Format Libraries.

Lam et al. synthesized a pentapeptide library and assayed it directly for streptavidin binding to the compounds displayed on the beads, sequencing to identify the positive compounds, which were then resynthesized and tested to confirm that binding was competitive with biotin. The consensus motif identified was HPQ, with a lesser number of HPM motif compounds. Furthermore, it did not matter if HPQ was N-terminal (e.g., HPQFV), internal (e.g., LHPQF), or C-terminal (e.g., MYHPQ); of course, the C-terminal position in the library was internal to the linker. The HPM motif containing peptides was displayed either as an internal motif (e.g., MHPMA) or C-terminal motif.

As a follow-up, Lam and Lebl explored the specificity of streptavidin and avidin using a library that lacked histidine.[108] A higher concentration of the streptavidin/alkaline phosphatase conjugate had to be used, but the result was the discovery of four motifs, WPAX or WXAX, RMDLY, and YMEXW, and two additional sequences, PPWPY and QYWQS. Representative compounds that were resynthesized demonstrated competitive binding with biotin, indicating they bound in the biotin binding pocket. This demonstrates the power of synthetic combinatorial libraries in discovering novel leads. Furthermore, when the apparently closely related avidin molecule was used as a target the authors found apparent overlap with streptavidin motif (HPYPX, HPFXX, and HPXPX motifs) as well as a different peptide specificity (HKXXX). Despite the apparent overlap, streptavidin did not bind to HPYPP, and avidin did not bind to the LHPQF, indicating that the HPYP motif is specific for avidin and the HPQ motif is specific for streptavidin, though both bind in the respective biotin binding pockets.

Following this study additional L- and D-amino acid peptides were discovered[109] The results show that the majority of the L-amino acid-containing peptides that bind streptavidin contain the HPQ motif, while avidin binds D-amino acid containing peptides that do not have this motif, but instead exhibit a vynxx, ltxsx, or vqsxw motif, again demonstrating the binding pocket specificity differences that existed between the two biotin binding proteins.

Protein–protein and protein–DNA binding interactions regulate signal transduction, transcription, and translation. SH3- and SH2-mediated protein–protein interaction is an important component in signal transduction, serving to interface selective ligand/receptor interactions into common cellular signal transduction pathways. Specific inhibition of SH3-dependent binding is therefore an attractive therapeutic target. Studies were undertaken utilizing a hexapeptide library and a cyclic heptapeptide library assayed on-bead using the fluoresceinated SH3 domains of phosphatidylinositol 3-kinase (P13KSH3) and Src presence of at first all 32 peptides. The set was divided into two pools of 16 and the active pool was identified, then this pool was split in half, and so on. In the end one high-affinity ligand was identified, Ac_2-L-Lys-D-Ala-D-ALa. The authors demonstrated the utility of this approach, but also evaluated the pool size that could be employed, concluding that the method was useful for pool sizes of 100–1000 compounds. Many methods can identify activity from a pool of compounds, but capillary electrophoresis gives a direct determination of binding affinity and does not require labeling of compound or a specific biological activity.

Identification of Enzyme Substrates

Peptide substrates have proven to be good starting points for developing substrate-based inhibitors of enzymes, and synthetic combinatorial chemistry peptide approaches should produce leads that can then be used for rational drug design. The identification of enzyme substrates from libraries represents a method to improve screening assays and should prove useful in creating screening assays for putative orphan enzymes, which have no known substrates. Such a tool should be instrumental in characterizing and validating new therapeutic targets through identification for substrates for the classification of orphan enzymes, identification of putative substrates from protein sequence database searches based on the sequences identified from library screening, and identification of putative substrates for kinases involved in critical functions such as signal transduction.

Multiple Synthesis Libraries. A multiple-synthesis mixture strategy was used to synthesize a library of 19 Kemptide-related peptides (format L(R/X)RASL, R is the native Kemptide residue) and a library of 19 analogs based on the v-Src autophosphorylation site (RRLIEDA(E/X)WAARG, E is the native v-Src residue). The cleaved peptides were incubated with [g-^{32}P] ATP plus either the cAMP-dependent kinase or the v-Abl protein kinase, and the phosphorylated compounds were identified by phosphopeptide-selective liquid LCMS.[93] When synthesis reached the point of randomization the resin was divided into four pools accounting for 16, 10.5, 31.5, and 42% of the total. Each pool was reacted with a different mixture of amino acids (I,V,T; Q,H; Y,D,E,K,R,P; F,G,A,L,M,S,H,W) so that the same proportion of reactive sites were occupied by equimolar amounts of each amino acid. The compounds were cleaved and the pools were recombined into one mixture of 19 compounds. Quality control was performed by MS, and the libraries were screened against their respective target kinases using [g-^{32}P] ATP. After removal of the enzyme by molecular weight centrifugal filtration (Microcon-10 unit), the peptide mixture was subjected to liquid chromatography. The recovered peaks were split for analysis of ^{32}P incorporation by g counting and specific MS detection of phosphopeptides.

Kemptide was the principle peptide phosphorylated by the cAMP-dependent kinase, but each analog was phosphorylated to some degree. The intensities of the negative ion (phosphorylated peptide, detecting the PO_2^- and PO_3^- ions) versus positive ion (peptide itself) permitted the authors to compute the percent conversion to phosphorylated peptide and rank them (R > H > G > Q = K > E > Y > H > A > P > T > T > S > (xle) > F > (xle) > W > V > M > A) according to suitability as substrates (MS cannot distinguish Ile and Leu, shown as xle). This ranking was consistent with that reported from a traditional study (R > H > K > A)[161] but contained much more information. The study with v-Abl protein kinase identified the Leu or Ile analog as the major substrate, with the Val analog as a minor substrate.

Resynthesis and testing of the Leu and Ile analog peptides compared with the Src substrate Glu peptide demonstrated a preference for the Ile analog as substrate (Glu: K_m = 1.56 mM, V_{max} = 27.8 nmol/min mg^{-1}; Leu: K_m = 0.727 mM, V_{max} = 20 nmol/min mg^{-1}; Ile: K_m = 0.67 mM, V_{max} = 73 nmol/min mg^{-1}). Thus libraries were shown to be effective means of defining substrates for both a receptor-dependent protein kinase (cSMP dependent kinase) and a receptor-independent protein kinase (v-Abl).

In a specialized variation of multiple synthesis, Songyang et al. synthesized a biased library MAXXXXYXXXXAKKK, where X represents 15 proteinogenic amino acids (excluding C, W, Y, S, and T), and then cleaved this library into solution.[162] The mixture was incubated with p60[c-src] protein tyrosine kinase and [g-^{32}P]ATP and the phosphorylated peptides were isolated from the mixture by ferric chelation chromatography and identified by sequencing. The authors identified as EEIYGEFF motif.

Split-and-Mix, One-Compound–One-Bead, Postassay Identification Format Libraries. Wu et al. identified peptide substrates serin/threonin cAMP-dependent kinase, known to exhibit specificity for peptides containing an RR(X)S motif.[113] A library of peptides displayed on the surface of beads were incubated with the enzyme in the presence [g-^{32}P]ATP. Substrate peptides were phosphorylated, all the beads were spread in agarose to immobilize them, and autoradiography was performed to locate the labeled beads. The ^{32}P-labeled beads were recovered and identified. Substrates that identified by this procedure had higher affinity than the commercial synthetic cAMP-dependent kinase substrate, Kemptide (LRRASLG, K_a = 1.33 mM). From approximately 500,000 compounds assayed, 55 positive compounds were selected from a pentapeptide library. Sequencing of two yielded the sequence RRYSV (K_a = 0.67 mM). Sequencing three compounds out of 60 positive beads from a hepta-peptide library produced SQRRFST (K_a = 0.83 mM), YRRTSLV (K_a = 1.43 mM), and IIRRKSE (K_a not determined). Two of these exhibited higher affinity for the enzyme than the commercial peptide.

The same approach was used subsequently to identify substrates for the p60[c-src] kinase.[114] The sequence YIYGSFK was identified from the library. The phosphorylation of YIYGSFK and several peptides derived from cellular proteins by Src and several other kinases was studied. YIYGSF was found to be a highly specific Scr-family kinase (Scr, Lyn, Fyn) substrate, exhibiting a K_m of 55 mM for Src, compared to a K_m of 353 mM for a peptide derived from cdc2 (residues 6–20, KVEKIGEGTYGVVYK), a native substrate of Src, but not a good substrate for non-Src family protein kinases (EGFR, Fes) compared to the cdc2 peptide. Thus both a higher affinity and more specific substrate was identified.

The next step would be to search sequence data bases to identify whether this motif appears in cellular proteins that may be putative substrates for Src. Information of this type can be used to design substrate based inhibitors.

More recently, by screening a secondary library (XIYXXXX) an even more potent and specific peptide substrate, GIYWHHY, was identified.[115]

In another study the substrate specificity of an endoproteinase was explored using an acetylated library of peptides on beads.[116] Two libraries were prepared, Ac-NleELRTQ(P'_1)(P'_2)SHR-NH$_2$, with P'_1 fixed as phenylalanine and P'_2 randomized in the first library, and P'_1 randomized with P'_2 fixed as serine in the second library. Incubation of the library beads with the 3C endoproteinase from heptatis A virus resulted in removal of the acetyl group from those peptides that were cleaved by the proteinase. The absence of the acetyl group permitted direct identification of the substrate sequence by Edman degradation. Thus there was no selection step, only incubation, assessment of the extent of reaction by measuring free amino groups using fluorescamine, and then sequencing of the entire batch of beads. The native P'_1/P'_2 residues inferred from other studies are serine/phenylalanine, which demonstrated a K_{cat}/K_m value of 0.004. The library results indicated a P'_1 preference for alanine, serine, and glycine, and a less selective P'_2 preference, but peptides containing a P'_2 arginine or proline were not cleaved. All the peptides that were cleaved were less active than the putative substrate domain, exhibiting K_{cat}/K_m values of 0.1 to 1.0. Thus this study confirmed the putative cleavage specificity to be serine/phenylalanine for P'_1/P'_2.

The substrate specificity of human fibroblast collagenase and stromelysin was evaluated using an on-bead library approach in which a fluorescent reporter was cleaved into solution, resulting in solution fluorescence in wells containing positive substrate beads.[117-119] Positive beads were recovered and the remaining material was microsequenced to determine the cleavage site. The first assumption was that the library should encompass a tripeptide window displayed in an alanine frame, AAXXXAA. To design the optimal method of covering this window an alanine scan was performed on the peptide $G_4P_3L_2A_1L_1,F_2$, cleaved by both collagenase and stromelysin. In terms of conventional nomenclature shown as subscripts to the amino acid residues, the P_3 proline, P_2 leucine, and P'_1 leucine, and to a lesser extent the P'_2 phenylalanine were important for either activity or selectivity. The authors concluded that PXXL was crucial for collagenase recognition as a substrate, and designed a library of three subforms, all synthesized on controlled pore glass (CPG):

Cop-$A_1A_2A_3X_4X_5B_6A_7$-(ε-aminocaproyl)$_5$-BAla-(3-aminopropyl)-CPG
Cop-$A_1A_2X_3A_4X_5B_6A_7$-(ε-aminocaproyl)$_5$-BAla-(3-aminopropyl)-CPG
Cop-$A_1A_2X_3X_4A_5B_6A_7$-(ε-aminocaproyl)$_5$-BAla-(3-aminopropyl)-CPG

Cop is a fluorophore (7-hydroxycoumarine-4-propanoyl) developed specifically for capping peptides synthesized on solid phase.[163] A represents fixed alanine residues, X represents positions varied one at a time between 20 amino acids such that the identity of the residue was known from the position in the synthetic/assay array, and B represents a degenerate position containing 20 different amino acids. Thus each subformat library consisted of 400 pools in which the identity of the X residues was known. The 20 proteinogenic amino acids were used, substituting S-methylcysteine for cysteine.

While a more powerful library design/identification strategy would have made it possible to simply synthesize and screen the complete tetrapeptide library, the results obtained with this approach were quite interesting. By defining cassettes of related compounds based on the shared identity of the N-terminal X residue, the relative importance of the C-terminal X residue could be compared. The cassettes defined as L, M, N, and P exhibited collagenase substrate activity. The L, M, and N cassettes exhibited an identical C-terminal X profile of $U > L > M > I > Q > Y$, while the P cassette had a different profile, $N > U > A > D = M > P$. Sequencing was used to identify the B residue identity and cleavage site. For instance, sequencing of the positive AAPANBA pool produced an abundance of and preference for B of $U > L > I > M$ in the first round of sequencing, indicating that cleavage occurred N-terminal to these residues, and identifying AAPANUA as a good substrate for collagenase. Additional analysis indicated the ablity to detect frame shifts in the point of cleavage. The authors also confirmed that there was a good correlation between cleavage of the peptides on the bead compared to cleavage of resynthesized peptides in solution. This approach should prove to be quite useful in the characterization of putative proteases identified from gene sequencing efforts and in the optimization of screening assays.

Another strategy used to identify protease substrates relied on an internally quenched fluorescent molecule where the fluorochrome (o-aminobenzamide, ABz) was separated from the quenching moiety (3-nitrotyrosine, $Y(NO_2)$) by the point of cleavage.[120] By synthesizing the library on macroporous beads and employing on-bead screening, the beads that displayed substrate peptides developed fluorescence and could be readily identified and selected. The format of the library was $Y(NO_2)X_5X_4PX_3X_2X_1K(ABz)$ bead, designed for the endopeptidase subtilisin Carlsberg. By inserting proline at the P-2 position the site of cleavage by this enzyme was directed between X_3 and X_2. By performing the assay at pH 8.5 and 5.2 the authors demonstrated that they could not only identify substrates, but they could optimize substrates for different reaction conditions. The substrates with the highest rates of cleavage on beads at pH 8.5 were $Y(NO_2)LQPFNEK(ABz)$, $Y(NO_2)FSPLQ$-$FK(ABz)$, and $Y(NO_2)IEPFFEK(ABz)$, while those with the highest rates of cleavage at pH 5.2 were $Y(NO_2)FQPLDVK(ABz)$, $Y(NO_2)LYPLDVK(ABz)$, and $Y(NO_2)ASPMGFK(ABz)$. However, the authors noted that while reflecting the rate of cleavage, this rank order did not hold in solution where accurate kinetic constants (K_{cat}/K_m) could be determined. Thus, final optimization had to be performed using compounds assayed in solution, producing $Y(NO_2)FQPLDEK(ABz)$ and $Y(NO_2)FQPLAEK(ABz)$ as the most active structures under pH 8.5 conditions.

Iterative Synthesis and Screening Library Methods. Phosphorylation of library peptides was also used to identify substrates of cAMP- and cGMP-dependent protein kinase.[20] The known consensus for cAMP-protein kinase substrates and inhibitors was RRX(S/A)X, but that for cGMP-protein kinase was less well defined. An iterative approach was used to identify octapeptides, incubating first a library Ac-Ac-XXXO$_1$O$_2$XXX-NH$_2$ with the kinase(s) plus [g-^{32}P]ATP, identifying Ac-XXXRRXXX for cAMP kinase and AcXXXRKXXX-NH$_2$ for cGMP kinase, respectively. Subsequent dependent sets of libraries (Ac-XXXRRO$_1$O$_2$X-NH$_2$, etc.) were synthesized until they had defined Ac-KRAEKASIY-NH$_2$ as the substrate for cAMP kinase (similar to the previously identified consensus motif RRX(S/A)X) and Ac-TQKARK-KSNA-NH$_2$ for cGMP kinase, a novel substrate series.

Validation of Library Design Concepts. An early study in which enzyme substrates were identified was based on the synthesis of several mixtures of peptides. Each mixture contained compounds of identical structure except at one position, which was randomized (e.g., Ac-GNSXR, where X = F, G, A, Y, L, V).[148] The mixtures were then exposed to an enzyme (angiotensin-I converting enzyme, atrial dipeptidyl carboxyhydrolase, bacterial dipeptidyl carboxyhydrolase, and meprin), and the products were analyzed by sequencing to determine which residues were important for binding and where the point of cleavage occurred. One cycle of sequencing revealed the preferred P$_1'$ residues, two cycles the preferred P$_2'$ residue, and so on. The authors made several new observations regarding substrate specificity, but the most interesting was perhaps the observation that meprin exhibited specificity for the Ala-Pro bond, while also cleaving Phe-Ala bonds. The authors speculate that bradykinin may be a physiological substrate of meprin.

Identification of Enzyme Inhibitors

Multiple Synthesis Libraries. Inhibitors of acetylcholinesterase activity were identified using a library of compounds synthesized and then spotted on a silica thin-layer chromatography plate and assayed.[94] Fifty-two compounds were synthesized and spotted in an array. The substrate acetylthiocholine iodide plus 5,5-dithiobis(2-nitrobenzoic acid) and then enzyme were evenly distributed over the plate by spraying. Inhibitors were detected by the clear zones that appeared because of the lack of enzymatic reaction (which produced a yellow background). Cellulose paper could also have been used as the support, in which case synthesis could have been carried out directly on the support. Positive control and several other inhibitors were identified, validating the method.

Synthesis of a library based on a noncleavable peptide inhibitor on pins was used to map the active site of and optimize inhibitor selectivity for human heart chymase, a chymotrypsin-like protease that converts angiotensin I to angiotensin II.[95] Two libraries were synthesized and the compounds were tested for inhibition of either chymotrypsin and chymase. One was based on use of 3-fluorobenzylpyruvamide (F-Phe-CO-) to bind in the P$_1$ pocket of the enzyme. It consisted of (F-Phe-CO)-X_1X_2-RG-pin, where X_1X_2 residues were each varied among the 20 different proteinogenic amino acids to produce 400 different peptides used to determine the specificity for the P_1' and P_2' residues. This library identified the X_1X_2 residue pair ED as selective for chymase. Synthesis and testing of (F-Phe-CO)-ED-RG(OMe) resulted in the determination of a K_i for chymase of 1 mM and for chymotrypsin of 100 mM. The other library was based on the use of an α-ketoamide (Phe-CO) in the P_1 position, producing the library Z-Ile-X_3X_2-(Phe-CO)-GG-pin, where X_3X_2 residues were each varied among 18 of the different proteinogenic amino acids (Met and Cys were omitted) to produce 324 different peptides used to determine the specificity for the P_3 and P_2 residues.

The results were the identification of chymase P_3P_2 preference for the amino acid pairs E_3P_2 and I_3Q_2. E_3D_2 was the most potent inhibitor, but it was not selective. The authors selected E_3P_2 for its acceptable potency and selectivity, synthesized the peptide Z-IEP-(Phe-CO_2Me) and established a K_i of 1 nM for chymase and 10 nM for chymotrypsin. Clearly, the potency of these two series were quite different, but selectivity of either 100- or 10-fold could be achieved. The authors then combined these series into one molecule, Z-IEP-(Phe-CO)-EDR-OMe, which exhibited a K_i for chymase of 100 nM and a K_i for chymotrypsin of 40 mM, a 400-fold selectivity for chymase with reasonable inhibitory potency. This demonstrates the power of rationally designed combinatorial peptide libraries to quickly optimize the selectivity of enzyme inhibitors.

Split-and-Mix, One-Compound–One-Bead, Postassay Identification Format Libraries. Meldal and Svendsen described a novel approach for identifying enzyme inhibitors.[121] The authors synthesized an intramolecularly quenched (by Ac-Y(NO_2)] fluorescent (Abz) substrate for subtilisin Carlsberg (Ac-Y(NO_2)FQPLAVK(Abz)-PEGA, cleaved at the L-A bond) inside macroporous beads composed of polyethylene glycol cross-linked polyamide (PEGA) together with a library of heptapeptides (XXXxXXVF, composed of the L (X) or D (x) stereoisomers). Enzyme inside the beads thus was in the presence of both a substrate and a potential inhibitor. Cleavage of the substrate present in each and every bead produced fluorescence. However, in those beads containing an inhibitory peptide the development of fluorescence was delayed due to the localized inhibitor. Novel inhibitors were identified, the most potent being AMMc(*tert*-butyl)MIVF (IC50 $\sim$ 3 mM).

The identification of inhibitors of protein tyrosine phosphatase PTP1B (implicated in breast and ovarian cancer) was used to validate a novel tagging method employing a radio-frequency (rf) encodable microchip to encode the synthetic steps to which a compound carrier system was exposed in a split-and-mix, one-compound–one-carrier approach to library synthesis (See Chapter 14).[122] A tripeptide library acetylated with *p*-carboxycinnamic acid (a nonspecific inhibitor) was synthesized on solid-phase resin contained in capsules that also held an rf microchip emitting a unique binary code. The objective was to confer specificity and additional binding affinity through the peptide interaction with enzyme residues surrounding the *p*-carboxycinnamic acid binding pocket. Each reaction pool was scanned and the identity of the microchips in that pool was recorded. The capsules from the various pools could then be mixed, split into the next set of reaction pools, and scanned again. Thus, at the end of the synthesis there was a record of which reactions each microchip-encoded capsule had been subjected to. In the actual example, a limited library of 125 compounds was synthesized, so the split-and-pooling strategy was also manually tracked to verify the ability of the rf tracking to accurately report the reaction history of each capsule.

Inhibitors of the glucosomal phosphoglycerate kinase (gPGK) from *Trypanosoma brucei* were identified from a pentapeptide library in an on-bead assay using fluorescently tagged or biotinylated gPGK and either manual selection of the fluorescent beads or selection onto steptavidin-coated magnetic beads.[123] Inhibitors of this gPGK are potential antiparasitic agents, because of the key role this enzyme has in the glycolytic pathway of the form of *Trypanosome brucei* found in the bloodstream, and because it is known that the *Trypanosome* is sensitive to compounds that interfere with glycolysis. Using this approach the authors found that NWMMF was a selective inhibitor of the parasite gPGK, IC_{50} of 80 mM, and was without effect on rabbit muscle gPGK at 500 mM. The next step toward a therapeutic agent is to optimize activity of this parasite-selective peptide and demonstrate anti-parasite activity.

Lowe and Quarrell started with a portion of sequence from a known 29-amino acid residue inhibitor of the serine proteases trypsin and chymotrypsin, CMT-1 (isolated from squash) to design an optimization library for leukocyte elastase.[124] The portion of sequence they used as a template defined a reactive loop in this conformationally constrained inhibitor. The authors synthesized a library containing mutants at positions P'_1 to P'_4 (C-terminal) to the presumed reactive site occurring between Arg-5 and Ile-6 relative to the N-terminus. Once cleaved this inhibitor remains bound to the enzyme acive site. Other investigators had shown that changing Arg-5 to Val-5 increased binding to human elastase. This led the authors to design two libraries, differing in the fifth residue, $RVCP(V_5/A_5)XXXXC_0KKDSDCLAECVCLEHGYCG$, and randomizing the P'_1, P'_2, P'_3, and P'_4 positions with the 19 proteinogenic amino acids. The residues that induced the tertiary conformation of the peptide through disulfide bond formation were all retained. Using this template library and fluorescently labeled human leukocyte elastase, they screened a therapeutically relevant serine protease in an on-bead assay using fluorescence microscopy to select positive beads. Disappointingly, the authors did not actually report on the success of screening or identify the peptides found using this library.

Inhibitors of thrombin are known to bind in both the active catalytic site as well as an exosite. An example where this was used to advantage is the inhibitor Hirulog, fPRPGGGG(NorLeu)GDFEEIPEEYL (K_i = 2 nM). Mirulog was designed on a rational basis from a known peptide that constitutes a pharmacophore that binds in the active site of the enzyme (fPRPG, K_i = 20 mM), a linker, and a sequence derived from a large molecular weight inhibitor, Hirudin, comprising a pharmacophore that binds in an exosite (DFEEIPEEYL, $K_i \sim 2$ mM). Pentapeptide libraries were assessed using on-bead screening to identify active site inhibitors.[125] From a library of format DLDLD, where a set of D- or L-amino acid isomers were randomized at every other position, several novel compounds were identified. Among them was wFrPf, which exhibited a K_i for thrombin of 3.5 mM, and dYaEw (K_i = 6 mM), compared to a K_i of 21 mM for the known pentapeptide inhibitor fPRPG (the *n*-terminal sequence of Hirulog, a thrombin inhibitor in clinical trials). From a library of the format LDLDL compounds such as TrFfP (K_i = 117 mM), LiFrP (K_i = 20 mM), and GfFfK (K_i = 23 mM) were identified, while from a library of the format DDDDD compounds such as ffrfh (K_i = 23 mM), ffrpt (K_i = 43 mM), ffr(norleu)(norleu) (K_i = 53 mM), and rlmrf (K_i = 83 mM) were identified.).

One benefit of combinatorial libraries in exploring binding domains around an initial starting point is to identify a novel series, as well as to improve activity of the original. With this in mind an on-bead library was designed with the format fPRPXXXXX-(linker)-bead, randomizing 19 proteinogenic amino acids at each of the X positions.[125] Screening this library identified SEL 1478, FPRPFGYRV-NH$_2$, which exhibited a K_i of 25 nM. Thus, using a single optimization library, the affinity of the peptide lead was optimized from 20 to 25 nM and a novel binding domain was identified. Multiple synthesis was employed in a combined alanine scan and truncation study to rapidly explore the SAR of SEL 1478, with the result that the compound FPRPFG(X)R-NH$_2$ was identified as necessary for maximal activity. Optimization using an expanded library of nonproteinogenic amino acids and peptide bond modifications would be the next logical step in optimization.

Factor Xa is another enzyme in the coagulation cascade. Drug discovery teams have searched for a specific low molecular weight, orally available inhibitor of this enzyme for the last 5–10 years without success. However, by screening an octapeptide library in an on-bead assay a novel lead was identified, YIRLAAFT-NH$_2$ (SEL 1691, K_i = 15 mM).[126,164] Many additional analogs were identified from other peptide libraries, including the consensus motif YIRLA, structurally different from any known peptide serine protease inhibitors. Furthermore, the series was specific for factor Xa. This factor Xa lead was optimized in a combined combinatorial chemistry and traditional medicinal chemistry one-at-a-time synthetic approach.

The first effort was the multiple synthesis of truncation and alanine substitution analogs, resulting in the identification YIRLA, as the structure necessary for activity. Libraries which allowed multiple substitutions within this motif were designed, synthesized, and tested. This provided valuable SAR and, as the most active compound, SEL 2060 (K_i = 150 nM). Traditional medicinal chemistry approaches based on SAR and molecular modeling ultimately produced a series of very active and specific inhibitors, among them the pentapeptide SEL 2711, Ac-*p*-amidonophenylalanylcyclohexylglycyl-3(methyl pyridinium)alanyl-L-P-NH2. This molecule exhibited a K_i of 3 nM for factor Xa, was >3000-fold selective compared to thrombin and a variety of other enzymes, was not cleaved by any serum enzymes, and exhibited antithrombotic activity in animals after intravenous, subcutaneous, and oral administration.

Iterative Synthesis and Screening Library Methods. Inhibitors of the HIV-1 protease were identified from a tetrapeptide library.[137] The peptides were screened free in solution in a microplate assay in which the cleavage of a fluorescent substrate by the HIV-1 protease was monitored. The library contained 22 amino acids at each position plus the aspartic acid protease transition-state analog Statine (Sta) at the second position. Twenty-two pools were synthesized to define the first position as Ac-(P/D)-XXX-NH2. Successive rounds of synthesis defined the rest of the peptide to be Ac-PI(Sta)D-Leu/D-Phe)-NH$_2$, the aspartic acid having lost activity/specificity on the second round of synthesis and screening. Thus, of the 23 residues incorporated at the second position, only the Statine-containing peptides were active, the activity (IC$_{50}$) of the most active peptide (Ac-PI(Sta)D-Leu)-NH$_2$) was 1.4 mM. Subsequent subtitution of the D-leucine with amino acids not in the original library yielded three leucine-substituted peptides (I, L, V) with 10-fold greater potency; substitution of isoleucine with valine produced a 2-fold enhancement of activity. Reassessment of the valine series indicated that tryptophan was more active than phenylalanine at the N-terminus, producing the tetrapeptide Ac-WV(Sta)D-Leu)-NH$_2$, IC$_{50}$ 200 nM. Further optimization from the initial lead produced a pentapeptide with a defined composition Ac-WV(Sta)VX (X not disclosed) with an IC$_{50}$ of 5 nM.

Eichler and Houghten identified inhibitors of trypsin from an interative set of 10 hexapeptide libraries with the P$_1$ position defined as K or R, synthesized in an array on cotton and screened in solution (Ac-KOXXXX-NH$_2$, Ac-XKOXXX-NH$_2$, Ac-XXKOXX-NH$_2$, Ac-XXXKOX-NH$_2$, Ac-XXXXKO-NH$_2$, Ac-ROXXXX-NH$_2$, Ac-XROXXX-NH$_2$, Ac-XXROXX-NH$_2$, Ac-XXXROX-NH$_2$, Ac-XXXXRO-NH$_2$).[24] This strategy, incorporating 19 of the proteinogenic amino acids, resulting in 190 pools containing 2,606,420 peptides, produced the sequence Ac-AKIYRP-NH$_2$, IC$_{50}$ = 46 mM, which was slowly cleaved by enzyme, but was a better inhibitor than the most active hexapeptide sequence derived from naturally occurring trypsin inhibitors (Ac-TKIYNP-NH$_2$, IC$_{50}$ = 102 mM). The authors went on to design a secondary library to optimize the activity of this novel lead.

Lead optimization was pursued using the iterative process described above[8] of synthesizing 400 pools of composition Ac-O$_1$O$_2$XXXX-NH$_2$, screening, and then synthesizing and screening subsequent iterative sets of 20 libraries each (Ac-Id$_1$Id$_2$O$_3$XXX-NH$_2$, Ac-Id$_1$Id$_2$Id$_3$O$_4$XX-NH$_2$, etc.) and proved unsuccessful in identifying any inhibitors of trypsin.[24,138] Eichler and Houghten synthesized an iterative set of 10 hexapeptide libraries with the P$_1$ position defined (biased) as K or R, as described above, and identified a novel hexamer sequence, Ac-AKIYRP-NH$_2$, IC$_{50}$ = 46 mM. The authors incorporated this into the design of a secondary dodecamer library to optimize the activity of this novel lead. Through another series of iterative synthesis/screening steps the most active dodecapeptide was identified. An example of one library is Ac-XXXAKIYRPOXX-NH$_2$. Altogether 19 libraries were synthesized in which the randomized position and the AKIYRPO motif was varied positionally throughout the template as the KO motif had been positionally varied. The final result was Ac-YYGAKIYRPDKM-NH$_2$, with an IC$_{50}$ of 10 mM.

Petithory et al. had demonstrated that endopeptidase substrates could be identified from peptide libraries. The identification of selective inhibitors of metalloproteases, the rat brain zinc endopeptidases 24-15 and 24-16, which metabolize neurotensin, was achieved using a phosphinic peptide library of the format Z-(L,D)PheY(PO$_2$CH$_2$)-(L,D) (G/A)$_1$-X$_2$X$_3$, where X was substituted by 20 amino acids (See Chapter 9). Phosphinic peptides (with in vivo activity) act as mixed inhibitors of the two peptidases, with nanomolar K_i, without affect on several other zinc peptidases, such as endopeptidase 24-11, angiotensin-converting enzyme, aminopeptidase M, leucine aminopeptidase, and carboxypeptidases A and B.[26] However by varying the P$'_1$ position between G and A, and randomizing the P$'_2$ (X$_2$) and P$'_3$ (X$_3$) residues the authors identified selective inhibitors from their optimization library.

A single-step deconvolution strategy was used to synthesize the peptides. X$_3$ was coupled to resin in 20 different reactions, the resin was mixed and split, and then X$_2$ was coupled in 20 different reactions. Each of these mixtures was split and then either Z-(L,D)PheY(PO$_2$CH$_2$)-(L,D)G-OH or Z-(L,D)PheY(PO$_2$CH$_2$)-(L,D)A-OH was coupled. This scheme produced 20 pools containing 40 Z-(L,D)PheY(PO$_2$CH$_2$)-G . . . (X$_3$) G/A-X$_2$ each, with G and X$_2$ defined, and 20 pools containing 80 Z-(L,D)PheY(PO$_2$CH$_2$)-(L,D)A . . . (X$_3$) peptides each, with A and X$_2$ defined. The peptides were cleaved from the resin and assayed. The X$_2$ selectivity determined for the Z-(L,D)PheY(PO$_2$CH$_2$)-G-X$_2$X$_3$ sublibrary was R/K for 24-15 and P for 24-16. Twenty Z-(L,D)PheY(PO$_2$CH$_2$)-GRX$_3$ peptides were synthesized, purified by HPLC, and tested, identifying X$_3$ selectivity for 24-15 to be F/M (~400-fold selective, K_i = 2.7/0.35 nM, respectively). Pursuing the Z-(L,D)PheY(PO$_2$CH$_2$)(L,D)A-X$_2$X$_3$ series the authors identified Pro, Nle as unselective, and Arg, Lys as 24-15-selective X$_2$ residues. From the deconvolution synthesis of X$_3$ peptides a nearly 2000-fold selective inhibitor was identified, Z-(L,D)PheY(PO$_2$CH$_2$)-(L,D)A-K-M, exhibiting a 24-15 K_i of 0.12 nM and a 1300-fold selective inhibitor was identified, Z-(L,D)PheY(PO$_2$CH$_2$)-(L,D)A-R-M, exhibiting a 24-15 K_i of 0.07 nM. This is a striking achievement of both selectivity and potency from an optimization library strategy.

Positional Scanning Format Library Methods. Inhibitors of chymotrypsin were discovered using a cyclic peptide template library.[72] A peptide backbone (KKKEG-resin) was synthesized using the tea-bag method of synthesis. Three different positional libraries were synthesized, requiring three different back-bone structures using orthogonally cleavable protecting groups on the lysine residues, depending on whether the residue being coupled was to be defined (Dde) or consisted of a mixture (Boc). The peptide backbone was cyclized between the N-terminal lysine and the glutamate residue before coupling of 19 proteinogenic and 10 carboxylic acids to the side chains of the three lysine residues to generate three positional libraries, cyclo[K(X)K(X)K(O$_1$)E]G-OH, cyclo[K(X)K(O$_2$)K(X)E]G-OH, and cyclo[K(O$_3$)K(X)K(X)E]G-OH. In these libraries O represents positions defined for each pool, X represents positions in which the mixture of side-chain acylating residues were attached, and the square brackets represent the cyclization between the N-terminal K and C-terminal E residues. A novel series of compounds was identified exhibiting side chains of piperonylic acid and 2-thiophenecarboxylic acid, with ~50 mM IC$_{50}$ values (compared to 1.6 mM for chymostatin and 35 mM for Ac-ygyyyr-NH$_2$).

Validation of Library Design Concepts. Using a more constrained and well-defined cyclic peptide scaffold, Domingo et al.[127] investigated the P$_1$ specificity for a series of different proteases. Their approach was based on the synthesis of an 11-mer peptide α-chymotrypsin inhibitor[165] mimicking the larger 6- to 9-kD Bowman–Birk inhibitors. The Bowman–Birk inhibitors present a nine-residue binding loop to proteases spanning the P$_3$ to P$'_6$ binding domain, while the peptide mimic presents a seven-residue loop to the enzyme. The authors synthesized a library (SCTX$_1$SIPPQCY) wherein each member had an identical sequence, randomized only at the P$_1$ position (X$_1$) by the 21 proteinogenic amino acids to generate a mixture of compounds. They screened this mixture for binding to porcine pancreatic elastase, subtilisin BPN', trypsin, α-chymotrypsin, and enzymatically inactive anhydro-α-chymotrypsin and anhydro-trypsin. Positive peptides were selected by passing the mixture over affinity columns composed of immobilized enzyme, and then the retained peptides were sequenced to identify the cleavage site and P$_1$ residues. In each case two sequences were obtained, one for the intact, bound peptide, and one for cleaved peptide. Trypsin exhibited a preference for K,R; α-chymotrypsin for F,Y,M,L,Nle; anhydro-α-chymotrypsin for F,Y; and elastase for L,M. K_i values ranged from 9000 to 10 nM.

Identification of Receptor Antagonists

Multiple-Synthesis Libraries. The method of synthesizing individual peptides on an array of pins and then either directly assaying for activity while the peptides are still bound to the pins or after their release into solution, lends itself to exploring structure activity relationships (SAR) of analogs based on the sequence of a lead molecule. The approach was applied to exploration of substance P analog binding to the NK1 receptor.[96] A radiolabeled binding assay was employed using a rat brain synaptosome tissue preparation. All the peptides synthesized had been reported previously, the purpose of the experiments described in the initial publication being to demonstrate that the crude material released from the pins could be used to establish SAR without further purification. The same group continued with the approach, applying it to study the SAR of substance P binding to the NK1 receptor through the simultaneous synthesis of 512 analogs to cover all possible D- and L-amino acid combinations of 9 of the 10 native residues of substance P (the C-terminal Met residue was not varied because it was directly incorporated into the support.[97]

This exhaustive SAR produced new insights and a rich database. Most interesting was that single changes were not predictive of dual changes. This is a fundamental difference in traditional medicinal chemistry and combinatorial medicinal chemistry approaches. In the traditional approach the parent molecule is frequently divided into domains and each domain is varied while the rest of the molecule is kept constant. Later on the best modifications of each domain are combined and the molecule is further refined. In the combinatorial approach, multiple changes can be introduced, and optimized simultaneously.

Much of the effort in synthetic combinatorial chemistry is now focused either on the synthesis of nonpeptide libraries or on the use of rational approaches to improve the design and utility of chemical libraries such as the design of a dipeptide library based on physicochemical properties of select amino acids.[166] After observing that the successful conversion of several peptides into nonpeptide peptidomimetics occurred through the discovery of millimolar dipeptides (angiotensin converting enzyme inhibitor CCK-B and CCK-A selective receptor antagonists), the authors considered that just using the 20 naturally occurring amino acids and four N-terminal protecting groups would generate 1600 possible peptides. They designed and synthesized an information-rich library of 256 N-protected dipeptides, encoding the greatest possible diversity of physicochemical information. The minimum analog peptide set was used,[167] plus various N-terminal blocking groups (the des-amino derivatives of three amino acids, A, G, F), and the C-terminus was amidated, providing structures with at least three potential pharmacophores, which were described by the z scales calculated using the physical properties of 55 amino acids. The set of 256 compounds was composed of dipeptides formed from of eight amino acids (D, K, S, F, L, W, T, and V).

This library was used to identify NK3 selective antagonists, followed by more traditional optimization incorporating multiple synthesis where appropriate.[98] The lead from the library was Boc(S)F(S)F-NH$_2$ (IC$_{50}$ = 1550 nM), where the S indicates the chirality of the Phe residues. This compound was somewhat specific for the NK3 receptor, exhibiting an IC$_{50}$ > 10 mM for NK1 and NK2 receptors. Of methylated analogs, only the C-terminal derivative retained activity, producing Boc(S)F(R,S)αMeF-NH$_2$ (IC$_{50}$ = 1520 nM). Exploration of the optimal chirality for the α-MeF and attempts to functionalize the N- and C-terminal groups produced Boc(S)F(R)αMeF-(CH$_2$)$_7$NHCONH$_2$ (IC$_{50}$ = 16 nM), exhibiting >700-fold selectivity for NK3 over NK1 and NK2, and greater potency than the reference compound; SuccDFNMeFGLM-NH$_2$.

Split-and-Mix, One-Compound–One-Bead, Postassay Identification Format Libraries. This method has been successfully applied to discover peptides that bind to the gpIIb/IIIa platelet integrin receptor.[55] A cyclic pentapeptide library containing both D- and L-amino acids was synthesized and assayed by competition of soluble gpIIb/IIIa binding to fibrogen-coated microplates. The peptide antagonists identified from the assay of only 50,000 compounds (a subportion of the library) were CRCDC and CRGdC.

The C5a receptor has been the target of drug discovery efforts for many years. For nearly 5 years a traditional peptide effort was conducted at Abbott Laboratories to discover receptor antagonists using a rational approach focused on the C-terminus of the C5a ligand. A series of peptides containing nonproteinogenic amino acids were identified, but all were agonists, typified by the structure FKA(D-cha)(L-Cha)r-OH (binding IC_{50} = 70 nM).[168–170] A subsequent effort carried out at another company did succeed in identifying an antagonist from this series after a similarly extensive effort.[171] This target thus represented a relevant model to compare the advantages of a combinatorial peptide library to traditional peptide approaches.

A library was designed based on an analog of the Abbott agonist (FKP(D-cha)(L-Cha)r-OH) in which the parent residue was coupled to 20% of the resin at each step of the synthesis, while the remaining 80% was split among the other residues (proteinogenic and nonproteinogenic amino acids) used for randomization and the C-terminal argenine was not varied. A library of 1,200,000 compounds was created on a double releasable linker designed to release the peptides as a free C-terminal OH group. Screened to identify only those compounds with a binding IC_{50} better than 50 nM, 37 compounds were identified (e.g., FL(Tic)(D-cha)(L-Cha)r-OH, (NMe-F)LY(D-cha)(L-Cha)-r-OH, FLY(D-cha)(L-Cha)r-OH), among them were close analogs to the antagonist described by Kawai et al.[169] In this case a single library not only provided new SAR, but was successful in converting an agonist into an antagonist. The library approach identified the same lead in a matter of a few weeks of synthesis and screening, which had taken years by a traditional approach.

Iterative Synthesis and Screening Library Methods. Three publications have addressed the discovery and optimization of endothelin receptor antagonists using combinatorial chemistry approaches. Exploring SAR, all possible (64) L- and D-amino acid combinations were assessed for a potent ETR_A hexapeptide ligand (Ac-f(Orn)DIIW-OH, IC_{50} = 230 nM), resulting in the identification of three analogs with millimolar activity, two containing 5 D-amino acid replacements, and with Ac-f(D-orn)Diiw-OH exhibiting a comparable IC_{50} as the parent of 250 nM.[139] The technique used was the split-and-mix synthesis of a 30,752-member tripeptide library (containing 31 or 32 residues consisting of proteinogenic or nonproteinogenic amino acids at each position) designed around a known ET antagonist, FR-139317.[77] Thirty-one pools (of mixtures of 992 compounds each) defined by the identity of the N-terminal residue, were screened. The authors then resynthesized and screened successive libraries to define the remaining two residues. Not only was FR-139317 identified, but extensive SAR around this molecule was obtained. Deconvolution of mixtures from a 19×19 combinatorial array, based on a similar but more potent hexapeptide ETR receptor antagonist (Ac-(D-BHG)LDIIW, $IC_{50} = 8$ nM) investigated by Spellmeyer et al.[139] not only produced extensive SAR, but produced several analogs with comparable affinity as the parent and one analog with higher affinity (Ac-(D-BHG)QDVIW-OH, $IC_{50} = 1.7$ nM).[140] Thus, all three independent studies demonstrated the power of combinatorial chemistry optimization libraries to identify potent compounds and to provide extensive SAR.

Positional Scanning Format Library Methods. To discover antagonists of the binding of IL-6 to its soluble receptor (sIL-6R), Wallace et al.[149] synthesized a positional scanning pentapeptide library. The ligand, IL-6, was bound to microtiter plates and then the library peptides were added with sIL-6R. Bound sIL-6R was detected with a specific anti-receptor antibody. No antagonists were discovered, so the investigators proceeded to synthesize the same library in a branched, polydentate configuration using lysine residues as the branching points to produce an octameric library. The branched lysine core is shown in Figure 3.13. The library was synthesized onto the eight free amino groups, such that each branch was identical. Thus, a multimeric or polydentate library was created in which each compound consisted of eight identical pentamers displayed off the branched lysine core. Again positional scanning was employed, but this time antagonists were identified, which produced a positional consensus (E/D/Y)(F/I/E)(Y/L/E/F/W)(F/I/W/Y)(W/Y/F) from which a possible 360 compounds could be derived. To simplify follow-up the degree of inhibition caused by certain pools was used to limit the residue selection, reducing the number of compounds synthesized in the follow-up to 36 peptides.

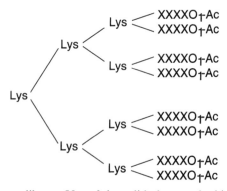

Figure 3.13. Polydentate library: Use of the solid phase and a highly branched linker complex of lysine residues to present multiple copies of the same peptide to the target, enhancing weak binding interactions. Shown is a multimeric "octodentate" library of compounds.

At a concentration of 45 mM, 20 of these produced greater than 50% inhibition, and dose–response curves were obtained for the 18 most active. These results indicated a consensus sequence of (E/D)(F/I)(L/Y)(I/F)W, with an IC_{50} of 75 nM for the six most active peptides (EFLIW, EFYIW, EILFW, EIYFW, EIYIW, DIYFW). Finally, the authors synthesized these as well as negative controls as tetramers and dimers. The tetramers exhibited similar activity as the octamers, but the dimers were inactive.

Ligands that bind the T-cell receptor were defined in order to assess the structural requirements for recognition of the MHC class 1/antigen complexes.[150,151] The mouse MHC class I amorphs H-2K and H-2L were chosen for study because both prefer octapeptides. The authors designed an octapeptide library, synthesizing eight libraries, each containing 19 pools with the position being defined occupied by 19 proteinogenic amino acids (OXXXXXXX, XOXXXXXX, XXOXXXXX, etc.). The ability to bind these peptides was assessed by the restoration of MHC expression. All the libraries contained peptides that bound and stabilized the two MHC class I molecules. In addition, the library pools sensitized target cells for cytolysis by T-lymphocytes, and the authors were able to define the specificity for several T-lymphocyte clones. In this functional assay the individual peptides identified were active in the pico- to femtomolar range. For instance, for one T-cell clone (4G3) that recognized the chicken ovalbumin-derived SIINFEKL epitope, they identified a novel epitope (DIKVGIEF), which was just as active. Even of greater interest was the identification of T-cell epitopes that induced cellular responses but that were inefficient in binding the MHC molecules. These latter molecules would appear to bypass the need for MHC interaction. This work demonstrates the potential of using synthetic combinatorial peptide libraries to understand the immune response and autoimmunity, and represents another approach to the development of synthetic vaccines.

Identification of Receptor Agonists

Iterative Synthesis and Screening Library Methods. A novel agonist of the m-opioid receptor, Ac-rfwink-NH$_2$, was identified from a hexapeptide library by testing the peptides in solution in a traditional radiolabel binding assay employing [^{3}H][D-Ala^2,MePhe^4,Gly^5-ol]enkephalin and rat brain tissue preparations.[142] The selectivity of this all-D-amino acid-containing peptide was high (K_i for m_1 = 16 nM and for m_2 = 41 nM), compared to the s receptor ($K_i >$ 1500 nM) and the k_1 ($K_i >$ 2000 nM), k_2 ($K_i >$ 5000 nM), and k_3 (K_i = 288 nM) receptor subtypes. Functional m-opioid receptor agonist activity (reversed by naloxone) was demonstrated on guinea pig ileum where Ac-rfwink-NH$_2$ exhibited an IC_{50} of 433 nM, compared to leu-enkephalin, which exhibited IC_{50} of 246 nM. In contrast, Ac-rfwink-NH$_2$ was functionally 100-fold less potent at the s receptor than leu-enkephalin, measured as the inhibition of electrically evoked contractions or mouse vas deferens tissue. Finally, intracerebroventricular injection of Ac-rfwink-NH$_2$ (3 nmol) showed analgesic properties in a nociception tail-flick model, exhibiting a long duration of action (120 min). Noteworthy is that the analgesic properties of this compound are more potent and longer lasting than those of morphine. The analgesic effect of Ac-rfwink-NH$_2$ was reversed by the irreversible m-selective antagonist β-funaltrexamine. Given intraperitoneal, Ac-rfwink-NH$_2$ also exhibited analgesic effects that were equipotent with morphine and reversed by intracerebroventricular naloxone. This observation suggests that the compound has bioavailability comparable to morphine. Due to its novel structure, bioavailability, potent activity, and good duration of effect, this compound or close analogs are being pursued as analgesics in an ongoing preclinical program.

L-Amino acid hexapeptides were also identified using a similar strategy.[141] While very potent at the m receptor, (Ac-RFMWMT-NH$_2$ K_i 0.8 nM, Ac-RFMWMK-NH$_2$ K_i 0.4 nM, Ac-RFMWMR-NH$_2$ K_i 0.5 nM), these compounds were less selective at the s receptor (K_i of 0.9, 5.6, and 7.4 nM, respectively) and the k$_3$ receptor (K_i of 0.6, 0.4, and 1.4 nM, respectively) than Ac-rfwink-NH$_2$. Subsequently, Dooley et al. identified several unrelated m-opioid agonists series, YPFGFX-NH$_2$ (X = R, IC$_{50}$ = 13 nM), WWPKHX-NH$_2$ (X = G, IC$_{50}$ = 9 nM), and Ac-FRWWYX-NH$_2$ (X = M, IC$_{50}$ = 33 nM), as well as antagonist, Ac-RWIGWR-NH$_2$ (IC$_{50}$ = 5 nM).[143] Interestingly, the truncated sequence WWPK-NH$_2$ exhibited an IC$_{50}$ of 17 nM, WWPR-NH$_2$ exhibited an IC$_{50}$ of 10 nM, and both retained >2000-fold selectivity for m- over s- and k-opioid receptors. These activities compare quite impressively with DAMGO (IC$_{50}$ = 3 nM; 253- and 844-fold selectivity over s and k, respectively) and YGGFL-OH (IC$_{50}$ = 5 nM; 0.7- and 257-fold selectivity over s and k, respectively. Relevant to the use of the iterative approach, the authors observed that at early stages in the iteration the YGG motif was more active, but because they pursued less active pools as well, they were able to define the YPFGFX-NH$_2$ and WWPKHX-NH$_2$ series, which were much more active and selective that the best YGGFMX-NH$_2$ molecules (IC$_{50}$ = 28 nM).

Split-and-Mix, One-Compound–One-Bead, Postassay Identification Format Libraries. A novel double-release assay was described for the identification of agonists and antagonists of G-protein coupled receptors.[129,130] After library synthesis, dry beads attached by surface effect to polyethylene film were exposed to gaseous TFA in a controlled manner to cleave ~4% of compound from each, and then neutralized with gaseous ammonium hydroxide. The released compound remained trapped in the microporous beads. For the cellular receptor assay, tissue culture dishes of confluent frog (*Xenopus laevis*) melanocytes, transiently expressing the bombesin receptor introduced by electroporation or expressing either the frog α-MSH receptor or the transfected human α-MSH receptor, were treated with 1 nM melatonin. They were then covered with a thin layer of agarose containing melatonin and, in the case of the α-MSH receptor studies, α-MSH. The polyethylene sheet displaying the library beads was inverted over this.

With the beads now in (agarose) solution the released compounds were free to diffuse out, contacting the underlying melanocytes. In this cellular system the pigmentation of the melanocytes is controlled through variations in the level of cAMP, which in turn is regulated by the G-coupled (bombesin or α-MSH) receptor. Agonists increase pigment darkening, whereas antagonists decrease pigmentation. In the bombesin studies, the plates with positive responses were frozen and the polyethylene film was removed (leaving the beads imbedded in the agarose). Image analysis was used to locate the positive beads, which were then collected by aspiration, washed, and subjected to a second round of release and screening.

The library synthesized for bombesin was derived from the sequence of bombesin, residues 8–14 (YAVGHLM-NH$_2$), and had the composition X_1X_2 VGHLX$_3$-NH$_2$. X_1 and X_2 were randomized with 19 L-amino acids, and X_3 was randomized with L, M, P, W to produce a 1444-member library. In the first round of screening 5000 beads were assayed and 500 were selected and reassayed. From this second round of screening 52 beads were selected and reassayed to produce four positive compounds. The beads producing these compounds were recovered and sequenced to determine their identity. The activities of agonists identified ranged from $\sim$100 nM (WFVGHLM and WAVGHLM) to 10 mM (AWVGHLM), compared to bombesin itself, which exhibited an EC$_{50}$ of $\sim$0.1 nM.

This study demonstrated an elegant method for identifying receptor agonists. The α-MSH study was directed toward the identification of antagonists. α-MSH is a 13-mer, while the library synthesized for screening was a tripeptide library released as 48 pools of Ac-$X_1X_2X_3$-NH$_2$ and 48 pools of $X_1X_2X_3$-NH$_2$ in which X_1 was defined for each pool. Iterative synthesis and screening of subpools was used to define X_2 and X_3, though it would also have been possible to recover the beads corresponding to underlying pigmentation changes and determine the identity of the compounds from information on the beads had a coding method been employed as well as either a multiple release pooling strategy or more sparse seeding of beads over the cell layer. From these studies the authors identified numerous antagonists, among them the most active was wRL-NH$_2$ (K_d = 63 nM). The potential for this approach is that any G-coupled receptor transfected into the melanocytes will be automatically coupled to the pigmentation response through stimulus-dependent effects on cAMP levels. Thus, this system represents a universal assay for a wide variety of therapeutically relevant receptors that is capable of identifying both agonists and antagonists.

Expanding on this work the same group synthesized three libraries based on the α-MSH-[5–13] sequence: a 7-mer (minux X_1X_2), 8-mer (minus X_1), and 9-mer library, with the format X_1X_2(D,L-F)$_3$(D,L-R)$_4$(D,L-W)$_5X_6K_7P_8V_9$-NH$_2$.[129] The approach used was split and mix, one compound–one bead, but split into 8 subpools depending on the chirality of the residues 3, 4, and 5 (e.g., LLL, DLL, LDL, etc.). Thus, by knowing the pool and sequencing the individual active peptide the authors could identify the compound including the chirality. Because their assay enabled them to assess the activity of the compounds from individual beads, as well as to identify both agonist and antagonist activity, they could assess individual responses within pools and not worry that compounds would interfere or cancel due to the simultaneous presence of an antagonist or agonist within the same pool. They identified an antagonist MP(D-F)(R)(D-W)FKPV-NH$_2$ (IC$_{50}$ = 11 nM), a partial agonist MP(D-F)(R)(D-W)WKPV-NH$_2$ (EC$_{50}$ = 690 nM), and a full agonist FH (D-F)(R)(L-W)QKPV-NH$_2$ (EC$_{50}$ = 0.4 nM); the agonist activity of α-MSH-[5-13] itself was EC$_{50}$ = 2 mM.

Identification of Functional Inhibitors

Iterative Synthesis and Screening Library Methods. Synthetic combinatorial peptide libraries have been screened to identify antimicrobial agents. In several reports of this application an iterative approach was used to identify L-amino acid-containing hexapeptides that exhibited antimicrobial activity,[8] as well as D-amino acid- and nonproteinogenic amino acid-containing peptides with or without acetylation of the N-terminus or other postsynthetic modification.[145] A liquid microbial growth assay was employed, and from these studies the peptide Ac-RRWWCR-NH$_2$ was identified as the most active among the L-amino acid-containing peptides, exhibiting a minimal inhibitory concentration (MIC) of 3.4 mg/mL against *Staphylococcus aureus*. From the libraries containing 56 different L, D, and nonproteinogenic amino acids at each position, four tetramers were identified: (aFmoc-elys)WfR-NH$_2$, (aFmoc-elys)Wfl-NH$_2$, (aFmoc-elys)Wf(aAIB)-NH$_2$, and (aFmoc-elys)Wfi-NH$_2$. The latter exhibited specific bacteriostatic activity for gram-positive bacteria.

Positional Scanning Format Library Methods. A hexapeptide positional scanning library was also used to identify peptides that inhibited the ability of melittin to be lyse red blood cells.[152] From a library containing 52,128,400 peptides, this study identified Ac-IVIFDC-NH2, with an IC$_{50}$ of 11 mg/mL. This compound had no similarity to melittin and inhibited hemolysis in a competitive manner, which was overcome by raising the concentration of melittin. The authors suggested that the effect of the peptide is through a specific interaction, perhaps with melittin itself, rather than through a nonspecific membrane effect.

Split-and-Mix, One-Compound–One-Bead, Postassay Identification Format Libraries. A novel antimicrobial assay has been described based on pH-sensitive single release.[131] The method was validated by use of analogs of the antimicrobial compound Ac-RRWWCR-NH$_2$, described originally by Houghten et al.[8,144] Multiple synthesis of an alanine scan series and N- and C-terminal truncations of the peptide rrwwcrc identified the importance of a C-terminal crc motif. There was only a marginal effect of deleting the N-terminal rr residues, and that the single replacements of the ww residues had the greatest effect on activity of any of the alanine replacements (the C-terminal cysteine was not replaced). Single replacements of the other residues had no affect on activity.

Once validated, the method was used to screen a pentapeptide library of D-proteinogenic and nonproteinogenic amino acids. The library peptides were synthesized on a linker that cleaved in aqueous solution at pH 7. The library beads were mixed in agarose with bacteria, poured into tissue culture plates (where the agar hardened), and incubated a sufficient period of time (12 h) for a lawn of bacteria to grow. Compounds were cleaved by the pH 7 medium and diffused out of the beads. Where the released peptides inhibited bacterial growth a clear ring formed around the bead of origin. The bead in the center of each clear zone was recovered and the identity of the compound was determined (either by directly sequencing or through sequencing of an encoding molecule). A series of compounds were identified, all analogs of the Houghten compound.

DISCUSSION

Peptide combinatorial chemistry has confronted the medicinal chemist with the problem of how to carry out challenging reactions in a routinely predictable, automatable manner. A variety of solid supports have been explored and are generally available for the solid-phase synthesis of the various types of library formats. Similarly, much work has been devoted to overcoming the synthetic issues that are specific to library synthesis. Good methods exist to assure the equimolar incorporation of multiple subunits and to monitor reactions at the level of single beads, a necessity in cases of library synthesis, with the exception of multiple synthesis. Newer methods of structure determination have also been described, and with the advent of automation and deconvolution methods, this is often more an iterative or deductive process rather than an analytical challenge.

There are growing bodies of literature addressing cyclization of peptides, the scaffolds created to confer structural rigidity onto library members, and the functionalization of side chains; N- and C-termini, and the peptide backbone. One successful approach has been the use of an N-substituted glycine backbone, referred to commonly as peptoids (see Chapter 6).

The three most common synthetic and screening combinatorial approaches being employed are multiple synthesis, iterative/deconvolution synthesis/identification, and one-compound–one-bead synthesis and postassay identification. All three general approaches work, though in specific circumstances one approach may be the most appropriate. Multiple synthesis plays an important role in interfacing lead generation with lead optimization, but it can also serve a valuable lead generation function in situations where there is basis for good design, just as the potentially larger iterative or one-compound–one-bead libraries can play a crucial role in lead optimization and SAR generation. The advantage of a sophisticated one-compound–one-bead/encoding/sequencing approach must be weighed against the investment that must be made in analytics for compound identification and in developing compatible chemical reactions and the expense of the linkers. Similarly, the elegance of a single synthesis must be compared with the need for iterative/recursive synthesis and screening, and the advantages of identifying an active compound versus elucidating the structure of a compound that should be active from the iterative algorithm. The synthesis and screening of mixtures is not just acceptable, but provides sound data and is in many instances crucial to success.

The research results that have been reviewed in this chapter demonstrate that combinatorial peptide approaches can identify peptide leads for many classes of targets that are of current therapeutic interest. For targets that have been extensively studied by traditional means, peptide combinatorial chemistry has frequently discovered novel leads and provided valuable new SAR. This in large part may be due to the ability to synthesize and screen so many compounds, and to incorporate many unnatural amino acids. It is also likely that such success is due to a difference between traditional and combinatorial approaches. In the former, medicinal chemists usually make one change at a time and assesss the affects, while in the latter multiple changes are made at one time, permitting the chemist for the first time to efficiently explore the role of and synergistic intramolecular interactions.

There are fewer reports demonstrating the use of combinatorial peptide chemistry to optimize lead structures, but those that are reviewed here indicate the power of the approach. Identifying more leads is not the solution to all problems of drug discovery; it simply moves the bottleneck upward in the process to optimization. Recognition that combinatorial chemistry approaches can assist in lead optimization is fundamentally important to the combinatorial chemistry/drug discovery revolution. To realize the real benefits of combinatorial chemistry, the drug discovery team must think and apply the combinatorial paradigm through all phases of optimization, including animal testing.

REFERENCES

1. Pavia MR, Sawyer TK, Moos WH (1993): The generation of molecular diversity. *Bioorg Med Chem Lett* 3:387–396.
2. Gallop MA, Barrett RW, Dower WJ, Fodor SPA, Gordon EM (1994): Applications of combinatorial technologies to drug discovery, 1: background and peptide combinatorial libraries. *J Med Chem* 37:1233–1251.
3. Gordon EM, Barrett RW, Dower WJ, Fodor SPA, Gallop MA (1994): Applications of combinatorial technologies to drug discovery, 2: combinatorial organic synthesis, library screening strategies, and future directions. *J Med Chem* 37:1385–1401.
4. Janda KD (1994): Tagged versus untagged libraries; methods for the generation and screening of combinatorial chemical libraries. *Proc Natl Acad Sci USA* 91:10779–10785.

5. Lebl M, Krchnak V, Sepetov NF, Seligmann B, Strop P, Felder S, Lam KS (1995): One-bead–one-structure combinatorial libraries. *Biopolymers (Pept Sci)* 37: 177–198.

6. Terrett NK, Gardner M, Fordon DW, Kobylecki RJ, Steele J (1995): Combinatorial synthesis: the design of compound libraries and their application to drug discovery. *Tetrahedron* 51:8135–8173.

7. Geysen HM, Meloen RH, Barteling SJ (1984): Use of peptide synthesis to probe viral antigens for epitopes to a resolution of a single amino acid. *Proc Natl Acad Sci USA* 81:3998–4002.

8. Houghten RA, Pinilla C, Blondelle SE, Appel JR, Dooley CT, Cuervo JH (1991): Generation and use of synthetic peptide combinatorial libraries for basic research and drug discovery. *Nature* 354:84–86.

9. Lam KS, Salmon SE, Hersh EM, Hruby VJ, Kasmierski WM, Knapp RJ (1991): A new type of synthetic peptide library for identifying ligand-binding activity. *Nature* 354:82–84.

10. Erb E, Janda KD, Brenner S (1994): Recursive deconvolution of combinatorial chemical libraries. *Proc Natl Acad Sci USA* 91:11422–11426.

11. Furka A, Sebestyen F, Asgedom M, Dibo G (1991): General method for rapid synthesis of multicomponent peptide mixtures. *Int J Pept Protein Res* 37:487–493.

12. Hortin, GL, Stantz WD, Santoro SA (1992): Preparation of soluble peptide libraries: application to studies of platelet adhesion sequences. *Biochem Int* 26:731–738.

13. Rapp W, Zhang L, Habich R, Bayer E (1989): Polystyrene–polyoxyethylene graft copolymer for high-speed peptide synthesis. In Jung G, Bayer E, eds. *Pept 88, Proc 20th Eur Pept Soc.* Berlin: de Gruyter, pp 199–201.

14. Barany G, Albericio F, Biancalana S, Bontems SL, Chang JL, Eritja R, Ferrer M, Fields CG, Fields GB, Lyttle MH, Sole NA, Tian Z, Van Abel RJ, Wright PB, Salipsky S, Hudson D (1992): Biopolymer synthesis on novel polyethylene glycol–polystyrene (PEG-PS) graft supports, In Smith JA, Rivier JE, eds. *Peptides: Chemistry and Biology.* Leiden: ESCOM Sci, pp 603–604.

15. Geysen HM, Meloen RH, Mason TJ (1986): A priori delineation of a peptide which mimics a discontinuous antigenic determinant. *Mol Immunol* 23:709–715.

16. Bray AM, Valerio RM, Dipasquale AJ, Greig J, Maeji NJ (1995): Multiple synthesis by the multipin method as a methodological tool. *J Pept Sci* 1:80–87.

17. Maeji NJU, Bray AM, Valerio RM, Wang W (1995): Larger-scale multipin peptide synthesis. *Pept Res* 8:33–38.

18. Bray AM, Chierfari DS, Valerio RM, Maeji NJ (1995): Rapid optimization of organic reactions on solid phase using multipin approach: synthesis of 4-aminoproline analogues by reductive amination. *Tetrahedron Lett* 36:5081–5084.

19. Frank R (1995): Simultaneous and combinatorial chemical synthesis techniques for the generation and screening of molecular diversity. *J Biotechnol* 41:259–272.

20. Tegge W, Frank R, Hofmann F, Dostmann WRG (1995): Determination of cyclic nucleotide-dependent protein kinase substrate specificity by the use of peptide libraries on cellulose paper. *Biochemistry* 34:10569–10577.

21. Pokorny V, Mudra P, Jehnicka J, Zenisek K, Pavlik M, Voburka Z, Rinnova M, Stierandova, A, Lucka AW, Eichler J, Houghten RA, Lebl M (1994): Compas 242: new type of multiple peptide synthesizer utilizing cotton and tea bag technology. In

Epton R, ed. *Innovative Perspect Solid Phase Syn Collects Pept Int Symp, 3d* Oxford, UK: Mayflower Worldwide, pp 643–648.

22. Eichler J, Houghten RA, Lebl M (1996): Inclusion volume solid-phase peptide synthesis. *J Pept Sci,* 2:240–244.

23. Eichler J, Bienert M, Stierandova A, Lebl M (1991): Evaluation of cotton as a carrier for solid phase peptide synthesis. *Pept Res* 4:296–307.

24. Eichler J, Houghten RA (1993): Identification of substrate-analog trypsin inhibitors through the screening of synthetic peptide combinatorial libraries. *Biochemistry* 32:11033–11041.

25. Stankova M, Wade S, Lam KS, Lebl M (1994): Synthesis of combinatorial libraries with only one representation of each structure. *Pept Res* 7:292–298.

26. Krchnak V, Vagner J (1990): Color-monitored solid-phase multiple peptide synthesis under low-pressure continuous flow conditions. *Pept Res* 3:182–193.

27. Houghten RA (1985): General method for the rapid solid-phase synthesis of large numbers of peptides: specificity of antigen–antibody interaction at the level of individual amino acids. *Proc Natl Acad Sci USA* 82:5131–5135.

28. Steward JM, Young CM (1984): *Solid Phase Peptide Synthesis.* Rockford, IL: Pierce Chemical.

29. Grant GA, Ed (1992): *Synthetic Peptides: A User's Guide.* New York: Freeman, pp 1–382.

30. Merrifield B (1995): Solid-phase peptide synthesis. In Gutte B, ed. *Pept: Synthesis Structures and Applications.* San Diego, CA: Academic, pp 94–169.

31. Lam KS, Lebl M (1994): Selectide technology: bead-binding screening. *Methods: A Companion to Methods in Enzymology* 6:372–380.

32. Lebl M, Krchnak V, Salmon SE, Lam KS (1994): Screening of completely random one-bead–one-peptide libraries for activities in solution. *Methods: A Companion to Methods in Enzymology* 6:381–387.

33. Carpino LA, El-Faham A (1995): Tetramethylfluoroformamidium hexafluorophosphate: a rapid-acting peptide coupling reagent for solution and solid phase peptide synthesis. *J Am Chem Soc* 117:5401–5402.

34. Krchnak V, Vagner J, Safar P, Lebl M (1988): Noninvasive continuous monitoring of solid-phase peptide synthesis by acid–base indicator. *Collect Czech Chem Commun* 53:2542–2548.

35. Kaiser E, Colescott RL, Bossinger CD, Cook PI (1969): Color test for detection of free terminal amino groups in the solid-phase synthesis of peptides. *Anal Biochem* 34:595–598.

36. Rutter WJ, Santi DV (1991): General method for producing and selecting peptides with specific properties. *US Patent 5,010,175.*

37. Pinilla C, Appel JR, Blanc P, Houghten RA (1992): Rapid identification of high-affinity peptide ligands using positional scanning synthetic peptide combinatorial libraries. *Bio Techniques* 13:901–905.

38. Pinilla C, Appel J, Blondelle SE, Dooley CT, Dorner B, Eichler J, Ostresh J, Houghten RA (1995): A review of the utility of soluble peptide combinatorial libraries. *Biopolymers (Pept Sci)* 37:221–240.

39. Kramer A, Volkmer-Engert R, Malin R, Reineke U, Schneider-Mergener J (1993): Simultaneous synthesis of peptide libraries on single resin and continuous

cellulose membrane supports: examples for the identification of protein, metal, and DNA binding peptide mixtures. *Pept Res* 6:314–319.

40. Andrews PC, Boyd J, Loo OR, Zhao R, Zhu CQ, Grant K, Williams S (1994): Synthesis of uniform peptide libraries and methods for physico-chemical analysis, In Crabb JW, ed. *Techniques in Protein Chemistry V*. San Diego: Academic, pp 485–492.

41. Frank R (1994): Spot synthesis: an easy and flexible tool to study molecular recognition. In Epton R, ed. *Innovative Perspect Solid Phase Syn Collect Pept Int Symp, 3d*. Birmingham, UK: Mayflower Worldwide, pp 509–512.

42. Zuckermann RN, Kerr JM, Siani MA, Banville SC (1992): Design, construction and application of a fully automated equimolar peptide mixture synthesizer. *Int J Pept Protein Res* 40:497–506.

43. Saneii HH, Shannon JD, Miceli RM, Fischer HD, Smith CW (1993): Comparison of a phage-derived peptide library with one synthesized chemically on the peptide librarian. *Pept Chem* 31:117–120.

44. Bartak Z, Bolf J, Kalousek J, Mudra P, Pavlik M, Pokomy V, Rinnova M, Voburka, Z, Zenisek K, Krchnak V, Lebl M, Salmon SE, Lam KS. (1994): Design and construction of the automatic peptide library synthesizer. *Methods: A Companion to Methods in Enzymology* 6:432–437.

45a. Fodor SPA, Read JL, Pierrung MC, Stryer L, Lu AT, Solas D (1991): Light-directed, spatially addressable parallel chemical synthesis. *Science* 251:767–773.

45b. Fodor SPA, Read JL, Pierrung MC, Stryer L, Lu AT, Solas D. (1991b): Light-directed, spatially addressable parallel chemical synthesis. *Science* 251:767–773.

46. Jacobs JW, Fodor SPA (1994): Combinatorial chemistry: applications of light-directed chemical synthesis. *Trends Biotechol* 12:19–26.

47. Han H, Wolfe MM, Brenner S, Janda KD (1995): Liquid-phase combinatorial chemistry. *Proc Natl Acad Sci USA* 92:6419–6423.

48. Dooley CT, Houghten RA (1995): A comparison of combinatorial library approaches for the study of opioid compounds. *Perspect Drug Disc Design* 2:287–304.

49. Pinilla C, Buencamino J, Appel JR, Hopp TP, Houghten RA (1995): Mapping the detailed specificity of a calcium-dependent monoclonal antibody through the use of soluble positional scanning combinatorial libraries: identification of potent calcium-independent antigens. *Mol Diversity* 1:21–28.

50. Deprez B, Williard X, Bourel L, Coste H, Hyafil F, Tartar A (1995): Orthogonal combinatorial chemical libraries. *J Am Chem Soc* 117:5405–5406.

51. Blake J, Litzi–Davis L (1992): Evaluation of peptide libraries: an interative strategy to analyze the reactivity of peptide mixtures with antibodies. *Bioconjugate Chem* 3:510–513.

52. Holmes CP, Rybak CM (1994): Peptide reversal on solid supports: a technique for the generation of C-terminal exposed peptide libraries. In Hodges RS, Smith JA, eds. *Pept: Proc 13 Am Pept Symp*. Leiden: ESCOM Sci, pp 992–994.

53. Kania RS, Zuckermann RN, Marlowe CK, (1994): Free C-terminal resin-bound peptides: reversal of peptide orientation via a cyclization/cleavage protocol. *J Am Chem Soc* 116:8835–8836.

54. Lebl M, Krchnak V, Sepetov NF, Nikolaev V, Stierandova A, Safar P, Seligmann

B, Strop P, Thorpe D, Felder S, Lake DF, Lam KS, Salmon SEL (1994): One-bead–one-structure libraries. In Epton R, ed. *Innovative Perspect Solid Phase Syn Collect Pept Int Symp, 3d.* Birmingham, UK: Mayflower Worldwide, pp 233–238.

55. Salmon SE, Lam KS, Lebl M, Kandola A, Khattri PS, Wade S, Patek M, Kocis P, Krchnak V, Thorpe D, Felder S (1993): Discovery of biologically active peptides in random libraries: solution-phase testing after staged orthogonal release from resin beads. *Proc Natl Acad Sci USA* 90:11708–11712.

56. Lebl M, Patek M, Kocis P, Krchnak V, Hruby VJ, Salmono SE, Lam KS (1993): Multiple release of equimolar amounts of peptides from a polymeric carrier using orthogonal linkage-cleavage chemistry. *Int J Pept Protein Res* 41:201–203.

57. Kocis P, Krchnak V, Lebl M (1993): Symmetrical structure allowing the selective multiple release of a defined quantity of peptide from a single bead of polymeric support. *Tetrahedron Lett* 34:7251–7252.

58. Jayawickreme CK, Graminski GF, Quillan JM, Lerner MR (1994): Creation and functional screening of a multi-use peptide library. *Proc Natl Acad Sci USA* 91:1614–1618.

59. Burbaum JJ, Ohlmeyer MHJ. Reader JC, Henderson I, Dillard LW, Li G, Randle TL, Sigal NH, Chelsky D, Balwin JJ (1995): A paradigm for drug discovery employing encoded combinatorial libraries. *Proc Natl Acad Sci USA* 92:6027–6031.

60. Madden D, Krchnak V, Lebl M (1995): Synthetic combinatorial libraries: views on techniques and their application. *Perspect Drug Disc Design* 2:269–286.

61. Sepetov N, Krchnak V, Stankova M, Wade S, Lam KS, Lebl M (1995): Library of libraries: approach to synthetic combinatorial library design and screening of "pharmacophore" motifs. *Proc Natl Acad Sci USA* 92:5426–5430.

62. Lebl M, Krchnak V, Stierandova A, Safar P, Kocis P, Nikolaev V, Sepetov NF, Ferguson R, Seligmann B, Lam KS, Salmon SE (1994): Nonsequencable and/or nonpeptide libraries. In Hodges RS, Smith JA, eds. *Pept: Proc 13 Am Pept Symp.* Leiden: ESCOM Sci, pp 1007–1008.

63. Lebl M, Krchnak V, Safar P, Stierandova A, Sepetov NF, Kocis P, Lam KS. (1994): Construction and screening of libraries of peptide and nonpeptide structures. In Crabb JW, ed. *Techniques in Protein Chem V.* San Diego, CA: Academic, pp 541–548.

64. Safar P, Stierandova A, Lebl M (1995): Amino acid like subunits based on iminodiacetic acid and their application in linear and DKP libraries. In Maia HLS, ed. *Pept 94, Proc 23d Eur Pept Soc.* Leider: ESCOM Sci pp 471–472.

65. Krchnak V, Weichsel AS, Cabel D, Lebl M (1995): Linear presentation of variable side-chain spacing in a highly diverse combinatorial library. *Pept Res* 8:198–205.

66. Simon RJ, Kaina RS, Zuckermann RN, Huebner VD, Jewell DA, Banville S, Ng S, Wang L, Hosenberg S, Marlowe CK, Spellmeyer DC, Tan R, Frnkel AD, Santi DV, Cohen FE, Bartlett PA (1992): Peptoids: a modular approach to drug discovery. *Proc Natl Acad Sci USA* 89:9367–9371.

67. Zuckermann RN, Martin EJ, Spellmeyer DC, Stauber GB, Shoemaker KR, Kerr JM, Figliozzi GM, Goff DA, Siani MA, Simon RJ, Banville SC, Brown EG, Wang L, Richter LS, Moos WH (1994): Discovery of nanomolar ligands for 7-transmembrane G-protein-coupled receptors form a diverse *N*-(substituted)glycine peptoid library. *J Med Chem* 37:2678–2685.

68. Zuckerman RN, Kerr JM, Kent SB:H, Moos WH (1992): Efficient method for the preparation of peptoids [oligo(N-substituted glycines)] by submonomer solid-phase synthesis. J Am Chem Soc 114:10646–10647.

69. Ellman JA, Thompson LA (1996): Synthesis and application of small molecule libraries. *Chem Rev* 96:555–600.

70. Rinnova M, Lebl M (1996): Molecular diversity and libraries of structures: synthesis and screening. *Collect Czech Chem Commun* 61:171–231.

71. Lam KS, Lebl M, Wade S, Stierandova A, Khattri PS, Collins N, Hruby VJ (1994): Streptavidin–peptide interaction as a model system for molecular recognition. In Hodges RS, Smith JA, eds. *Pept: Proc 13 Am Pept Sym* Leiden: ESCOM Sci, pp 1005–1006.

72. Eichler J, Lucka AW, Houghten RA (1994): Cyclic peptide template combinatorial libraries: synthesis and identification of chymotrypsin inhibitors. *Pept Res* 7:300–307.

73. Spatola AF, Darlak K, Romanovskis P (1996): An approch to cyclic peptide libraries: reducing epimerization in medium sized rings during solid phase synthesis. *Tetrahedron Lett* 37:591–594.

74. Virgilio AA, Ellmann JA (1994): Simultaneous solid-phase synthesis of β-turn mimetics incorporating side-chain functionality. *J Am Chem Soc* 116:11580–11581.

75. Blondelle SE, Takahashi E, Houghten RA, Perez–Paya E (1996): Rapid identification of compounds with enhanced antimicrobial activity by using conformationally defined combinatorial libraries. *Biochem J* 313:141–147.

76. Jiracek J, Yiotakis A, Vincet B, Lecoq A, Nicolaou A, Checler F, Dive V (1995) Development of highly potent and selective phosphinic peptide inhibitors of zinc endopeptidase 24–15 using combinatorial chemistry. *J Biol Chem* 270:21701–21706.

77. Terrett NK, Bojanic D, Brown D, Bungay PJ, Gardner M, Gordon DW, Mayers CJ, Steele J (1995): The combinatorial synthesis of a 30,752-compound library: discovery of SAR around the endothelin antagonist, FR-139,317. *Bioorg Med Chem Lett* 5:917–922.

78. Combs AP, Kapoor TM, Feng S, Chen JK, Daude–Snow LF, Schreiber SL (1996): Protein structure-based combinatorial chemistry: discovery of non-peptide binding elements to Src SH3 domain. *J Am Chem Soc* 118:287–288.

79. Gennari C, Nestler HP, Salmon B, Still WC (1995): Solid-phase synthesis of vinylogous sulphonyl peptides. *Angew Chem Int Ed* 34:1763–1765.

80. Gennari C, Nestler HP, Salmon B, Still WC (1995): Synthetic receptors based on vinylogous sulfonyl peptides. *Angew Chem Int Ed* 34:1765–1768.

81. Burger MT, Still WC (1995): Synthetic ionophores: encoded combinatorial libraries of cyclen-based receptors for Cu^{2+} and Co^{2+}. *J Org Chem* 60:7382–7383.

82. Boyce R, Li G, Nestler HP, Suenaga T, Still WC (1994): Peptidosteroidal receptors for opioid peptides: sequence-selective binding using a synthetic receptor library. *J Am Chem Soc* 116:7955–7956.

83. Hiroshige M, Hauske JR, Zhou P (1995): Palladium-mediated macrocyclization on solid support and its applications to combinatorial synthesis. *J Am Chem Soc* 117:11590–11591.

84. Wang GT, Li S, Wideburg N, Krafft GA, Kempf DJ (1996): Synthetic chemical

diversity: solid phase synthesis of libraries of C(2) symmetric inhibitors of HIV protease containing diamino diol and diamino alcohol cores. *J Med Chem* 38:2995–3002.

85. Wang Z, Laursen RA (1992): Multiple peptide synthesis on polypropylene membranes for rapid screening of bioactive peptides. *Peptide Res* 5:275–280.

86. Frank R, Guler S, Krause S, Lindenmaier W (1990): Facile and rapid "spot-synthesis" of large numbers of peptides on membrane sheets. In Giralt E, Andreu D, eds. *Peptides: Proc 21st Eur Pept Symp.* Leiden: ESCOM Sci, pp 151–152.

87. Holmes CP, Adams CL, Kochersperger LM, Mortensen RB, Aldwin LA (1995): The use of light-directed combinatorial peptide synthesis in epitope mapping. *Biopolymers (Pep Sci)* 37:199–211.

88. Kerr JM, Banville SC, Zuckermann RN (1993): Encoded combinatorial peptide libraries containing nonnatural amino acids. *J Am Chem Soc* 115:2529–2531.

89. Bergmann C, Stohlmann SA, McMillan M (1993): An endogenously synthesized decamer peptide efficiently primes cytotoxic T cells specific for the HIV-1 envelope glycoprotein. *Eur J Immunol* 23:2777–2781.

90. Kramer A, Vakalopoulou E, Schleuning W-D, Schneider–Mergener J (1995): A general route to fingerprint analysis of peptide-antibody interactions using a clustered amino acid peptide library: comparison with a phage display library. *Mol Immunol* 32:459–465.

91. Cass R, Dreyer ML, Giebel LB, Hudson D, Johnson CR, Ross MJ, Schaeck J, Shoemaker KR (1994): Pilot, a new peptide lead optimization technique and its application as a general library method. In Hodges RE, Smith JA, eds. *Pep: Chem, Struct, Biol,; Proc Am Pept Symp 13.* Leiden: ESCOM Sci, pp 975–977.

92. Chu Y-H, Avila LZ, Biebuyck HA, Whitesides GM (1993): Using affinity capillary electrophoresis to identify the peptide in a peptide library that binds most tightly to vancomycin. *J Org Chem* 58:649–652.

93. Till JH, Annan RS, Carr SA, Miller WT (1994): Use of synthetic peptide libraries and phosphopeptide-selective mass spectrometry to probe protein kinase substrate specificity. *J Biol Chem* 269:7423–7428.

94. Kiely JS, Moos, WH, Pavia MR, Schwarz RD, Woodard GL (1991): A silica gel plate-based qualitative assay for acetylcholinesterase activity: a mass method to screen for potential inhibitors. *Anal Biochem* 196:438–442.

95. Bastos M, Marji NJ, Abeles RH (1995): Inhibitors of human heart chymase based on a peptide library. *Proc Natl Acad Sci USA* 92:6738–6742.

96. Wang J-X, Bray AM, Dipasquale AJ, Maeji NJ, Geysen HM (1993): Systemic study of substance P analogs, I: evaluation of peptides synthesized by the multipin method for quantitative receptor binding assay. *Int J Pept Protein Res* 42:384–391.

97. Wang J-X, DiPasquale AJ, Bray AM, Maeji NJ, Geysen HM (1993): Study of stereo-requirements of substance P binding to NK1 receptors using analogues with systematic D-amino acid replacements. *Bioorg Med Chem Lett* 3:451–456.

98. Boden P, Eden JM, Hodgson J, Horwell DC, Howson W, Hughes J, McKnight AT, Meecham K, Pritchard MC, Raphy J, Ratcliffe GS, Suman–Chauhan N, Woodruff GN (1994): The rational development of small molecule tachykinin NK3 receptor selective antagonists: the utilisation of a dipeptide chemical library in drug design. *Bioorg Med Chem Lett* 4:1679–1684.

99. Kassarjian A, Schellenberger V, Turck CW (1993): Screening of synthetic peptide libraries with radiolabeled acceptor molecules. *Pept Res* 6:129–133.

100. Needels MC, Jones DG, Tate EH, Heinkel GL, Kochersperger LM, Dower WJ, Barrett RW, Gallop MA (1993): Generation and screening of an oligonucleotide-encoded synthetic peptide library. *Proc Natl Acad Sci USA* 90:10700–10704.

101. Ohlmeyer MHJ, Swanson RN, Dillard LW, Reader JC, Asouline G, Kobayashi R, Wigler M, Still WC (1993): Complex synthetic chemical libraries indexed with molecular tags. *Proc Natl Acad Sci USA* 90:10922–10926.

102. Youngquist RS, Fuentes GR, Lacey MP, Keough T (1994): Matrix-assisted laser desorption ionization for rapid determination of the sequences of biologically active peptides isolated from support-bound combinatorial peptide libraries. *Rapid Commun Mass Spectrosc* 8:77–81.

103. Youngquist RS, Fuentes GR, Lacey MP, Keough T (1995): Generation and screening of combinatorial peptide libraries designed for rapid sequencing by mass spectroscopy. *J Am Chem Soc* 117:3900–3906.

104. Steward MW, Stanley CM, Obeid OE (1995): A mimotope from a solid-phase peptide library induces a measles virus-neutralizing and protective antibody response. *J Virol* 69:7668–7673.

105. Lam KS, Lou Q, Zhao Z-G, Smith J, Chen M-L, Pleshko E, Salmon SE (1995): Idiotype specific peptides bind to the surface immunoglobulins of two murine B-cell lymphoma lines, inducing signal transduction. In *Biomedical Pept, Protein & Nucleic Acids, I.* Birmingham, UK: Mayflower Worldwide, pp 205–210.

106. Lam KS, Lake D, Salmon SE, Smith J, Chen M-L, Wade S, Abdul-Latif F, Leblova Z, Ferguson RD, Krchnak V, Sepetov NF, Lebl M (1996): Application of a one-bead–one-peptide combinatorial library method for B-cell epitope mapping. *Immunomethods,* in press.

107. Smith MH, Lam KS, Hersh EM, Lebl M, Grimes WJ (1994): Peptide sequences binding to MHC class I proteins. *Mol Immunol* 31:1431–1437.

108. Lam KS, Lebl M (1992): Streptavidin and avidin recognize peptide ligands with different motifs. *Immunomethods* 1:11–15.

109. Gissel B, Jensen MR, Gregorius K, Elsner HI, Svendsen I, Mouritsen S (1995): L- and D-peptide ligands to streptavidin and avidin found in a synthetic peptide library. In Maia HLS, ed. *Pept 94; Proc 23 Eur Pept Soc* Leiden: ESCOM Sci, pp 1–6.

110. Fassina G, Bellitti MR, Cassani G (1994): Screening synthetic peptide libraries with targets bound to magnetic beads. *Protein Pept Lett* 1:15–18.

111. Chen JK, Lane WS, Brauer AW, Tanaka A, Schreiber SL (1993): Biased combinatorial libraries: novel ligands for the SH3 domain of phosphatidylinositol 3-kinase. *J Am Chem Soc* 115:12591–12592.

112. Yu H, Chen JK, Feng S, Dalgarno DC, Brauer AW, Schreiber SL (1994): Structural basis for the binding of proline-rich peptides to SH3 domains. *Cell* 76:933–945.

113. Wu J, Ma QN, Lam KS (1994): Identifying substrate motifs of protein kinases by a random library approach. *Biochemistry* 33:14825–14833.

114. Lam KS, Wu J, Lou Q (1995): Identification and characterization of a novel

synthetic peptide substrate specific for Src-family protein tyrosine kinases. *Int J Pept Protein Res* 45:587–592.

115a. Lou Q, Wu J, Salmon SE, Lam KS (1995): Structure–activity relationship of a novel peptide substrate for p60[c-src] protein tyrosine kinase. *Lett Pept Sci* 2:289–296.

115b. Lou Q, Leftwich ME, Lam KS (1996): Identification of GIYWHHY as a novel peptide substrate for human p60[c-src] protein tyrosine kinase. *Bioorg Med Chem, Bioorg Med Chem* 4:677–682.

116. Petithory JR, Masiarz FR, Kirsch JF, Santi DV, Malcolm BA (1991): A rapid method for determination of endoproteinase substrate specificity: specificity of the 3C proteinase from hepatitis A virus. *Proc Natl Acad Sci USA* 88:11510–11514.

117. Ator MA, Beigel S, Dankanich TC, Echols M, Gainor JA, Gilliam CL, Gordon TD, Koch D, Koch JF, Kruse LI, Morgan BA, Krupinski–Olsen R, Siahaan TJ, Singh J, Whipple DA (1994): Immobilized peptide arrays: a new technology for the characterization of protease function In Hodges RE, Smith JA, eds. *Pep: Chem, Struct, Biol, Proc Am Pept Symp 13*. Leiden: ESCOM Sci, pp 1012–1015.

118. Singh J, Allen M, Solowiej, Ator M, Olsen R, Whipple D, Gainor J, Gilliam C, Treasurywala A, Echols M, Petersen J, Kruse L, Upson D (1995): Screening of "unbiased" immobilized peptide library versus proteases: an approach for rapid determination of substrate specificity and selectivity. In Maia HLS, ed. *Pept 94: Proc 23rd Eur Pept Symp*. Leiden: ESCOM Sci.

119. Singh J, Allen M, Ator M, Gainor J, Whipple D, Solowiej JE, Treasurywala A, Morgan BA, Gordon TD, Upson D (1995): Validation of screening immobilized peptide libraries for discovery of protease substrates. *J Med Chem* 38:217–219.

120. Meldal M, Svendsen I, Breddam K, Auzanneau F-I (1994): Portion-mixing peptide libraries of quenched fluorogenic substrates for complete subsite mapping of endoprotease specificity. *Proc Natl Acad Sci USA* 91:3314–3318.

121. Meldal M, Svendsen I (1995): Direct visualization of enzyme inhibitors using a portion mixing inhibitor library containing a quenched fluorogenic peptide substrate, I: inhibitors for subtilisin Carlsberg. *J Chem Soc Perkin Trans* 1:1591–1596.

122. Moran EJ, Sarshar S, Cargill JF, Shahbaz MM, Lio A, Mjalli AMM, Armstrong RW (1995): Radio frequency tag encoded combinatorial library method for the discovery of tripeptide-substituted cinnamic acid inhibitors of the protein tyrosine phosphatase PTP1B. *J Am Chem Soc* 117:10787–10788.

123. Samson I, Kerremans L, Rozenski J, Samyn B, Van Beeumen J, Van Aerschot A, Herdewijn P (1995): Identification of a peptide inhibitor against glycosomal phosphoglycerate kinase of *Trypanosoma brucei* by a synthetic peptide library approach. *Bioorg Med Chem* 3:257–265.

124. Lowe G, Quarrell R (1994): Screening peptide bead libraries for enzyme inhibitors. *Methods: A Comparison to Methods in Enzymology* 6:411–416.

125. Seligmann, B, in press.

126. Seligmann, B, in press.

127. Domingo GJ, Leatherbarrow RJ, Freeman N, Patel S, Weir M (1995): Synthesis of a mixture of cyclic peptides based on the Bowman–Birk reactive site loop to screen for serine protease inhibitors. *Int J Pept Res* 46:79–87.

128. Vlattas I, Sytwn II, Dellureficio J, Stanton J, Braunwalder AF, Galakatos N, Kramer R, Seligmann B, Sills MA, Wasvary J (1994): Identification of a receptor-

binding region in the core segment of the human anaphylatoxin C_{5a}. *J Med Chem* 37:2783–2790.

129. Jayawickreme CK, Quillan JM, Graminski GF, Lerner MR (1994): Discovery and structure–function analysis of *a*-melanocyte-stimulating hormone antagonists. *J Biol Chem* 47:29846–29854.

130. Quillan JM, Jayawickreme CK, Lerner MR (1995): Combinatorial diffusion assay used to identify topically active melanocyte-stimulating hormone receptor antagonists. *Proc Natl Acad Sci USA* 92:2894–2898.

131. Seligmann, B. in press.

132. Lam KS, Zhao Z-C, Wade S, Krchnak V, Lebl M (1994): Identification of small peptides that interact specifically with a small organic dye. *Drug Dev Res* 33:157–160.

133. Pinilla C, Appel JR, Houghten RA (1993): Synthetic peptide combinatorial libraries (SPLCs): identification of the antigenic determinant of β-endorphin recognized by monoclonal antibody 3E7. *Gene* 128:71–76.

134. Zuckermann RN, Kerr JM, Siani MA, Banville SC, Santi DV (1992): Identification of highest-affinity ligands by affinity selection from equimolar peptide mixtures generated by robotic synthesis. *Proc Natl Acad Sci USA* 89: 4505–4509.

135. Bianchi E, Folgori A, Wallace A, Nicotra M, Acali S, Phalipon A, Barbato G, Bazzo R, Cortese R, Felici F, Pessi A (1995): A conformationally homogenous combinatorial peptide library. *J Mol Biol* 247:154–160.

136. Sila U, Mutter M (1995): Topical templates as a tool in molecular recognition and peptide mimicry: synthesis of a TASK library. *J Mol Recog* 8:29–34.

137. Owens RA, Gesellchen PD, Houchins BJ, DiMarchi RD (1991): The rapid identification of HIV protease inhibitors through the synthesis and screening of defined peptide mixtures. *Biomed Biophys Res Commun* 181:402–408.

138. Eichler J, Houghten RA (1993): Identification of substrate-analog trypsin inhibitors through the screening of synthetic peptide combinatorial libraries. *Biochemistry* 32: 11035–11041.

139. Spellmeyer DC, Brown S, Stauber GB (1993): Endothelin receptor ligands. multiple D-amino acid replacement net approach. *Bioorg Med Chem Lett* 3:1253–1256.

140. Neustadt B, Wu A, Smith EM, Nechuta T, Fawzi A, Zhang H, Ganguly AK (1995): A case study of combinatorial libraries: endothelin receptor antagonist hexapeptides. *Bioorg Med Chem Lett* 5:2041–2044.

141. Dooley CT, Chung NN, Schiller PW, Houghten RA (1993): Acetalins: opioid receptor antagonists determined through the use of synthetic peptide combinatorial libraries. *Proc Natl Acad Sci USA* 90:10811–10815.

142. Dooley CT, Chung NN, Wilkes BC, Schiller PW, Bidlack JM, Pasternak GW, Houghten RA (1994): An all D-amino acid opioid peptide with central analgesic activity from a combinatorial library. *Science* 266:2019–2022.

143. Dooley CT, Kaplan RA, Chung NN, Schiller PW, Bidlack JM, Houghten RA (1995): Six highly active mu-selective opioid peptides identified from two synthetic combinatorial libraries. *Pept Res* 8:124–137.

144. Houghten RA, Appel JR, Blondelle SE, Cuervo JH, Dooley CT, Pinilla C (1992): The use of synthetic peptide combinatorial libraries for the identification of bioactive peptides. *Biotechniques* 13:412–421.

145. Blondelle SE, Takahashi E, Weber PA, Houghten RA (1994): Identification of antimicrobial peptides by using combinatorial libraries made up of unnatural amino acids. *Antimic Agents Chemother* 38:2280–2286.

146. Pinilla C, Appel JR, Milich D, Houghten RA (1994): Identification of an antigenic determinant for the surface antigen of hepatitis B virus using peptide libraries. In Hodges RE, Smith JA, eds. *Pep: Chem, Struct, Biol, Proc Am Pept Symp 13.* Leiden: ESCOM Sci, pp 1016–1017.

147. Pinilla C, Appel JR, Houghten RA (1994): Investigation of antigen–antibody interactions using a soluble, non-support-bound synthetic decapeptide library composed of four trillion (4×10^{12}) sequences. *Biochem J* 301:847–853.

148. Birkett AJ, Soler DF, Wolz RL, Bond JS, Wiseman J, Berman J, Harris RB (1991): Determination of enzyme specificity in a complex mixture of peptide substrates by N-terminal sequence analysis. *Anal Biochem* 196:137–143.

149. Wallace A, Altamura S, Toniatti C, Vitelli A, Bianchi E, Delmastro P, Ciliberto G, Pessi A (1994): A multimeric synthetic peptide combinatorial library. *Peptide Res* 7:27–31.

150. Walden P, Wiesmuller KH, Jung G (1995): Elucidation of T-cell epitopes: a synthetic approach with random peptide libraries. *Biochem Soc Trans* 23:678–681.

151. Udaka K, Wiesmuller K-H, Kienle S, Jung G, Walden P (1995): Decrypting the structure of major histocompatability complex class I-restricted cytotoxic T lymphocyte epitopes with complex peptide libraries. *J Exp Med* 181:2097–2108.

152. Blondelle SE, Simpkins LR, Houghten RA (1992): Inhibition of melittin's hemolytic activity using synthetic peptide combinatorial libraries. In Schneider CH, Eberle AN, eds. *Peptides: Proc 22nd Eur Pept Symp 1.* Leiden: ESCOM Sci, pp 761–762.

153. Stigler R-D, Ruker F, Katinger D, Elliott G, Hohne W, Henklein P, Ho, JX, Keeling K, Carter DC, Nugel E, 'Kramer A, Psrstmann T, Schneider–Mergener J (1995): Interaction between a Fab fragment against gp41 of human immunodeficiency virus 1 and its peptide epitope: characterization using a peptide epitope library and molecular modeling., *Protein Eng* 8:471–479.

154. Maeji NJ, Bray AM, Geysen HM (1990): Multipin peptide synthesis strategy for T cell determinant analysis. *J Immunol Methods* 134:23–33.

155. Kozlowski S, Corr M, Takeshita T, Boyd LF, Pendleton CD, Germain RN, Berzofsky JA, Margulies DH. (1992): Serum angiotensin-1 converting enzyme activity processes a human immunodeficiency virus 1 gp160 peptide for presentation for presentation by Major Histocompatability Complex Class I molecules. *J Exp Med* 175:1417–1422.

156. Shirai M, Pendleton D, Berzofsky JA (1992): Broad recognition of cytotoxic T cell epitopes form the HIV-1 envelope protein with multiple class I histocompatability molecules. *J Immunol* 148:1657–1667.

157. Wilson IA, Niman HL, Houghten RA, Cherenson ML (1984): The structure of an antigenic determinant in a protein. *Cell* 37:767–778.

158. Appel JR, Pinilla C, Niman H, Houghten RA (1990): Elucidation of discontinuous linear determinants in peptides. *J Immunol* 144:976–983.

159. Wong WY, Sheth HB, Holm A, Irvin RT, Hodges RS (1994): Investigation of antibody binding diversity utilizing representative combinatorial peptide libraries

In Hodges RE, Smith JA, eds. *Pep: Chem, Struct, Biol, Proc Am Pept Symp 13.* Leiden: ESCOM Sci, pp 995–997.

160. Wong WY, Sheth HB, Holm A, Hodges RS, Irvin RT (1994): Representative combinatorial peptide libraries: an approach to reduce both synthesis and screening efforts. *Methods: A Companion to Methods in Enzymology* 6:404–410.

161. Kemp BE, Graves DJ, Benjamini E, Krebs EG (1977): Role of multiple basic residues in determining the substrate specificity of cyclic AMP-dependent protein kinase. *J Biol Chem* 252:4888–4894.

162. Songyang Z, Carraway KL, Eck MJ, Harrison SC, Feldman RA, Mohammadi M, Schlessinger J, Hubbard SR, Smith DP, Eng C, Lorenso MJ, Ponder BAJ, Mayer BJ, Cantley LC (1995): Catalytic specificity of protein–tyrosine kinase is critical for selective signalling. *Nature* 373:591–594.

163. Gainor JA, Gordon TD, Morgan BA (1994): The synthesis and coupling efficiency of 7-hydroxycoumarin-4-propionic acid, a fluorescent marker useful in immobilized substrate libraries. In Hodges RE, Smith JA, eds. *Peptides: Chemistry Structure and Biology: Proceedings of the 13th American Peptide Symposium.* Leiden: ESCOM Sci, pp 989–991.

164. Al-Obeidi F, Lebl M, Ostrem JA, Safar P, Stierandova A, Strop P, Walser A (1995): Factor Xa Inhibitors. *WO PCT 95/29189.*

165. Maeder DL, Sunde M, Botes DP (1992): Design and inhibitory properties of synthetic Bowman–Birk loops. *Int J Protein Res* 40:97–102.

166. Horwell DC, Howson W, Ratcliffe GS, Rees DC (1993): The design of a dipeptide library for screening at peptide receptor sites. *Bioorg Med Chem Lett* 3:799–802.

167. Horwell DC, Hughes J, Hunter JC, Pritchard MC, Richardson RS, Roberts E, Woodruff GN (1991): Rationally designed "dipeptide" analogs of CCK: a-methyltryptophan derivatives as highly selective and orally active gastrin and CCK-B antagonists with potent anxiolytic properties. *J Med Chem* 34:404–414.

168. Drapeau G, Brochu S, Godin D, Levesque L, Rioux R, Marceau F (1993): Synthetic Cra receptor agonists: pharmacology, metabolism and in vivo cardiovascular and hematologic effects. *Biochem Pharm* 45:1289–1299.

169. Kawai M, Wiedeman PE, Luly JR, Or YS (1992): *Publication WO92/12168-A1.*

170. Luly JR, Megumi K, Wiedeman PE (1994): *US Patent 005 190922A.*

4

LIGHT-DIRECTED CHEMICAL SYNTHESIS OF POSITIONALLY ENCODED PEPTIDE ARRAYS

JEFFREY W. JACOBS, DINESH V. PATEL, ZHENGYU YUAN
Versicor, Inc., Fremont, California

CHRISTOPHER P. HOLMES, and JOHN SCHULLEK
Affymax Research Institute, Palo Alto, California

VALERY V. ANTONENKO, J. RUSSELL GROVE, NICOLAY KULIKOV, DEREK MACLEAN, MARC NAVRE, CINDY NGUYEN, LIHONG SHI, ARATHI SUNDARAM, AND STEVEN A. SUNDBERG
Affymax Research Institute, Santa Clara, California

Positional encoding may be broadly defined as the construction of a library on a solid support using either parallel or combinatorial synthesis techniques, but in a fashion whereby each member of the library resides in a known location. This approach has several advantages, particularly in keeping track of each compound as one manipulates the library during synthesis, and the facile way in which active "hits" are identified upon assay. This chapter highlights work in this area performed on plastic surfaces[1] and cotton sheets,[2] but the focus will be on the light-directed chemical synthesis method for creating positionally encoded arrays.[3]

Combinatorial Chemistry and Molecular Diversity in Drug Discovery, Edited by
Eric M. Gordon and James F. Kerwin, Jr.
ISBN 0-471-15518-7 Copyright © 1998 by Wiley-Liss, Inc.

POSITIONAL ENCODING VIA PHYSICAL SEPARATION

Plastic PINS

Geysen and coworkers pioneered the pin method of parallel, positionally encoded synthesis in the early 1980s.[1] This technique utilizes a fixed array of 96 polyethylene pins, grafted at their tips with a porous, polymeric material such as polyacrylic acid to form the actual synthesis area. These pins are arranged such that they occupy a footprint identical to that of a standard microtiter plate. The plates can then contain individual reagents into which one immerses the tips of the array to effect chemical synthesis, biological assays, or cleavage from the support. Compound locations are positionally defined using conventional microtiter plate nomenclature in which the eight rows are designated alphabetically by A–H and the 12 columns are designated numerically.

The pin method has distinct advantages in facilitating the logistics of a given synthesis in that each compound is positionally encoded by its fixed position in the array. By mechanically manipulating a single rack of pins, one is simultaneously manipulating 96 compounds. After a given synthesis, the prepared compounds can either be assayed while still immobilized to the pins, or, provided that the synthesis was performed using a cleavable linker, the compounds can be cleaved into solution and assayed in that form.[4,5] Cleavage from the solid support is facilitated in that it requires only a single manipulation to immerse the pin array into a deep-well microtiter plate containing the cleavage cocktail, thus performing 96 individual cleavage reactions. The use of a microtiter plate as the cleavage vessel not only allows the liberated compounds to reside in their original locations—in essence one replicates the footprint of the formerly immobilized compounds—but it also facilitates subsequent assay, because most high throughput screens utilize microtiter plates as source vessels for the compounds of interest.

A variety of synthetic methods have been applied to the pin format, including peptide chemistry[1,6] and solid-phase organic synthesis.[7,8] In general, chemical syntheses on plastic surfaces appear to proceed as easily as syntheses on other solid supports. The pin method remains a medium-throughput approach, because it would be impractical to increase the density of the pin array. However, the existing approach is sufficiently convenient that a skilled operator can work with several racks of pins simultaneously. Libraries prepared in this fashion have been screened for Substance P analogs[9] as well as for cholecystokinin A receptor-specific ligands.[8]

Cotton Sheets

Frank has utilized cotton sheets as solid supports for the preparation of positionally addressable libraries.[2] In this approach, reagents are precisely applied to a specific region of the sheet to construct the desired compounds. The density of synthesis sites is determined by the precision with which one can add reagents to a unique position on the sheet, the volume added, and its relative diffusion within the support. This technique can be performed using common laboratory hand tools, but higher-precision arrays require the use of robotic liquid handling devices. A typical array consists of 96 synthesis sites (spots) occupying the dimensional area of a conventional microtiter plate; however, it has been proposed that precision liquid handling equipment capable of delivering nanoliter volumes of reagents could allow the density of synthesis sites to approach $100/cm^2$.[10] Arrays of this complexity would require a solid-phase assay as only subanalytical amounts of material could be prepared. This technique has been used primarily in the preparation of arrays of peptides, and has found application in the epitope mapping of anti-peptide antibodies, as recently reviewed.[10]

LIGHT-DIRECTED, POSITIONALLY ENCODED SYNTHESIS

Researchers at Affymax have developed a chemistry-driven technology that combines the tools of solid-phase peptide chemistry, photolabile protecting groups, and photolithography to construct extremely compact arrays of compounds on glass substrates, or chips.[3] The synthesis area measures only 1.28×1.28 cm^2, yet it is possible to prepare thousands of compounds within these dimensions. This miniaturization serves to reduce reagent consumption, particularly when performing postsynthesis biological assays. The entire array of immobilized compounds is assayed by a single incubation of the molecular-binding agent of interest, such as an antibody, enzyme, or receptor, typically requiring less than 1 mL of the recoverable assay solution. Detection is achieved via high-sensitivity fluorescence microscopy, such that relative fluorescence intensities are obtained for every compound synthesized, with both positive and negative binding-events being scored.

We will first describe the process development that was required to make this technique practical, then outline a series of array synthesis strategies, describe some specific applications of this technology to lead discovery, and conclude with a description of the nonpeptidic chemistries that have been performed in this format.

Process Development

Conventional solid-phase peptide synthesis is an iterative process in which a suitable solid support is first acylated with an N-protected amino acid, the protecting group is removed, and then the process is repeated with the next amino acid until the desired peptide is assembled.[11] To enable light-directed peptide synthesis we first had to develop several new materials and reagents, including a solid support suitable for both chemistry and photolithography, photolabile N-protecting groups with physicochemical properties orthogonal to the ligands being prepared, and extremely sensitive analytical techniques to assess the quality of the materials due to the miniaturization inherent in this approach. The results of some of this process development are described below.

Substrate Preparation. Substrate, or chip development is critical since the same surface is used for both ligand synthesis and subsequent detection. The detection of an array is performed using optical methods—the surface is literally imaged—thus the fidelity and homogeneity of the chip is extremely important. In a typical batch procedure, borosilicate glass wafers are cleaned with strong acid, washed, silanized with the heterobifunctional organosilane 3-aminopropyl-1-triethoxysilane, and then heat-cured. The organosilane provides an amine functionality for subsequent peptide synthesis, and is incorporated at a site density of 25 to 50 Å (approximately 10 pmol of amines/cm^2) (unpublished). The chip is then either used immediately in a synthesis or acylated with a suitable protecting group or linker, and then stored.

Photocleavable Protecting Groups. Central to our approach is a protecting group that can be removed with light. Early work focused on the nitroveratryl group (**1**; Figure 4.1A) first described by Patchornik et al.[12] This was converted to its chloroformate derivative (Nvoc-Cl) and used to acylate α-amino acids to afford the corresponding Nvoc-protected derivatives for use in light-directed synthesis. The optimal photolysis solvent was determined to be 5 mM sulfuric acid in dioxane; however, during the course of our studies, the photolysis of some Nvoc-protected amino acids was not found to yield free amines quantitatively. These observations spurred us to examine the photochemical pathway of nitroveratryl alcohol in detail as described below.[13]

The photolysis of nitroveratryl alcohol (**1**) proceeds initially to afford a nitrosoaldehyde (**2**), which rearranges to an isoxazolone (**3**) via an acid-catalyzed, thermal process (Figure 4.1A). Further photolysis dimerizes this compound to a mixture of *cis*- and *trans*-azo derivatives (**4**). A similar analysis of other nitroveratryl derivatives developed in-house revealed one, α-methylnitropiperonyloxycarbonyl (MeNPoc, **5**),[14] that photolyzed much more cleanly (Figure 4.1B). The photolysis of MeNPoc alcohol (**5**) afforded a relatively unreactive nitrosoketone by-product (Figure 4.1B), proceeded most efficiently under neutral conditions, and had a shorter half-life than Nvoc (average $t_{1/2}$ = 19 versus 110 s). The photodeprotection of MeNPoc-protected amino acids in solution phase was determined to proceed quantitatively.

Analytical Tools. Each substrate contains approximately 10 pmol of amines/cm^2, which is insufficient to directly characterize the purity of the synthesis products using conventional methods such as HPLC. We therefore needed to develop a sensitive analytical method to characterize the purity of peptides cleaved from the surface of a chip. The reagent *N*-α-fluorenylmethyloxycarbonyl-*N*-ε-dabsyllysine (Fmoc-Lys(Dbs); Figure 4.2) was developed as a chro-

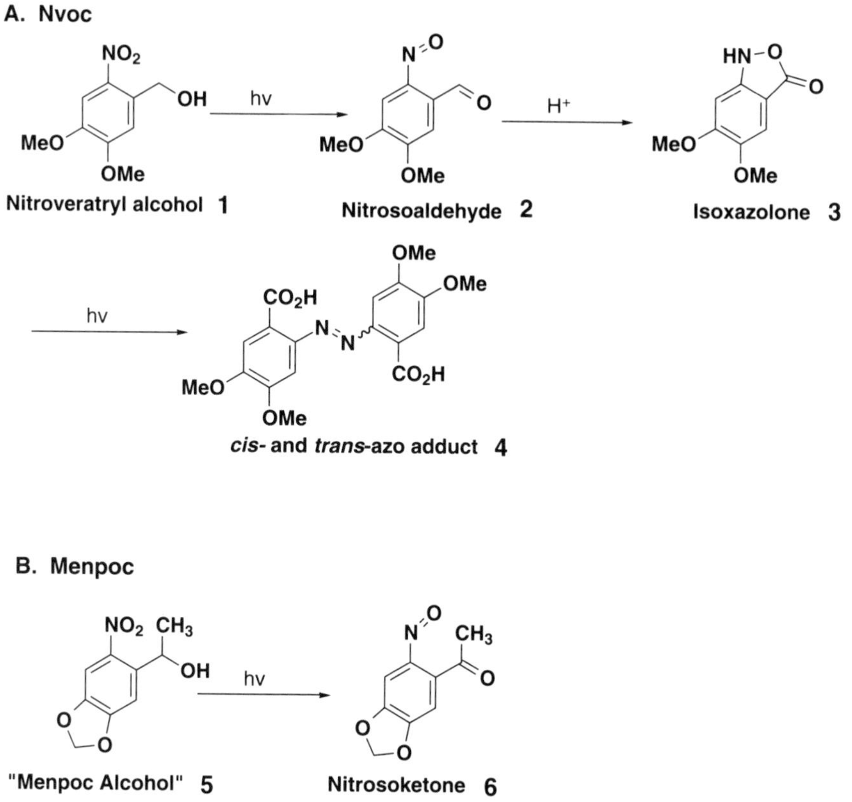

A. Nvoc

Nitroveratryl alcohol **1**

Nitrosoaldehyde **2**

Isoxazolone **3**

cis- and trans-azo adduct **4**

B. Menpoc

"Menpoc Alcohol" **5**

Nitrosoketone **6**

Figure 4.1. Photochemical pathways of the benzyl alcohols derived from (A) Nvoc and (B) MeNPoc.

mophoric tag for the HPLC detection of peptides cleaved from the surface of the substrate.[15] The HPLC trace of such products could then be compared with that of fully characterized authentic samples prepared using conventional methods. Figure 4.2 shows a representative HPLC trace of a peptide, Fmoc-YINQK(dbs)-NH$_2$, synthesized on a chip using conventional Fmoc chemistry.

SYNTHESIS AND DETECTION OF ARRAYS

A typical synthesis procedure for assembling an array is illustrated in Figure 4.3. A glass substrate modified with photolabile protecting groups (X) is selectively illuminated through a lithographic mask (M1). The lithographic mask consists of optically opaque and optically transparent regions, the pattern of which determines those sites on the chip surface that are deprotected. After photolysis, the entire surface of the chip is then flooded with an activated

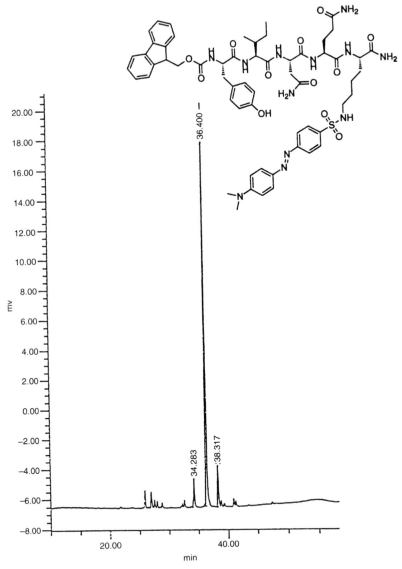

Figure 4.2. High-sensitivity HPLC analysis of a dabsyl-tagged peptide synthesized on a glass chip using Fmoc chemistry. Approximately 100 pmol injection of Fmoc-YINQK(Dbs)-NH$_2$. Fmoc, 9-fluorenylmethyloxycarbonyl; K(Dbs), N-ε-dabsyllysine. Uppercase letters refer to conventional single-letter codes for amino acids.

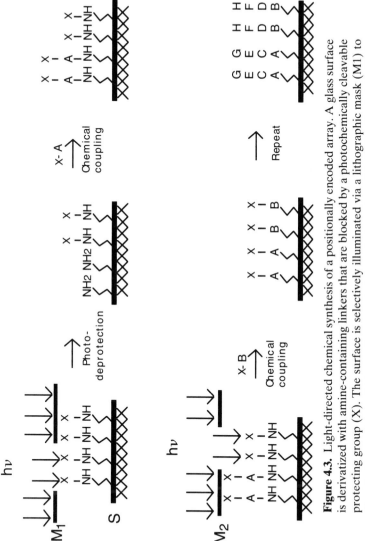

Figure 4.3. Light-directed chemical synthesis of a positionally encoded array. A glass surface is derivatized with amine-containing linkers that are blocked by a photochemically cleavable protecting group (X). The surface is selectively illuminated via a lithographic mask (M1) to liberate free amines, which are then coupled to a photoprotected building block (A). Light is directed to a different region of the substrate through a new mask (M2) and the process is repeated until the desired array of compounds is prepared. The patterns of photolysis and the order of addition of building blocks define the products and their position within the array.

amino acid (A); however, acylation occurs only in those regions that were photodeprotected. After the surface is washed, a second mask (M2) is moved into position and the process is repeated. The pattern(s) of the lithographic mask(s) and the order of addition of activated building blocks determines the location and sequence of the synthesized peptides.

Masking Strategies

Peptide arrays can be assembled using a variety of strategies.[3] For example, the desired series of n monomers can be anchored to the surface of the glass iteratively by photolyzing a consecutive series of adjoining stripes of a width $1/n$th that of the synthesis surface (1.28 cm), with one monomer residing in each stripe. Dimers are then created in the same fashion, but this second set of building blocks is added by photolyzing stripes orthogonal to the first set. As in the divide-and-pool method,[16] for a given number of monomers n, one creates n^2 compounds in $2n$ chemical steps.[17] This orthogonal stripe synthesis strategy can be continued to make longer peptides by simply subdividing previous stripes into n synthesis regions until the limit of photolithographic resolution is achieved (currently 10–20 μm).[18] Note that one needs to incorporate photoprotected building blocks only at positions immediately to the C-terminus of the point of modification; conventionally protected amino acids can be incorporated prior to conserved positions in the array.

A more powerful method for creating molecular diversity in this synthesis format is via a binary masking strategy.[3] With this technique, each building block is added after 50% of the available synthesis space is photolyzed. Subsequent additions are performed such that each new mask illuminates exactly one-half of the previous synthesis area. Since each monomer addition divides the synthesis region into two different populations, the number of compounds in the array increases by a factor of two with a single coupling cycle. In a 16-step binary synthesis, one would create 2^{16} compounds (65,536) in only 16 chemical steps (photoprotected building blocks are required at every step).

Detection of Peptides

To assay individual compounds present in matrices of the complexity described above requires an extremely sensitive and high-resolution detection system. Our detection system comprises a scanning epifluorescence confocal microscope with a laser source for excitation. This system provides strong background rejection from the unbound target, the glass, and other parts of the system; a large, dynamic range; high resolution over a broad area; and the ability to collect both kinetic and equilibrium data.[18] In a typical procedure, the peptide array is mounted inverted in a flow cell on the microscope stage. The array is incubated with a fluorescently labeled receptor, washed, and then scanned with the epifluorescence microscope. This system measures the fluorescent signal from receptor bound at the surface of the array. The fluorescent intensity at each synthesis site is a function of the affinity of the receptor for the compound, the concentration of receptor, and the number and density of interacting sites on the array. After performing the assay, the merger of two positionally related data sets—fluorescent signal versus location and peptide sequence versus location—decodes the array, associating compound identity with corresponding fluorescent signal for every member of the array.

APPLICATIONS OF LIGHT-DIRECTED CHEMICAL SYNTHESIS

Epitope Mapping

In an early model system we examined the binding specificity of the anti-peptide monoclonal antibody D32.39, which binds to the C-terminal region of the opioid peptide dynorphin B (YGGFLRRQFKVVT; K_d = 8 nM). To determine the minimal active fragment, a simple binary synthesis of the 10-mer FLRRQFKVVT was performed, in which 2^{10} (1024) different peptides representing all the possible N-terminal truncations, C-terminal truncations, and internal deletion peptides were prepared. This synthesis scheme also afforded all the overlapping 9-mers, 8-mers, 7-mers, etc., providing data for a frame-shift analysis. The peptides were assayed for activity by a single incubation with fluorescein-labeled antibody D32.39, and the relative fluorescent intensities of the peptides were determined by fluorescence microscopy. The results of this experiment clearly identified RQFKVVT as the minimal active fragment.[19]

In a second chip synthesis, all 20 amino acids encoded by genes were inserted into each position of the minimal active fragment to afford all 140 singly substituted peptides. The results of this analysis did not reveal any mutants with statistically higher activity, but did indicate that both Arg and Phe in RQFKVVT are intolerant of substitutions.[19]

Pharmacophore Libraries

One can incorporate into a peptide array specific building blocks that contain unique functional groups, or pharmacophores, that can impart proscribed biological activities to these compounds. For example, pharmacophore libraries could include functional groups such as free carboxylates, mercaptans, or hydroxamates to bias the library toward the inhibition of metalloproteases. A second example would include the incorporation of phosphotyrosine or a phosphotyrosine isostere (PTI) into an array of peptides to search for compounds capable of binding phosphopeptide-specific proteins, such as Src homology 2 (SH2) domains or tyrosine kinases. Examples of such pharmacophore libraries are presented below.

Proteases

Stromelysin is a matrix metalloprotease (MMP) and has been implicated in the pathological loss of cartilage proteoglycans associated with osteo- and rheumatoid arthritis.[20] We have studied the positional preference of building blocks for both substrates and inhibitors of stromelysin, and the following is a brief summary of this work.[21]

Protease Substrate Arrays. An array of substrates of the form Ac-PL(P_1)(P_1')(P_2')A was prepared as outlined in Figure 4.4 (P_1 refers to that amino acid one residue to the N-terminus of the scissile bond, P_1' refers to that amino acid one residue to the C-terminus of the scissile bond). In this experiment, eight different amino acids were inserted at P_1, four were incorporated at P_1', and eight at P_2' to afford an array of 256 variants. After chemically "capping" the chip with acetic anhydride, the wafers were exposed to stromelysin for varying periods of time, and the degree of proteolysis was determined by subsequently staining the array with the amine-specific probe fluorescein isothiocyanate (FITC). Upon fluorescence imaging, synthesis sites containing peptides that were efficiently proteolyzed will be extensively labeled with the fluorophore and will give strong signals, unproteolyzed regions will not be labeled, while substrates of intermediate activity will display a corresponding level of fluorescent intensity. By inspection, one can see that the incorporation of the negative control D-Ala at P_1' results in the preparation of 64 poor substrates, as anticipated; however, sequences in the array that contain Leu, Val, or cyclohexylalanine (CHA) at P_1' yield substrates with a spectrum of proteolytic susceptibilities. From this analysis, the most productive substitutions at P_1 include Ala and Gly, at P_1' Leu and Val appear best, while there is not a dramatic preference at P_2'.[21]

Protease Inhibitor Assays. Many metalloendoproteases primarily recognize those residues in regions to the C-terminus of the scissile bond of their substrates. Thus, using conventional C-to-N peptide synthesis, one can prepare arrays of immobilized peptides and then cap them with metalloprotease inhibitor pharmacophores to create protease inhibitor libraries (PILs). These pharmacophores could include chelators such as hydroxamates, carboxylates, and mercaptans, or transition state analogs such as phosphoramidates and phosphonates.[22] Similar strategies have been used for the solid-phase synthesis of inhibitors of the aspartyl protease from the human immunodeficiency virus (HIV)[23,24] as well as inhibitors of the metalloprotease thermolysin.[25] We have successfully prepared and screened pharmacophore libraries of this nature against the MMP stromelysin, and some of this work is described below.

In this study, we utilized the light-directed chemical synthesis technology to deconvolute active pools of stromelysin inhibitors, which were identified by screening a large bead-based library of peptidyl succinic acids. These peptidyl succinic acids contain free carboxylates at P_1'—a functional group known to

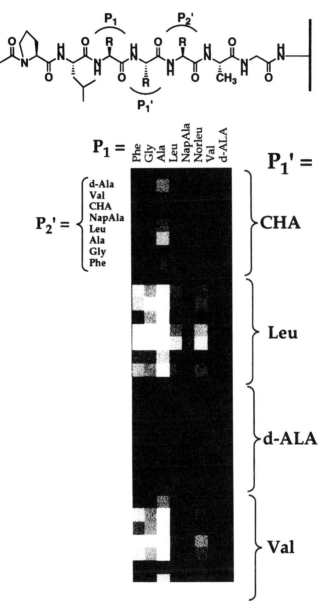

Figure 4.4. Light-directed synthesis of an array of stromelysin substrates. Synthesis sites with a high fluorescent signal are red and indicate efficient proteolysis, those that are dark blue or black indicate little proteolysis, and those of intermediate signal are colored accordingly. The naturally occurring amino acids in this array are designated by conventional three-letter code. Unnatural amino acids are defined as follows: cyclo-hexylalanine (CHA), 1-naphthylalanine (NapAla), and norleucine (Norleu).

be a metalloprotease inhibitor pharmacophore.[22] First, a large bead mass consisting of 1156 dipeptides was synthesized from a set of 34 natural and unnatural amino acids using the split/pool method of Furka et al.[16] These dipeptides were portioned into 16 different reaction vessels, and each aliquot was capped with a substituted succinic acid, resulting in a library of 18,496 peptidyl succinic acids. Each pool was incubated with fluorescently labeled stromelysin and then sorted using a fluorescence-activated cell sorter (FACS). Beads displaying high-affinity inhibitors should be coated with fluorescently labelled stromelysin and as such will be counted during FACS analysis, whereas pools containing inactive beads will score poorly. This analysis identified the preferred succinates at P_1', while the preferred residues at P_2' and P_3' remained unknown. This final deconvolution was accomplished by replicating the active pools via light-directed synthesis on chips.

Arrays of dipeptides were prepared lithographically using the orthogonal stripe method, capped with the succinic acids of interest, and then assayed against stromelysin. Figure 4.5 shows an array of stromelysin inhibitors that contain an n-heptyl-substituted succinic acid at P_1'. The results from this assay indicated that those residues at P_2' that afforded the highest fluorescent intensities include branched, aliphatic residues such as Ile, norleucine, and $tert$-butylglycine, while a strong preference was seen for Leu and Ile at P_3'. Compounds with low fluorescent intensities contained residues such as Gly, Ala and Pro at P_2', and Pro and Thr at P_3'.[21]

Because we performed a positionally addressable synthesis, we were able to obtain semiquantitative SAR for every compound in the array and use it to select a cross section of compounds for individual resynthesis and IC_{50} testing in solution phase.[21] Were we to continue the bead-based deconvolution analysis we would have obtained little information on compounds of intermediate or poor activity, since, by definition, a deconvolution is a positive-selection strategy.

SH2 Domains

Src homology 2 (SH2) domains appear in many kinases as well as other proteins and participate in intracellular signaling events and other protein/protein interactions by recognizing specific phosphotyrosine-containing reporter sequences on other proteins. Many SH2 domains have been cloned and shown to specifically recognize phosphotyrosine-containing peptide fragments in vitro.[26] In an approach similar to that described above for proteases, we have prepared SH2-specific pharmacophore libraries by derivatizing peptide arrays with phosphotyrosine or a phosphotyrosine isostere (PTI). Described below is a library of phosphopeptides used to screen for inhibitors of the SH2 domain derived from the tyrosine kinase Fyn.

Fyn is a member of the Src family of tyrosine kinases, and its SH2 domain is known to recognize the phosphopeptide sequence EPQpYEEIPIYL (pY = phosphotyrosine; pTyr) with affinity and selectivity similar to that of

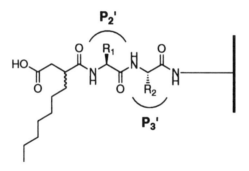

Figure 4.5. An array of peptidyl succinic acid inhibitors of stromelysin in which P_1' was fixed as a 2-*n*-heptyl succinic acid, and P2' and P3' were varied.

other members of the Src family.[27] The cocrystal structure of the Src family SH2 domain Lck in complex with the above sequence suggests that these proteins should be highly selective for hydrophobic residues at pY + 3 (that amino acid three residues to the C-terminus of pY) of the phosphopeptide sequence, and moderately selective for Glu at pY + 1.[28] We therefore synthesized a series of derivatives of the parent sequence Ac-pYEEIPI in which pY + 1 (Glu) and pY + 3 (Ile) were varied.

Linker Library. In any solid-phase assay it is critical that the immobilized ligands are displayed such that they can be recognized by the protein of interest. Before proceeding with the large-scale synthesis of arrays, we first synthesized a linker library, consisting of an array of positive and negative controls in which we varied the linkers on which these ligands were displayed (Figure 4.6).[29] The array of linkers was capped with the positive control EPQpYEEIPIYL as well as three mutants of this sequence, in which pTyr was replaced with Tyr or Phe or simply deleted (Figure 4.6A). The results from this experiment are highlighted in the bar graph in Figure 4.6B, from which one can clearly see that tetra-Lys is an unsuitable linker, affording poor discrimination among the controls. Others, such as tetra-Ser, APS, and a 15-atom polyethyleneglycol-based linker (PEG15),[30] all afford the expected relative signals. Subsequent preparation of arrays was performed on top of PEG15.

Synthesis of Arrays. An array of 512 variants of the parent sequence Ac-pYEEIPI was prepared in which 16 replacements of Y + 1 (biased toward negatively charged residues, or residues isosteric with Glu) and 32 replacements of Y + 3 (biased toward hydrophobic residues) were incorporated (Figure 4.7). Compounds with fluorescent intensities in the top 5% of the data generally contained residues consistent with predictions made from the crystal structure: Preferred residues at pY + 1 included Glu as well as the negatively charged residues pTyr, Tyr(SO3H2), and Asp. Preferred residues at pY + 3 included the aromatic amino acids Phe, pCl-Phe, pF-Phe, and pBz-Phe as well as the D-amino acids D-Val and D-Leu.[29]

ALTERNATIVE CHEMISTRIES

Light-directed chemical synthesis should be applicable to any solid-phase synthesis technique where the synthesis proceeds linearly and in which light can be used to deprotect the appropriate functional groups. Two nonpeptidyl systems are described below.

Figure 4.6. (A) A library of display linkers. The positive control Ac-EPQpYEEIPIYL (pY) and three negative controls (pY replaced with Y or F and deleted) were synthesized on top of a series of linkers to determine the optimal spacer for the solid-phase assay of phosphopeptide inhibitors of this SH2 domain. (B) A bar graph showing data selected from the linker library assay, in which the linkers tetraserine (SSSS), 3-aminopropylsilane (APS), and a 15-atom polyethyleneglycol linker (PEG15) afford excellent signal discrimination between the positive and negative controls, while the linker tetralysine (KKKK) afforded poor signal discrimination.

Linkers

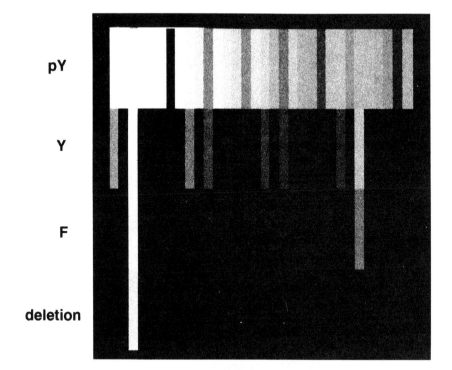

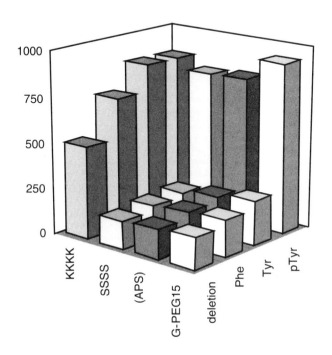

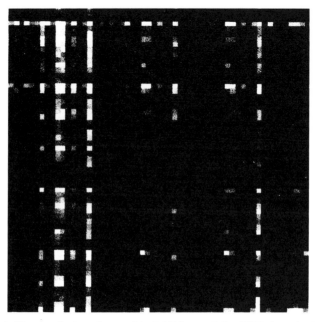

Figure 4.7. A library of phosphopeptide inhibitors of the SH2 domain from Fyn, in which 512 variants of the parent sequence Ac-pYEEIPI were synthesized such that each compound is present in eight different locations in the array (4096 synthesis sites). Synthesis sites with high fluorescent signals are red and indicate sequences to which the SH2 domain binds with high affinity; those that are dark blue or black indicate little recognition, and those of intermediate signal are colored accordingly.

Oligonucleotide Synthesis

Solid-phase oligonucleotide synthesis consists of the stepwise coupling/depro-tection of 5′-dimethoxytrityl (DMT) protected nucleoside monomers in the 3′ to 5′ direction. In a typical coupling procedure, the 5′ hydroxyl of an immobilized nucleoside is deprotected with mild acid, the liberated hydroxyl is phosphitylated with a DMT-protected deoxynucleoside 3′-phosphoramidite, and the resulting phosphite is oxidized to a phosphotriester. The process is continued until the desired oligonucleotide is prepared. This technique was adapted to light-directed parallel chemical synthesis by replacing the 5'-protecting group DMT with MeNPoc and incorporating a MeNPoc-protected hydroxyl linker into the synthesis substrate.[31] Hydroxyl groups are selectively photodeprotected as described previously, and arrays of oligonucleotides are assembled using standard amidite chemistry.

Using the orthogonal stripe method one could potentially assemble all 65,566 possible octanucleotides (4^8) in only 32 chemical steps ($4n$; $n = 8$). Such miniaturized arrays of densely packed oligonucleotides could serve as hybridization probes for DNA sequencing by hybridization (SBH). To test this approach using light-directed synthesis, Fodor and coworkers prepared an array of 256 different octanucleotide probes and screened them against the target sequence 5'-GCGTAGGC-fluorescein. The fluorescence signals from complementary probes were 5–35 times stronger than those containing hybridization mismatches.[31] More sophisticated arrays have been reported, and a large body of results suggest that high-density oligonucleotide probe array technology has significant power in enabling de novo sequencing.[18]

Oligonucleotide arrays have also been generated using a solution-channeling device to direct the oligonucleotide probe synthesis.[32] A potential advantage of this technique is that spatial resolution is achieved using a barrier method, such that conventional chemical deprotection strategies can be used. However, the need for physically separating the reaction sites limits the utility of this method since the resolution limit of approximately 1×1 mm may restrict the number of probes one can prepare in a convenient surface area.[32]

Carbamate Backbone

Libraries of unnatural polymers provide unique vehicles for drug discovery in that they may probe sequence and conformational space not accessible to naturally occurring biopolymers. These compounds could also possess desirable characteristics such as proteolytic resistance, high serum stability, and oral bioavailability. In recent years several such unnatural polymers have been reported, including N-alkylated glycines or "peptoids,[33]" poly ureas,[34] and poly vinylogous sulfonamides.[35] Schultz and coworkers reported the light-directed chemical synthesis of a peptide-like polymer that differs from native peptides in that a carbamate linkage replaces the amide bond[36] (see also Chapter 12).

The building blocks consisted of Nvoc-protected chiral aminocarbonates prepared from optically active amino alcohols or synthesized from N-protected amino acids via reduction of the carboxylic acid. Acylation of the alcohol with nitrophenyl chloroformate affords activated carbonates that couple readily and with high efficiency to free amines to afford a carbamate linkage. Using light-directed parallel chemical synthesis, libraries of oligocarbamates were prepared and used to epitope map anti-oligocarbamate monoclonal antibody 20D6.3, which was generated against $AcY^cK^cF^cL^c$.[25,26] A binary synthesis of $AcY^cF^cA^cS^cK^cI^cF^cL^c$ afforded an array of 256 oligocarbamates containing all possible deletions and truncations of the parent sequence. This was treated with 20D6.3 and antibody–oligocarbamate complexes were detected by scanning fluorescence microscopy with a fluorescein-conjugated goat α-mouse secondary antibody. The oligocarbamates AcK^{c}F^{c}L^{c}G-OH, AcF^{c}K^{c}F^{c}L^{c}G-OH, AcY^{c}K^{c}F^{c}L^{c}G-OH, AcA^{c}K^{c}F^{c}L^{c}G-OH, and AcI^{c}F^{c}L^{c}G-OH were among the 10 highest affinity ligands based on fluorescence intensities. These compounds were synthesized preparatively and tested in solution and found to have IC$_{50}$ values that ranged between 60 and 180 nM. Comparison of these structures suggests that -F^{c}L^c- is the dominant epitope of 20D6.3.

We have synthesized small organic molecules on glass chips (unpublished), but the photolithographic approach is best suited for linear syntheses such as the preparation of polymeric materials. Organic chemistry typically involves the stepwise generation of reactive intermediates during the course of a specific transformation. Since the entire array is flooded with each reagent, spatial addressability is lost unless one can photoprotect these intermediates. Second, combinatorial organic chemistry requires a diverse set of building blocks, with unique classes of reagents typically used in each step of the synthesis. The preparation of photoprotected building blocks—with each class possessing functional groups that require chemically different protection strategies— would be impractical.

FUTURE DIRECTIONS

Light-directed chemical synthesis is a powerful way to generate positionally encoded arrays of peptides, oligonucleotides, and other linear polymers. Upon assay, binding events are detected simultaneously for every compound in the array via high-sensitivity fluorescence detection. Since the location of every compound is known, one is able to obtain both positive and negative information for each ligand without having to either deconvolute or decode the library. Photolithography permits the synthesis of extremely compact arrays because there is no need for the physical separation of reaction sites, as exists for the pin,[1] spot,[2] and solution-channeling methods.[32]

Light-directed synthesis technology has the potential to become an important diagnostic tool for sequencing DNA via the SBH approach because oligonucleotide synthesis and screening complements the strengths of light-directed chemical synthesis: Oligonucleotide chemistry is well adapted to solid-phase synthesis and requires a minimum of photoprotected building blocks, subsequent assays work well in the solid phase, and photolithography is a practical technique for the synthesis of the massive arrays of probes required for SBH. Other applications could include materials science. Xiang and Schultz and coworkers prepared a binary array of solid-state materials using physical masking strategies, and these materials were screened for their ability to behave as high-temperature superconductors.[37] Their lithographic synthesis represents an extension of the combinatorial approach from biological and organic molecules to the remainder of the periodic table.

REFERENCES

1. Geysen HM, Meloe RH, Barteling SJ (1984): Use of peptide synthesis to probe viral antigens for epitopes to a resolution of a single amino acid. *Proc Natl Acad Sci USA* 81:3998–4002.

2. Frank R (1992): Spot-Synthesis: an easy technique for the positionally addressable, parallel chemical synthesis on a membrane support. *Tetrahedron* 48:9217–9232.

3. Fodor SPA, Read JL, Pirrung MC, Stryer L, Lu AT, Solas D (1991): Light-directed, spatially addressable parallel chemical synthesis. *Science* 251:767–773.

4. Bray AM, Maeji NJ, Valerio RM, Campbell RA, Geysen HM (1991): Direct cleavage of peptides from a solid support into aqueous buffer: application in simultaneous multiple peptide synthesis. *J Org Chem* 56:6659–6666.

5. Valerio RM, Bray AM, Maeji NJ (1994): Multiple peptide synthesis on acid-labile handle derivatized polyethylene supports. *Int J Pept Protein Res* 44:158–65.

6. Maeji MJ, Bray AM, Valerio RM, Wang W (1995): Larger scale multipin peptide synthesis. *Pept Res* 8:33–38.

7. Bray AM, Chiefari DS, Valerio RM, Maeji NJ (1995): Rapid optimization of organic reactions on solid phase using the multipin approach: synthesis of 4-aminoproline analogues by reductive amination. *Tetrahedra Lett* 36:5081–5084.

8. Bunin BA, Plunkett MJ, Ellman JA (1994): The combinatorial synthesis and chemical and biological evaluation of a 1,4-benzodiazepine library. *Proc Natl Acad Sci USA* 91:4708–4712.

9. Wang JX, Bray AM, Dipasquale AJ, Maeji NJ, Geysen HM (1993): Systematic study of substance P analogs, I: evaluation of peptides synthesized by the muyltipin method for quantitative receptor binding assay. *Int J Pept Protein Res* 42:384–391.

10. Frank R (1995): Simultaneous and combinatorial chemical synthesis techniques for the generation and screening of molecular diversity. *J Biotech* 41:259–272.

11. Merrifield RB (1963): Solid phase peptide synthesis, I: the synthesis of a tetrapeptide. *J Am Chem Soc* 85:2149–2154.

12. Patchornik A, Amit B, Woodward RB (1970): Photosensitive protecting groups. *J Am Chem Soc* 92:6333–6335.

13. Dong L-C, Holmes CH, Jacobs JW (1998): Characterization of photoprotecting groups for light-directed combinatorial synthesis. In preparation.

14. Holmes CP, Kiangsoontra B (1994): Development of a new photo-removable protecting group for the amino and carboxyl groups of amino acids. In Hodges RS, Smith JA, eds. *Peptides, Chemistry, Structure and Biology: Proceedings of the 13th American Peptide Symposium, 20–25 June 1993, Edmonton, Alberta, Canada.* Leiden: ESCOM, pp 110–112.

15. Antonenko VV, Kulikov N, Sundaram A (1998): Analytical tools for combinatorial chemistry: HPLC detection of picomole quantities of peptides. In preparation.

16. Furka A, Sebestyen F, Asgedom M, Dibo G (1991): General method for rapid synthesis of multicomponent peptide mixtures. *Int J Pept Protein Res* 37:487–493.

17. Jacobs JW, Fodor SPA (1994): Combinatorial chemistry: applications of light-directed chemical synthesis. *Trends Biotech.* 12:19–26.

18. Lipshutz RJ, Morris D, Chee M, Hubbell E, Kozal MJ, Shah N, Shen N, Yang R, Fodor SPA (1995): Using oligonucleotide probe arrays to access genetic diversity. *BioTechniques* 19:442–447.

19. Holmes CP, Adams CL, Kochersperger LM, Mortensen RB (1995): The use of light-directed combinatorial peptide synthesis in epitope mapping. *Biopolymers* 37:199–211.

20. Flannery CR, Lark MW, Sandy JD (1992): Identification of a stromelysin cleavage site within the interglobular domain of human aggrecan. *J Biol Chem* 267:1008–1014.

21. Yuan Z, et al. (1998): Combinatorial synthesis of substrates and inhibitors of matrixmetalloproteases. In preparation.

22. Petrillo EW, Ondetti MA (1982): Angiotensin-converting enzyme inhibitors: medicinal chemistry and biological actions. *Med Res Rev* 2:1–41.

23. Owens RA, Gesellchen PD, Houchins BJ, DiMarchi RD (1991): The rapid identification of HIV protease inhibitors through the synthesis and screening of defined peptide mixtures. *Biochem Biophys Res Commun* 181:402–408.

24. Wang GT, Li S, Wideburg N, Krafft GA, Kempf DJ (1995): Synthetic chemical diversity: solid phase synthesis of libraries of C2 symmetric inhibitors of HIV protease containing diamino diol and diamino alcohol cores. *J Med Chem* 38:2995–3002.

25. Campbell DA, Bermak JC, Burkoth TS, Patel DV (1995): A transition state analogue inhibitor combinatorial library. *J Am Chem Soc* 117:5381–5382.

26. Songyang Z, Shoelson SE, Chaudhuri M, Gish G, Pawson T, Haser WG, King F, Roberts T, Ratnofsky S, Lechleider RJ, Neel BG, Birge RB, Fajardo JE, Chou MM, Hanafusa H, Schaffhausen B, Cantley LC (1993): SH2 domains recognize specific phosphopeptide sequences. *Cell* 72:767–778.

27. Payne G, Stolz LA, Pei D, Band H, Shoelson SE, Walsh CT (1994): The phosphopeptide-binding specificity of Src family SH2 domains. *Chem Biol* 1:99–105.

28. Eck M, Shoelson SE, Harrison SC (1993): Recognition of a high-affinity phosphotyrosyl peptide by the Src homology 2 domain of p56[lck]. *Nature* 362:87–91.

29. Jacobs JW, Antonenko VV, Grove JR, Kulikov N, MacLean D, Sundberg SA (1998): The phosphopeptide-binding specificity of the SH2 domain from Fyn. In preparation.

30. Sundberg SA, Holmes CP (1998): A novel display linker for solid-phase assays. In preparation.

31. Pease AC, Solas D, Sullivan EJ, Cronin MT, Holmes CP, Fodor SPA (1994): Light-generated oligonucleotide arrays for rapid DNA sequence analysis. *Proc Natl Acad Sci USA* 91:5022–5026.

32. Southern EM, Maskos U, Elder JK (1992): Hybridization with oligonucleotide arrays. *Genomics* 13:1008–1017.

33. Simon RJ, Kania RS, Zuckermann RN, Huebner VD, Jewell DA, Banville S, Ng S, Wang L, Rosenberg S, Marlowe CK, Spellmeyer DC, Tan R, Frankel AD, Santi DV, Cohen FE, Bartlett PA (1992): Peptoids: a modular approach to drug discovery. *Proc Natl Acad Sci USA* 89:9367–9371.

34. Hutchins SM, Chapman KT (1994): A general method for the solid-phase synthesis of ureas. *Tetrahedron Lett* 35:4055–4058.

35. Gennari C, Nestler HP, Salom B, Still WC (1995): Solid-phase synthesis of vinylogous sulfonyl peptides. *Angew Chem Int Ed Engl* 34:1763–1765.

36. Cho CY, Moran EJ, Cherry SR, Stephans JC, Fodor SPA, Adams CL, Sundaram A, Jacobs JW, Schultz PG (1993): An unnatural biopolymer. *Science* 261:1303–1305.

37. Xiang X-D, Sun X, Briceno G, Lou Y, Wang K-A, Chang H, Wallace–Freedman WG, Chen S-W, Schultz PG (1995): A combinatorial approach to materials discovery. *Science* 268:1738–1740.

5

CONFORMATIONALLY RESTRICTED PEPTIDE AND PEPTIDOMIMETIC LIBRARIES

ALEX A. VIRGILIO AND JONATHAN A. ELLMAN
Department of Chemistry, University of California at Berkeley, Berkeley, California

A major goal in the treatment of human disease has been the development of compounds that reproduce or block the activity of specific bioactive peptides or proteins.[1] The most direct approach has been to employ the peptide or simple analogs as a therapeutic agent, and several important peptide-based therapeutic agents have been developed.[2] Unfortunately, peptides have several undesirable properties that limit their utility for most therapeutic applications.[3]

Cyclic peptides or peptidomimetics that display side-chain functionality from a constrained template offer several improvements over the corresponding linear peptide. Stability to enzymatic degradation is almost always observed.[4] In addition, increased rigidity often results in improved selectivity between receptor subtypes and increased binding affinity. Constraints have been directly incorporated into peptides through either the peptide backbone or side chains by a number of different strategies, such as amide,[5-14] disulfide,[15,16] and thioether[17,18] cyclization reactions. Alternatively, amino acids or amino acid side chains have been displayed from cyclic templates.

An overview of chemical methods used to obtain libraries of constrained peptides or peptidomimetics is provided, and several case examples are described. The scope of this review is confined to those compounds that have been designed to display peptide side-chain functionality in a constrained

Combinatorial Chemistry and Molecular Diversity in Drug Discovery, Edited by
Eric M. Gordon and James F. Kerwin, Jr.
ISBN 0-471-15518-7   Copyright © 1998 by Wiley-Liss, Inc.

manner and that incorporate a number of amino acid residues as building blocks.

CYCLIC PEPTIDE LIBRARIES

One of the most direct routes to obtain libraries of constrained peptides would be through peptide cyclization. Several investigators have pursued this approach and have reported the synthesis of collections of cyclic peptides.[19,20] For example, Mihara and coworkers employ a *p*-nitrobenzophenone oxime resin that is loaded with an ε-aminocaproic acid residue.[21] Linear pentapeptides are then prepared by standard Boc chemistry followed by cyclization of the N-terminal amine onto the C-terminal ε-aminocaproic acid residue to provide the cyclic peptides free in solution. Unfortunately, the generality of these approaches is limited by the variable yields of the cyclization reactions, which are highly sequence-dependent.[22]

GENERAL SCAFFOLDS FOR THE DISPLAY
OF AMIDE LINKED FUNCTIONALITY

Several researchers have focused on the design and synthesis of general scaffolds from which to display diverse amide-linked functionality. These general scaffolds offer two advantages over cyclic peptides. First, since the scaffolds are prepared before the introduction of diverse functionality, the problems associated with the cyclization of linear peptides are circumvented. Second, these compounds can mimic protein surfaces and discontinuous binding sites in a manner not attainable by short, linear, or cyclic peptides.

Houghten and coworkers have identified novel chymotrypsin inhibitors from combinatorial libraries based on a cyclic peptide template.[23] A positional scanning format[24] was utilized to prepare three different positional libraries (O_1XX, XO_2X, XXO_3) based on the cyclic peptidic scaffold (Figure 5.1). Each

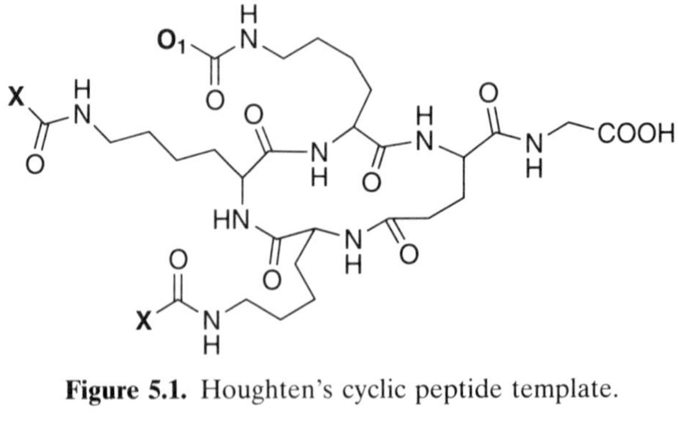

Figure 5.1. Houghten's cyclic peptide template.

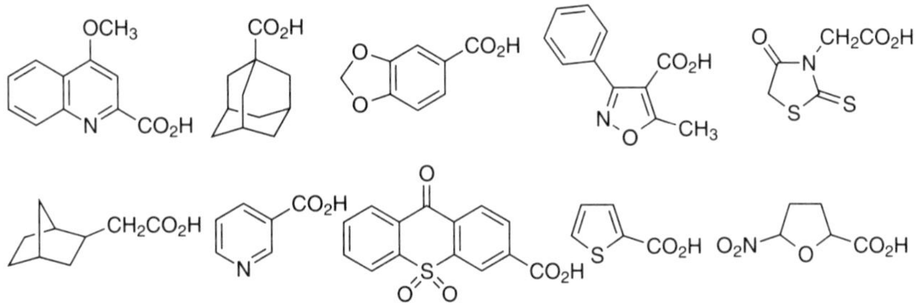

Figure 5.2. The 10 carboxylic acids used in conjunction with amino acids as building blocks for Houghten's library.

position, X, represents an approximately equimolar mixture of 19 of the 20 proteinogenic amino acids (cysteine excluded) along with 10 diverse carboxylic acids (Figure 5.2). The fixed (O_i) position represents individually defined positions occupied by one of the 20 amino acids or one of the 10 carboxylic acids. Thus, each positional libary consisted of 30 separate peptide mixtures comprising 841 (29^2) compounds. In the synthesis of the positional libraries, the side chains at the fixed (O_i) position in each library were introduced through a Lys side chain protected with 1-(4,4-dimethyl-2,6-dioxocyclohex-1-ylidene)ethyl (Dde) group, while the mixed (X) positions were derived from Boc-protected Lys residues.

The template was synthesized on Boc-Gly-PAM polystyrene resin by first attaching Fmoc-Glu(Oallyl)-OH. The peptide chain was then extended in the C to N direction by incorporating an Fmoc-Lys(Dde) and two Fmoc-Lys(Boc) residues. After cleavage of the Glu side chain allyl ester and removal of the N-terminal Fmoc group, the support-bound peptide was cyclized to form the 14-membered lactam. High-performancee liquid chromatography (HPLC) analysis revealed the major product to be desired cyclic monomer (42% of total peak area) contaminated with some cyclic dimer (26% of total peak area). The Boc protecting groups were removed and the resin, in 90 separate resin bags (30 fixed positons × 3 positional libraries), was coupled to an equimolar mixture of the 19 Fmoc-protected amino acids and 10 carboxylic acids. Only 1.2 equivalents of the acylation mixture relative to support-bound amine was used to attain approximate equal representation of each amino acid and carboxylic acid at each position X.[25] After removal of the Fmoc protecting group of the newly introduced amino acids, the free amines were capped to form N-terminal acetamides. The Dde protecting group was removed by treatment with hydrazine, and the resin was divided into 30 groups of three bags. Each set of three resin bags was immersed in a solution of one of the 20 Fmoc-amino acids, or one of the 10 carboxylic acids to couple the defined side chain O_i. The Fmoc protecting groups in each of the 60 bags to which an Fmoc-amino acid was coupled were removed and the amines capped

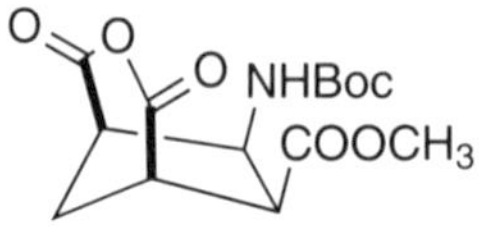

Figure 5.3. The two templates used by Selectide for the display of amide-linked functionality.

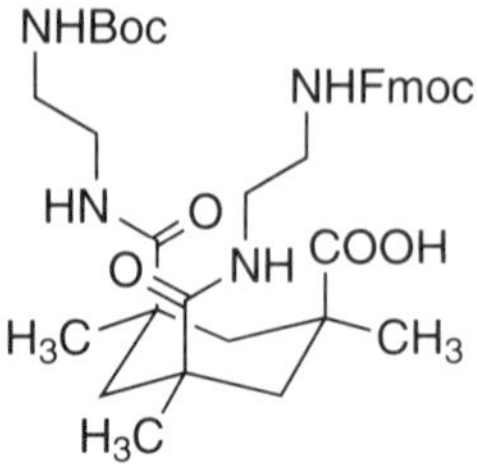

as before. Side-chain deprotection followed by cleavage of the peptide mixtures from the resin provided the three positional libraries in solution.

The peptide mixtures were then evaluated for chymotrypsin inhibition. At all three positions, two carboxylic acids—piperonylic acid and 2-thiophenecarboxylic acid—were found to be the most effective at chymotrypsin inhibition. Resynthesis of the eight compounds corresponding to all combinations of these two carboxylic acids at all three positions provided inhibitors with an average K_d of 60 μM. The inhibitors were approximately 30-fold less active than chymostatin ($K_d = 1.6$ μM) and 10-fold more active than the corresponding linear analogs. Furthermore, a control peptide based on the cyclic peptidal template was found to be completely resistant to trypsin, chymotrypsin, and proteinase K.

Researchers at Selectide have utilized two different templates for the display of amide-linked functionality (Figure 5.3). Early work involved a scaffold derived from *cis*-5-norbornene-*endo*-2,3-dicarboxylic anhydride **1**.[26] More recently, a related template based on 1,3,5-trimethyl-1,3,5-cyclohexanetricarboxylic acid (Kemp's triacid) **2** was employed.[27]

The racemic acid **2** was attached to TentaGel resin through a β-Ala-Gly-β-Ala-Gly linker derivatized at the N-terminus with a set of 30 amino acids to introduce R$_1$ (Scheme 5.1). After deprotection of the Fmoc group, a set of

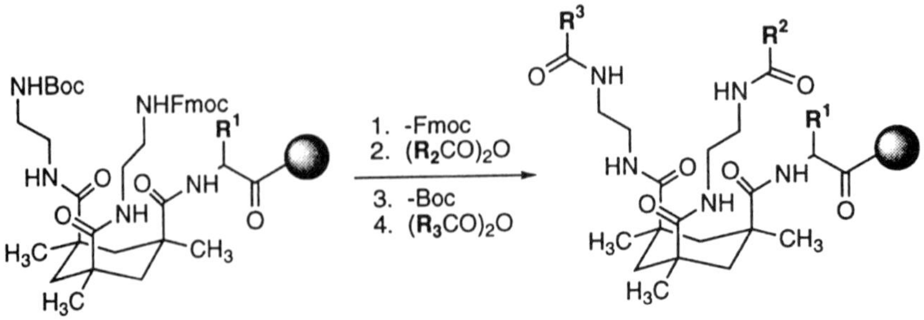

Scheme 5.1.

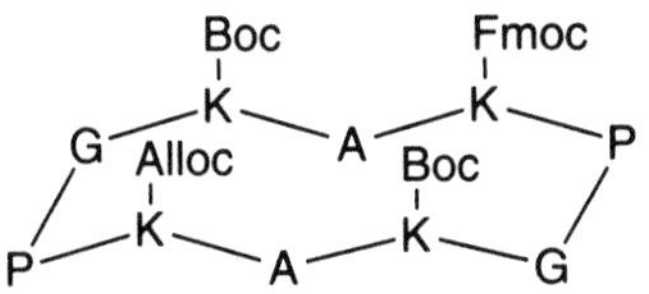

3

Figure 5.4. The cyclic decapeptide template used by Sila and Mutter.

50 carboxylic acids were coupled to introduce the second element of diversity, R_2. Next, the Boc group was removed and the set of 50 carboxylic acids were coupled at the third position, R_3. After side-chain deprotection, the library was evaluated against various biological targets while still attached to the solid support. The structures of the active compounds were determined by comparing the fragmentation patterns from LC/MS/MS of cleaved material (one compound per bead[28]) to that of model compounds. The screening targets and active compounds were not provided, although the authors reported that these are forthcoming.

Sila and Mutter used another template-based library to mimic the discontinuous epitope of lysozyme, which interacts with the monoclonal antibody HyHEL-5.[29] A collection of 32 compounds was assembled in solution based on the cyclic decapeptide template **3** with four differentially protected Lys residues (Figure 5.4). Two of the Lys residues were used to introduce a set of eight amino acid side chains, while the other two Lys residues were reserved for Arg side chains, which had been implicated as important for antibody binding based on an X-ray crystal structure of the epitope–antibody complex.[30]

The cyclic decapeptide template **3**, c(Lys(Fmoc)-Pro-Gly-Lys(Boc)-Ala-Lys(Alloc)-Pro-Gly-Lys(Boc)-Ala), was assembled on support, cleaved, diluted to 1 mM, and cyclized in solution (Scheme 5.2). The desired cyclic monomer was obtained in 58% yield (relative to acyclic precursor) after diethyl ether precipitation and purification by RP-HPLC. The two Boc-protected Lys side chains of **3** were deprotected and AcArg(Pmc) was coupled to each side chain. The Fmoc group of **4** was removed with diethylamine and the lyophilized residue was divided into eight portions. Each portion was coupled with one of eight N-acetyl amino acids (Leu, Gly, Pro, Thr(tBu), Tyr(tBu), Asp(tBu), Asn(Trt), and Gln(Trt)), and the products were again isolated by diethyl ether precipitation, lyophilization, and purification by RP-HPLC. The eight compounds were recombined to give mixture **5**, the Alloc group was removed, and the dividing, coupling, and purification procedures were repeated to afford an equimolar pool of 32 compounds **6**. Preliminary binding tests using a HyHEL-5 affinity column implicated a Tyr side chain as crucial for binding to the monoclonal antibody HyHEL-5.

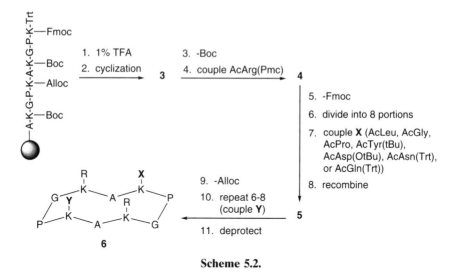

Scheme 5.2.

SCAFFOLDS DESIGNED TO DISPLAY FUNCTIONALITY IN A β-TURN CONFORMATION

Many researchers have focused on the design and synthesis of general pepti-domimetic scaffolds that are based on common secondary structural motifs of bioactive peptides. The design of peptidomimetics based on the β-turn structure (Figure 5.5) has been the most heavily investigated due to the large number of bioactive peptides that have been shown to contain this turn struc-ture.[31-33] Unfortunately, a single peptidomimetic scaffold cannot be applied to all β-turn types due to the wide variation in structures that are included in the turn motif. The wide variation in side-chain functionality and also in side-chain orientation is particularly significant since the relative orientation of the side chains is a critical determinant of activity.[34] Furthermore, the bioactive conformation has not been determined for most bioactive peptides upon which turn mimetics would be based. Therefore, the identification of a

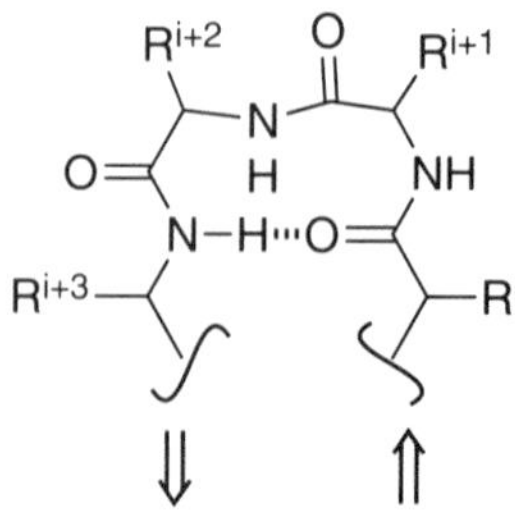

Figure 5.5. General structure of a β-turn.

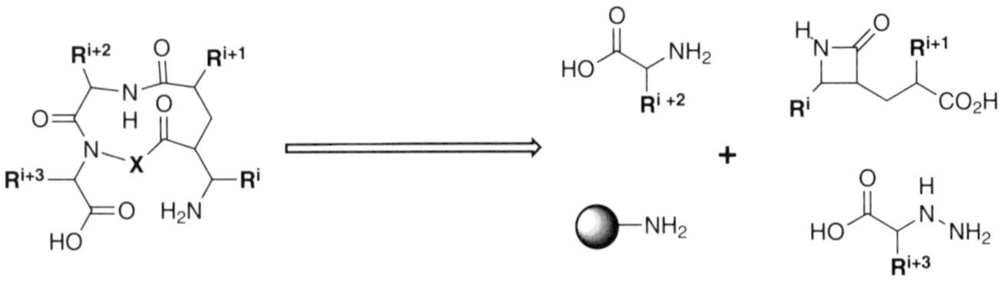

Figure 5.6. Retrosynthesis of Kahn's turn mimetic.

bioactive turn mimetic against a receptor target will require the synthesis of multiple derivatives to identify the key side chains present in the mimetic, as well as to determine the necessary orientations of those side chains.

β-Turn mimetics have been extensively reviewed and many types of turn mimetics exist.[35–37] This section focuses on general turn mimetic scaffolds, which are well-suited for the construction of libraries and which can display functionality at a number of positions, especially at the highly exposed $i + 1$ and $i + 2$ positions.

Kahn and coworkers have designed a number of different β-turn mimetics. In particular, Kahn has reported a general turn mimetic scaffold that has the potential to be synthesized with side chains corresponding to all four of the positions of a β-turn (Figure 5.6). Furthermore, the syntheses of turn mimetics based on this scaffold have been performed in solution and on solid phase. The mimetics are prepared from an α-hydrazino acid ($i + 3$ side chain), an azetidinone ($i + 1$ and i side chains), and α-amino acids ($i + 2$ and other side chains). The hydrazine moiety can also be substituted with diamino alkyl spacers to provide larger rings.

Employing this synthesis strategy Kahn has prepared several bioactive turn mimetics that contain side-chain functionality. Each turn mimetic was incorporated into a peptide and included either no side chain or a serine side chain at the $i + 2$ site and no side chain at the $i + 1$ site. One of the turn mimetics, **7** (Figure 5.7), was modeled after one of the complementary determining loop regions of an antibody that binds to the retrovirus type 3 cellular receptor.[38] The mimetic has an IC_{50} of 55–75 μM for the antibody–ligand complex. The second mimetic, **8**, was modeled after a critical loop in CD4 and binds to the gp120 glycoprotein of the HIV virus with low micromolar K_d values.[39] Lastly, a series of four mimetics, **9**, was synthesized to probe the receptor-bound conformation of leucine enkephalin.[40] Each mimetic contains a different spacer, X, resulting in rings of both different sizes and rigidity. One of the series of mimetics displays modest affinity toward the μ opiod receptor ($IC_{50} = 8$ μM). All of Kahn's mimetics were designed based on either X-ray crystal structures or extensive SAR data.

Recently Hermkens and coworkers have reported a strategy to prepare β-turn mimetics that is capable of introducing all four of the side chains of the

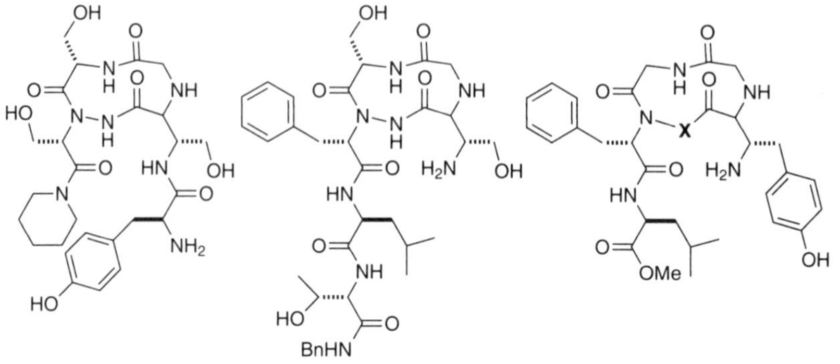

Figure 5.7. Turn mimetics developed by Kahn.

β-turn.[41] Employing this strategy, they prepared a potential ligand to the GPIIb/IIIa receptor **10** (Figure 5.8). Unfortunately, little binding was observed, which the authors ascribe to incorrect assignment of the bioactive conformation of the naturally occurring peptide ligand. Aubé and coworkers have also recently described a method to construct β-turn mimetics **11** that should allow the straightforward introduction of both the *i* + 1 and *i* + 2 side chains.[42]

β-Turn Peptidomimetics Displaying the *i* + 1 and *i* + 2 Side Chains

We have developed two general peptidomimetic templates for library synthesis. Each displays amino acid side-chain functionality so as to emulate peptides that adopt β-turn conformations. The first generation mimetic is constrained in a turn structure by replacing the hydrogen bond between the *i* and *i* + 3 residue with a covalent aminoalkylthiol backbone linkage (Figure 5.9). The secondary amide between the *i* + 1 and *i* + 2 residue is retained from the β-turn structure, since this is the most exposed amide of the turn and is the most likely to be involved in receptor interactions. The flexibility of the turn mimetic and the relative orientations of the side chains can be varied by

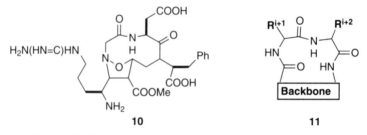

Figure 5.8. Hermken's and Aubé's β-turn mimetic scaffolds.

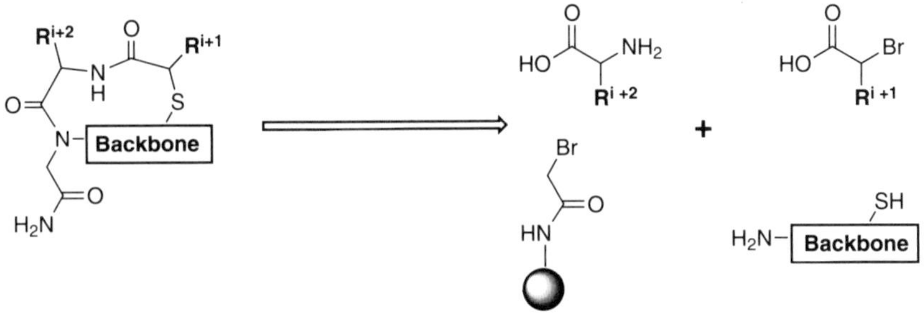

Figure 5.9. Retrosynthesis of first generation turn mimetics.

introducing different backbone linkages to provide 9- or 10-membered ring turn mimetics with different backbone substitutions. In addition, the relative orientations of the side chains can also be varied by incorporating different combinations of the absolute configurations at each of the stereocenters intro- duced by the $i + 1$ and $i + 2$ side chains of the turn mimetic.

The peptidomimetic is constructed from three readily available compo- nents. The $i + 2$ side chain is derived from an α-amino acid, many of which are commercially available in both (R) and (S) configurations in protected form. The $i + 1$ side chain is derived from an α-bromo acid, which can be prepared with complete retention of configuration from the corresponding commercially available amino acid employing sodium nitrite and HBr.[43,44] Finally, a number of aminoalkylthiols are commercially available or can be prepared in relatively few steps.

Before initiating construction of the first-generation turn mimetic **17,** p- nitrophenylalanine is coupled to the Rink amide support employing standard solid-phase peptide synthesis conditions to provide precursor **12** (Scheme 5.3).[45] Bromoacetic acid is then coupled to the support-bound p-nitrophenylal- anine by activation with diisopropylcarbodiimide. The backbone element is introduced by treatment of α-bromo amide **13** with either 2-aminoethanethiol *tert*-butyl disulfide or 3-aminopropanethiol *tert*-butyl disulfide in DMSO to provide the secondary amine **14,** which is then coupled with the appropriate Fmoc-protected amino acid employing O-(7-azabenzotriazol-1-yl)-1,1,3,3-tet- ramethyluronium hexafluorophosphate (HATU) to provide **15.** After coupling the Fmoc-protected amino acid to introduce the $i + 2$ side chain to obtain **16,** the Fmoc protecting group is removed and the free amine is acylated with the α-bromo acid to incorporate the $i + 1$ side chain. Reduction of the mixed *tert*-butyl disulfide **16** and rapid cyclization with tributylphosphine and tetra- methylguanidine (TMG) affords the 9- or 10-membered thioether **17.** Cleavage from support is then accomplished by treatment with a 1:1:18 water/dimethyl sulfide/trifluoroacetic acid comixture. For all of the turn mimetics synthesized, cyclization provided the desired cyclic monomer with no detectable amount of dimer. This included mimetics incorporating (R) and (S) α-bromo acids,

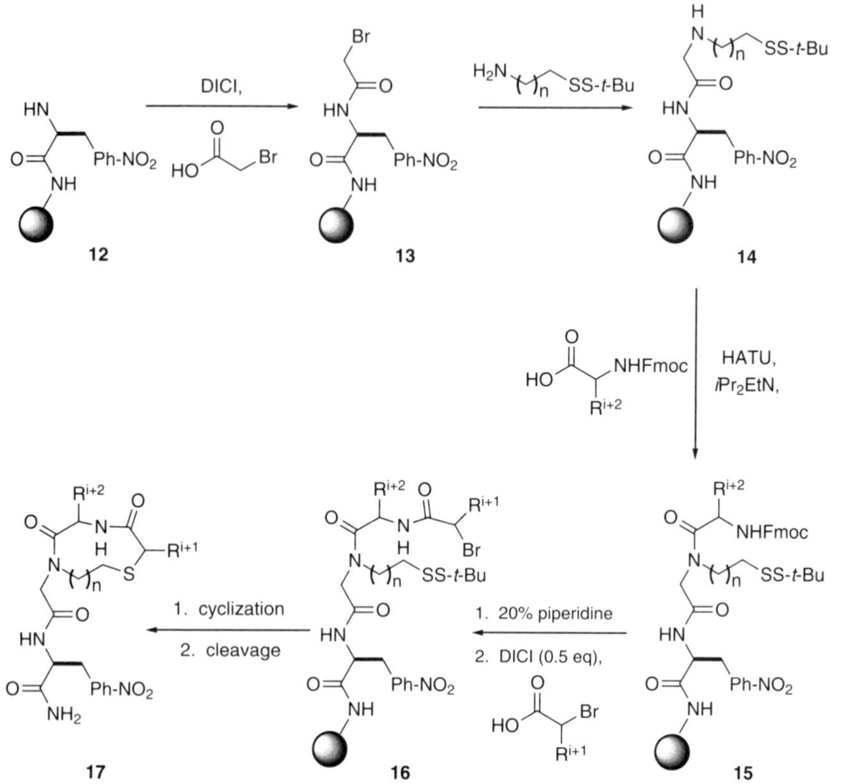

Scheme 5.3.

α-bromoisovaleric acid (corresponding to the sterically hindered amino acid, valine), and α-chloroacetic acid (corresponding to the least sterically hindered amino acid, glycine) at the $i + 1$ site. Turn mimetics containing a variety of side-chain functionality were also synthesized to demonstrate the compatibility of a range of side-chain functionality with the synthesis sequence.[46]

Once the synthetic sequence was optimized and shown to be general, a library of 1152 turn mimetics was generated using 18 α-halo acids, 32 Fmoc-protected amino acids, and two backbone elements (Figure 5.10). A phenylalanine residue was incorporated in all derivatives prior to construction of the turn mimetic, at the putative $i + 4$ position, to provide an additional hydrophobic interaction since the library would primarily be screened against membrane-bound receptors. A modified library was also prepared from the initial library by oxidation of the thioethers to sulfoxides. The α-bromo acids and amino acids were selected to generate a small library with diverse functionality representative of the naturally occurring amino acid sidechains. Side-chain protection was accomplished using the TFA-labile Boc and t-Bu protecting groups.

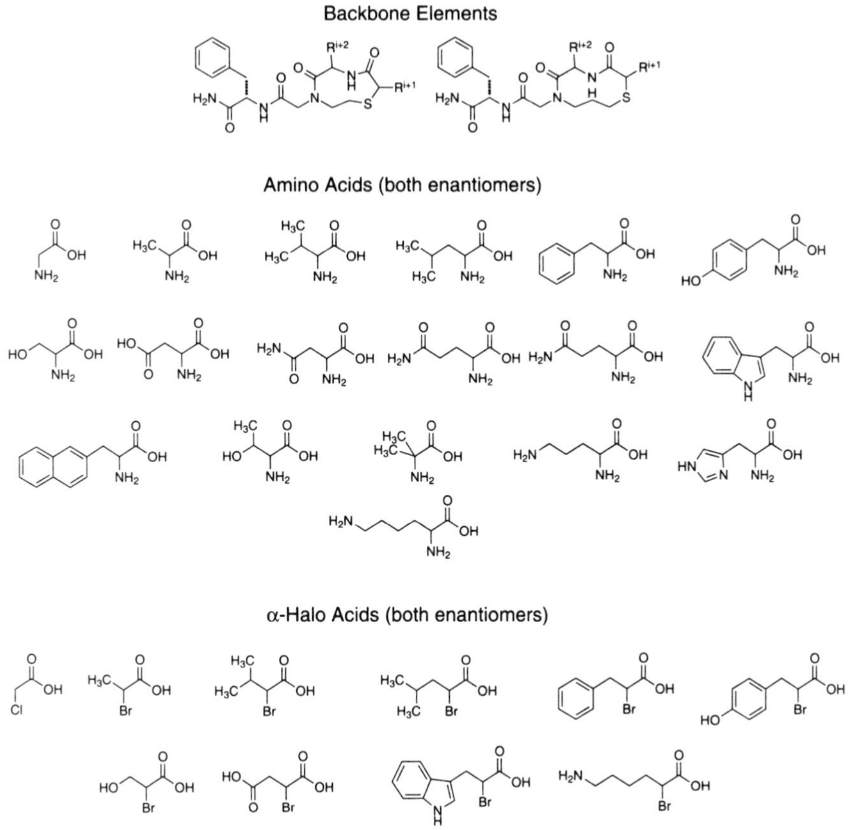

Backbone Elements

Amino Acids (both enantiomers)

α-Halo Acids (both enantiomers)

Figure 5.10. Library of first generation turn mimetics.

The 1152 turn mimetics were synthesized simultaneously in less than two weeks employing the Chiron Mimotopes pin apparatus[47] following the previously described reaction sequence. Synthesis was performed on polyethyl-enepoly(N,N-dimethylacrylamide/methacrylic acid) graft copolymer pins, pre-derivatized with the acid-cleavable Rink linker.[48] All pins were derivatized as one batch until the i + 2 side chains were introduced.[49] At this point the pins were affixed to pin holders such that each pin fit into a well of a 96-well microtiter plate. Upon completion of the synthesis, isolation of the spatially separate mimetics in microtiter plates was accomplished by cleavage of the mimetics off of the pins with concomitant side-chain deprotection, followed by concentration in vacuo.

Seven percent of the library was evaluated by mass spectrometry using electrospray ionization; for all of the derivatives that were tested the expected molecular ions were observed. A subset of the library, corresponding to those mimetics containing only Trp and Lys side chains at the i + 1 and i + 2

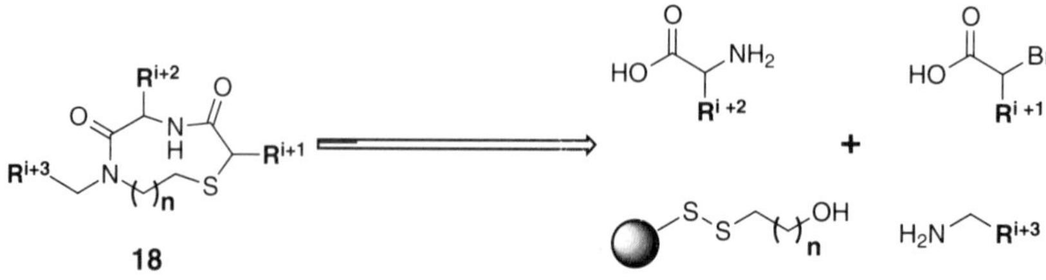

Figure 5.11. Retrosynthesis of second generation turn mimetics.

positions, was screened against three murine somatostatin receptor subtypes (msr1, msr2, and msr3).[50] Low micromolar competitive inhibitors of somatostatin were found, several of which displayed selectivity for a particular receptor subtype. The complete library of 1152-turn mimetics was evaluated in a series of one-point competitive radioligand binding assays at a concentration of 5 μM against three seven-transmembrane G-protein coupled receptors. Active compounds that displayed greater than 50% inhibition were not observed when the library was screened against the dopamine-D_2 receptor, but 5 were found for the cytokine, f-Met-Leu-Phe, receptor and 15 for the NK1 receptor.[51,52] The library has also yielded lead compounds in proprietary assays.

β-Turn Peptidomimetics Displaying the i +1, i + 2, and i + 3 Side Chains

A second-generation turn peptidomimetic **18** (Figure 5.11) was developed to provide additional display of functionality and improved solubility by incorporating the i + 3 side chain in place of the primary amide functionality present in the first-generation peptidomimetic (Figure 5.3). The second-generation turn mimetics are attached to the solid support through a disulfide linkage instead of the Rink amide linkage, and the i + 3 side chain is introduced by means of a non-α-branched primary amine.

Support-bound backbone component **20** is prepared by methanolysis of thioester **19**[53] with NaOMe in 3:1 THF/MeOH followed by disulfide interchange with the thiol backbone component activated as the 2-benzothiazolyl (BT) mixed disulfide (Scheme 5.4).[54] The 4,4'-dimethoxytrityl (DMT) ether is not required as a protecting group, but, rather, it enables the spectrophotometric determination of the loading level from which the overall yields are calculated.[55] The DMT group is removed by treatment with 3% trichloroacetic acid in CH_2Cl_2 and the alcohol is then mesylated. Any nonmesylated alcohol is capped to prevent formation of ester side products, which could interfere with biological evaluation. The support-bound mesylate is treated with a concentrated solution of the appropriate primary amine to provide intermediate **21** that incorporates the i + 3 side chain. The i + 2 and i + 1 side chains are then introduced as described previously for the synthesis of the first-generation turn mimetics.

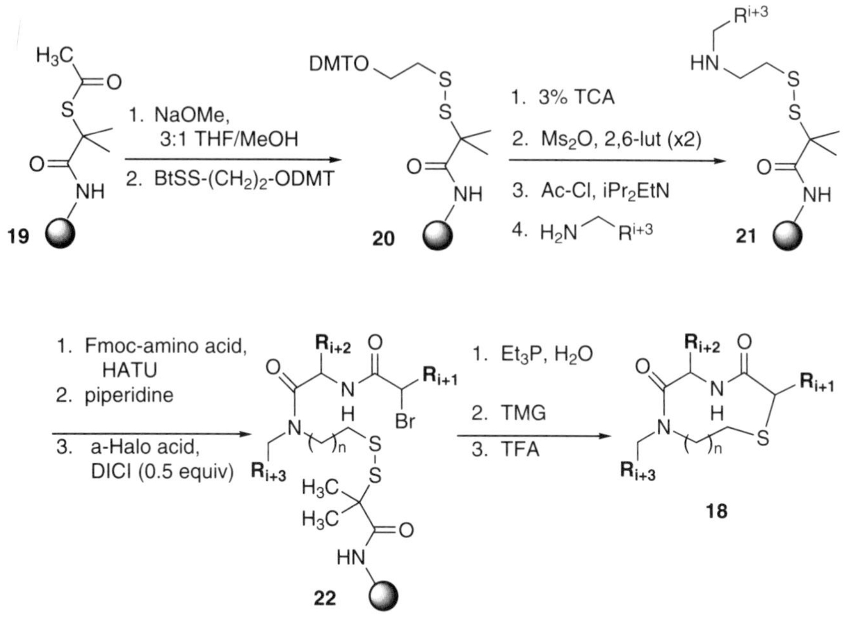

Scheme 5.4.

The cyclization precursor **22** is cleaved from the support and allowed to cyclize in solution. The lower boiling reagents, triethylphosphine and N-methylmorpholine, are employed so that excess of these reagents can be removed after cleavage by concentration in vacuo. Tentagel was initially employed as the support due to difficulties encountered previously concerning the swelling of polystryrene (PS) supports in aqueous solvents, but it was found that by increasing the proportion of DMF in the solvent comixture, PS supports could be utilized. The cyclic product **18** is obtained in a high level of purity for different turn mimetics; however, the rate of cyclization is dependent on both the size of the ring $(9 > 10)$ and the stereochemistry at the two α-carbons in the final product $((S, R)$ and $(R, S) > (S, S)$ and $(R, R))$. Furthermore, the concentration of the acyclic mimetic must be kept ≤ 1 mM to prevent the formation of oligomers or cyclic dimer. After cleavage from the support and cyclization in solution, side-chain protecting groups are removed by treatment with a $TFA/Me_2S/H_2O$ comixture. Reconcentration to remove the volatile deprotection cocktail then provides the desired turn mimetics.

The above synthesis sequence has been used to prepare a library of 176 second-generation mimetics **18** biased toward the somatostatin receptor subtypes. All combinations of both enantiomers of tryptophan and lysine were incorporated at either the $i + 1$ and $i + 2$ positions based on the binding data from the first-generation library and based on previous researchers' studies with cyclic peptides.[56,57] Twenty-two different amines were incorporated at the

$i + 3$ position. Fourteen amines were selected to display a maximum of diverse functionality at this site,[58] and the remaining eight amines were selected based on the sequences of bioactive somatostatin analogs. A number of subtype selective ligands have been identified and the binding data and functional activities will be reported in due course.[59]

REFERENCES

1. Hirschmann R (1991): Medicinal chemistry in the golden age of biology: lessons from steroid and peptide research. *Angew Chem Int Ed Engl* 30:1278–1301.
2. Schmidt G (1986): Recent developments in the field of biologically-active peptides. *Top Curr Chem* 109:109–159.
3. Plattner J, Norbeck DW, eds. (1990): *Obstacles to Drug Development From Peptide Leads.* Chichester, UK: Ellis Horwood, pp 92–126.
4. Geysen HM, Mason TJ (1993): Screening chemically synthesized peptide libraries for biologically-relevant molecules. *Bioorg Med Chem Lett* 3:397–404.
5. Al-Obeidi F, Hruby VJ, Hadley ME, Sawyer TK, Castrucci AMD (1990): Design, synthesis, and biological activities of a potent and selective alpha-melanotropin antagonist. *Int J Pept Protein Res* 35:228–234.
6. Crusi E, Huerta JM, Andreu D, Giralt E (1990): Gly/Lys-containing peptide macrocycles—synthesis and cyclization studies. *Tetrahedron Lett* 31:4191–4194.
7. Osapay G, Profit A, Taylor JW (1990): Synthesis of tyrocidine-A. use of oxime resin for peptide-chain assembly and cyclization. *Tetrahedron Lett* 31:6121–6124.
8. Trzeciak A, Bannwarth W (1992): Synthesis of head-to-tail cyclized peptides on solid support by Fmoc chemistry. *Tetrahedron Lett* 33:4557–4560.
9. Kates SA, Sole NA, Johnson CR, Hudson D, Barany G, Albericio F (1993): A novel, convenient, 3-dimensional orthogonal strategy for solid-phase synthesis of cyclic peptides. *Tetrahedron Lett* 34:1549–1552.
10. Marlow CK (1993): Peptide cyclization on TFA labile resin using the trimethysilyl (TMSE) ester as an orthogonal protecting group. *Bioorg Med Chem Lett* 3:437–440.
11. McMurray JS, Lewis CA (1993): The synthesis of cyclic peptides using Fmoc solid-phase chemistry on the linkage agent 4-(4-Hydroxymethyl-3-methoxyphenoxy)-butyric acid. *Tetrahedron Lett* 35:8059–8062.
12. Alsina J, Rabanal F, Firalt E, Albericio F (1994): Solid-phase synthesis of "head-to-tail" cyclic peptides via lysine side-chain anchoring. *Tetrahedron Lett* 35:9633–9636.
13. Aletras A, Barlos K, Gatos D, Koutsogianni S, Mamos P (1995): Preparation of the very acid-sensitive Fmoc-Lys (MH)-OH. Application in the synthesis of side-chain to side-chain cyclic peptides and oligolysine cores suitable for the solid-phase assembly of MAPs and TASPs. *Int J Prept Protein Res* 45:488–496.
14. Dumy P, Eggleston IM, Cervigni S, Sila U, Sun X, Mutter M (1995): A convenient synthesis of cyclic peptides as regioselectively addressable functionalized templates (RAFT). *Tetrahedron Lett* 36:1255–1258.
15. Albericio F, Hammer RP, Garciaecheverria C, Molins MA, Chang JL, Munson MC, Pons M, Giralt E, Barany G (1991): Cyclization of disulfide-containing peptides in solid-phase synthesis. *Int J Pept Protein Res* 37:402–413.

16. Camarero JA, Giralt E, Andreu D (1995): Cyclization of a large disulfide peptide in the solid phase. *Tetrahedron Lett* 36:1137–1140.

17. Barker PL, Bullens S, Bunting S, Burdick DJ, Chan KS, Deisher T, Eigenbrot C, Gadek TR, Gantzos R, Lipari MT, Muir CD, Napier MA, Pitti RM, Padua A, Quan C, Stanley M, Struble M, Tom JYK, Burnie JP (1992): Cyclic RGD peptide analogues as antiplatelet antithrombotics. *J Med Chem* 35:2040–2048.

18. Polinsky A, Cooney MG, Toypalmer A, Osapay G, Goodman M (1992): Synthesis and conformational properties of the lanthionine-bridge opioid peptide [D-AlaL2, AlaL5] enkephalin as determined by NMR and computer simulations. *J Med Chem* 35:4185–4194.

19. Darlak K, Romanovskis P, Spatola AF (1994): Cyclic peptide libraries, In Peptides: Chemistry, Structure and Biology. Hodges R; Smith J (eds.) Leiden: ESCOM, pp 981–983.

20. Lyttle MH, Berry COA, Hocker MH, Kauvar LM (1994): Novel concepts in the preparation and use of planar arrays of peptides with unnatural amino acids and cyclic structures. Leiden: ESCOM, pp 1009–1011.

21. Mihara H, Yamabe S, Niidome T, Aoyagi H, Kumagai H (1995): Efficient preparation of cyclic peptide mixtures by solid-phase synthesis and cyclization cleavage with oxime resin. *Tetrahedron Lett* 36:4837–4840.

22. Nishino N, Xu M, Mihara H, Fujimoto T, Ueno Y, Kumagai H (1992): Sequence dependence in solid-phase-synthesis-cyclization-cleavage for Cyclo(-arginyl-glycyl-aspartyl-phenylglycyl-). *Tetrahedron Lett* 33:1479–1482.

23. Eichler J, Lucka AW, Houghten RA (1994): Cyclic peptide template combinatorial libraries: synthesis and identification of chymotrypsin inhibitors. *Pept Res* 7:300–307.

24. Dooley CT, Houghten RA (1993): The use of positional scanning synthetic peptide combinatorial libraries for the rapid determination of opioid receptor ligands. *Life Sci* 52:1509–1517.

25. Kramer A, Volkmerengert R, Malin R, Reineke U, Schneidermergener J (1993): Simultaneous synthesis of peptide libraries: examples for the identification of protein, metal and DNA binding peptide mixtures. *Pept Res* 6:314–319.

26. Patek M, Drake B, Lebl M (1994): All-cis cyclopentane scaffolding for combinatorial solid-phase synthesis of small nonpeptide compounds. *Tetrahedron Lett* 35:9169–9172.

27. Kocis P, Issakova O, Sepetov NF, Lebl M (1995): Kemps triacid scaffolding for synthesis of combinatorial nonpeptide uncoded libraries. *Tetrahedron Lett* 36:6623–6626.

28. Lam KS, Salmon SE, Hersh EM, Hruby VJ, Kazmierski WM, Knapp RJ (1991): A new type of synthetic peptide library for identifying ligand-binding activity. *Nature* 354:82–84.

29. Sila U, Mutter M (1995): Topological templates as tool in molecular recognition and peptide mimicry: synthesis of a TASK library. *J Mol Recognition* 8:29–34.

30. Novotny J, Bruccoleri RE, Saul F (1989): On the attribution of binding energy in antigen-antibody complexes McPC 603, D1.3, and HyHEL-5. *Biochemistry* 28:4735–4739.

31. Smith JA, Pease LG (1980): Reverse turns in peptides and proteins. *CRC Crit Rev Biochem* 8:315–400.

32. Rose GDG, Gierasch LM, Smith JA (1985): Turns in peptides and proteins. *Adv Protein Chem* 37:1–109.

33. Gierasch LM, Rizo J (1992): Constrained peptides: models of bioactive peptides and protein substructures. *Annu Rev Biochem* 61:387–418.

34. Hruby VJ (1982): Conformational restrictions of biologically active peptides via amino acid side chain groups. *Life Sci* 31:189–199.

35. Ball JB, Alewood PF (1990): Conformational constraints: nonpeptide beta-turn mimics. *J Mol Recognition* 3:55–64.

36. Kahn M, ed. (1993): vol 49, pp 3433–3677.

37. Kahn M (1993): Peptide secondary structure mimetics: recent advances and future challenges. *Synlett,* 821–826.

38. Saragovi HU, Fitzpatrick D, Raktabutr A, Nakanishi H, Kahn M, Greene MI (1991): Design and synthesis of a mimetic from an antibody complementary-determining region. *Science* 253:792–795.

39. Chen SX, Chrusciel RA, Nakanishi H, Raktabutr A, Johnson ME, Sato A, Weiner D, Hoxie J, Saragovi HU, Greene MI, Kahn M (1992): Design and synthesis of a CD4 beta-turn mimetic that inhibits human-immunodeficiency-virus envelope glycoprotein GP120 binding and infection of human-lymphocytes. *Proc Natl Acad Sci USA* 89:5872–5876.

40. Gardner B, Nakanishi H, Kahn M (1993): Conformationally constrained nonpeptide beta-turn mimetics of enkephalin. *Tetrahedron* 49:3433–3448.

41. Hermkens PHH, Vondinther TG, Joukema CW, Wagenaars GN, Ottenheijm HCJ (1994): Peptide backbone cyclization as an avenue to beta-turn mimics. *Tetrahedron Lett* 35:9271–9274.

42. Kitagawa O, Velde DV, Dutta D, Morton M, Takusagawa F, Aubé J (1995): Structural-analysis of beta-turn mimics containing a substituted 6-aminocaproic acid linker. *J Am Chem Soc* 117:5169–5178.

43. Dener JM, Zhang LH, Rapoport H (1993): An effective chirospecific synthesis of (+) -pilocarpine from L-aspartic acid. *J Org Chem* 58:1159–1166.

44. Kwack H, Virgillio AA, Ellman JA (1996): Preparation of optically pure α-bromo acids containing mild acid labile sidechain protecting groups, vol 61.

45. Virgilio AA, Ellman JA (1994): Simultaneous solid-phase synthesis of beta-turn mimetics incorporating side-chain functionality. *J Am Chem Soc* 116:11580–11581.

46. Maeji NJ, Valerio RM, Bray AM, Campbell RA, Geysen HM (1994): Grafted supports used with the multipin method of peptide-synthesis. *React Polym* 22:203–212.

47. The Fmoc-Rink-Amide-Handle-Gly-HMD-MA/DMA pins (5 μmol/pin) were supplied by Chiron Mimotopes (Victoria, Australia).

48. The backbone components were introduced by reaction of the support-bound α-bromo amide **13** with an equimolar mixture of 2-aminopropanethiol and 3-aminopropanethiol *tert*-butyl disulfide so that there would be two compounds generated per pin.

49. The assays were performed by Daniel Fitzpatrick at Genentech (San Francisco, CA).

50. The assays were performed under the supervision of Mike Morrissey at Berlex (Richmond, CA).

51. Virgilio AA, Bray AM, Ellman JA (1995): Abstracts of Papers of the American Chemical Society 209: p. 263. Synthesis and evaluation of a library of β-turn mimetics.

52. Mery J, Granier C, Juin M, Brugidou J (1993): Disulfide linkage to polyacrylic resin for automated Fmoc peptide-synthesis—immunochemical applications of peptide resins and mercaptoamide peptides. *Int J Pept Protein Res* 42:44–52.

53. Brzezinska E, Ternay AL (1994): Disulfides .1. synthesis using 2,2'-dithiobis (benzothiazole). *J Org Chem* 59:8239–8244.

54. Caruthers MH, Barone AD, Beaucage SL, Dodds DR, Fisher EF, McBride LJ, Matteucci M, Stabinsky Z, Tang J-Y (1987): Chemical synthesis of deoxyoligonucleotides by the phosphoramidite method. In Methods in Enzymology vol 154. San Diego, CA: Academic, pp 287–313.

55. Veber D, Freidinger R, Perlow D, Paleveda WJ, Holly F, Strachan R, Nutt R, Arison B, Homnick C, Randall W, Glitzer M, Saperstein R, Hirschmann R (1981): A potent cyclic hexapeptide analog of somatostatin. *Nature* 292:55–58.

56. Veber D, Saperstein R, Nutt R, Freidinger R, Brady S, Curley P, Perlow D, Palveda W, Colton C, Zacchei A, Tocco D, Hoff D, Vandlen R, Gerich J, Hall L, Nadarino L, Cordes E, Anderson P, Hirschmann R (1984): A super active cyclic hexapeptide analog of somatostatin. *Life Sci* 34:1371.

57. Muskal S: MDL Information Systems.

58. Virgilio A, Bray AA, Zhang W, Trinh L, Snyder M, Morrissey MM, Ellman J (1997): Synthesis and evaluation of a library of peptidomimetics based upon the beta-turn *Tetrahedron* 53:6635–6644.

6

SUBMONOMER APPROACHES FOR THE GENERATION OF MOLECULAR DIVERSITY: NONNATURAL OLIGOMER AND ORGANIC TEMPLATE LIBRARIES

LUTZ S. RICHTER, DANE A. GOFF, KERRY L. SPEAR, ERIC J. MARTIN, AND RONALD N. ZUCKERMANN
Chiron Corporation, Emeryville, California

Progress in molecular biology has made purified receptors, enzymes, and other biological reagents available for the high-throughput screening of chemical compound collections in a tremendous variety of new assay systems. Consequently, there has been an increased demand for small organic molecules in order to speed up the discovery process of new lead structures for drug discovery and development. In an attempt to keep pace with the increased demand for synthetic compounds, a growing number of methods for the rapid generation of diverse chemical libraries have been developed.[1-3] Straightforward and efficient submonomer approaches for the automated, high-throughput synthesis of nonnatural oligomer[4,5] and organic template libraries are summarized in this chapter.

Combinatorial Chemistry and Molecular Diversity in Drug Discovery, Edited by
Eric M. Gordon and James F. Kerwin, Jr.
ISBN 0-471-15518-7   Copyright © 1998 by Wiley-Liss, Inc.

N-SUBSTITUTED GLYCINE OLIGOMERS

Oligomers of N-substituted glycines[6] (NSG peptoids) are conceptually and structurally related to peptides. They possess comparable spacing of the side chains and the backbone amide bonds (Figure 6.1) (Scheme 6.1). In contrast to these formal analogies, a number of other features clearly distinguish N-substituted glycines as a separate and entirely different class of molecules. NSG peptoids are nonnatural oligomers that are devoid of chirality in the backbone. Moreover, peptoids lack amide protons, which decreases their polarity and renders them stable to degradation by most common proteases.[7,8]

The conformational profile of an NSG peptoid monomer differs significantly from every proteinogenic amino acid, including glycine and proline.[4,9] Since the α-carbon is devoid of any substituent, rotation about the ϕ and ψ angles should be almost unhindered. Moreover, the difference in energy between cis and trans isomers of the amide bond is significantly reduced, allowing for access to both conformations at room temperature. Depending on the nature of the side chain, the substituent at the nitrogen will confer structural properties on peptoids that should limit their flexibility. Since the backbone of NSG peptoids is overall more flexible and less polar than the peptide backbone, a driving force for the preferred conformation of many peptoids in aqueous solution should be minimization of the hydrophobic surface ("hydrophobic collapse"[10]).

NSG peptoids are easily accessible using a modification of conventional solid-phase peptide synthesis. Peptoid oligomers were synthesized in good yields and high purity by condensation of N^{α}-Fmoc-protected, N-substituted glycine monomers.[6] However, this approach requires the synthesis of individual monomers prior to assembly of the oligomers. Monomers for NSG peptoid synthesis were obtained via alkylation of primary amines with haloacetic acids and acrylamides and by reductive amination of glyoxylic acid.[6]

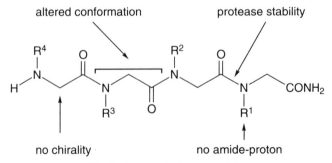

Figure 6.1. Structure of an *N*-substituted glycine peptoid tetramer (C-terminal amide) and some differences as compared to a peptide.

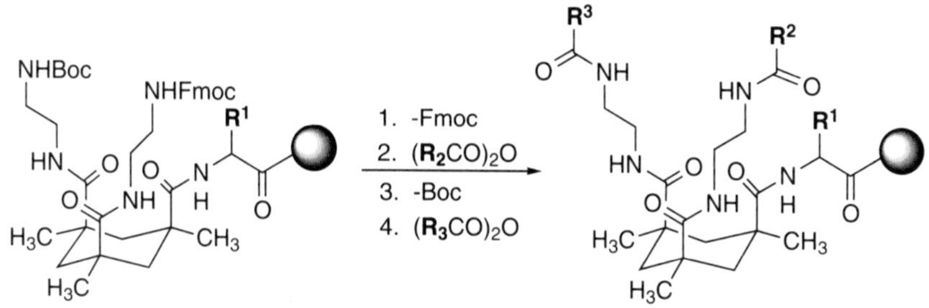

Scheme 6.1.

"Submonomer" Synthesis of N-Substituted Glycine Oligomers

In a straightforward and efficient process, NSG peptoids are obtained by submonomer solid-phase synthesis.[11] Using this method, the N-substituted glycine monomers are assembled on solid support from two submonomers: bromoacetic acid and a primary amine (Figure 6.2).

The submonomer route to NSG peptoids takes advantage of the spatial separation (pseudo-dilution[12]) of the substrate molecules on an insoluble matrix, thereby preventing bisalkylation of the primary amine. As a consequence, NSG peptoids are accessible in excellent purity without the need to synthesize N^α-Fmoc-protected, N-substituted glycine monomers.

A wide variety of commercially available, structurally diverse primary amines are suitable for submonomer peptoid synthesis. Resin and synthesis equipment used in peptide synthesis can be used directly in the submonomer chemistry. Thus, highly chemically diverse libraries are accessible without substantial cost or time investment. Reactive side-chain functionalities like

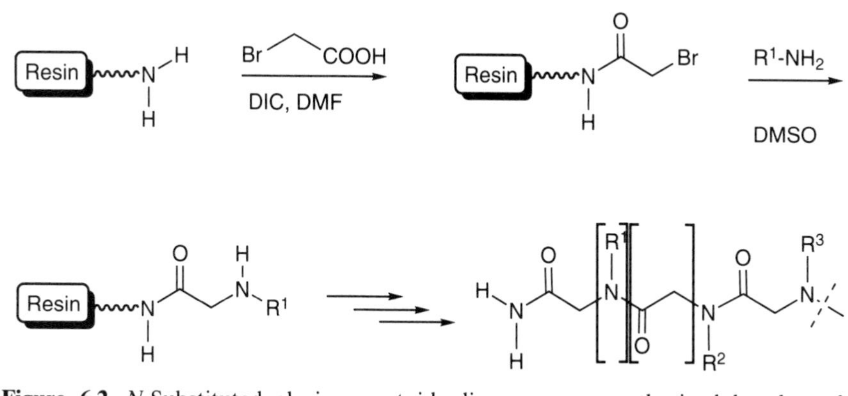

Figure 6.2. *N*-Substituted glycine peptoid oligomers are synthesized by the submonomer method, which consists of a very simple and efficient two-step monomer addition cycle. The method uses low-cost, easily obtained starting materials, and requires no main-chain protecting groups.

carboxyl, hydroxyl, amino, and thiol groups require protection. Bifunctional amines bearing reactive side chains (such as carboxyl and hydroxyl) can be used without protection when they are used to install the last (N-terminal) submonomer.

The efficiency of the submonomer method was demonstrated by the synthesis of a variety of homo- and heterooligomers in excellent purity and yield.[11] Since structurally and functionally diverse oligomers can be synthesized reproducibly and all reactions essentially go to completion,[13] this peptoid approach is ideally suited for the automated generation of molecular diversity.

NSG Peptoid Ligands for 7-Transmembrane G-Protein Coupled Receptors

Therapeutically relevant drugs that interact with 7-transmembrane G-protein coupled receptors (7TM) frequently contain similar pharmacophores. A peptoid trimer library was designed to bind to the 7TM receptor superfamily by incorporation of structural elements that repeatedly occur in small molecule agonists and antagonists. These structural features included hydrophobic, aromatic groups and hydrogen bond donors/acceptors like hydroxyl or phenol groups.[14] The complete library contained 4500 members and was screened for inhibition of binding of high-affinity radioligands to 7TM receptors.

After deconvolution of the most inhibitory pool for [3H]prazosin binding[15] to an α_1-adrenergic receptor preparation, the most active compounds were resynthesized. CHIR 2279 (Figure 6.4) was identified as a highly potent α_1-receptor antagonist (K_i = 5 nM).[14] Similarly, CHIR 4531 was shown to inhibit [3H]DAMGO (μ-specific) binding to opiate receptors[16] with a K_i = 6 nM. The newly discovered peptoid ligands show little structural homology with either the endogenous ligands or therapeutically used drugs (e.g., epinephrine/norepinephrine and prazosin for the α_1-adrenergic receptor, Met-enkephalin and morphine for opiate receptors) (Figure 6.3), and represent a new class of 7-transmembrane G-protein coupled receptor ligands.

SUBMONOMER APPROACHES FOR NON-NSG OLIGOMERS

N-Substituted Alanine Oligomers

Introduction of a substituent at the α-carbon of NSG peptoids provides chiral and sterically more constrained oligomers. After some modification of the acylation conditions, N-substituted alanine (NSA peptoid) oligomers are accessible from ($\pm$)-2-bromopropionic acid and primary amines, yielding mixtures of all possible stereoisomers (Figure 6.4).[17]

Enantiomerically enriched, N-substituted alanines are obtained in good optical yields (average enantiomeric ratio = 80:20) when enantiopure α-bromopropionic acid and an optimized combination of activating agent and base are used.[17] The introduction of an α-methyl group has profound effects on

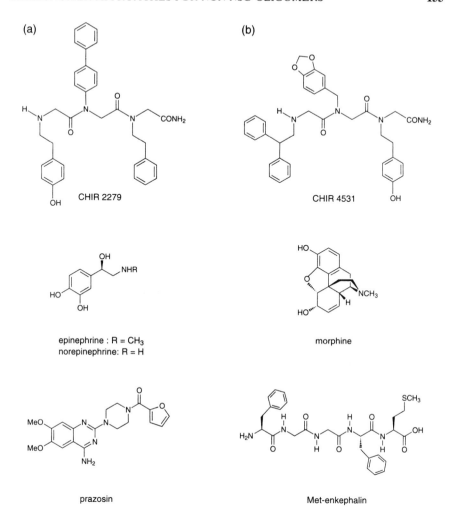

Figure 6.3. Peptoid trimers have been shown to be a new class of high-affinity 7-transmembrane G-protein coupled receptor ligands. (a) CHIR 2279 has a 5 nM affinity for the α_1-adrenergic receptor and is distinct from its natural ligand epinephrine or a commercially available antagonist prazosin. (b) CHIR 4531 has a 6 nM affinity for the (μ-specific) opiate receptor and is distinct from the known classes of ligands, morphine, and the enkephalins. Both ligands were discovered from a 4500-component library.

the conformational profile of the oligomers. Overall, NSA peptoids possess much less flexibility, and the energy minima are better defined.[9]

Polyamide Nucleic Acids by Submonomer Solid-Phase Synthesis

When a different type of oligomeric backbone is used for the side-chain display of nucleobases, "peptoid" nucleic acids are obtained from bromoacetic acid,

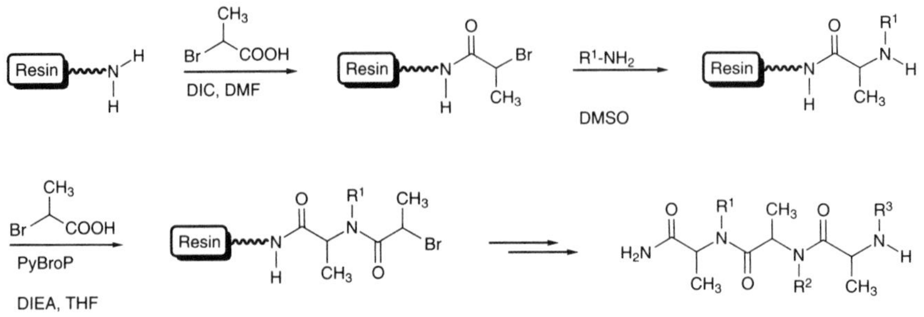

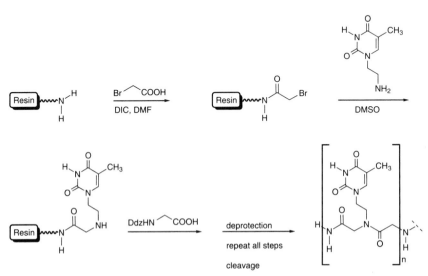

Figure 6.4. *N*-Substituted alanines can also be prepared by the submonomer method by using 2-bromopropionic acid instead of bromoacetic acid in the acylation step. These oligomers are more conformationally constrained in the backbone than *N*-substituted glycine peptoids, and are chiral.

an aminoethylated nucleobase, and Ddz-protected glycine (Figure 6.5).[18] Although these polyamides are closely related to peptide nucleic acids,[19,20] peptoid nucleic acids have significantly less binding affinity for either DNA or RNA molecules.[18]

Peptide nucleic acids (PNS) are also accessible by submonomer solid-phase synthesis, but some optimization of the synthetic conditions is required[21] (Fig-

Figure 6.5. Peptoid nucleic acids can be synthesized by the submonomer method using a 1-(2-aminoethyl)nucleobase building block.

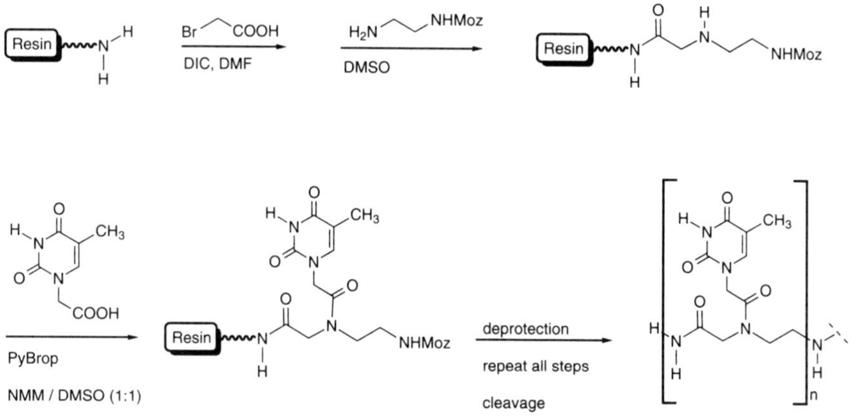

Figure 6.6. Peptide nucleic acids can be synthesized by the submonomer method using a 1-carboxymethyl nucleobase building block.

ure 6.6). The method allows for assembly of PNA molecules from inexpensive precursors and for structural variation of both the backbone and the nucleobase side chain. A PNA pentamer T_5 has been synthesized in >60% purity by this method.[21]

The backbones shown here (and substituted variations thereof) can be used for the display of other diverse, nonnucleotide-derived amine or carboxylic acid submonomers. All non-NSG peptoid oligomer libraries reviewed in this chapter can be assembled with essentially quantitative yields for the individual reaction steps. The automation developed for combinatorial library synthesis was successfully employed for high-throughput synthesis.

ORGANIC TEMPLATES VIA SUBMONOMER SOLID-PHASE SYNTHESIS

Encouraged by the successful synthesis and testing of diverse, nonnatural oligomer libraries, we became interested in the generation of molecular diversity using constrained organic templates. Although conformationally constrained libraries may sample less conformational space than oligomer libraries, active ligands from these libraries could be more easily incorporated into structure-based drug design strategies. The submonomer concept has been applied to the high-throughput, automated synthesis of organic heterocycles displaying diverse side chains.

Diketopiperazines and Diketomorpholines

The organic template that is most closely related to NSG peptoids is probably a N,N'-disubstituted diketopiperazine. Formally "cyclic dipeptoids," diketo-

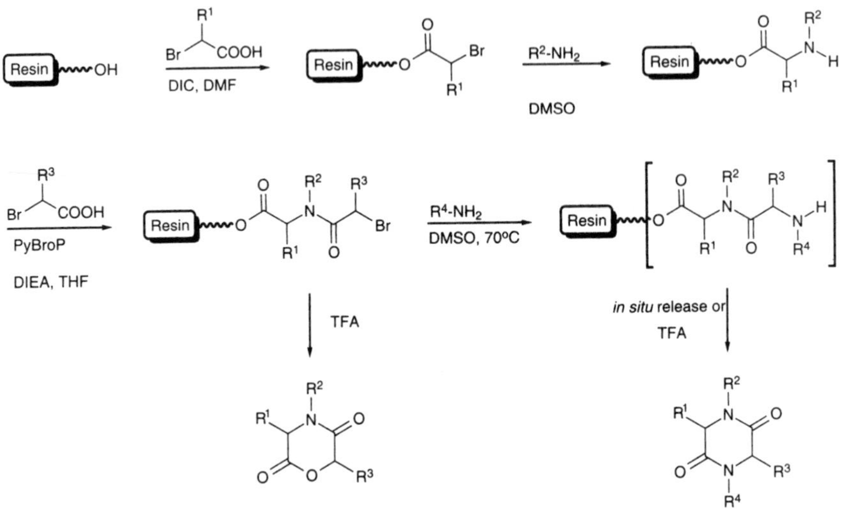

Figure 6.7. Diketopiperazines and diketomorpholines can be synthesized by the submonomer method using α-bromo acid and amine submonomers.

piperazines are easily accessible by submonomer solid-phase synthesis.[22] Similarly, defined mixtures of diketomorpholines can be synthesized (Figure 6.7) Highly substituted diketopiperazines and diketomorpholines bearing diverse substituents at the α-carbons are obtained by using α-substituted bromoacetic acids.

Libraries containing 22,000 diketopiperazines and 1000 diketomorpholines have been synthesized from readily available, diverse building blocks. Diketomorpholines and diketopiperazines are also formed from linear precursors like N-aryldipeptoid amides and the corresponding alcohols upon exposure to trifluoroacetic acid.[23] When the N-terminal aromatic amine is replaced by an aliphatic amine, trifluoroacetic acid-induced ring closure occurs generally slower and in a sequence-dependent manner.

1(2H)-Isoquinolinones

The activated double bond of vinylogous peptoids has been used for intramolecular Heck reactions.[24] Diverse (1(2H)-isoquinolinones are obtained in excellent purity and yield using a palladium-induced arylation of a crotonamide for ring closure (Figure 6.8). Prior to library synthesis, various o-iodo- or o-bromoarylcarboxylic acids have been synthesized from anthranilic acids and heteroarylcarboxylic acids. The nitrogen atom in the heterocycle stems from the amino group of a primary amine and is used for the display of diverse side chains.

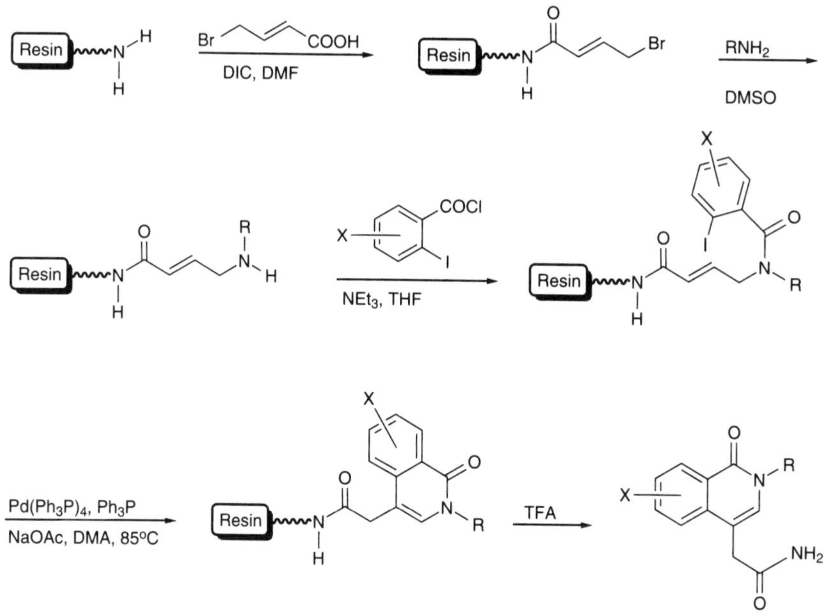

Figure 6.8. Peptoid 1(2*H*)-isoquinolinones can be prepared by an intramolecular Heck reaction from a terminal iodobenzoyl group to backbone double bond.

1,4-Benzodiazepine-2,5-diones

Cyclization at the N-terminus of NSG peptoids has been accomplished using the Staudinger (aza-Wittig) reaction.[25] When an α-amino acid ester is used for the final displacement step, the resulting secondary amine can be aroylated using α-azidobenzoyl chlorides. Diverse 1,4-benzodiazepine-2,5-diones are obtained when the resulting α-azidobenzamide is reacted with tributylphosphine in toluene, followed by heating (Figure 6.9). Prior to library construction, a set of various 2-azidoaroylchlorides was synthesized individually from anthranilic acids. Again, the substituent at the side-chain nitrogen atom originates from readily available primary amines, allowing for the display of highly dissimilar side chains.

AUTOMATED SYNTHESIS OF COMBINATORIAL LIBRARIES

We have modified a Zymate XP robot to perform automated syntheses of diverse combinatorial libraries.[26] The workstation allows for high-throughput synthesis with precise control over the composition of the mixtures. All synthetic manipulations like delivery of solvents and reagents as well as distribution and recombining of the resin particles ("resin splitting") are performed by the robotic apparatus.

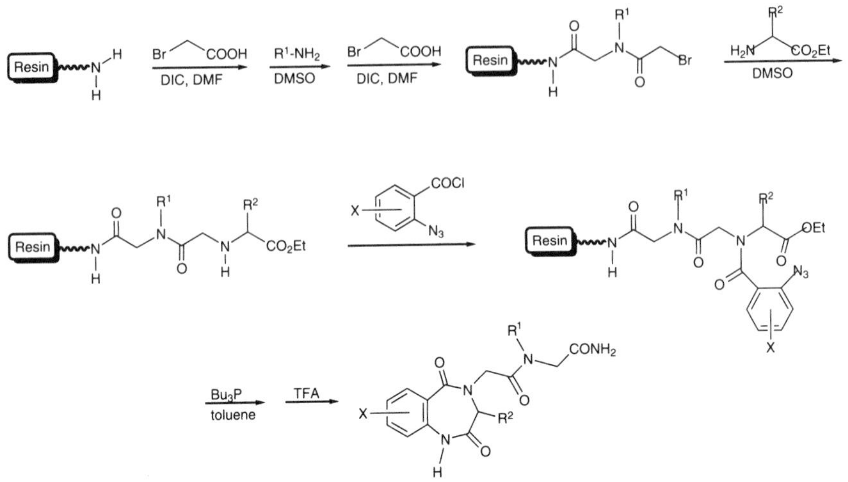

Figure 6.9. Peptoid 1,4-benzodiazepine-2,5-diones can be prepared via an aza-Wittig reaction between a terminal benzoylazide and a sidechain carboxylate ester.

The chemical reactions are carried out in fritted, cylindrical glass reaction vessels that are equipped with an aluminum heating block. Syntheses are performed on polystyrene beads with acid-labile linkers. The robotic arm adds reagents and solvents through an opening in the top of the reaction vessels. Bubbling with argon (to mix the bead slurry) and solvent removal are accomplished by applying argon pressure or vacuum to the bottom of the reaction vessels.

After completion of all synthetic manipulations, the equimolar, combinatorial libraries are cleaved from the solid support by exposure to trifluoroacetic acid. Using an automated cleavage station of our own design,[27] library throughput is greatly increased.

RATIONAL APPROACHES FOR MAXIMIZING STRUCTURAL AND FUNCTIONAL DIVERSITY

More than 13,000 chemically distinct primary amines are listed in the Available Chemicals Database.[28] Even for automated synthesis of short oligomers like peptoid trimers, the number of possible combinations is too great to handle. We have developed computational strategies and methods for the rational selection of subsets containing dissimilar monomers. By using building blocks that are maximally diverse, structural and functional redundancy of combinatorial libraries can be suppressed, increasing efficiency and decreasing deconvolution problems.

To measure similarity and dissimilarity between monomeric building blocks, we compute a variety of structural and functional properties, such as shape, branching, receptor-recognition properties, and lipophilicity.[29] Maximally diverse subsets are identified with D-optimal design.[30] The method allows for the intuitive selection of an initial subgroup of building blocks. Subsequently, the algorithm adds the desired number of additional compounds, ensuring that the combined set of building blocks is maximally diverse.[29,31] Although not every pharmaceutically relevant feature will be taken into account, this strategy minimizes redundancy and should enhance the probability to discover novel, bioactive molecules.

SUMMARY AND DISCUSSION

Using a multidisciplinary approach that combines organic synthesis, computational chemistry, and robotic automation, we have developed methods and tools for the generation of molecular diversity. Our initial focus was the establishment of a reproducible method for the automated, high-throughput synthesis of novel chemical entities from inexpensive, readily available starting materials. Diverse random libraries and nonredundant, biased libraries of N-substituted glycine oligomers are easily accessible by submonomer solid-phase synthesis. NSG peptoid trimers have provided some of the first examples of structurally novel, low-molecular-weight ligands for pharmaceutically relevant targets that were discovered using combinatorial chemistry techniques.

The submonomer approaches developed for the synthesis of diverse, nonnatural oligomer libraries were successfully applied to the synthesis of conformationally constrained heterocycles. As described above, computational methods were used for maximizing diversity of the substituents. Many of the templates used for library synthesis are found in commercially available pharmaceuticals or drug leads.

The organic template libraries reviewed here are currently being evaluated in several biological assays. As the initial successes with nonnatural oligomer libraries are repeated with constrained organic templates, combinatorial libraries of small organic molecules will play an ever increasing role in the discovery process of novel lead molecules for drug development.

ACKNOWLEDGMENT

The authors thank W. Moos, G. Stauber, D. Spellmeyer, J. Blaney, and our other colleagues in the Drug Discovery Research team at Chiron.

REFERENCES

1. Gordon EM, Barrett RW, Dower WJ, Fodor SPA, Gallop MA (1994): Applications of combinatorial technologies to drug discovery, 2: combinatorial organic synthesis, library screening strategies, and future directions. *J Med Chem* 37:1385–1401.

2. Gallop MA, Barrett RW, Dower WJ, Fodor SPA, Gordon EM (1994): Applications of combinatorial technologies to drug discovery, 1: background and peptide combinatorial libraries. *J Med Chem* 37:1233–1251.

3. Desai MC, Zuckermann RN, Moos WH (1994): Recent advances in the generation of chemical diversity libraries. *Drug Dev Res* 33:174–188.

4. Richter LS, Spellmeyer DC, Martin EJ, Figliozzi GM, Zuckermann RN (1995): Automated synthesis of nonnatural oligomer libraries: the peptoid concept. In Jung G, ed. *Peptide and Nonpeptide Libraries: A Handbook for Organic, Medicinal, and Biochemists.* Weinheim: Verlag Chemie.

5. Zuckermann RN (1993): The chemical synthesis of peptidomimetic libraries. *Curr Op Struct Biol* 3:580–584.

6. Simon RJ, Kania RS, Zuckermann RN, Huebner VD, Jewell DA, Banville S, Ng S, Wang L, Rosenberg S, Marlowe CK, Spellmeyer DC, Tan R, Frankel AD, Santi DV, Cohen FE, Bartlett PA (1992): Peptoids: a modular approach to drug discovery. *Proc Natl Acad Sci USA* 89:9367–9371.

7. Miller SM, Simon RJ, Ng S, Zuckermann RN, Kerr JM, Moos WH (1994): Proteolytic studies of homologous peptide and N-substituted glycine peptoid oligomers. *Bioorg Med Chem Lett* 4:2657–2662.

8. Miller SM, Simon RJ, Ng S, Zuckermann RN, Kerr JM, Moos WH (1995): Comparison of the proteolytic susceptibilities of homologous L-amino acid, D-amino acid and N-substituted glycine peptide and peptoid oligomers. *Drug Dev Res* 35:20–32.

9. Spellmeyer DC, unpublished results.

10. Wiley RA, Rich DH (1993): Peptidomimetics derived from natural products. *Med Res Rev* 13:327–384.

11. Zuckermann RN, Kerr JM, Kent SBH, Moos WH (1992): Efficient method for the preparation of peptoids[oligo(N-substituted glycines)] by submonomer solid-phase synthesis. *J Am Chem Soc* 114:10646–10647.

12. Kates SA, Sole NA, Johnson CR, Hudson D, Barany G, Albericio F (1993): A novel, convenient, three-dimensional orthogonal strategy for solid-phase synthesis of cyclic peptides. *Tetrahedron Lett* 34:1549–1552.

13. Figliozzi GN, Goldsmith R, Ng SC, Banville SC, Zuckermann RN (1996): Synthesis of N-(substituted)glycine peptoid libraries. *Methods Enzymol,* 267:437–447.

14. Zuckermann RN, Martin EJ, Spellmeyer DC, Stauber GB, Shoemaker KR, Kerr JM, Figliozzi GM, Goff DA, Sianti MA, Simon RJ, Banville SC, Brown EG, Wang L, Richter LS, Moos WH (1994): Discovery of nanomolar ligands for 7-transmembrane G-protein coupled receptors from a diverse (N-substituted)glycine library. *J Med Chem* 37:2678–2685.

15. Timmermans, PBMWM, Ali FK, Kwa HY, Schoop AMC, Slothorst–Grisdijk FP, Zwieten PAv (1981): Identical antagonist selectivity of central and peripheral alpha-1-adrenoceptors. *Mol Pharmacol* 20:295–301.

16. Gillan MGC, Kosterlitz HW (1982): Spectrum of the mu, delta, and kappa binding sites in homogenates of rat brain. *Br J Pharm* 77:461–469.

17. Richter LS, Zuckermann RN (1997): Enantioselective solid-phase synthesis of N-alkyl- and N-aryl-substituted amino acids. *Tetrahedron Lett,* in press.

18. Almarsson O, Bruice TC, Kerr J, Zuckermann RN (1993): Molecular mechanics calculations of the structures of polyamide nucleic acid DNA duplexes and triple-helix hybrids. *Proc Natl Acad Sci USA* 90:7518–7522.

19. Egholm M, Buchardt O, Nielsen PE, Berg RH (1992): Peptide nucleic acids (PNA): oligonucleotide analogues with an achiral peptide backbone. *J Am Chem Soc* 114:1895–1897.

20. Egholm M, Nielsen PE, Buchardt O, Berg RH (1992): Recognition of guanine and adenine in DNA by cytosine and thymine containing peptide nucleic acids (PNA). *J Am Chem Soc* 114:9677–9678.

21. Richter LS, Zuckermann RN (1995): Synthesis of peptide nucleic acids (PNA) by submonomer solid-phase synthesis. *Bioorg Med Chem Lett* 5:1159–1162.

22. Scott BO, Siegmund AC, Marlowe CK, Pei Y, Spear KL (1995): Solid-phase organic synthesis (SPOS): a novel route to diketopiperazines and diketomorpholines. *Mol Diversity*, 1:125–134.

23. Richter LS, Zuckermann RN, unpublished results.

24. Goff DA, Zuckermann RN (1995): Solid-phase synthesis of highly substituted peptoid 1(2*H*)-isoquinolinones. *J Org Chem* 60:5748–5749.

25. Goff DA, Zuckermann RN (1995): Solid-phase synthesis of defined 1,4-benzodiaze-pine-2,5-dione mixtures. *J Org Chem* 60:5744–5745.

26. Zuckermann RN, Kerr JM, Siani MA, Banville SC (1992): Design, construction, and application of a fully automated equimolar peptide mixture synthesizer. *Int J Pept Protein Res* 40:497–506.

27. Zuckerman RN, Banville SC (1992): Automated peptide-resin deprotection/cleavage by a robotic workstation. *Pept Res* 5:169–174.

28. *Available Chemicals Directory* (1993). San Leandro, CA: MDL.

29. Martin EJ, Blaney JM, Siani MA, Spellmeyer DC, Wong AK, Moos WH (1995): Measuring diversity: experimental design of combinatorial libraries for drug discovery. *J Med Chem* 38:1431–1436.

30. Federov (VV (1972): *Theory of Optimal Experiments.* New York: Academic.

31. Simon RJ, Martin EJ, Miller SM, Zuckermann RN, Blaney JM, Moos WH (1994): Using peptoid libraries [oligo-N-substituted glycines] for drug discovery. *Tech Protein Chem* 5:533–539.

7

OLIGONUCLEOTIDE LIBRARIES AS A SOURCE OF MOLECULAR DIVERSITY

HOUNG-YAU MEI

Parke–Davis Pharmaceutical Research, Division of Warner–Lambert Company, Ann Arbor, Michigan

ANTHONY W. CZARNIK

IRORI
La Jolla, California

While the advantage of oligonucleotide diversity for the storage of genetic information was instantly apparent, the same advantage applied to the discovery of selective receptors was not obvious. Biological function requires that nucleic acids provide for recognition of cognate proteins, for which the docking of extended surfaces is anticipated to yield a zipper-like interaction. Until quite recently, biological function did not seem to require similar recognition of small organic ligands. Presumably, such complexes would require preorganization of the nucleic acid to form a binding pocket. The discovery of ribozymes that bind guanosine or guanosine triphosphates prior to self-splicing has erased this misconception. In retrospect, it should have been apparent that the self-folding of RNA into structures less symmetrical than the double helix would engender more nearly concave binding sites than does double-stranded DNA.

In this chapter, we provide an overview of the methods reported to date for the preparation of oligonucleotide libraries. Because the sequencing of oligonucleotides can be accomplished on a vanishingly small amount of mate-

Combinatorial Chemistry and Molecular Diversity in Drug Discovery, Edited by
Eric M. Gordon and James F. Kerwin, Jr.
ISBN 0-471-15518-7   Copyright © 1998 by Wiley-Liss, Inc.

rial, oligonucleotide libraries allow the preparation of enormous numbers of compounds (estimated up to 10^{18}), each in very small amounts. The first part of this chapter discusses such libraries and the methods used to select receptors. The second part of the chapter summarizes our group's work on the parallel synthesis of cyclic oligonucleotides using solid-phase synthesis methods.

LARGE LIBRARIES CONTAINING SMALL AMOUNTS OF EACH COMPONENT

Darwinian evolution, as embodied in the phrase "survival of the fittest," explains the biological selection process embraced by nature. Variants of parent DNA molecules were an infrequent side-product of reproduction. Some of these variants were corrected through self-editing systems, while others escaped; a few even led to viable organisms and survived in the form of mutant progeny. In most cases such mistakes occurred randomly, but the "fittest" was usually selected due to pressure imposed by the environment. The outcome of this selection process was passed along to the progeny until another round of selection occurred. The natural-selection process usually extends far longer than the average lifetime of a scientist. In theory, it is possible to expedite this evolutionary process, at least in some instances, by establishing a simplified reproduction system in the test tube. The prospect of performing evolution in a test tube is exciting, because such experiments might lead to (1) the discovery of something that is not otherwise observable due to the low probability of a particular event, and/or (2) the creation of a completely novel molecule or system within an artificial environment. This chapter surveys the unique opportunities to test this theory in the world of oligonucleotides.

The reduction-to-practice of in vitro evolution was enabled by the development of two technologies: solid-phase oligonucleotide synthesis[1] and the polymerase chain reaction.[2] By combining the power of these techniques, a population of oligonucleotides with random sequences at selected (or all) positions can be generated. This pool of combinatorially synthesized DNA or RNA molecules can then be screened for biological activity. As shown in Figure 7.1, a twin cycle of selection and amplification is repeated until the most active molecule(s) in the pool are discovered. The strength of this combinatorial approach to oligonucleotide libraries relies on a juxtaposition of facts: (1) An oligonucleotide library can be synthesized with high yield and accuracy; (2) only small quantities of each component need to be present to accomplish biological screening; (3) high degeneracy (as many as 10^{18} molecules in a micromole of oligonucleotide library) in the population is obtained from combinatorial synthesis; (4) the selected sequence can be amplified with high fidelity which is impossible with any other combinatorial approach, and thus can be identified without any tagging or deconvolution procedures; (5) the secondary or even tertiary structure of the selected sequence can be predicted with high confidence using experimentally derived thermodynamic parameters; and (6) iterative synthesis and selection can be achieved under *in vitro*

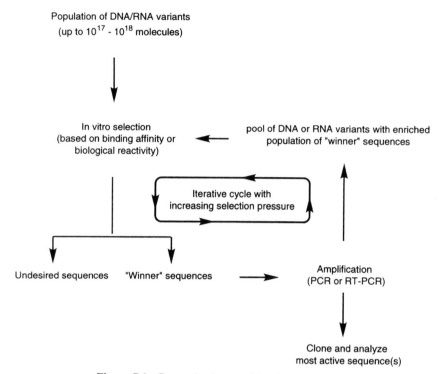

Figure 7.1. General scheme of in vitro selection.

or sometimes even *in vivo* conditions. *In vitro* selection using oligonucleotide libraries has been applied broadly to understanding the evolutionary process, inventing novel catalysis for biochemical reactions, and discovering therapeutic drug candidates.

In biological systems, enzymes such as Qβ replicase serendipitously catalyze the synthesis of a reservoir of RNA variants from a single RNA template.[3] By comparison, the *in vitro* chemical synthesis of oligonucleotide libraries offers the opportunity to make variants of oligonucleotides, either DNA or RNA, by design. The *in vitro* selection procedure begins by synthesizing a population of oligonucleotide analogs. A library of oligonucleotides can be generated by randomizing sequences at selected or at all positions. For example, when the conserved sequence in an active site of ribozyme is determined to be essential for its activity, the nonconserved domains can be mutated to select for a new generation of ribozyme with improved activity. The synthesis of an oligonucleotide library is then followed by iterative selection and amplification until a winning sequence is determined. This is analogous to the generation of antibodies launched by our immune system. When the immune system detects a foreign object (antigen), an enormous population of antibodies (ca. 10^{10} in the mouse) is generated. For each antibody in the population, a specific interaction occurs between the antigen and the hypervariable complementarity determining regions (CDR) in the Fab region. Similarly, high-affinity, oligonucleotides can also be selected from a completely randomized pool of sequences to target one specific ligand without any information of the ligand's structure. Specific sequences of oligonucleotide will be selected solely by the interaction between the oligonucleotide and the target ligand. In most reported examples, the selected winner sequence usually adopts a stable, well-defined structure that survives the iterative selection process. This conformation is in the absence of any chaperon molecules, which are usually required for protein folding in the cellular environment. Given the substantial chemical diversity in oligonucleotide libraries, the probability of finding a selective oligonucleotide receptor from a pool of billions is considerable.

In 1990, three research groups independently reported the invention and application of a method that is now known either as *in vitro* selection (evolution) or SELEX (systematic evolution of ligands by exponential enrichment). *In vitro* selection of DNA/RNA molecules with specific biological function or binding ability provides a revolutionary tool for modern biotechnology. The selection method has since become a powerful laboratory method for basic research as well as for applications such as drug discovery.

Beaudry and Joyce reported the first successful demonstration of *in vitro* selection of an RNA enzyme (ribozyme) that catalyzes sequence-specific cleavage of a DNA substrate.[4] Under physiological conditions, wild-type *Tetrahymena* group I intron RNA catalyzes a sequence-specific cleavage reaction on an RNA substrate but not on any DNA molecules. By creating a population estimated at 10^{13} RNA variants of the *Tetrahymena* ribozyme and repeating the selection process 10 times under increasing selection pressure, one RNA molecule was found that cleaves a DNA substrate with a 100-fold rate enhancement. In a later report, the catalytic efficiency was further improved up to 10^5-fold.[5] This was the first example of the directed evolution of an RNA enzyme. Although it is the same chemical reaction, the discovery of a ribozyme that works on an unnatural substrate (in this case, a DNA molecule) has stimulated a completely novel area of research. Numerous efforts have been made to engineer or discover catalytic oligonucleotides from iterative selection and amplification processes. In one case, scientists have found ribozymes in bacteria[6] and *Tetrahymena*[7] that catalyze the hydrolysis of the ester linkage. Others have been trying to discover ribozymes derived from a variety of sources that catalyze chemical reactions potentially of interest to organic chemists.[8] One of the most intriguing areas is to find a DNA or RNA molecule that will bind a transition-state analog and increase the rate of the corresponding reaction, analogous to that found in the catalytic antibody area. Research to discover catalysts from DNA or RNA libraries has only just begun. One important, yet unrealized goal is to identify catalysts with usefully large rate accelerations.

While oligonucleotide libraries found use in evolving new ribozymes, Tuerk and Gold reported a highly efficient way to select RNA ligands for bacteriophage T4 DNA polymerase.[9] The SELEX method was used to study the interaction between bacteriophage T4 DNA polymerase and the ribosome binding site of the mRNA that encodes for it. A pool estimated at 65,536 RNA sequences, containing a consensus five base-pair stem sequence from the bacteriophage mRNA and eight randomized bases in the loop region, underwent alternate cycles of selection under increasing pressure of binding affinity. Two different sequences with similar binding affinity survived after eight rounds of the selection cycle. One of the sequences selected was the wild-type sequence, while the other varied from the wild type at four positions at which two extended base pairs were found. The reported protocols can presumably be used to obtain high-affinity ligands for any protein that binds nucleic acids. SELEX offers an attractive approach for studying RNA–protein interactions that are usually too complex or time-prohibitive to address otherwise. It is also conceivable that the same method can be applied to the development of high-affinity nucleic acid ligands for other target molecules. Nucleic acid ligands that have high affinity for reverse transcriptase of human immunodeficiency virus,[10] blood clotting human thrombin,[11] autoantibodies,[12] and many other protein targets[13] have been reported.

Before 1990, there was little or no appreciation for the fact that some DNA or RNA molecules that form stable three-dimensional structures can serve as selective receptors for small molecules. Ellington and Szostak showed that some RNA molecules selected from an RNA library with completely random sequences bind specifically to a variety of organic dyes.[14] The planar aromatic rings and hydrogen-bonding substituents on these ligands provide interaction sites with the folded RNA molecules. It was estimated that every one molecule in 10^{10} random sequences, folds in a conformation that recognizes this class of small organic ligand. This was the first demonstration that oligonucleotides can recognize targets other than biological macromolecules. These individual RNA sequences with high affinities for a specific target were termed "aptamers" by the authors. DNA or RNA molecules that recognize amino acids such as L-arginine[15] and theophylline,[16] important biological cofactors such as adenosine triphosphate,[17] and many other small molecular targets have been selected from random pools of oligonucleotides. These findings further support the proposal of selecting nucleic acid molecules that bind transition-state analogs and catalyze their corresponding reactions. They also provide some plausible hypotheses for the roles of RNA molecules in the early stages of the evolution.

As discussed above, the combinatorial approach of oligonucleotide *in vitro* selection has made a significant impact on our basic understanding of what has been termed "the RNA world." Success in finding RNA molecules that possess biological activities beyond our traditional thinking has stimulated exciting hypotheses on the origin of life. Oligonucleotide libraries also portend potential utility in pharmaceutical research. The efficient iterative selection process ensures quick turnaround in looking for drug leads, although the issue of whether nucleotide-based drugs have therapeutic value is yet unresolved.

The examples given above suggest that oligonucleotide libraries offer con-

TABLE 7.1. Deconvolution Strategies of SURF

Round	Sequences	Different Sequences per Subset	Activity[a] when X = A	G	C	T
1	NNNXNNN	4096	+++			
2	NNNANXN	1024		+++		
3	NXNANGN	256			+++	
4	NCNAXGN	64				+++
5	NCXATGN	16		+++		
6	NCGATGX	4			+++	
7	XCGATGC	1	+++			

Note. Selected sequence: ACGATGC.

[a] For clarity sake, only the most active ones are indicated by +++.

siderable potential for studying some fundamental issues of evolution. However, oligonucleotides incorporating only the four natural nucleotides as building blocks are limited in their ability to display diverse chemical properties. Due to the chemical nature of the phosphate backbone, low bioavailability is expected for these molecules. Oligonucleotides of natural monomers have been found to be extremely unstable against nuclease degradation and are slowly taken up by the cell membrane. Only limited success has been had in converting biological macromolecules of any kind to small molecule mimetics. These issues must be addressed before oligonucleotides can be considered useful in pharmaceutical applications. A combinatorial approach known as synthetic unrandomization of randomized fragments (SURF) has been developed that facilitates the identification of stable oligonucleotide analogs potentially useful in therapeutic, diagnostic, or basic research applications.[18-20] SURF is an iterative synthetic and screening approach for identifying novel molecules from oligonucleotide libraries composed of either natural or unnatural subunits. Because the library of oligonucleotides is synthesized chemically, incorporation of modified nucleotide subunits offers the advantage of generating stabilized oligonucleotides at predetermined positions. The pool of oligonucleotide analogs can be used in either *in vitro* or cellular environments. In addition, the synthesis of medium to short randomized oligonucleotide libraries is possible since no additional primer sequences for enzymatic amplification is required.

An example of using SURF to select an optimized sequence from a pool of 7-mer oligonucleotides is illustrated in Table 7.1. Although only the four natural nucleotides—A, G, C, and T—were used in this example, a wide variety of analogs can be included as monomers. SURF deconvolution begins with the synthesis of a nonoverlapping set of mixtures by incorporating a unique monomer at a common position of each subset. The subsets are tested separately and the one with the greatest activity is identified. A second set of compound mixtures is prepared, with each subset containing the fixed monomer showing the greatest activity from the previous round. In addition, another position is fixed with each of the unique monomers to give another set of subsets. The complexity of the mixture is reduced and the process is repeated seven times until a unique molecule is identified. Although the deconvolution profile depends on the order of unrandomization, only limited effect was found on which molecule was finally selected.[18]

The SURF selection strategy has been employed successfully to screen libraries of oligonucleotides for several drug discovery programs. Phosphorothioate DNA sequences displaying submicromolar activities have been selected from cellular screens of oligonucleotide libraries and show anti-human herpes simplex virus and anti-human immunodeficiency virus activity in cell culture.[19,20] Another 9-mer oligonucleotide, selected from a library estimated to contain 65,536 2'-O-methyl RNA molecules, recognizes a hairpin-loop sequence from *Ha-ras* RNA with binding affinity of 0.01 μM.[19] Recently, a model system based on oligonucleotide hybridization has been used to simulate the ability of iterative SURF to identify the most active molecule in the mixture.[17] The results suggested that SURF could generally find the best molecules (or one with activity very close to the best). It was also suggested that a proper analysis of the first few rounds of the winning sets will allow the prediction of the ultimate outcome of the iterative selection. In the same report, it was further noted that SURF analysis could be applied more generally to other synthetic combinatorial libraries as well.

SMALL LIBRARIES CONTAINING LARGE AMOUNTS OF EACH COMPONENT

In our group at Parke–Davis, we have taken a different synthetic approach to the generation of oligonucleotide libraries, in particular, focused libraries of cyclic oligonucleotides incorporating natural and/or unnatural monomers. The biological functions of naturally occurring cyclic oligonucleotides are largely unexplored. When compared with their linear analogs, it is believed that these low molecular weight cyclic oligonucleotide molecules (Figure 7.2) may offer some unique structural and functional properties. For example, cyclic dinucleotides such as c-r(UpUp) and c-r(ApUp) have been found to inhibit the DNA-dependent RNA polymerase of *Escherichia coli* during the initiation phase of transcription.[21] Cyclic dinucleotide phosphorothioates have been found to inhibit the replication of human immunodeficiency virus-1 at both the protein and the RNA levels.[22] Endogenous cyclic ribodiguanylic acid, c-r(GpGp), has been reported to regulate the biological synthesis of cellulose in the gram-negative bacterium *Acetobacter xylinum.*[23] Activation of the nitrocellulose synthase is highly specific for c-r(GpGp). It is believed that c-r(GpGp) exerts its function by binding to membrane-bound synthase through intercalation and inducing a conformational change in the protein. This was supported by structural studies of a series of cyclic dinucleotides and

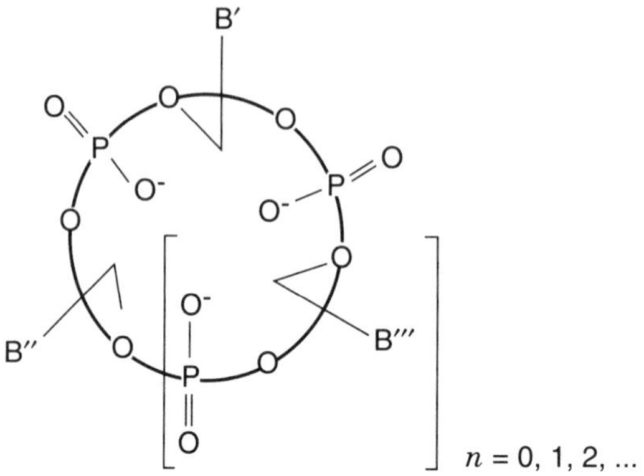

Figure 7.2. Cyclic oligonucleotides.

trinucleotides.[24–28] c-r(GpGp) forms a self-intercalated dimer in both solution and solid phase.[24] A model of high-order aggregates that consist of tetrameric c-r(GpGp) molecules was also proposed. Other cyclic dinucleotides such as c-r(ApAp) and c-(TpTp), which are not synthase regulators, however, have a lesser tendency to adopt this self-intercalative form.[25–27] This can be explained by the fact that adenine or thymine bases have less of a tendency to stack or to form hydrogen bonds among themselves. Interestingly, an analog much less active in regulating cellulose synthase, c-r(GpGpGp), also adopts a completely different conformation.[28] The NMR solution structure proposed for c-r(GpGpGp) indicates no signs of stacking interaction among the three guanine bases. These variations in structure and sequence may have a direct effect on the biological functions of these cyclic oligonucleotides.

In general, cyclic oligonucleotides are considered to be more stable against enzymatic degradation, specifically from exonuclease activity. It is also likely that, due to their cyclic nature, these compounds have more distinct conformations than their linear analogs. A library of cyclic oligonucleotides, therefore, should offer diversity in both chemical properties and shape. To further examine the biological functions of the cyclic oligonucleotides, a rapid synthetic method must be available for generating samples of these molecules. Toward this end, we have applied Parke–Davis' Diversomer approach[29,30] to the generation of cyclic oligonucleotide libraries. Diversomer technology combines solid-phase organic synthesis, automation, informatics, and custom equipment to facilitate the parallel synthesis of discrete compounds in significant (i.e., milligram) quantities. The approach that we designed and present here includes several features: (1) The library of compounds is presented in spatially addressable fashion; (2) sufficient quantities are prepared for biological testing and for archival purposes; (3) chemical modification of the bases of phosphate backbone are accessible, permitting the generation of stable compounds for direct testing in in vitro or cellular assays; (4) variations in conformation and in sequence coexist in a library of cyclic oligonucleotides; and (5) since

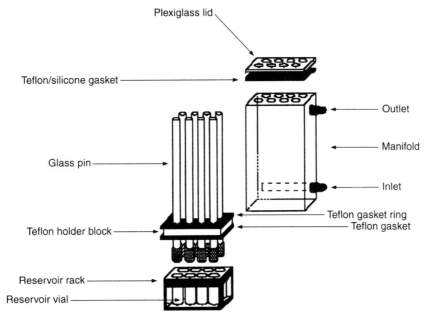

Figure 7.3. An 8-array version of the Diversomer apparatus.

unnecessary sequences are not present, relatively low molecular weight molecules can be synthesized and tested for application in drug discovery, drug
delivery, and molecular recognition. Success in the preparation of libraries of
the commercially available drugs Dilantin and Valium have been reported
using this technology.[29] The extension of the approach for the preparation of
a library of cyclic oligonucleotides is described here.

An 8-array version of the Diversomer apparatus is shown in Figure 7.3.
This apparatus consists of gas dispersion tubes (pins) as reaction vessels, a
holder block for parallel operation of 8 pins as one unit, a reservoir block
where arrays of vials are to receive the pins at any step of the reactions, and
a manifold that permits reactions to be run under inert atmosphere. For
reactions run at elevated temperatures, the upper portions of the pins serve
as reflux condensers when a chilled gas is circulated through the manifold.
Gaskets on either side of the holder block provide for sealing and a controlled
environment. A third gasket at the top of the manifold allows for the addition
of reagents or withdrawal of samples during the reaction cycles. In addition
to this unique apparatus, the filling of resin slurries in the pins and all liquid
sample handling is achieved using a Tecan Robotic Sample Processor (RSP).

As a demonstration, the preparation of a small library of cyclic dinucleotides
using the 8-array Diversomer apparatus was performed.[31] The synthetic strategy for the preparation of cyclic dinucleotides is illustrated in Scheme 7.1.
Solid-phase phosphotriester methods for oligonucleotide synthesis were

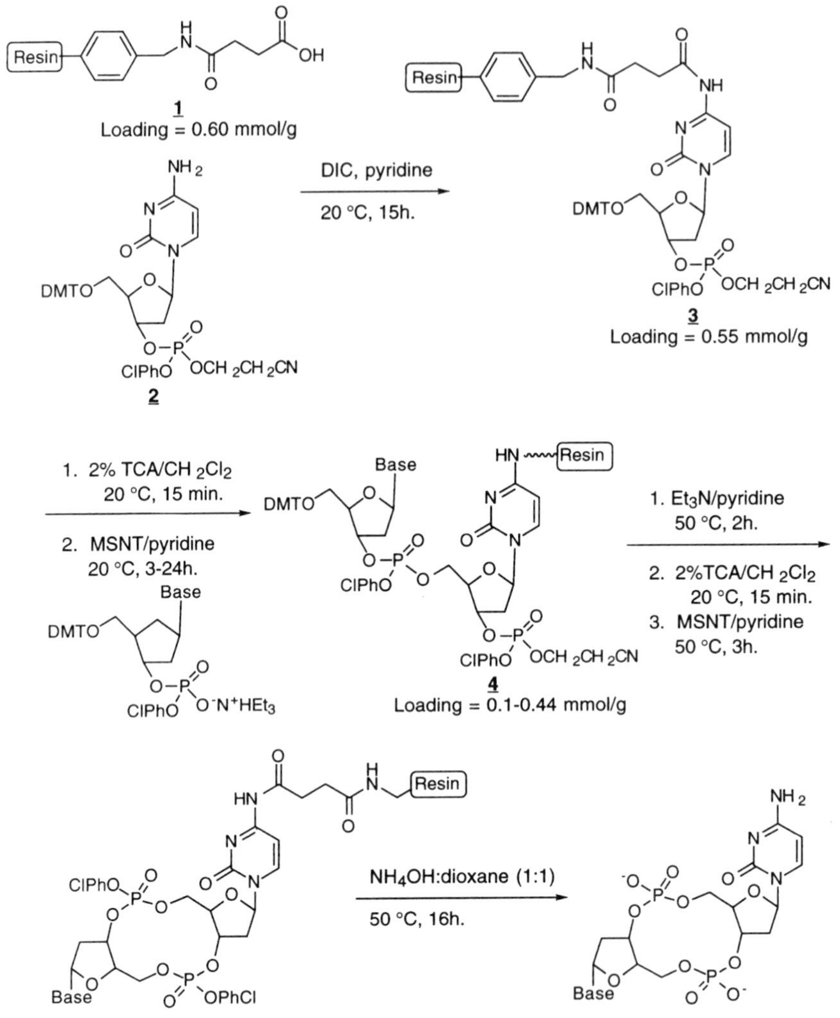

Scheme 7.1. Synthetic route of cyclic oligonucleotides.

adapted.[32] Commercially available aminomethyl polystyrene resin was selected as the solid support for the preparation of an 8-unit array of cyclic dinucleotides. The resin was first functionalized with a succinyl linker (1) to enable attachment of the first nucleotide through the exocyclic amino group of each base. For convenience, cytidine, **2,** was chosen as the first nucleotide to be attached to the solid support through the exocyclic amino group of the base to provide **3.** Following immobilization and deprotection of **3,** a set of eight linear dinucleotides (4) were constructed using both natural and modified nucleotides as the second building blocks. The 5'- and 3'-ends of the linear

TABLE 7.2. Synthesis Yields for Cyclic Dinucleotides Using Diversomer Technology

Sample[a] c(CpXp) X =	Theoretical Yield[b] (mg)	Crude Yield[c] (mg)	Purified Yield[d]	
			(mg)	(%)
5-Methyl cytosine	178	95.0	3	2
Uridine	175	52.2	8.5	5
Cytosine	175	47.3	7	4
Thymine	178	40.8	7.2	4
5-Iodocytosine	212	33.7	NA[e]	NA
Inosine	183	86.5	2	1
Adenine	183	44.5	2	1
Guanine	187	42.9	2	1

[a] Each reaction was performed with ≈500 mg of resin.
[b] Based on loading of aminomethyl polystyrene (0.6 mmol/g).
[c] Weight of products after purification.
[d] Yield calculated by ^{1}H NMR/internal standard.
[e] Due to impure starting material (5-iodocytidine triethyl ammonium salt from Glen Research) was used for the synthesis. A compound having similar spectral data to c(CpCp) was isolated.

dimers were then deprotected and cyclized to afford the penultimate resin-bound intermediates, **5.** Final cleavage of the cyclic compounds and concurrent removal of the chlorophenyl protecting group provided a set of eight cyclic dinucleotides of structure **6.**

A large-scale (retaining up to 800 mg of resin in each pin), 8-unit Diversomer apparatus was used for the parallel synthesis of eight cyclic dinucleotides. After synthesis and complete deprotection, the eight cyclic dinucleotides were further purified in parallel utilizing automation and solid-phase extraction (SPE) technology. SPE cartridges prepacked with C-18 silica were attached to a customized vacuum box at the Tecan RSP workstation. The Tecan RSP was programmed to dispense solvent to condition the cartridges, load the crude products onto cartridges, and then dispense the selected solvent to elute the product. After final purification, milligrams of the purified products were obtained over the eight-step synthesis starting with aminomethyl polystyrene. Yields of the eight cyclic oligonucleotides are listed in Table 7.2. The overall low yields for the eight products could be attributed to the inefficient cyclization step in the reaction sequence and/or lack of optimization of the SPE purification procedures.

The combination of robust phosphotriester chemistry, high loading of polystyrene-based solid supports, customized equipment amenable to a wide range of chemistries, semiautomated parallel synthesis, and purification methods enables the rapid generation of cyclic nucleotides. One may envision extension of these tools to the syntheses of other modified cyclic nucleotides, including trinucleotides and other cyclic polynucleotides.

SUMMARY

It is now clear that rather selective receptors for small organic compounds can be discovered within very large libraries of oligonucleotides. Recognition of theophylline versus caffeine serves as an illuminating example. It is fair to say that small molecule interactions with such receptors are not often very strong, and this will be an important issue to resolve if applications are to result. Both chemical and enzymatic methods of library generation are capable of yielding small amounts of large numbers of compounds. Smaller, focused libraries, such as those that might be included in a chemical archive for routine screening applications, are probably better made via parallel synthesis. The utility of any of these compounds in drug discovery is yet to be established, but the principles demonstrated via library synthesis and evolutionary screening have provided new ways to think about the search for molecules with desirable properties.

REFERENCES

1. For a review, see Gait MJ, ed (1990): *Oligonucleotide Synthesis: A Practical Approach.* New York: Oxford University Press.
2. For a review, see Erlich H (1989). *PCR Technology: Principles and Applications for DNA Amplification.* New York: Stockton.
3. Mills DR, Peterson RL, Spiegelman S (1967): An extracellular Darwinian experiment with a self duplicating nucleic acid molecule. *Proc Natl Acad Sci USA* 58:217–224.
4. Beaudry AA, Joyce GF (1990): Directed evolution of an RNA enzyme. *Science* 257:635–641.
5. Tsang J, Joyce GF (1994): Evolutionary optimization of the catalytic properties of a DNA-cleaving ribozyme. *Biochemistry* 33:5966–5973.
6. Noller HF, Hoffarth V, Zimniak L (1992): Unusual resistance of peptidyl transferase to protein extractions procedures. *Science* 256:1416–1419.
7. Piccirilli JA, McConnell TS, Zaug AJ, Noller HF, Cech TR (1992): Aminoacyl esterase activity of the tetrahymena ribozyme. *Science* 256:1420–1424.
8. Morris KN, Tarasow TM, Julin CM, Simons SL, Hilvert D, Gold L (1994): Enrichment for RNA molecules that bind a Diels-Alder transition state analog. *Proc Natl Acad Sci USA* 91:13028–13032.
9. Tuerk C, Gold L (1990): Systematic evolution of ligands by exponential enrichment: RNA ligands to bacteriophage T4 DNA polymerase. *Science* 249:505–510.
10. Tuerk C, MacDougal S, Gold L (1992): RNA pseudoknots that inhibit human immunodeficiency virus type 1 reverse transcriptase. *Proc Natl Acad Sci USA* 89:6988–6992.
11. Bock LC, Grissin LC, Latham JA, Vermaas EH, Toole JJ (1992): Selection of single-stranded DNA molecules that bind and inhibit human thrombin. *Nature* 355:564–566.

12. Kenan DJ, Tsai DE, Keene JD (1994): Exploring molecular diversity with combinatorial shape libraries. *Trends Biol Sci* 19:57–64.

13. For example, (a) Jellinek D, Green LS, Bell C, Janjic N (1994): Inhibition of receptor binding by high-affinity RNA ligands to vascular endothelial growth factor. *Biochemistry* 33:10450–10456; (b) Schneider D, Gold L, Platt T (1993): Selective enrichment of RNA species for tight binding to Escherichia coli rho factor *FASEB* 7:201–207.

14. Ellington AD, Szostak JW (1990): In vitro selection of RNA molecules that bind specific ligands. *Nature* 346:818–822.

15. Connell GJ, Illangesekare M, Yarus M (1993): Three small ribooligonucleotides with specific arginine sites. *Biochemistry* 32:5497–5502.

16. Gold L, Pieken W, Tasset D, Janjic N, Kershenheuter GP, Polisky B, Jayasena S, Biesecker G, Smith D, Jenison RD (1995): *PCT Int Appl CODEN: PIXXD2, WO 9,507,364 A1 950316.*

17. Sassanfar M, Szostak JW (1993): An RNA motif that binds ATP. *Nature* 364:550–553.

18. Freier SM, Konings DAM, Wyatt JR, Ecker DJ (1995): Deconvolution of combinatorial libraries for drug discovery: a model system. *J Med Chem* 38:344–352.

19. Ecker DJ, Vickers TA, Hanecak R, Driver V, Anderson K (1993): Rational screening of oligonucleotides combinatorial libraries for drug discovery. *Nucleic Acids Res* 21:1853–1856.

20. Wyatt JR, Vickers TA, Roberson JL, Buckheit RW Jr, Klimkat T, Debaets E, Davis PW, Rayner B, Imbach JL, Ecker DJ (1994): Combinatorially selected guanosine-quartet structure is a potent inhibitor of human immunodeficient virus envelope-mediated cell fusion. *Proc Natl Acad Sci USA* 91:1356–1360.

21. Hsu C-YJ, Dennis D (1984): RNA polymerase: linear competitive inhibition by bis-(3′ to 5′)-cyclic dinucleotides, NpNp. *Nucleic Acids Res* 10:5637–5647.

22. Battistini C, Fustinoni S, Brasca MG, Ungheri D (1993): *UK patent GB 2,257,704.*

23. Ross P, Weinhouse H, Aloni Y, Michaeli D, Weinberger–Ohana P, Mayer R, Braun S, de Vroom E, van der Marel GA, van Boom JH, Benziman M (1987): *Nature* 325:279–281.

24. Guan Y, Gao TG, Liaw YC, Robinson H, Wang AH-J (1993): Molecular-structure of cyclic diguanylic acid at 1 angstrom resolution of 2 crystal forms—self association, interactions with metal ion/planar dyes and modeling studies. *J Biomol Struct Dyn* 11:253–276.

25. Frederick CA, Coll M, van der Marel GA, van Boom JH, Wang AH-J (1988): Regulation of cellulose synthesis in *Acetobacter xylinum* by cyclic diguanylic acid. *Biochemistry* 27:8350–8361.

26. Blommers MJJ, Haasnoot CAG, Walters JALI, van der Marel GA, van Boom JH, Hilbers CW (1988): Solution structure of the 3′-5′ cyclic binucleotide d(pApA). A combined NMR, UV melting, and molecular mechanics study. *Biochemistry* 27:8361–8369.

27. Hamoir G, Sonveaux E (1993): 3′-5′ cyclic oligothymidylic acids—conformation and complexation of intercalating agents. *Bull Soc Chim Belg* 102:335–342.

28. Mooren MMW, Wijmenga SS, van der Marel GA, van Boom JH, Hilbers CW (1994): The solution structure of the circular trinucleotide CR(GPGPGP) deter-

mined by NMR and molecular mechanics calculation. *Nucleic Acids Res* 22:2658–2666.

29. Dewitt SH, Kiely JS, Stankovic CJ, Schroeder MC, Reynolds DM, Pavia MR (1993): "Diversomers": an approach to nonpeptide, nonoligomeric chemical diversity. *Proc Natl Acad Sci USA* 90:6909–6913.

30. Dewitt SH, Schroeder MC, Stankovic CJ, Strode JE, Czarnic AW (1994): Diversomer (TM) technology—solid-phase synthesis, automation, and integration for the generation of chemical diversity. *Drug Dev Res* 33:116–124.

31. Preliminary results have been presented: Ghosh S, Dewitt SH, Mei H-Y, Sanders KB, Czarnik AW (1994): Multiple synthesis of cyclic oligonucleotides using DIVERSOMER technology. *208th National Meeting of the American Chemical Society, Washington, DC, August 1994.*

32. Barbato S, Denapoli L, Mayol L, Piccialli G, Santacroce C (1989): A polymer-nucleotide linkage useful for the solid-phase synthesis of cyclic oligodeoxyribonucleotides. *Tetrahedron* 45:4523–4536.

PART II

SMALL MOLECULE LIBRARIES

8

SMALL MOLECULE LIBRARIES: OVERVIEW OF ISSUES AND STRATEGIES IN LIBRARY DESIGN

JOHN J. BALDWIN

Pharmacopeia, Inc., Princeton, New Jersey

The number of new therapeutically relevant targets provided by molecular biology has grown substantially over the past 10 years and is expected to increase further as proteins from the human genome project are defined. Since many, if not most of these targets do not lend themselves to rational de novo design, lead identification has come to rely on high-volume screening capabilities. Such a strategy makes even the most difficult targets, such as orphan receptors and novel subtypes of known receptors, and channels approachable. These high-capacity bioassays are ideally based on a discrete biochemical target using the human form of the macromolecule, whether an enzyme, receptor, or channel protein.

Once established, such assays require large numbers of diverse compounds. This need has been one of the major driving forces behind the evolution of combinatorial chemistry, a strategy that allows hundreds or thousands of compounds to be rapidly synthesized. Peptide combinatorial libraries served as the earliest examples and established a paradigm for discovery.[1-3] However, the recognition of the fundamental bioavailability problem associated with peptides and their intrinsic instability in animals shifted the focus from oligomeric libraries to small molecules. This change was reinforced by two publications that demonstrated that pharmacologically interesting small molecules such as benzodiazepines could be efficiently synthesized on solid support.[4,5]

Combinatorial Chemistry and Molecular Diversity in Drug Discovery, Edited by
Eric M. Gordon and James F. Kerwin, Jr.
ISBN 0-471-15518-7   Copyright © 1998 by Wiley-Liss, Inc.

FACTORS INFLUENCING LIBRARY DESIGN

The design and construction of small molecule libraries posed entirely different challenges from those of peptides. With peptides, the chemistry on solid support was fully developed and structure determination by microsequencing was available. The questions to be addressed with oligomeric libraries were primarily statistical ones, that is, establishing efficient screening methods to identify the most active compound in a large collection. The development of deconvolution and positional scanning evolved as successful strategies for the lead identification problem.[6-8]

In contrast, the design of a nonoligomeric collection is influenced by a wide range of issues, all of which affect the ultimate composition of the library. The principal factors that influence design are chemistry adaptable to the solid phase, the range of commercially available synthons, and screening strategies as they affect linking group selection. The overriding influence for producing a small molecule library is chemistry. This critical role is based on the observation that for library construction, solid-phase chemistry offers significant advantages over more classical solution chemistry; therefore, most libraries described to date use solid-phase synthesis. The need to execute library synthesis on solid support and thus the requirement to convert solution chemistries to the solid phase is the most significant challenge in the combinatorial exercise. As practiced, combinatorial chemistry encompasses not only the assembly of the building blocks selected for the library, but also the type of solid support and the linking strategy used for the attachment and ultimate release of the library member. Although a number of supports are theoretically available for small molecule construction, the resin composed of 1% divinylbenzene cross-linked polystyrene is evolving as one of the most preferred. This solid support, especially in the polyethylene glycol-grafted version, offers properties, including swelling characteristics across a range of solvents, that can be critical in optimizing nonamide chemistries from solution to the solid phase. The development of solid supports, especially those that are designed for the variable chemistry encountered in nonoligomer synthesis, is a rapidly evolving field and improvements in loading capacity and the pseudosolution characteristics are to be expected.

SOLID SUPPORT CHEMISTRY

Any synthetic method adapted to the solid phase must be reliable, predictable, and capable of being optimized for yield efficiency. Of the chemistries used to date, amine manipulation has been the most exploited and small molecule libraries of ureas, hydantoins, carbamates, and sufonamides have been described.[4,9-13] Alkylation chemistry has converted easily to the solid phase with both enolate and N-alkylation reactions successfully performed.[12,14-18] Functional group modifications have also been optimized for solid support, including oxidation and reductions,[19–22] reductive amination,[23–25] and Mitsunobu reaction[26–29] Aldol condensation,[30–32] Michael additions,[33] and the Horner Emmons reaction[20–42] serve to illustrate the solid-phase adaptation of carbon–carbon bond formation followed by heteroatom addition to the enone.

Organometallic reactions appear to adapt well to solid support, and several examples of the Heck, Stille, and Suzuki reactions have been reported.[36–39,41] Multiple examples of 1,3-dipolar addition on solid support have appeared[30–32,43,44] and progress is being made toward the solid-phase synthesis of carbohydrates.[45]

It is the chemical limitation, narrowed by any newly optimized transformations, that serves as the design boundary for library constructs. If specific reactions must be converted to the solid phase to accomplish a library design, then additional time should be allocated to complete the synthesis.

The design itself may evolve from a variety of approaches. It may be based simply on doable chemistry, since combinatorial chemistry is attempting to optimize serenedipity; that is, it may be viewed as an approach designed to discover a random event. Libraries based solely on chemistry and screened over a wide range of assays may be the ideal way to find entirely new bioactive small molecules. Alternatively, libraries may be built around a known pharmacophoric scaffold or a bioactive small molecule. For example, the utilization of the Horner–Emmons followed by Michael addition of thiols by Kurth serves as a library example whose design rested entirely on chemical methodology development.[33] Libraries built around the diketopiperazine,[15,25] and benzopyran[46] themes illustrate the use of the preferred scaffold concept.

LIBRARY SIZE

The size of the library is a strategic decision that depends on a number of factors, including the projected use, the number of readily available synthons or building blocks, and access to a robust reliable encoding method. When a library is intended for random screening, large size with maximum structural diversity among the members is highly desirable. For example, a library constructed in only four combinatorial steps by split synthesis[47] would yield a family of 160,000 members using only 20 synthons at each step. In split synthesis each particle of solid support or bead contains multiple copies of a single compound. Each bead is anonymous during synthesis and therefore the division of supports among reaction vessels is a random event. These anonymous beads are divided statistically according to a hypergeometric distribution pattern. Certain members will be over represented, others missing. To avoid the possibility of missing members, libraries are usually prepared in high redundancy such that the number of solid supports used is 100 to 300-fold more than the number of library members.

Since the number of individual compounds in a library is a multiple of the number of synthons at each step, very few synthetic steps with only a few synthons at each step can lead to large libraries. Therefore, a judicious selection of reactants is required. Various methods capable of multivariant analysis using, for example, structural parameters such as shape, lipophilicity, dipole moment, and molecular weight, are being increasingly employed in the design phase. Diversity is also reflected in the various conformations available to the members. Random libraries, directed toward lead discovery, benefit by conformational flexibility in the design where significant advantages may be gained from the greater number of allowed conformations maximizing the probability for interaction with an association site on the macromolecule. With more biased libraries, conformationally constrained designs may offer a preferred approach to potency and selectivity.

More directed or biased libraries are usually smaller because of the design constraints implied above. When a design is based on an active lead compound or on structural or mechanistic knowledge of the target, the point of ligand attachment to the solid support must be carefully selected. From model compounds made in solution the selected point of attachment and the resulting functionality must be such that it does not adversely affect bioactivity on release. In the design exercise the need to develop new chemistries should be minimized and, since it is likely that optimal building blocks will not be commercially available, intermediates that require synthesis should be carefully selected to cover as wide a diversity range as allowed by the strategy. Whether the library be random or biased, its size, in large measure, is dependent on the method chosen for synthesis.

LIBRARY CONSTRUCTION

Small libraries can be generated by parallel synthesis where individual compounds are prepared by automated or semiautomated methods. This approach provides multimilligram quantities for more detailed in vitro analysis. Large libraries are most efficiently prepared by the split synthesis technique. Although only relatively small libraries have been reported thus far, once the chemistry has been optimized there is no reason why libraries of 100,000 or more members cannot be produced. Such large libraries can stress the capacity of the most automated assays; therefore, they are usually tested as mixtures of detached members. Structure determination of an active mixture usually depends on deconvolution strategies[48] or relies on encoding that defines the reaction history of the bead from which the active compound was derived. Four encoding strategies have been described: peptide encoding strands,[49,50] oligonucleotide encoding,[51] binary encoding with electrophoric tags,[52] and radio-frequency tag encoding.[53]

The quantity of a compound prepared on a single bead by split synthesis or the one-compound–one-bead method is limited. Usually 200–300 pmol can be released from a single bead, which allows two tests to be performed at the 1-μM level. The first test can be run as a mixture in 96-well plate format, usually with 10–30 compounds per well. The second 100 pmol is released from each bead associated with the active well but now rearrayed in a one bead per well format. The reaction history and implied structure of any active rearrayed compound can be determined by the encoding method.

SUMMARY

Library design, as executed on solid support, must be based on chemistry optimized for the solid phase or on a commitment to develop the appropriate synthetic methods. Libraries designed for lead discovery should be large and based on either synthetically allowable constructs or on pharmacologically relevant scaffolds. The library should be synthesizable from commercially available synthons with their selection based on maximizing diversity. The selection should be such that toxic or metabolically labile substituents are eliminated and the average molecular weight of the library members is controlled to the 350- to 500-dalton range. Considerations in synthon selection should also reflect the aim of having the final library members possess an aqueous solubility and lipophilicity that would suggest oral availability.

Biased or directed libraries should be smaller. The hit rate in bias libraries will be higher and assay follow-up could become rate limiting. This suggests that synthons be carefully selected to provide the maximum amount of information. It is likely that the most appropriate intermediates may not be available and more time will be devoted to their preparation.

Although considerable amount of thought has been directed toward the optimization by deconvolution strategies, encoding methods or highly sensitive analytical techniques offer significant advantages and will likely replace deconvolution. Through such approaches, structural information is rapidly obtained, providing significant structure activity information even in the screening mode. It is this information derived from large numbers of compounds that should assist the medicinal chemist in the final design and thus speed the overall drug discovery process.[54]

REFERENCES

1. Houghten RA (1994): Combinatorial libraries: finding the needle in the haystack. *Curr Biol* 4:564–567.

2. Houghten RA (1993): The broad utility of soluble peptide libraries in drug discovery. *Gene* 137:7–11. (a) Houghten RA (1993): Peptide libraries: criteria and trends. *Trends Genet* 9:235–239. (b)

3. Gordon EM, Barrett RW, Dower WJ, Fodor SPA, Gallop M (1994): Applications of combinatorial technologies to drug discovery, 1: background and peptide combinatorial libraries. *J Med Chem* 37:1233.

4. DeWitt SH, Kiely JS, Stankovic CJ, Schroeder MC, Reynolds Cody DM, Pavia MR (1993): "Diversomers": An approach to nonpeptide, nonoligomeric chemical diversity. *Proc Natl Acad Sci USA* 90:6909–6913.

5. Bunin BA, Plunkett MJ, Ellman JA (1994). The combinatorial synthesis and chemical and biological evaluation of a 1,4-benzodiazepine library. *Proc Natl Acad Sci USA* 91:4708–4712.

6. Geysen HM, Mason T (1993): Screening chemically synthesized peptide libraries for biologically relevant molecules. *Bioorg Med Chem Lett* 3:397–404. (a) Geysen MH, Meloen RH, Barteling SJ (1984): Use of peptide synthesis to probe viral antigens for epitopes to a resolution of a single amino acid. *Proc Natl Acad Sci USA* 81:3998–4002. (b)

7. Blake J, Litzidavis L (1993): Evaluation of peptide libraries: an iterative strategy to analyze the reactivity of peptide mixtures with antibodies. *Bio-Conj Chem* 3:510–513.

8. Pinilla C, Appel JR, Houghton RA (1994): Investigation of antigen–antibody interactions using a soluble, non-support-bound synthetic decapeptide library composed of four trillion (4×10^{12}) sequences. *Biochem J* 301:847–853.

9. Baldwin JJ, Burbaum JJ, Henderson I, Ohlmeyer MHJ (1995): Synthesis of a small molecule combinatorial library encoded with molecular tags. *J Am Chem Soc* 117:5588.

10. Burgess K, Linthicum DS, Shin H (1995): Solid-phase synthesis of unnatural biopolymers containing repeating urea units. *Angew Chem Int Ed Engl* 34:907.

11. Hutchins SM, Chapman KT (1994): A general method for the solid phase synthesis of ureas. *Tetrahedron Lett* 35:4055–4058.

12. Kick EK, Ellman JA (1995): Expedient method for the solid-phase synthesis of aspartic acid protease inhibitors directed toward the generation of libraries. *J Med Chem* 38:1427–1430.

13. Cho CY, Moran EJ, Cherry SR, Stephans JC, Fodor SPA, Adams CL, Sundaram A, Jacobs JW, Schultz PG (1993): An unnatural biopolymer. *Science* 261:1303–1305.

14. Zuckermann RN, Kerr JM, Kent SBH, Moos WH (1992): Efficient method for the preparation of peptoids [oligo(N-substituted glycines)] by submonomer solid-phase synthesis. *J Am Chem Soc* 114:10646–10647.

15. Dankwardt SM, Newman SR, Krstenansky JL (1995): Solid phase synthesis of aryl and benzylpiperazines and their application in combinatorial chemistry. *Tetrahedron Lett* 36:4923–4926.

16. Bunin BA, Ellman JA (1992): A general and expedient method for the solid-phase synthesis of 1,4-benzodiazepine derivatives. *J Am Chem Soc* 114:10997–10998.

17. Backes BJ, Ellman JA (1994): Carbon–carbon bond-forming methods on solid support: utilization of Kenner's "safety-catch" linker. *J Am Chem Soc* 116:11171–11172.

18. Moon H-S, Schore NE, Kurth MJ (1994): A polymer-supported C_2-symmetric chiral auxiliary: preparation of non-racemic 3,5-disubstituted-γ-butyrolactones. *Tetrahedron Lett* 35:8915–8918.

19. Kurth MJ, Randall LAA, Chen C, Melander C, Miller RB, McAlistèr K, Reitz G, Kang R, Nakatsu T, Green C (1994): Library-based lead compound discovery: antioxidants by an analogous synthesis/deconvolution assay strategy. *J Org Chem* 59:5862.

20. Ellman JA, Virgilio AA (1994): *Int. Patent WO94/06291.*

21. Schlatter JM, Mazur RH (1977): Hydrogenation in solid-phase peptide synthesis, I: removal of product from the resin. *Tetrahedron Lett* 2851–2852.

22. Ojima I, Tsai C-Y, Zhang Z (1994): Catalytic asymmetric synthesis of peptides on polymer support. *Tetrahedron Lett* 35:5785–5788.

23. Green J (1995): Solid-phase synthesis of lavendustin A and analogues. *J Org Chem* 60:4287–4290.

24. Bray AM, Chiefari DS, Valerio RM, Maeji NJ (1995): Rapid optimization of organic reactions on solid phase using the multipin approach: synthesis of 4-aminoproline analogues by reductive amination. *Tetrahedron Lett* 36:5081–5084.

25. Gordon DW, Steele J (1995): Reductive alkylation on a solid phase: synthesis of a piperazinedione combinatorial library. *Biomed Chem Lett* 5:47–50.

26. Stanley M, Tom JYK, Burdick DJ, Struble M, Burnier JP (1991): Poster presentation at the *12th American Peptide Symposium Cambridge, MA.*

27. Richter LS, Gadek TR (1994): A surprising observation about Mitsunobu reactions in solid-phase synthesis. *Tetrahedron Lett* 35:4705–4706.

28. Rano TA, Chapman KT (1995): Solid-phase synthesis of aryl ethers via the Mitsunobu reaction. *Tetrahedron Lett* 36:3789–3792.

29. Campbell DA, Bermak JC (1994): Solid-phase synthesis of peptidylphosphonates. *J Am Chem Soc* 116:6039–6040.

30. Beebe X, Schore NE, Kurth MJ (1992): Polymer-supported synthesis of 2,5-disubstituted tetrahydrofurans. *J Am Chem Soc* 114:10061.

31. Beebe X, Chiappari CL, Olmstead MM, Kurth MJ, Schore NE (1995): Polymer-supported synthesis of cyclic ethers: electrophilic cyclization of tetrahydrofuroisoxazolines. *J Org Chem* 60:4204–4212.

32. Beebe X, Schore NE, Kurth MJ (1995): Polymer-supported synthesis of cyclic ethers: electrophilic cyclization of Isoxazolines. *J Org Chem* 60:4196–4204.

33. Chen C, Randall LAA, Miller RB, Jones AD, Kurth MJ (1994): "Analogous" organic synthesis of small-compound libraries: validation of combinatorial chemistry in small molecule synthesis. *J Am Chem Soc* 116:2661–2662.

34. Look GC, Murphy MM, Campbell DA, Gallop MA (1995): Trimethylorthoformate: a mild and effective dehydrating reagent for solution- and solid-phase imine formation. *Tetrahedron Lett* 36:2937–2940.

35. Look GC, Holmes CP, Chinn JP, Gallop MA (1994): Methods for combinatorial organic synthesis: the use of fast ^{13}C NMR analysis for gel phase reaction monitoring. *J Org Chem* 59:7588–7590.

36. Yu KL, Deshpande MS, Vyas DM (1994): Heck reactions in solid-phase synthesis. *Tetrahedron Lett* 35:8919–8922.

37. Hiroshige M, Hauske JR, Zhou P, (1995): Formation of CC bond in solid-phase synthesis using the Heck reaction. *Tetrahedron Lett* 36:4567–4570.

38. Deshpande MS (1994): Formation of carbon–carbon bond on solid support: Application of the Stille reaction. *Tetrahedron Lett* 35:5613–5614.

39. Plunkett MJ, Ellman JA (1995): Solid-phase synthesis of structurally diverse 1,4-benzodiazepine derivatives using the Stille coupling reaction. *J Am Chem Soc* 117:3306–3307.

40. Forman FW, Sucholeiki I (1995): Solid-phase synthesis of biaryls via the Stille reaction. *J Org Chem* 60:523–528.

41. Frenette R, Friesen RW (1994): Biaryl synthesis via Suzuki coupling on a solid support. *Tetrahedron Lett* 35:9177–9180.

42. Williard R, Jammalamadaka V, Zava D, Benz CC, Hunt CA, Kushner PJ, Scanlan TS (1995): Screening and characterization of estrogenic activity from a hydroxystilbene library. *Chem Biol* 2:45–51.

43. Smith PW, Lai JYQ, Whittington AR, Cox B, Houston JG, Stylli CH, Banks MN, Tiller PR (1994): Synthesis and biological evaluation of a library containing potentially 1600 amides/esters: a strategy for rapid compound generation and screening. *BioMed Chem Lett* 4:2821–2824.

44. Pei Y, Moos WH (1994): Postmodification of peptoid side chains: [3 + 2] cycloaddition of nitrile oxides with alkenes on the solid-phase. *Tetrahedron Lett* 35:5825–5828.

45. Danishefsky SJ, McClure KF, Randolph JT, Ruggeri RB (1993): A strategy for the solid-phase synthesis of oligosaccharides. *Science* 260:1307–1309.

46. Burbaum JJ, Ohlmeyer MHJ, Reader JC, Henderson I, Dillard LW, Li G, Randle TL, Sigal NN, Chelsky D, Baldwin JJ (1995): A paradigm for drug discovery employing encoded combinatorial libraries. *Proc Natl Acad Sci USA* 92:6027–6031.

47. Furka A, Sebestyen F, Asgedom M, Dibo G (1989): *14th International Congress of Biochemistry, Prague, Czechoslovakia, 10–15 July 1988*. Berlin: Gruyter, vol 5, p 47 (abstract); Furka A, Sebestyen F, Asgedom M, Dibo G (1989): *10th International Symposium on Medicinal Chemistry, Budapest, Hungary, 15–19 August 1988*. Amsterdam: Elsevier, p 288 (abstract).

48. Janda KD (1994): Tagged versus untagged libraries: methods for the generation and screening of combinatorial chemical libraries *Proc Natl Acad Sci USA* 91:10779–10785.

49. Kerr JM, Banville SC, Zuckermann RN (1993): Encoded combinatorial peptide libraries containing nonnatural amino acids. *J Am Chem Soc* 115:2529.

50. Nikolaiev V, Stierandova A, Krchnak V, Seligmann B, Lam KE, Salmon SE, Lebl M (1993): Peptide-encoding for structure determination of nonsequenceable polymers within libraries synthesized and tested on solid-phase supports. *Pept. Res* 6:161–170.

51. Brenner S, Lerner RA (1992): Encoded combinatorial chemistry. *Proc Natl Acad Sci USA* 89:5381–5383.

52. Ohlmeyer MHJ, Swanson RN, Dillard LW, Reader JC, Asouline G, Kobayashi R, Wigler M, Still WC (1993): Complex synthetic chemical libraries indexed with molecular tags. *Proc Natl Acad Sci USA* 90:10922–10926.

53. Morgan EJ, Sarshar S, Cargill JF, Shahbaz MM, Lio A, Mjalli AMM, Armstrong RW (1995): Radio frequency tag encoded combinatorial library method for the discovery of tripeptide-substituted cinnamic acid inhibitors of the protein tyrosine phosphatase PTPIB *J Am Chem Soc* 117:10787–10788.

54. Gordon EM, Barrett RW, Dower WJ, Fodor SPA, Gallop M (1994): Applications of combinatorial technologies to drug discovery, 2: combinatorial organic synthesis, library screening strategies, and future directions. *J Med Chem* 37:1385–1401.

9

PROTEASE INHIBITOR LIBRARIES

DINESH V. PATEL
Versicor Inc., Fremont, California

DAVID A. CAMPBELL
Affymax Research Institute, Santa Clara, California

Conceptual validation of combinatorial chemistry as an efficient technique for generating large numbers of compounds commenced with the synthesis of homooligomeric peptide molecules. While this achievement generated tremendous interest from the drug industry, the limited potential of peptides as orally bioavailable medicines could not be ignored. To overcome this practical limitation, organic and medicinal chemists have tried to develop solid-phase methodologies that can be applied for combinatorial syntheses of nonpeptidic, small, heterocyclic molecules.[1] Going a step further, researchers have begun utilizing mechanistic and structural information available on certain biological targets in the rational design of targeted, pharmacophore-based, libraries. Incorporation of a pharmacophoric group in combinatorial synthesis whose selection is based on the target of interest provides quick access to libraries well suited for lead discovery and optimization. To date, the discovery of such pharmacophoric units has been mainly restricted to enzyme targets. As an example, a clear understanding of the mechanism of hydrolysis of peptides has resulted in the discovery of numerous transition-state analog templates as the critical pharmacophoric unit in novel and potent protease inhibitors. In this chapter, we describe several specific examples of solid-phase synthesis of such protease inhibitor-based pharmacophores and their utility in the generation of protease inhibitor libraries (PILs).

Combinatorial Chemistry and Molecular Diversity in Drug Discovery, Edited by
Eric M. Gordon and James F. Kerwin, Jr.
ISBN 0-471-15518-7 Copyright © 1998 by Wiley-Liss, Inc.

PHOSPHONIC ACID TRANSITION-STATE ANALOG LIBRARIES

Phosphinyl acid derivatives are transition-state analogs and metal chelators that have found utility in the rational design of potent inhibitors of metalloproteases such as thermolysin and angiotensin converting enzyme (ACE).[2,3] Combinatorial libraries of molecules possessing this pharmacophore are expected to be an excellent new tool for discovery of metalloprotease inhibitors of various known and novel metalloproteases. A key requirement for generation of quality libraries is the development of mild reaction conditions that can be used reliably in a repetitive fashion with a diverse set of starting materials. Thus, central to the development of phosphonate pharmacophore-based libraries was identification of a mild phosphorous–oxygen (P-O) bond-forming reaction. Traditional methods such as reaction of alcohol with phosphonochloridate seemed inappropriate because of inconsistent yields and limited stability of chloridate intermediates. This prompted Campbell to investigate alternative methods, and led to the development of a Mitsunobu-based coupling reaction for general synthesis of phosphonates (Scheme 9.1).[4]

Successful bond formation involved treatment of a solution of monomethyl alkyl phosphonic acid, triphenylphosphine, and appropriate alcohol in THF

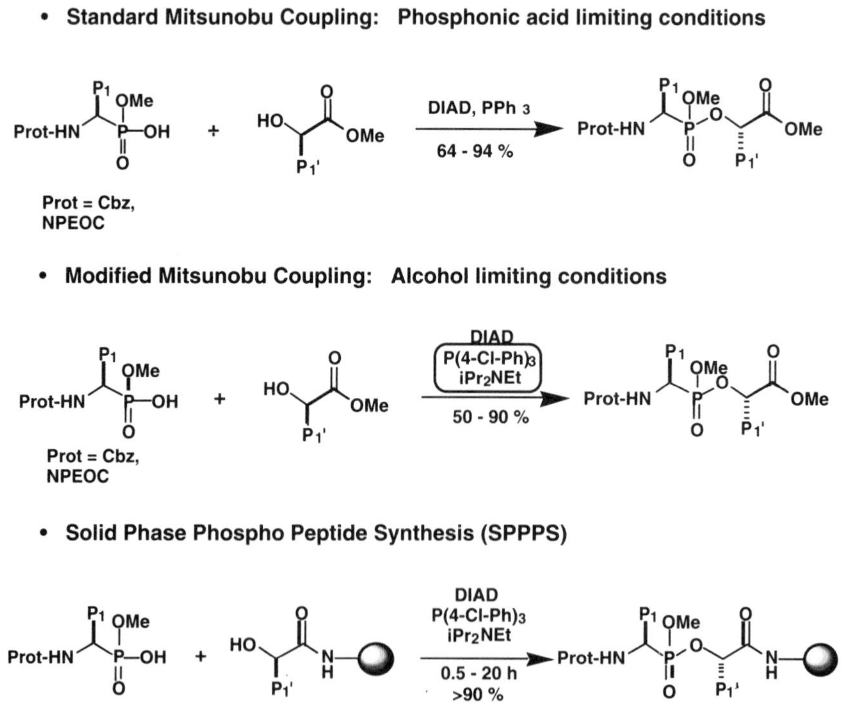

- **Standard Mitsunobu Coupling: Phosphonic acid limiting conditions**

- **Modified Mitsunobu Coupling: Alcohol limiting conditions**

- **Solid Phase Phospho Peptide Synthesis (SPPPS)**

Scheme 9.1. Developing and rehearsing solid-phase "P-O" coupling chemistry.

with diisopropyl azodicarboxylate (DIAD). The reaction is general, works well for α-aminophosphonic acids, and is relatively insensitive to steric bulk of phosphonic acids. To capitalize on the split-and-pool protocol, it became necessary to develop a solid-phase synthesis version of this method. This required further modification, since unlike in the solution phase, an alcohol on solid phase is the limiting reagent. Reaction proceeded between the alcohol on solid support, a phosphonic acid in solution, DIAD, tris(4-chlorophenyl)phosphine, which is a more electron-deficient phosphine than the typically used triphenylphosphine, and an exogenous base such as triethylamine as a general base catalyst. The procedure is versatile in scope and yields satisfactory results even with most hindered coupling partners.[5]

As proof of principle, we decided to synthesize a library of trimer peptidyl phosphonates and screen it for activity against the zinc metalloprotease thermolysin. Such compounds have been reported previously by Bartlett as potent inhibitors of thermolysin.[2] Retrosynthetic analysis (Scheme 9.2) reveals the necessity for several specific chemistries and building blocks (BBs) for successful execution of this strategy: (1) The first and foremost requirement of an efficient method for synthesizing phosphonic acids on solid support was to be fulfilled by the Mitsunobu conditions described above. (2) A collection of α-aminophosphonic acid building blocks (BBs) bearing an appropriate protecting group on nitrogen that will tolerate the reagents needed for P-O bond forming reaction was necessary for representing adequate diversity at the P_1 position. These were prepared following the method of oxidative decarboxylation of amino acids reported in the literature (Scheme 9.3). The more stable 4-nitrophenethyloxycarbonyl (NPEOC) was chosen as the amino protecting group since the betaine formed between phosphine and DIAD during Mitsunobu coupling was basic enough to cause β-elimination of the commonly used Fmoc group. (3) Fmoc-protected α-hydroxy carboxylic acid building blocks (Bbs) were prepared starting from amino acids by diazotization followed by hydrolysis (Scheme 9.3). Since Mitsunobu coupling involves inversion of chirality at the α center, D-amino acids were employed for synthesis of hydroxy acid building blocks. The necessary chemistry was rehearsed first by individual synthesis of a few literature-based peptidyl phosphonate thermolysin inhibitors.[6] Reproducibly high yields (>95%) were obtained for these compounds so validating this synthetic route. The stage was now set for constructing the peptidyl phosphonate library.

The split–pool method was used to prepare a 540-member Cbz-$X_{(6)^p}$ – $^{o}Y_{(5)}$-$Z_{(18)}$-NH-resin library on a noncleavable solid support.[7] The Z residue (P_2 position) consisted of 18 natural amino acids, excluding cysteine and asparagine. Five α-hydroxy acids were used as the ^{o}Y component at P_1', and the X^p residue comprised six α-amino phosphonic acid Bbs (P_1 position). Double couplings were performed for amino acid and hydroxy acid condensations, first with HBTU/HOBt and then with PyBroP (Scheme 9.4). The modified Mitsunobu protocol was employed for amino phosphonic acid couplings. After the N-termini were capped with Cbz groups, appropriate deprotections were

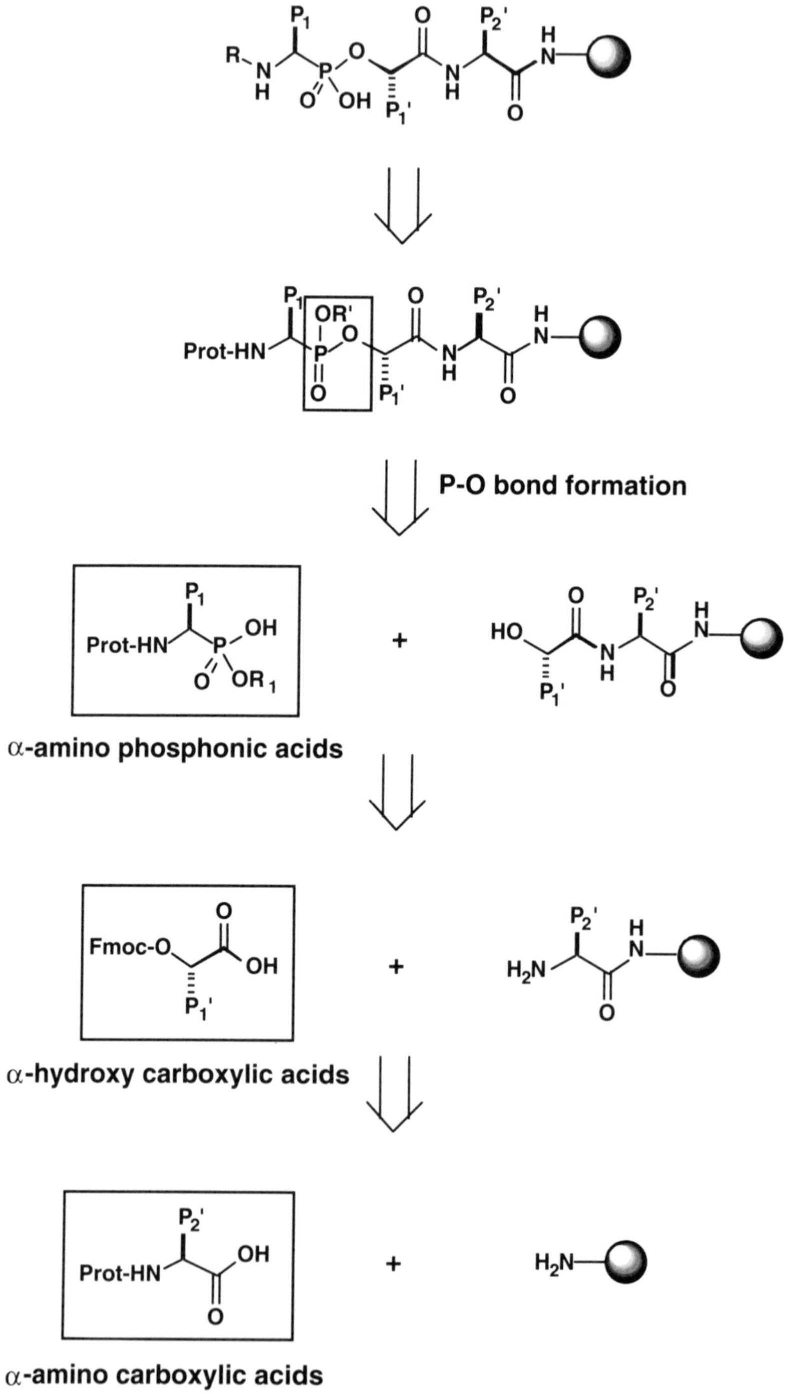

Scheme 9.2. Solid-phase peptidyl phosphonate synthesis: retro-synthetic analysis.

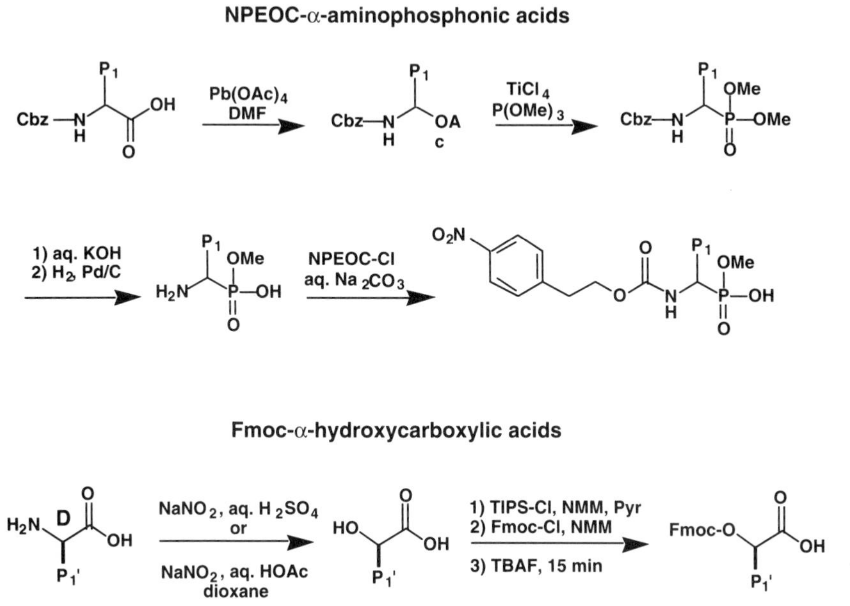

NPEOC-α-aminophosphonic acids

Fmoc-α-hydroxycarboxylic acids

Scheme 9.3. Synthesis of building blocks required for P-O bond construction.

performed to generate a 540-member (6 × 5 × 18) tripeptidyl phosphonate library.

The library was assayed for thermolysin inhibition by rank ordering of pools of compounds in an iterative fashion employing a depletion assay (Scheme 9.5). For deconvoluting the optimal Xp residue for the P_1 position, six pools composed of 90 members per pool were assayed. While all pools inhibited thermolysin to some extent, a definite rank ordering consistent with literature data (Phe[p] the best, Gly[p] the worst) was observed. Next, Phe was held constant at P_1, and the P_1' position was studied with five pools (18 compounds per pool) of Cbz-Phe[p] − [o]Y$_{(5)}$-Z$_{(18)}$-NH-resin for uncovering the optimal [o]Y residue. In agreement with the literature data, the leucine side chain emerged as the most inhibitory at P_1'. For the P_2' position, it was now necessary to evaluate the 18 individual Cbz-Phe[p] − [o]Leu$_{(5)}$-Z$_{(18)}$-NH-resin phosphopeptides. Besides uncovering the most potent peptidylphosphonate reported in literature (P_2' = Ala, K_i = 49 nM), the assay procedure also identified additional active sequences comprised of basic [P_2' = arg and his] and neutral H-bonding (P_2' = gln).

Inhibitory potency and relative rank ordering was confirmed by individual synthesis and in vitro evaluation of a number of analogs (Scheme 9.6). The discovery of P_2' arg, his, and gln analogs was unexpected, since all peptidyl-phosphonate inhibitors reported previously possessed hydrophobic residues at that position.

Overall, this study illustrates various aspects of pharmacophoric libraries and their application to drug discovery. First, it highlights the fact that combi-

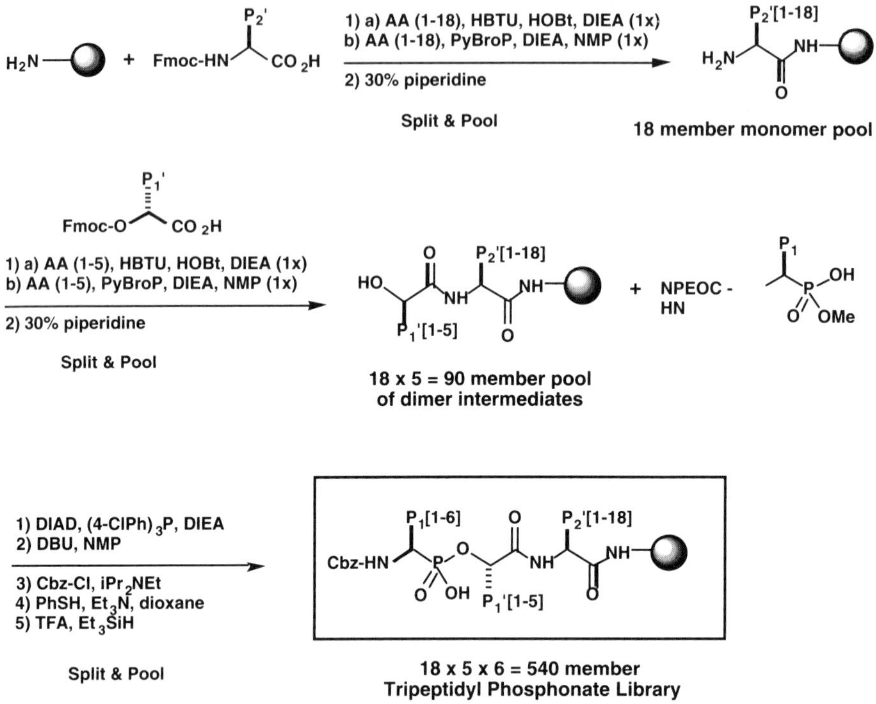

Scheme 9.4. Construction of a 540-member tripeptidyl phosphonate library.

natorial chemistry creates a different set of demands on medicinal and organic chemists than those traditionally encountered, and that these will require development of new connection chemistries. In this example, the objective of preparing a library of rationally designed phosphonic acid pharmacophore-based metalloprotease inhibitors created a necessity for new solid-phase chemistry that would permit the assembly of this core unit on solid support. This led to the development of a modified Mitsunobu-type P-O bond-forming reaction for both solution- and solid-phase chemistry. In general, a predictable consequence of the need for combinatorial libraries of nonpeptidic, small drug-like templates will be the discovery and development of new solid-phase chemistries.

Second, this drug discovery tool can meet minimal expectations, as illustrated by the successful identification of previously reported tripeptidyl phosphonate Cbz-Phe^p − ^{o}Leu-Ala-NH$_2$ (K_i = 49 nM) as a good thermolysin inhibitor from the library. Third, the surprising identification of additional potent analogs with not hydrophobic but polar P$_2'$ side chains demonstrates an important advantage of the combinatorial drug discovery approach over the classical medicinal chemistry approach. Individual analog synthesis may lead to prematurely biased approaches based on initial findings (e.g., study

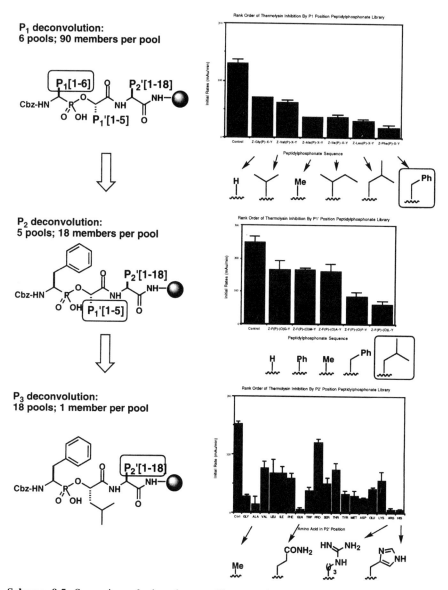

Scheme 9.5. Screening of phosphonate library—deconvolutions and lead identification.

of thermolysin inhibitors with only hydrophobic P_2' side chains), whereas a combichem approach will be less biased and more versatile and will provide more extensive SAR data. Thus, there will be more options to choose from, which may lead to an early selection of compounds with better pharmacokinetic properties.

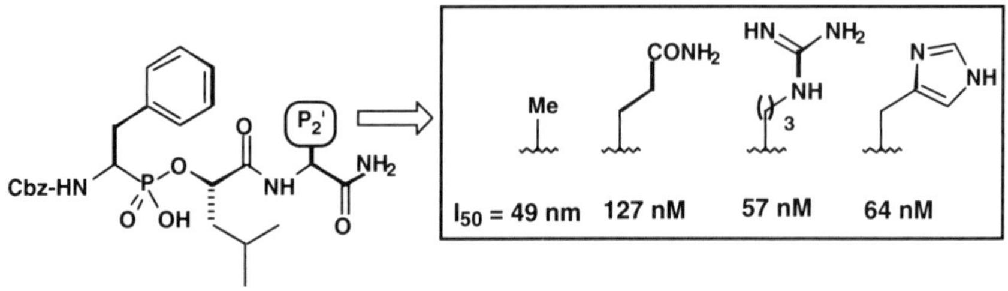

Scheme 9.6. Individual analog synthesis and lead confirmation.

TRANSITION-STATE ANALOG HIV PROTEASE INHIBITORS

Extensive efforts toward the rational design of aspartyl protease inhibitors such as renin and HIV have led to the discovery of several transition-state analog mimics. These templates can serve as the central unit around which molecular diversity can be generated by application of appropriate chemistries. Recently, solid-phase synthesis of hydroxyethylamine and 1,2-diol transition-state analog pharmacophore units and their utility for synthesis of HIV protease inhibitors have been reported by two different groups.[8,9]

In the first instance, bifunctional linkers are used by Wang and coworkers to serve the dual purpose of protecting the hydroxyl groups of these BBs and providing points for attachment on solid support (Scheme 9.7).[8] Thus, one linker possesses a vinyl ether group at one end and a free carboxylate group at the other. The vinyl ether moiety is reacted with diamino alcohol BB **1** under acid-catalyzed conditions to form an acetal protecting group, and the carboxylic acid group is used for ester-type linkage to the solid support. The other linker possesses a methyl ketone and carboxylic groups at the two ends, with the ketone group forming a ketal with diol **3**. Resulting intermediates **2** and **4** are now well suited for a bidirectional solid-phase synthesis strategy for preparing C-2 symmetric HIV protease inhibitors. The two terminal amino groups of **2** and **4** are deprotected and reacted with a variety of carboxylic acids, sulfonyl chlorides, isocyanates, and chloroformates to extend the core units in both directions and generate a wide variety of aspartyl protease inhibitors. The authors claim (no data reported) that a library of >300 discrete analogs was prepared and screened against HIV protease to identify several potent inhibitors ($I_{50} < 100$ nM).

In the method reported by Kick and Ellman,[9] a masked aminodiol pharmacophoric unit is attached through its hydroxyl group onto a dihydropyran functionalized polystyrene support (Scheme 9.8). The tosyl and azido groups of this unit provide convenient handles for bidirectional derivatization. The tosyl group can be displaced with primary amines, and the resulting secondary amine can be converted to ureas or acylated to give amides. The azido group of the pharmacophore can then be reduced to an amine, which is now available for further functionalization. The versatility of this approach was demonstrated

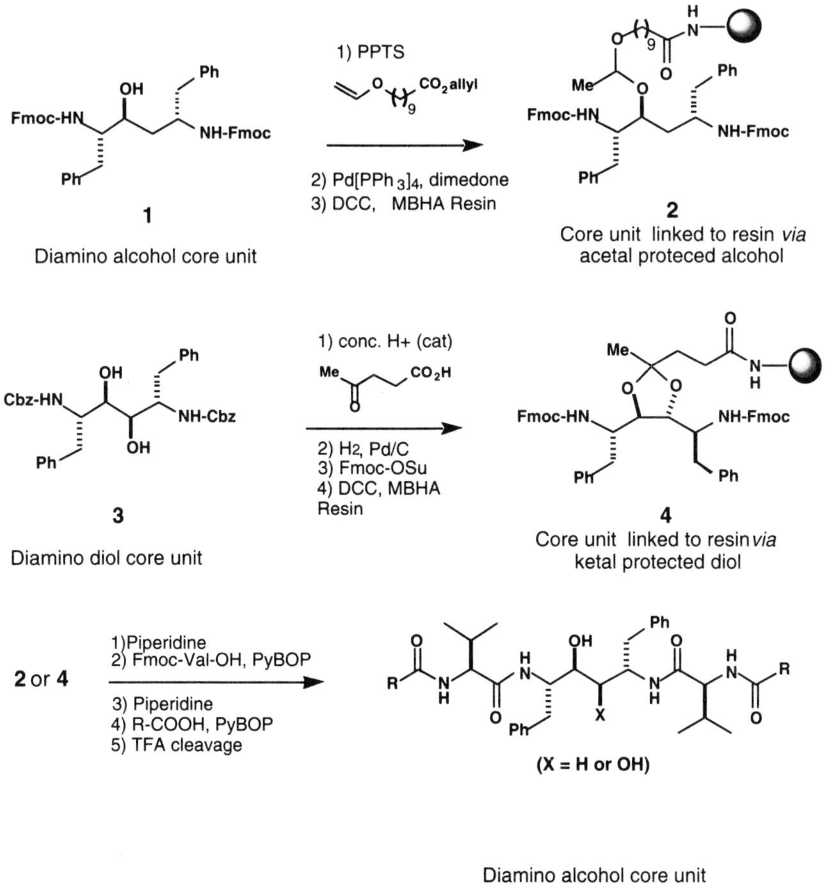

Scheme 9.7. Solid-phase synthesis of C2 symmetric HIV protease inhibitors.

by synthesis of various known HIV protease inhibitors in good yields (47–86%).

CONCLUSIONS

The field of drug discovery now spans from synthesis and screening of individual compounds to random libraries of millions of analogs. Somewhere within these two extremes lies an opportunity to rationally design and synthesize focused libraries biased by the presence of a pharmacophoric group. The choice of such functionality can be based on an understanding of the mechanism of action or an adequate degree of 3D structural information regarding the target. Since the pharmacophoric group is expected to provide critical binding interactions with the specific target, the need for very large libraries

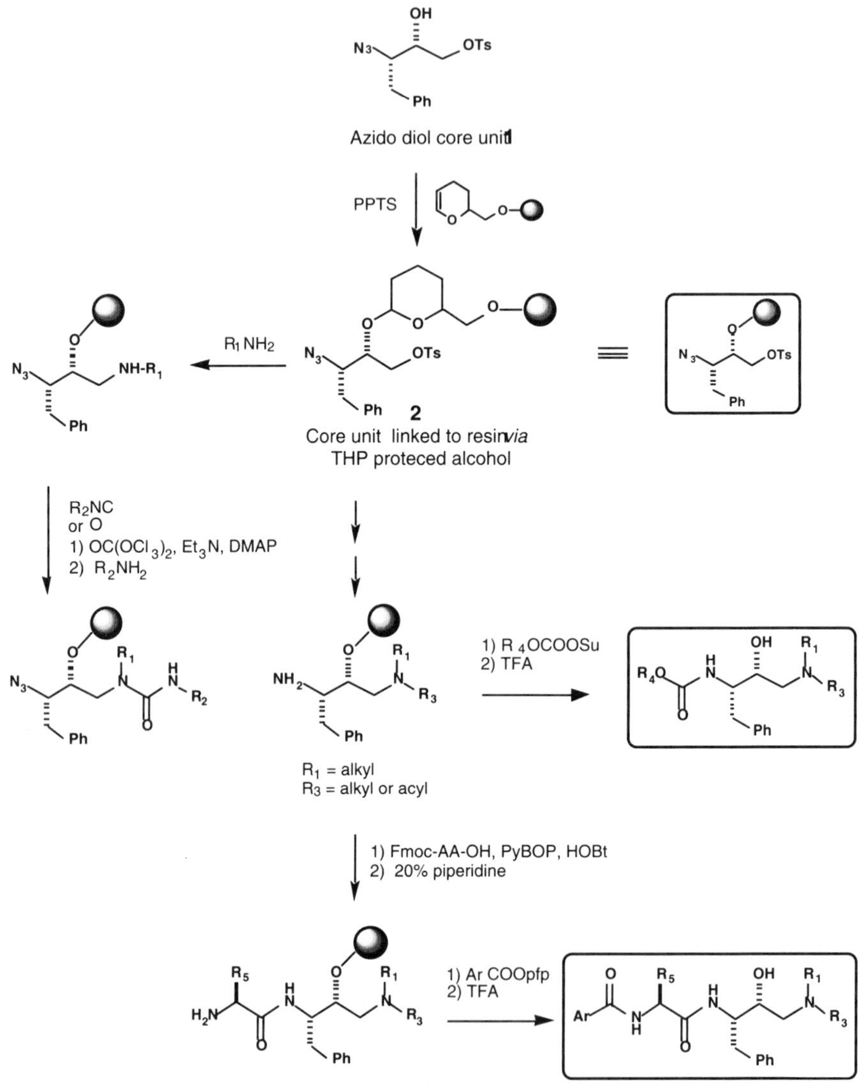

Scheme 9.8. SPS of hydroxyethyl amine TS analog HIV protease inhibitors.

is avoided, and medium-sized pharmacophoric libraries of 100 to several thousand compounds should be adequate. As one progresses further along the lead optimization pathway, a gradual transition from pools to discrete analogs is a logical choice. Besides furnishing rapid and valuable SAR information necessary for lead optimization, such approaches may also result in the discovery of novel and clinically relevant drug-like entities.

In this chapter, some current examples that fall in the category of pharmacophoric libraries are described. Phosphonic acid-based PILs provided good inhibitors of the metalloprotease thermolysin,[4-7] and hydroxy ethyl amine or 1,2-diol-type transition-state mimicking pharmacophore libraries[8,9] were prepared to generate good inhibitors against aspartyl proteases such as HIV.

REFERENCES

1. (a) Gordon EM, Barret RW, Dower WJ, Fodor SPA, Gallop MA (1994): Applications of combinatorial chemistry to drug discovery. *J Med Chem* 37:1385–1401; (b) Terrett NK, Gardner M, Gordon DW, Kobylecki RJ, Steele J (1995): Combinatorial synthesis: the design of compound libraries and their application to drug discovery. *Tetrahedron* 51:8135–8173; (c) Patel DV, Gordon EM (1996): Applications of small molecule combinatorial chemistry to drug discovery. *Drug Discovery Today* 1:134–144.

2. (a) Bartlett PA, Marlowe CK (1987): Possible role for water dissociation in the slow binding of phosphorus-containing transition transition-state-analogue inhibitors of thermolysin. *Biochemistry* 26:8553–8561; (b) Morgan BP, Scholtz JM, Ballinger MD, Zipkin ID, Bartlett PA (1991): Differential binding energy—A detailed evaluation of the influence of hydrogen-bonding and hydrophobic groups of the inhibition of thermolysin by phosphorus-containing inhibitors. *J Am Chem Soc* 113:297–307.

3. Karanewsky DS, Badia MC, Cushman SW, DeForrest JM, Dejneka T, Loots MJ, Perri MG, Petrillo EW, Powell JR (1988): (Phosphinyloxy) Acyl Amino-Acid inhibitors of angiotensin converting enzyme (ACE). 1. discovery of (S)-1-[6-amino-2-[hydroxy (4-phenylbuty)phosphinyl] oxy]-1 oxohexyl]-L proline, a novel orally active inhibitor of ACE. *J Med Chem* 31:204.

4. Campbell DA (1992): The synthesis of phosphonate esters: an extension of the Mitsunobu reaction. *J Org Chem* 57:6331–6335.

5. Campbell DA, Bermak JC (1994): Phosphonate ester synthesis using a modified Mitsunobu condensation. *J Org Chem* 59:658–660.

6. Campbell DA, Bermak JC (1994): Solid-phase synthesis of peptidylphosphonates. *J Am Chem Soc* 116:6039–6040.

7. Campbell DA, Bermak JC, Burkoth TS, Patel DV (1995): A transition-state analogue inhibitor combinatorial library. *J Am Chem Soc* 117:5381–5382.

8. Wang GT, Li S, Wideburg N, Krafft GA, Kempf DJ (1995): Synthetic chemical diversity: solid-phase synthesis of libraries of C_2 symmetric inhibitors of HIV protease containing diamino diol and diamino alcohol cores. *J Med Chem* 38:2995–3002.

9. Kick EK, Ellman JA (1995): Expedient method for the solid-phase synthesis of aspartic acid protease inhibitors directed toward the generation of libraries. *J Med Chem* 38:1427–1430.

10

HETEROCYCLIC COMBINATORIAL CHEMISTRY: AZINE AND DIAZEPINE PHARMACOPHORES

MIKHAIL F. GORDEEV AND DINESH V. PATEL
Versicor, Fremont, California

Heterocyclic compounds hold a special place among pharmaceutically important natural and synthetic materials. The remarkable ability of heterocyclic nuclei to serve both as biomimetics and reactive pharmacophores has largely contributed to their unique value as traditional key elements of numerous drugs. Heterocyclic derivatives such as morphine alkaloids, β-lactam antibiotics, and benzodiazepines are just a few familiar examples from various pharmaceuticals featuring a heterocyclic component.[1]

Recently, combinatorial chemistry has exerted a tremendous influence on the process of drug discovery. The technology clearly possesses the potential of rapidly making available large numbers of diverse compounds—more molecules than have existed before in the databases of the entire pharmaceutical industry. Besides reducing the time required to discover drugs, this process should make available larger numbers and types of leads for current targets, thereby offering more choices for efficient lead progression. Combinatorial chemistry can also be expected to greatly improve chances of discovering leads against difficult targets, where screening of currently available collections of compounds may have been unsuccessful. While methods for the generation of combinatorial libraries of peptides and oligonucleotides are now well established,[2] preparation of libraries of small organic molecules remains a rapidly evolving area of research.[3] Significantly, because of limited bioavailability

Combinatorial Chemistry and Molecular Diversity in Drug Discovery, Edited by
Eric M. Gordon and James F. Kerwin, Jr.
ISBN 0-471-15518-7   Copyright © 1998 by Wiley-Liss, Inc.

of peptides and oligonucleotides, combinatorial chemistry of small organic molecules may hold the most promise for efficient discovery of new drug-like molecules.

Although some researchers practice combinatorial syntheses in solution,[2e] far greater attention is being paid to the development of combinatorial syntheses on solid supports. The latter approach permits application of Furka's[4] "split-and-pool" methodology for efficient construction of large and diverse libraries employing appropriate connection chemistries and building blocks. Thus, a critical prerequisite to combinatorial drug discovery is to develop solid-phase routes to bioactive molecules on solid supports and to explore the utility of such synthetic methodologies for preparation of corresponding libraries.

In this short chapter, it is impossible to cover all significant developments in the area of heterocyclic combinatorial chemistry. Rather, we discuss combinatorial synthesis of representative pharmacophoric benzodiazepine and azine derivatives. Equally interesting examples of heterocyclic scaffolds are discussed elsewhere in this book, and a number of interesting developments in this field occurred after this chapter was written. While in certain cases combinatorial libraries can be utilized in a tethered format, it is generally preferable to screen compounds in solution,[2a] and reported examples of heterocyclic syntheses on solid supports for drug discovery purposes fall in this last category.

AZINE HETEROCYCLIC PHARMACOPHORES

In our laboratories, we have developed efficient routes to immobilized enamino ester and β-keto ester reactive intermediates. Our efforts have concentrated on expanding the utility of these versatile intermediates for generation of combinatorial heterocyclic libraries. This has resulted in methodologies for solid-phase synthesis (SPS) of heterocyclic scaffolds such as 1,4-dihydropyridine, pyridine, and pyrimidine, and their application toward preparation of the corresponding heterocyclic libraries.

1,4-Dihydropyridines

The 1,4-dihydropyridine (DHP) nucleus is a well-studied traditional pharmacophoric scaffold that has emerged as a core structural unit of various vasodilator, antihypertensive, bronchodilator, antiatherosclerotic, hepatoprotective, antitumor, antimutagenic, geroprotective, and antidiabetic agents.[5] We have recently developed a general solid-phase synthesis of DHPs involving two- or three-component condensation of N-immobilized enamino esters with pre-formed 2-arylidene β-keto esters, or with β-keto esters and aldehydes, respec-

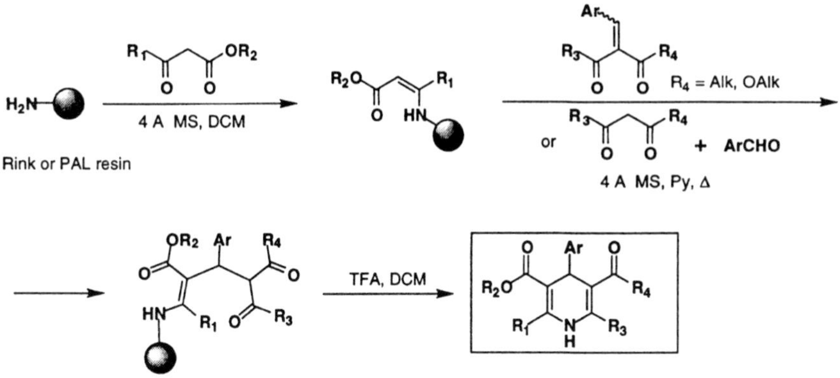

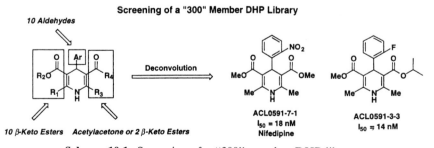

Figure 10.1. Synthesis of a dihydropyridine library.

tively (Figure 10.1). The mechanism of this transformation has been studied on solid support using ^{13}C NMR and IR spectroscopy.[6]

Beginning with the synthesis of Nifedipine on solid support as a model study, an extensive chemical rehearsal using various reaction parameters, such as solvent, temperature, and resin, was performed on a set of building blocks (BBs) with diverse steric and electronic properties. This study has enabled us to define the scope and limitations of the SPS methodology, a critical prerequisite to construction of libraries with adequate degree of yield and purity. Following this protocol, a 300-member focused library of DHPs was synthesized and screened using a [^{3}H]Nimodipine competition binding bioassay on rat cortex membranes to identify potent calcium channel blockers (Scheme 10.1). Several active pools were identified, and subsequent deconvolutions led to the identification of cardiovascular drug Nifedipine (**ACL0591-7-1**, I_{50} = 18 nM) along with other novel compounds as high affinity binders (e.g., **ACL0591-3-3**, I_{50} = 14 nM).[7] This study has served to demonstrate the feasibility of preparing and screening DHP scaffold libraries for discovery of novel bioactive molecules.

Scheme 10.1. Screening of a "300"-member DHP library.

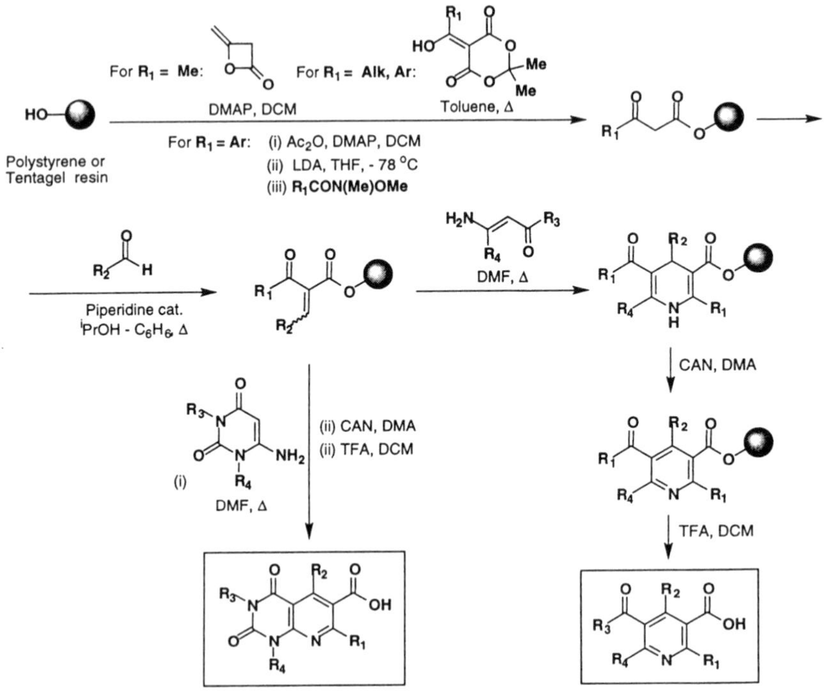

Figure 10.2. SPS of substituted pyridines and pyrido[2,3-*d*]pyrimidines.

Pyridines

The pyridine nucleus is a key feature of various drugs, including numerous antihistamines, antiseptic, antiarrhytmic, antirheumatic, and other pharmaceutical agents.[2] 1,4-Dihydropyridines can serve as convenient penultimate precursors in the Hantzsch pyridine synthesis. The N-tethered route to DHPs (Figure 10.1) described above would require cleavage of DHPs from the solid support for subsequent conversion to pyridines. Indeed, the library-to-library oxidation of DHPs cleaved from the solid support using ceric ammonium nitrate (CAN) as the oxidant readily afforded the corresponding pyridine derivatives.

A complementary procedure that permits direct construction of a pyridine nucleus on solid support has also been developed (Figure 10.2).[6b,c] This strategy commences with immobilized β-keto esters and leads to O-tethered dihydropyridine intermediates that can be smoothly oxidized to the corresponding pyridines by CAN treatment prior to the cleavage from solid support. Depending on the structure of the desired β-keto esters, three routes to such immobilized intermediates have been developed. Thus, while O-tethered acetoacetic ester could be readily prepared from alcohol resin with diketene,[6c] aliphatic analogs (R_1 = Alk) were made using acyl Meldrum's acid building blocks as substituted diketene synthons, whereas aromatic and heteroaromatic (R_1 = Ar) substituents were successfully introduced via the reaction of lithiated acetoxy-resin with appropriate Weinreb's amides (Figure 10.2).[6b] Knoevenagel condensation of immobilized β-keto esters followed by Hantzsch-type condensation with α-oxo enamines gives the DHPs. The latter, depending on the nature of the linker employed, can either be cleaved from resin to soluble DHPs (e.g., via keto amides derived from immobilized amino acids), or be oxidized to pyridines by CAN and then cleaved to yield the corresponding functionalized nicotinic acids.

The synthetic scope of this transformation is further enhanced by the reactivity of immobilized keto ester intermediates toward various α-oxo enamines.[6c] These react efficiently with 6-aminouracils to form fused DHPs, which can be oxidized and cleaved in usual manner to afford pyrido[2,3-d]pyrimidines (Figure 10.2). Additional functionalization in the pyridine scaffold has been achieved using ketene aminal derivatives as the enamine component (e.g., R_3 = Me, R_4 = PhCONH; Figure 10.2) to access corresponding 2-aminopyridine derivatives.[6c] Such molecules can be subjected to further synthetic transformations to generate more extensively functionalized heterocycles with enhanced diversity.

Pyrimidine Derivatives

Examples of pyrimidine-based bioactive molecules include vitamin B_1 (thiamine), CNS drugs, and antibacterial and antitumor agents.[1] We have successfully applied the cyclocondensation of immobilized β-keto esters with S-alkylisothioureas followed by CAN oxidation and cleavage to produce substituted 2-alkylthiopyrimidine-5-carboxylic acids in high yields and purity (Figure 10.3). The alkylthio group in such derivatives can be displaced with appropriate nucleophiles prior to the cleavage from solid support, thereby providing further opportunity for combinatorial diversification.

SPS of a structurally related and pharmaceutically attractive class of heterocycles, namely 2-aminopyrimidines, has also been accomplished, as shown in Figure 10.3.[6b] The two key steps in this synthetic transformation include guanylation of immobilized amino esters and cyclization of resulting guanidines with reactive β-diketones. Given the wide choice of readily available amino acid and β-diketone building blocks, this route is well suited for generation of pyrimidine pharmacophore libraries.

Diketopiperazines

Gordon and Steele[8] have reported an SPS of a 1000-membered diketopiperazine (DKP) library employing the synthetic protocol shown in Figure 10.4. The three key steps of this methodology are (1) reductive alkylation of immobilized amino acids, (2) acylation with a second N-Boc amino acid, and (3) deprotection and cyclization with release of final products from support. The chemistry

PCA = 1-Pyrazolecarboxamidine hydrochloride
ASA = 1-Amidinosulfonic acid

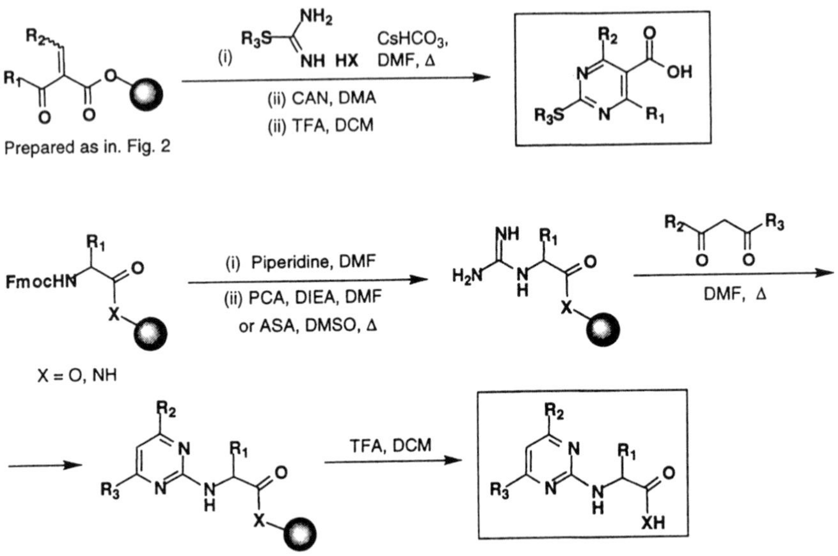

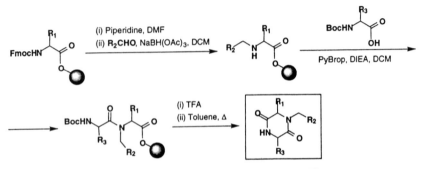

Figure 10.3. Synthesis of pyrimidine-5-carboxylic acids and 2-aminopyrimidines.

is reliable and versatile, as demonstrated by correct LC-MS identification of 96 out of the theoretically anticipated 100 members resulting from a combination of 10 amines and 10 aldehydes. Iterative screening and deconvolution of the DKP libraries has identified several bioactive compounds, including a high-affinity binder for the neurokinin-2 receptor.[9]

2*H*-Isoquinoline-1-ones

En route to structurally novel rigid heterocyclic scaffolds, researchers have begun translating efficient organometallic solution chemistries into combinato-

Figure 10.4. Synthesis of a diketopiperazine library.

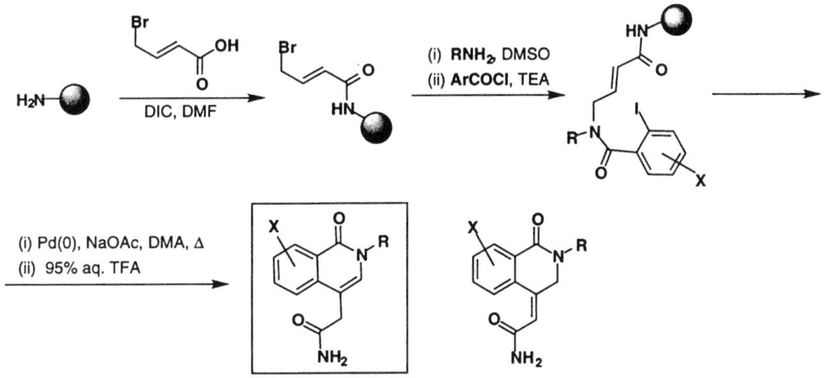

Figure 10.5. SPS of 2*H*-isoquinoline-1-ones.

rial synthesis on solid supports. An application of homogeneous transition metal mediated reaction on solid support was reported by Goff and Zuckermann[10a] in their SPS of 2*H*-isoquinoline-1-ones (Figure 10.5). The key step of this synthesis involves an intramolecular Heck reaction to afford, after cleavage, desired 2*H*-isoquinoline-1-ones in good yields (Figure 10.5) (in some cases, kinetic *exo*-methylene products have also been isolated). Intermolecular versions of solid-phase Heck reactions have also been reported recently.[10b]

1,4-DIAZEPINE HETEROCYCLIC PHARMACOPHORES

1,4-Benzodiazepine-2-ones

One of the first and most elegant examples of SPS of heterocycles was provided by Ellman and coworkers in their preparation of libraries of benzodiazepines (BDPs).[3a] These were prepared on solid support by a sequence of transformations (Figure 10.6) utilizing three types of building blocks: 2-aminobenzophenones, amino acids, and alkylating reagents. This synthetic protocol enabled high-yielding preparations of fully derivatized benzodiazepines. The original SPS has been extended using pin methodology to perform a parallel synthesis of 192 diversely functionalized BZDs.[2c] From the cholecystokinin A receptor binding assay, detailed structure activity relationship (SAR) data were acquired and an indole-substituted BZD was identified as a potent CCK ligand.

Since only few appropriately functionalized 2-aminobenzophenones are readily available, an original synthesis of such immobilized reagents based on palladium-mediated Stille coupling between immobilized 2-aminoarylstannane and an acid chloride as the solution coupling partner was developed (Figure 10.7).[11a] More recently, this methodology has been employed to generate an 11,200-analog library from 20 acid chlorides, 35 amino acids, and 16 alkylating agents.[11b]

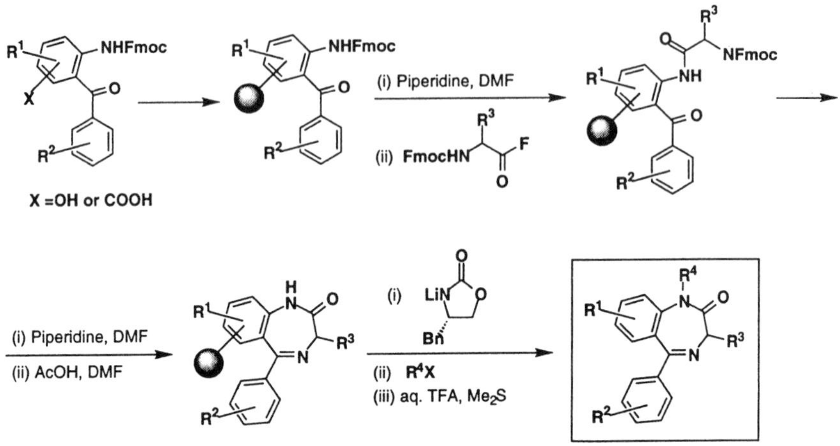

Figure 10.6. Combinatorial synthesis of 1,4-benzodiazepine-2-ones.

Another preparation of 1,4-benzodiazepine-2-ones, described by Hobbs De Witt et al.,[12] commences by condensation of resin-bound α-amino ester with 2-aminobenzophenone imines followed by TFA treatment of the intermediate to effect cleavage and cyclization (Figure 10.8). In this approach, solid-phase chemistry was performed using a multipin Diversomer apparatus to simultaneously but separately to synthesize 40 discrete structurally related compounds. The crude products were assayed employing bovine cortical membranes with [³H]fluoronitrazepam as the radioligand, and the experimental IC_{50} SAR data was reported to be within an order of magnitude of literature

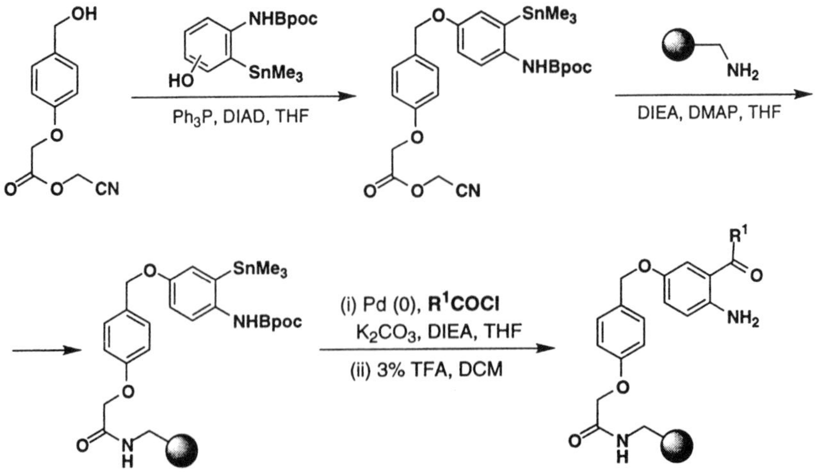

Figure 10.7. Synthesis of 2-aminobenzophenone derivatives on a solid support.

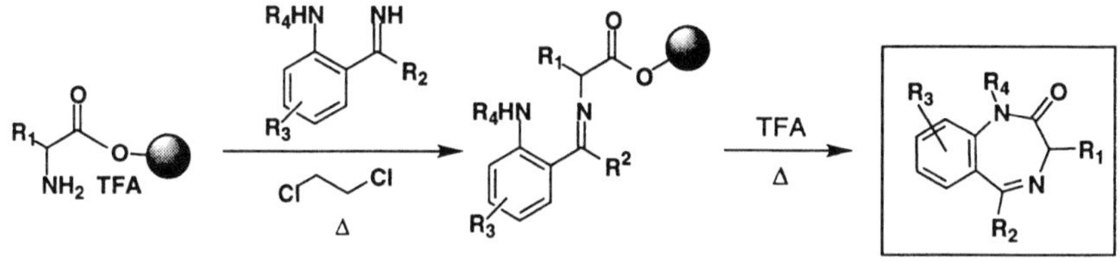

Figure 10.8. Preparation of a soluble 1,4-benzodiazepine-2-one library.

reports. An advantage of this approach is the semiautomated laboratory setup at all stages, including synthesis, product isolation, and characterization.[12]

1,4-Benzodiazepine-2,5-diones

Two different routes have been simultaneously reported for synthesis of structurally related scaffolds; 1,4-benzodiazepine-2,5-diones. One method (Figure 10.9), reported by Ellman et al.,[13] involves N-immobilization of amino acid esters via reductive alkylation with an aryl aldehyde linker followed by acylation with functionalized anthranilic acid building blocks. Racemization-free cyclization of functionalized *ortho*-amino benzamide intermediates with lithiated acetanilides and in situ alkylation affords the target compounds in good yields.

An alternative procedure, reported by Goff and Zuckermann,[14] generates

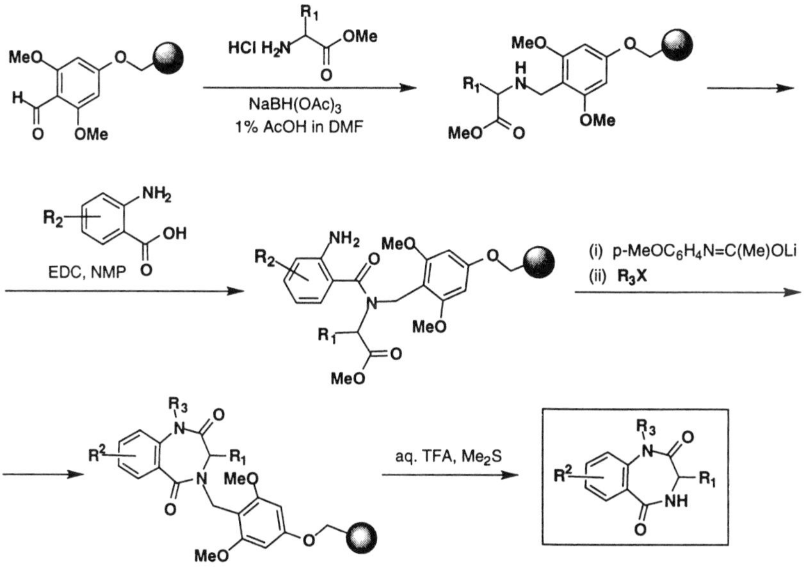

Figure 10.9. SPS of 1,4-benzodiazepine-2,5-diones.

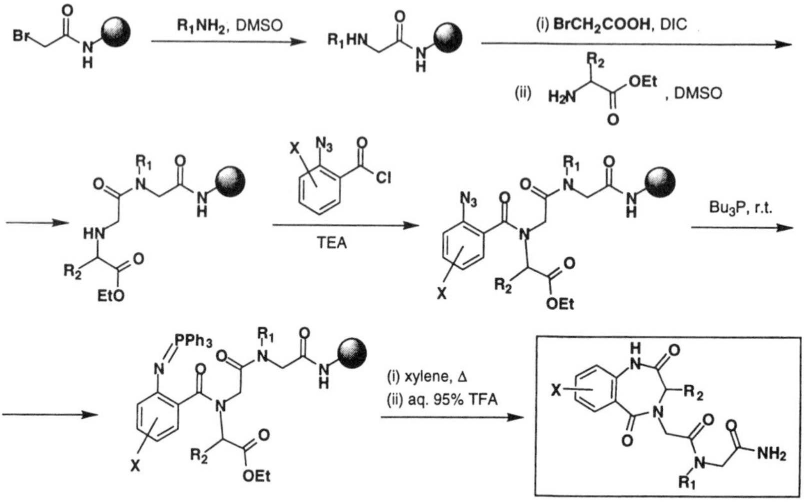

Figure 10.10. SPS of peptoid 1,4-benzodiazepine-2,5-diones.

a BZPD scaffold on a "peptoid" scaffold and is outlined in Figure 10.10. The key step in this transformation involves an intramolecular aza-Wittig-type reaction of iminophosphorane intermediates generated in situ on solid support with amino acid-derived carboxyl esters (see also Chapter 6).

CONCLUSION

Heterocyclic combinatorial chemistry is rapidly emerging as a powerful tool for drug discovery. One can foresee further explosive activity in this field leading to significant advances in medicinal chemistry. Libraries of larger size and diversity will be best suited for novel lead discovery, whereas smaller, focused libraries can be expected to facilitate target-specific lead development. Integration of rational drug design with combinatorial techniques may prove to be a particularly powerful strategy, enabling a medicinal chemist to quickly generate novel chemical entities with desired pharmacological properties. Future developments in this field will be closely associated with new solid-phase synthetic methodologies, automation techniques to speed up production of libraries, miniaturized assays to facilitate high throughput screening, and establishment of database management systems to store, retrieve, and capitalize on all the useful information generated in these processes.

ACKNOWLEDGMENT

We thank Eric M. Gordon for helpful discussions, Bruce England and Jesse Combs for bioassay support, and Supriya Jonnalagadda, Miles She, Jie Wu, Jiang Zhu, and

Seifu Tadesse for technical assistance with some of the research activities described in this chapter.

REFERENCES

1. Roth HJ, Kleemann A (1988): In *Pharmaceutical Chemistry,* vol 1: *Drug Synthesis.* New York: Wiley.

2. For reviews on the use of combinatorial technology for drug discovery, see (a) Gallop MA, Barrett RW, Dower WJ, Fodor SPA, Gordon EM (1994): Application of combinatorial technologies to drug discovery, 1: background and peptide combinatorial libraries. *J Med Chem* 37:1233–1251; (b) Gordon EM, Barrett RW, Dower WJ, Fodor SPA, Gallop MA (1994): Application of combinatorial technologies to drug discovery, 2: combinatorial organic synthesis. *J Med Chem* 37:1385–1401; (c) Moos WH, Green GD, Pavia MR (1993): Recent advances in generation of molecular diversity. In Bristol JA, ed. *Annual Reports in Medicinal Chemistry,* vol 28. San Diego, CA: Academic, pp 315–324; (d) Desai MC, Zuckermann RN, Moos WH (1994): Recent advances in the generation of chemical diversity libraries. *Drug Dev Res* 33:174–188; (e) Ecker DJ, Crooke ST (1995): Combinatorial drug discovery: which methods will produce the greatest value? *Biotechnology* 13:351–360; (f) Terrett NK, Gardner M, Gordon DW, Kobylecki RJ, Steele J (1995): Combinatorial synthesis: the design of compound libraries and their application to drug discovery. *Tetrahedron* 51:8135–8173.

3. (a) Bunin BA, Ellman JA (1992): A general and expedient method for the solid-phase synthesis of 1,4-benzodiazepine derivatives. *J Am Chem Soc* 114:10997; (b) Hobbs De Witt S, Kiely JS, Stankovic CJ, Schroeder MC, Reynolds Cody DM, Pavia MR (1993): "Diversomers": an approach to nonpeptide, nonoligomeric chemical diversity. *Proc Natl Acad Sci USA* 90:6909–6913; (c) Bunin BA, Plunkett MJ, Ellman JA (1994): The combinatorial synthesis and chemical and biological evaluation of 1,4-benzodiazepine library. *Proc Natl Acad Sci USA* 91:4708–4712; (d) Kurth MJ, Randall LAA, Chen C, Melander C, Miller RB, McAlister K, Reitz G, Kang R, Nakatsu T, Green C (1994): Library-based lead compound discovery: antioxidants by an analogous synthesis/deconvolutive assay strategy. *J Org Chem* 59:5862–5864; (e) Yu K-L, Deshpande MS, Vyas DM (1994): Heck reaction in solid-phase synthesis. *Tetrahedron Lett* 35:8919; (f) Backes BJ, Ellman JA (1994): Carbon–carbon bond-forming methods on solid support: utilization of Kenner's "safety-catch" linker. *J Am Chem Soc* 116:11171–11172; (g) Patek M, Drake B, Lebl M (1994): *All-cis*-Cyclopentane scaffolding for combinatorial solid-phase synthesis of small nonpeptide compounds. *Tetrahedron Lett* 35:9169–9172; (h) Smith PW, Lai JYQ, Whittington AR, Cox B, Houston JG, Stylli CH, Banks MN, Tiller PR (1994): Synthesis and biological evaluation of a library containing potentially 1600 amides/esters: a strategy for rapid compound generation and screening. *Bioorg Med Chem Lett* 4:2821–2824; (i) Kick EK, Ellman JA (1995): Expedient method for the solid-phase synthesis of aspartic acid protease inhibitors directed toward the generation of libraries. *J Med Chem* 38:1427–1430; (j) Holmes CP, Chinn JP, Look GC, Gordon EM, Gallop MA (1995): Strategies for combinatorial organic synthesis: solution and polymer-supported synthesis of 4-thiazolidinones and 4-metathiazanones derived from amino acids. *J Org Chem* 60:7328–7333;

(k) Campbell DA, Bermak JC, Burkoth TS, Patel DV (1995): A transition state analogue inhibitor combinatorial library. *J Am Chem Soc* 117:5381–5382; (l) Murphy MM, Schullek JR, Gordon EM, Gallop MA (1995): Combinatorial organic synthesis of highly functionalized pyrrolidines: identification of a potent angiotensin converting enzyme inhibitor from a mercaptoacyl proline library. *J Am Chem Soc* 117:5381–5382.

4. Furka A, Sebestyen F, Asgedom M, Dibo G (1991): General method for rapid synthesis of multicomponent peptide mixtures. *Int J Pept Protein Res* 37:487.

5. For reviews, see (a) Godfraind T, Miller R, Wibo M (1986): Calcium antagonism and calcium entry blockade. *Pharmacol Rev* 38:321–416; (b) Janis RA, Silver PJ, Triggle DJ (1987): Drug action and calcium cellular regulation. *Adv Drug Res* 16:309–590; (c) Sausins A, Duburs G (1988): Synthesis of 1,4-dihydropyridines by cyclocondensation reactions. *Heterocycles* 27:269–289.

6. (a) Gordeev MF, Patel DV, Gordon EM (1986): Approaches to combinatorial synthesis of heterocycles: a solid-phase synthesis of 1,4-dihydropyridines. *J Org Chem* 61:924–928; (b) Gordeev MF, Patel DV (1995): Methods for synthesizing diverse collections of pyridines, pyrimidines, and 1,4-derivatives thereof. *US patent application 08/431,083*, filed 28 April 1995; (c) Gordeev MF, Patel DV, Gordon EM (1996): Approaches to combinatorial synthesis of heterocycles: solid-phase synthesis of pyridines and pyrido[2,3-*d*]pyrimidines. *Tetrahedron Lett* 37:4643–4646.

7. Gordeev MF, England B, Patel DV, Gordon EM, unpublished results.

8. Gordon DW, Steele J (1995): Reductive alkylation on a solid phase: synthesis of a piperazinedione combinatorial library. *Biorg Med Chem Lett* 5:47–50.

9. Terret NK, Gardner M, Gordon DW, Kobylecki RJ, Steele J (1995): Combinatorial synthesis: the design of compound libraries and their application to drug discovery. *Tetrahedron Lett* 51:8135–8173.

10. (a) Goff DA, Zuckermann RN (1995): Solid-phase synthesis of highly substituted peptoid 1(2*H*)-isoquinolines. *J Org Chem* 60:5748–5749; (b) Yu K-L, Deshpande MS, Vyas DM (1994): Heck reactions in solid-phase synthesis. *Tetrahedron Lett* 35:8919–8922.

11. (a) Plunkett MJ, Ellman JA (1995): Solid-phase synthesis of structurally diverse 1,4-benzodiazepine derivatives using the Stille coupling reaction. *J Am Chem Soc* 117:3306–3307; (b) Bunin BA, Plunkett MJ, Ellman JA (1996): Synthesis and evaluation of 1,4-Benzodiazepine libraries. *Methods Enzymol,* 267:448–465.

12. Hobbs De Witt S, Schroeder MC, Stankovic CJ, Strode JE, Czarnik AW (1994): DIVERSOMER technology: solid-phase synthesis, automation, and integration for generation of chemical diversity. *Drug Dev Res* 33:116–124.

13. Boojamra CG, Burow KM, Ellman JA (1995): An expedient and high-yielding method for the solid-phase synthesis of diverse 1,4-benzodiazepine-2,5-dione mixtures. *J Org Chem* 60:5742–5743.

14. Goff DA, Zuckermann RN (1995): Solid-phase synthesis of defined 1,4-benzodiazepine-2,5-dione mixtures. *J Org Chem* 60:5744–5745.

11

SCAFFOLDS FOR SMALL MOLECULE LIBRARIES

MICHAEL R. PAVIA

Millenium Pharmaceuticals, Inc., Cambridge, Massachusetts

New lead discovery in the pharmaceutical industry is now taking place more effectively by generating large libraries of nonpeptide molecules to be evaluated using high-throughput screening paradigms.[1-4] The concept of preparing nonpeptide, drug-like libraries was first reported just five years ago. Since that time we have seen steady development of new chemistries applied to library generation to the point where it is now possible to synthesize almost endless varieties of structural libraries using these methodologies.

But what structural class of nonpeptide molecule(s) will be most useful in a new lead discovery library? Scaffolding approaches have been exploited by a number of groups over the years for both library and nonlibrary use. The concept in simplest terms is to prepare a backbone or scaffold to hold, in the correct alignment, the important functional groups for interaction with a biologically important ligate. Herein is presented a brief overview of scaffolding approaches and how they may be used to generate diversity libraries.

LINEAR, REPEATING SCAFFOLDS

One approach to scaffold design is to assemble, in linear fashion, a series of structurally related monomers, each containing a functional group or side chain that is displayed for recognition by the biological target. In the three

Combinatorial Chemistry and Molecular Diversity in Drug Discovery, Edited by
Eric M. Gordon and James F. Kerwin, Jr.
ISBN 0-471-15518-7   Copyright © 1998 by Wiley-Liss, Inc.

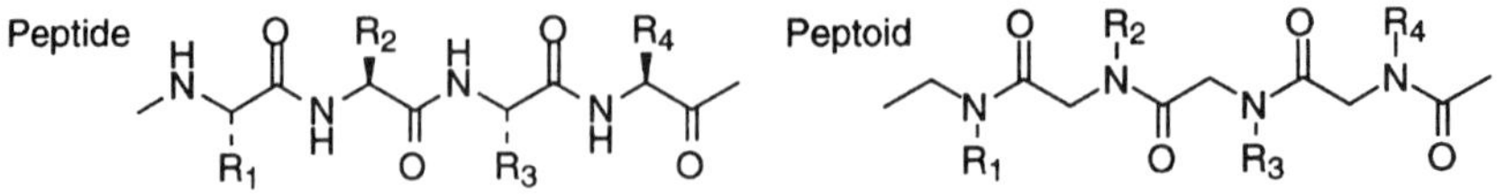

Figure 11.1. Structures of an oligopeptide and the corresponding peptoid.

approaches described below the readily hydrolyzable peptide backbone is replaced with one more suited to pharmaceutical use.

Possibly the most interesting of the linear scaffolds are the "peptoids," (Figure 11.1) developed at Chiron.[5] Peptoids are oligomers of N-substituted glycines, with the side chains attached to the amide nitrogen. Computational analysis indicates that peptoids have available a greater diversity of potential conformational states than do peptides. Furthermore, replacement of the peptide backbone with the N-substituted monomers results in increased absorption characteristics and increased metabolic stability to proteases. Among the advantages of this approach for library generation is that the monomers are achiral; they are readily available with a wide variety of functional groups presented as side chains; and the linking chemistry is high yielding and amenable to automation (see Chapter 6).[6,7]

Another approach to repeating monomer peptide mimics can be found in the oligocarbamate libraries described by Schultz et al.[8] (Figure 11.2) (see Chapter 12). Finally, an alternative to the peptide backbone, the aminimide backbone (Figure 11.3), has recently been shown to display side-chain functional groups in much the same way as the peptide.[9] In each of these cases, the final molecules are linear with limited conformational constraints. While they appear to perform satisfactorily in addressing the issue of chemical and biological stability, they lack the rigidity that is usually required for a high affinity ligand.

β-TURN AND β-SHEET MIMETICS

The design of β-turn mimetics has been an active area of medicinal chemistry research for years and has been recently reviewed.[10] This theme addresses the issue of rigidity of structure in biological activity. Recent work described below illustrates the need for β-strand mimetics.

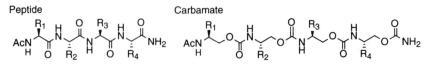

Figure 11.2. Structures of an oligopeptide and the corresponding oligocarbamate.

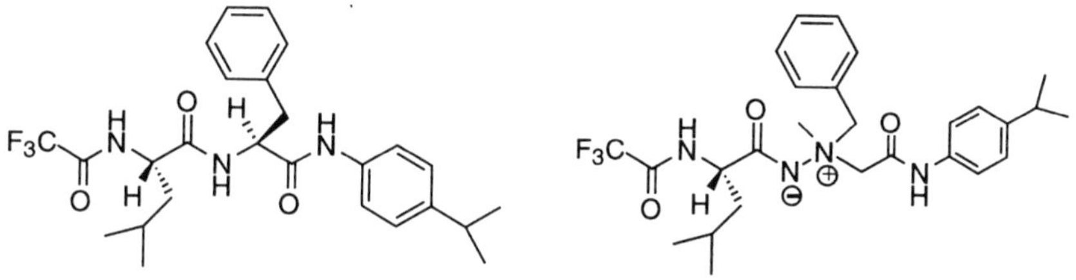

Figure 11.3. Structure of a dipeptide enzyme inhibitor and the corresponding aminimide inhibitor.

Ellman has recently described the simultaneous solid-phase synthesis of a library of β-turn mimetics (Figure 11.4) incorporating side-chain functionality (see Chapter 5).[11] Such libraries can easily contain many side-chain combinations as well as a large number of relative orientations of the side chains. It is anticipated that by having this level of variation, the difficulties normally encountered in designing one optimal compound will be circumvented. The required constraint is built by replacing the hydrogen bond normally found in β-turns between the *i* + 1 and *i* + 3 residues with a covalent backbone linkage. The flexibility and relative orientations of the side chains can be varied by introducing different backbone linkages or altering the absolute configurations at various stereocenters. The target molecules are constructed on solid support using a Mimotopes Pin Approach[12] from readily available materials (an α-halo acid and an α-amino acid).

In addition to the work being done on β-turn mimetics, a pyrrolinone-based β-pleated sheet mimetic has recently been reported, which may be applied to library construction.[13,14] These molecules mimic the zigzag shape of a peptide β-pleated sheet without containing any amide bonds, thus protecting the compounds from cleavage by proteolytic enzymes. There is some evidence that in cellular assays these unusual molecules might also possess enhanced cell transport properties. The mimetic structure was designed by computer modeling. After comparing the proposed pyrrolinone structure with the X-ray structure of an angiotensinogen fragment, a tetrapeptide with a β-pleated sheet conformation, it was predicted that carbonyl groups and side chains in the pyrrolinone would be oriented similarly to the corresponding

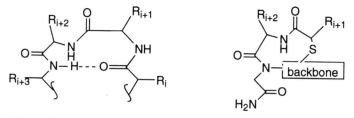

Figure 11.4. Structure of a β-turn and the corresponding mimic.

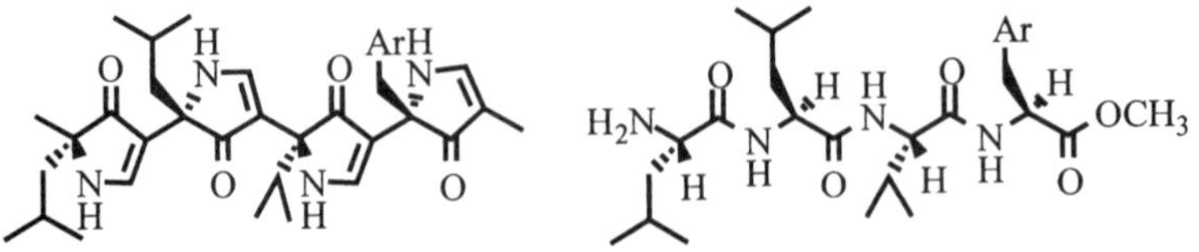

Figure 11.5. Structure of an angiotensinogen fragment and its corresponding pyrrolinone-based mimic.

structures in the tetrapeptide. The fact that the pyrrolinone NH groups are comparable to amide NH groups in basicity suggests that they might stabilize a β-pleated sheet conformation through intrastrand hydrogen bonding. Synthesis of the proposed target and analysis of its X-ray structure demonstrated that such β-sheet mimicry was indeed occurring.

The repeating 3,5,5-trisubstituted pyrrolinone units are synthesized by an iterative synthetic approach via cyclization of metalated imino esters. Interestingly, the presence or absence of a nitrogen protecting group on the polymeric pyrrolinones controls whether the molecules exist in an antiparallel or a parallel sheet (Figure 11.5).

Since the β-pleated sheet motif constitutes a recognition element for enzymes such as renin and HIV protease, mimics of this basic structure might act as inhibitors of those proteolytic enzymes. The ability to mimic a β-pleated sheet may also have implications for the design of anti-amyloid agents for the treatment of Alzheimer's disease and for the design of DNA binders.

CONFORMATIONALLY RESTRAINED PEPTIDE SCAFFOLD

The TASK (template-assembled side chains) concept was introduced by Mutter and coworkers[15,16] for mimicking bioactive conformations of peptide ligands. In this approach, amino acid side chains representing discontinuous epitopes are assembled on a conformationally restrained peptide (or peptide-like) scaffold mimicking the topological features of the original peptide.

The synthetic strategy employed allows for the selective and independent functionalization of each attachment point thereby affording spatially well-defined functional sites. Scaffolds carrying up to eight independently addressable sites are possible. By varying the structural parameters of the template as well as the orientational flexibility and chemical nature of the functional groups this class of molecules can be applied to the rational design of biologically active compounds or alternatively, TASK libraries can be used in high-throughput screening.

SMALL ORGANIC MOLECULE SCAFFOLDS

The vast majority of marketed pharmaceuticals are low molecular weight, nonpeptide, nonpolymeric entities. Therefore, it is logical that small organic

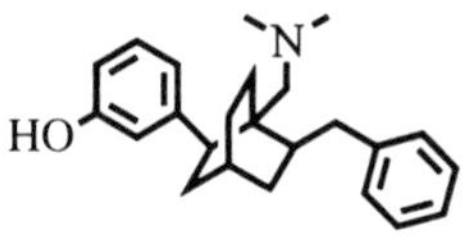

Figure 11.6. Enkephalin mimic.

molecules that can display functional groups would surface as a scaffold-ing approach. The examples that follow have, in general, not been used in library approaches for the reasons described at the end of this section. How-ever, they are included here to illustrate the promise of small organic molecule scaffolds and to set the stage for the introduction of library ap-proaches.

Probably the first use of a small, nonpeptide-like rigid scaffold to hold functional groups important for biological activity was reported in 1986.[17] The target molecule was designed after the proposed active conformation of methionine enkephalin using computer modeling (Figure 11.6). A bi-cyclo[2.2.2]octane was derivatized with three functional groups designed to bind to the opiate receptor. Testing revealed an IC_{50} of 225 nM in a [^{3}H] naloxone binding assay. This activity is relatively weak when one considers that morphine and methionine enkephalin bind with IC_{50} values of 4.5 and 9.0 nM, respectively.

In an attempt to eliminate the peptide backbone in peptide hormone and neurotransmitter agonists/antagonists the use of both carbohydrate and ste-roid-based scaffolds has been reported. By utilizing a carbohydrate-based scaffold it was shown that a designed nonpeptidyl peptidomimetic can act as a scaffold for molecules that bind to seven-transmembrane G-protein-coupled receptors.[18,19] In this case, Merck scientists were attempting to mimic a hexa-peptide somatostatin agonist discovered in house. Four amino acid side chains were selected that were felt to be important for binding to the somatostatin receptor (Figure 11.7). Molecular modeling was used to orient the functional groups in the appropriate places around the carbohydrate ring through attach-

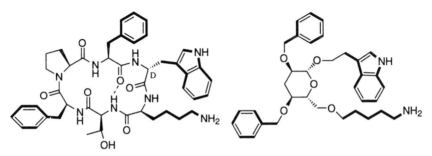

Figure 11.7. Structure of a cyclic hexapeptide somatostatin agonist and the correspond-ing glucose-based mimic.

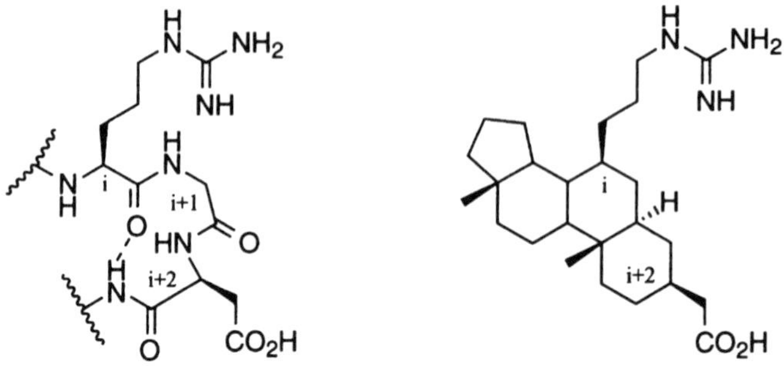

Figure 11.8. Structure of a type 1 β-turn and the corresponding steroid mimic.

ment to the hydroxyl groups. Glucose was selected for the scaffold since its pyranoside structure affords a well-defined conformation, the side chains can be arranged in an equatorial disposition around the pyran by attachment through the hydroxy groups, and a wealth of synthetic knowledge has been reported in the literature. The target compound bound to the somatostatin receptor with an IC_{50} of 1.3 μM and bound to the substance P receptor with an IC_{50} of 0.18 μM. A closely related analog bound with an IC_{50} of 60 nM and exhibited no significant interaction with 50 other targets. The authors postulate that the weak binding of this analog may be due to the ability of the carbohydrate molecule to undergo hydrophobic collapse as well as the need for further optimization in the placement of the side chains.

A steroid-like scaffold has been used in the design of mimics of the binding of the Arg-Gly-Asp (RGD) sequence at the fibrinogen receptor.[20] While effectively mimicking the size of the cyclic hexapeptides of interest this scaffold has the further advantage that the rigid nucleus should reduce the tendency for hydrophobic collapse hypothesized, for example, for side chains attached to the glucose scaffold. In addition, steroids are a well-known class of drugs with excellent oral bioavailability and well-understood chemistry. The target molecule was designed around the hypotheses that the Arg and Asp residues are found in a β-turn with the Gly as the $i + 1$ residue (Figure 11.8). Computer modeling suggested that equatorial substituents at the 3 and 7 positions of an allopregnane scaffold have the same spacing and geometry as the i and $i + 2$ amino acid side chains of a β-turn. The compound bound to the GPIIb/IIIa receptor with an IC_{50} of 100 μM.

A final example of a rigid scaffold displaying functional groups proposed important for biological activity has been reported by Kalindjian et al.[21] To design a CCK_B/gastrin receptor antagonist they began with a model shape of the tetragastrin molecule derived from molecular mechanics calculations and concluded that tetragastrin could exist in a 3_{10} helix with two aromatic rings of the tryptophan and the phenylalanine side chains interacting in a π-stacking arrangement with a separation of 5–7 Å. The authors utilized dibenzobicyclo[2.2.2]octane as a rigid scaffold that could be used to replace the peptide

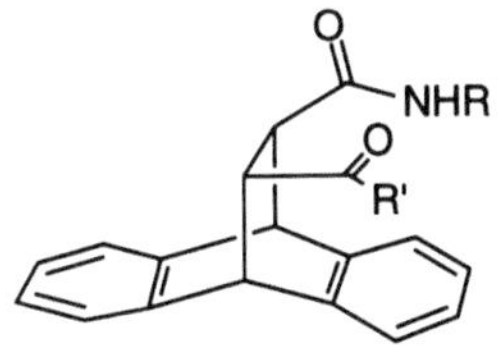

Figure 11.9. CCK_B/gastrin mimic.

backbone of tetragastrin while maintaining the stereoelectronic features. One of the fused rings represents the peptide side-chain aromatic groups. Construction of the proposed target afforded a compound with reasonable binding affinity ($pK_B' = 7.0$) (Figure 11.9).

Each of the above examples relies on the ability to rationally design a molecule (or library) with a defined binding conformation through the use of computer modeling. This is a very difficult design problem that has met with only limited success. While this approach may be useful for testing a particular hypothesis regarding possible active conformations and functional groups important for binding, it is unlikely that it can be utilized to rapidly optimize a compound's activity through a library approach.

The major reason for the inability to pursue a library approach is due to the considerable synthetic effort needed to prepare these compounds. If one wishes to reorient the functional groups in different positions or with alternative stereochemical orientation a novel chemical route must be devised in most cases. Therefore, it is very difficult to imagine from a chemical standpoint how one could easily place the important functional groups in enough different ways to explore a large number of configurations to achieve optimal binding.

LIBRARY APPROACHES WITH SMALL ORGANIC MOLECULES

Recently, several examples have been reported of libraries based on rigid or semirigid scaffolds to which functional groups postulated to be important for biological activity can be attached. In each case the conformation and position of the functional groups were fixed. It is important to note that these scaffolds were designed in most cases to be denovo drug discovery libraries, and not designed using computer modeling techniques to target any specific biological target.

Patek et al. prepared racemic, all-cis-substituted cyclopentanes.[22] Four functional groups could then be attached to this scaffold by attachment through the three carboxylic acids and one amine group displayed in cis fashion around the cyclopentane. The functional groups are attached through formation of an amide bond. By clever use of a cyclic anhydride, a methyl ester, and a Boc protected amine it is a relatively straightforward task to sequentially introduce

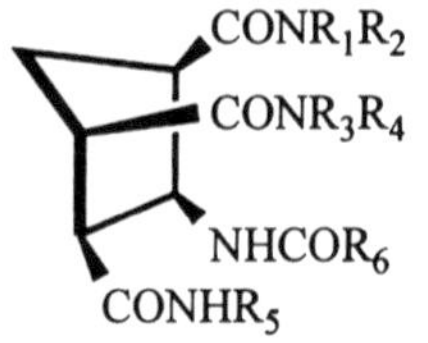

CONR₁R₂
CONR₃R₄
NHCOR₆
CONHR₅

Figure 11.10. Cyclopentane scaffold.

four different functional groups as desired (Figure 11.10). The synthesis could easily be carried out in solution or through initial attachment to a solid support, functionalization, cleavage from the solid support, and finally functionalization to afford the tetrasubstituted scaffold. No biological activity was reported by these authors.

The generation of three structurally unique libraries of compounds has been described by Rebek and coworkers.[23-25] A rigid scaffold possessing multiple reactive functional groups is reacted with a mixture of functional groups that can form a covalent bond with the scaffold. The scaffolds chosen were 9,9-dimethylxanthene-2,4,5,7-tetracarboxylic acid tetrachloride, cubane 1,3,5,7-tetracarboxylic acid tetrachloride, and 1,3,5-*tris*-carboxymethylbenzene tricarboxylic acid trichloride (Figure 11.11). Relatively rigid core structures from which the functional groups would radiate outward were chosen, presuming that lead compounds should have convex shapes able to compliment the concave surfaces of macromolecular targets such as enzyme active sites, receptor cavities, or nucleic acid grooves. The highly symmetric and compact cubane derivative displays the groups in a tetrahedral array, the xanthane in a less symmetric arrangement that is more planar, and the benzene derivative occupies a still different space with less rigidity.

The scaffolds were reacted in solution with a set of 21 amines consisting of amino acid derivatives and small heterocycles. By attaching the building blocks on the scaffold at random, every possible combination is theoretically possible. Comparable reactivity toward the scaffold acid chloride was the main criteria for selection. One equivalent of the scaffold was reacted with four equivalents of the mixture of amines. Electrospray mass spectroscopy and high-performance liquid chromatography (HPLC) suggested that in the

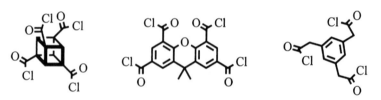

Figure 11.11. Rebek library scaffolds.

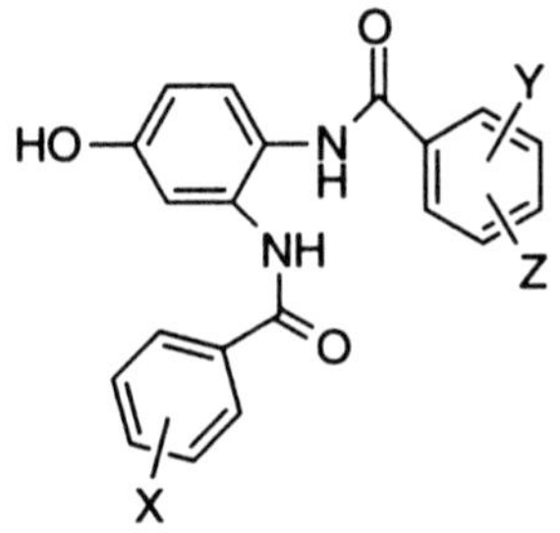

Figure 11.12. Example of a Sphinx library backbone.

case of the smaller libraries (up to 55 components) that most of the expected compounds are in fact present. An iterative deconvolution procedure was used to identify a low micromolar inhibitor of trypsin-catalyzed hydrolysis.

The Sphinx research group has been very active in preparing a large number of rigid or semirigid scaffolds to which a wide variety of functional groups could be attached. We have reported a facile, multistep array synthesis of hundreds of small molecule compounds with good purity levels and in sufficient quantities for direct screening in various biological assays.[26] An example of one such scaffold is shown in Figure 11.12.

Scaffolds have been designed to mimic a specific molecule, and rigid scaffolds with fixed display orientations have been used to generate large libraries. However, for effective lead discovery one may need a facile method to functionalize a small molecule scaffold with a wide variety of functional groups in a wide variety of spatial configurations. In other words, is there a way of easily varying the functional groups and scaffold simultaneously in a chemically simple fashion?

THE UNIVERSAL LIBRARY CONCEPT

Any biological macromolecule (receptor, enzyme, antibody, etc.) recognizes binding substrates through a number of precisely oriented physicochemical interactions described by parameters such as size, charge, hydrogen bonding ability, and hydrophobic interactions. We are attempting to systematically explore this multiparameter space by designing libraries that orient groups responsible for these binding interactions at many unique locations in space through a scaffolding approach. Each scaffold has the ability to display three or four functional groups in a large number of spatial orientations, and, furthermore, the scaffolds are easily modified to change their size, shape, and physical properties.

To rapidly prepare the required library using the smallest number of building blocks, the universal library is generated using a double-combinatorial

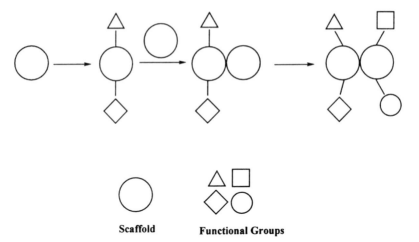

Figure 11.13. The double combinatorial approach.

approach developed in the Sphinx laboratories (Figure 11.13). In this scheme, functional groups, representing various physicochemical interacting properties, are introduced onto the first solid support-bound scaffold building block. Then the second scaffold building block is added, followed by an additional round of functional group introduction. The completed target molecule is then cleaved from the solid support to afford the desired product in solution and ready for screening. By applying the double-combinatorial approach a very large number of highly functionalized low molecular weight target molecules can be rapidly produced from a small collection of building blocks.

A major challenge was to select a general structural class of drug-like target molecules of sufficient generality to allow a wide variation in substitution patterns as well as a method to readily effect changes in overall size, shape, and physical properties. The biphenyl scaffold was selected as our initial class of target molecules. This scaffold allows for facile introduction of three or four functional groups in a large number of spatial arrangements simply by altering the substitution pattern on each aromatic ring. Furthermore, we have built into our design a simple method of changing the biphenyl scaffold to easily change the size, shape, and physical properties of the final products. While the products are prepared via a multistep procedure on solid support, the final products are cleaved from the resin for biological evaluation. It is important to note that the final cleavage reaction results in a pendant methyl group. This is significant since we desired not to be limited to an invariant OH or CO_2H group in our final products.

The biphenyl libraries (single compounds in a spatially addressable format) can be generated through a solution coupling reaction or through a solid-phase biphenyl coupling. The scaffold is functionalized with appropriate side

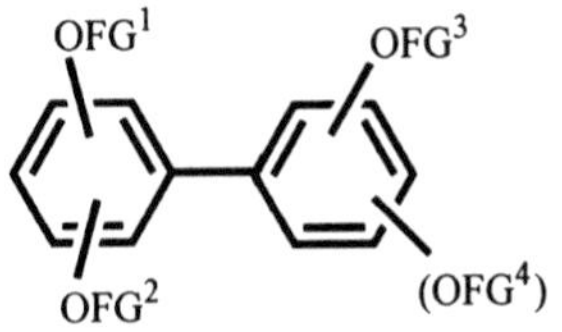

Figure 11.14. Generic structure of the biphenyl library.

chains on the solid support using the Mitsunobu reaction and the final products are then cleaved from the solid support for testing (Figure 11.14).[27]

The double-combinatorial approach described for the biphenyl libraries can easily be used to prepare vast numbers of target molecules. Still, this represents a formidable challenge in materials and labor and is probably unnecessary to explore the relevant space. Therefore, use of computational tools to design a subset of molecules that will most effectively explore multiparameter space is being pursued.

In conclusion, we have devised a versatile class of molecules that allows for the rapid display of multiple functional groups in large numbers of spatial arrangements and also allows simple modifications to significantly change the size, shape, and physical properties of the target molecules. We believe that this approach will effectively explore a large segment of multiparameter space. These universal libraries will be screened against biological targets of interest (receptors, enzymes, ion channels, antibodies, etc.) to quickly afford novel chemical leads.

REFERENCES

1. Moos WH, Green GD, Pavia MR (1993): Recent advances in the generation of molecular diversity. *Annu Rep Med Chem* 28:315–324.
2. Pavia MR, Sawyer TK, Moos WH (1993): The generation of molecular diversity. *Bioorg Med Chem Lett* 3:387–396.
3. Gallop MA, Barrett RW, Dower WJ, Fodor SPA, Gordon EM (1994): Applications of combinatorial technologies to drug discovery, 1: background and peptide combinatorial libraries. *J Med Chem* 37:1233–1251.
4. Gordon EM, Barrett RW, Dower WJ, Fodor SPA, Gallop MA (1994): Applications of combinatorial technologies to drug discovery, 2: combinatorial organic synthesis, library screening strategies, and future directions. *J Med Chem* 37:1385–1401.
5. Simon RJ, Kania RS, Zuckermann RN, Huebner VD, Jewell DA, Banville S, Ng S, Wang L, Rosenberg S, Marlowe CK, Spellmeyer DC, Tan R, Frankel AD, Santi DV, Cohen FE, Bartlett PA (1992): Peptoids: a modular approach to drug discovery. *Proc Natl Acad Sci USA* 89:9367–9371.
6. Zuckermann RN, Kerr JM, Kent SBH, Moos WH (1992): Efficient method for the preparation of peptoids [oligo(N-substituted glycines)] by submonomer solid-phase synthesis. *J Am Chem Soc* 114:10646–10647.

7. Zuckermann RN, Martin EJ, Spellmeyer DC, Stauber GB, Shoemaker KR, Kerr JM, Figliozzi GM, Goff DA, Siani MA, Simon RJ, Banville SC, Brown EG, Wang L, Richter LS, Moos WH (1994): Discovery of nanomolar ligands for 7-transmembrane G-protein-coupled receptors from a diverse N-(substituted)glycine peptoid library. *J Med Chem* 37:2678–2685.

8. Cho CY, Moran EJ, Cherry SR, Stephans JC, Fodor SPA, Adams CL, Sundaram A, Jacobs JW, Schultz PG (1993): An unnatural biopolymer. *Science* 261:1303–1305.

9. Peisach E, Casebier D, Gallion SL, Furth P, Petsko GA, Hogan JC Jr, Ringe D (1995): Interaction of a peptidomimetic aminimide inhibitor with elastase. *Science* 269:66–69.

10. Kahn M, Guest Editor (1993): Symposia 50. *Tetrahedron* 49:3433–3677.

11. Virgilio AA, Ellman JA (1994): Simultaneous solid-phase synthesis of β-turn mimetics incorporating side-chain functionality. *J Am Chem Soc* 116:11580–11581.

12. Geysen HM, Meloen R, Barteling S (1984): Use of peptide synthesis to probe viral antigens for epitopes to a resolution of a single amino acid. *Proc Natl Acad Sci USA* 81:3998–4002.

13. Smith AB III, Hirschmann R, Pasternak A, Akaishi R, Guzman MC, Jones DR, Keenan TP, Sprengeler PA, Darke PL, Emini EA, Holloway MK, Schleif WA (1994): Design and synthesis of peptidomimetic inhibitors of HIV-1 protease and renin: evidence for improved transport. *J Med Chem* 37:215–218.

14. Smith AB III, Guzman MC, Sprengeler PA, Keenan TP, Holcomb RC, Wood JL, Carroll PJ, Hirschmann R (1994): Denovo design, synthesis, and x-ray crystal structures of pyrrolinone-based β-strand peptidomimetics. *J Am Chem Soc* 116:9947–9962.

15. Tuchscherer G, Domer B, Sila U, Kamber B, Mutter M (1993): The TASP concept: mimetics of peptide ligands, protein surfaces and folding units. *Tetrahedron* 49:3559–3575.

16. Sila U, Mutter M (1995): Topological templates as tool in molecular recognition and peptide mimicry: synthesis of a TASK library. *J Mol Recognit* 8:29–34.

17. Belanger PC, Dufresne C (1986): Preparation of *exo*-6-benzyl-*exo*-2-(*m*-hydroxyphenyl)-1-dimethylaminomethylbicyclo[2.2.2]octane: a nonpeptide mimic of enkephalins. *Can J Chem* 64:1515–1519.

18. Hirschmann R, Nicolaou KC, Pietranico S, Salvino J, Leahy EM, Sprengler PA, Furst G, Smith AB III, Strader CD, Cascieri MA, Candelore MR, Donaldson C, Vale W, Maechler L (1992): Nonpeptidal peptidomimetics with a β-glucose scaffolding: a partial somatostatin agonst bearing a close structural relationship to a potent, selective substance P antagonist. *J Am Chem Soc* 114:9217–9218.

19. Hirschmann R, Nicolaou KC, Pietranico S, Leahy EM, Salvino J, Arison B, Cichy MA, Spoors PG, Shakespeare WC, Sprengler PA, Hamly P, Smith AB III, Reisine T, Raynor K, Maechler L, Donaldson C, Vale W, Freidinger RM, Cascieri MA, Strader CD (1993): Denovo design and synthesis of somatostatin nonpeptide peptidomimetics utilizing β-D-glucose as a novel scaffolding. *J Am Chem Soc* 115:12550–12568.

20. Hirschmann R, Sprengler PA, Kawasaki T, Leahy JW, Shakespeare WC, Smith AB III (1992): The first design and synthesis of a steroidal peptidomimetic: the potential value of peptidomimetics in elucidating the bioactive conformation of peptide ligands. *J Am Chem Soc* 114:9699–9701.

21. Kalindjian SB, Bodkin MJ, Buck IM, Dunstone DJ, Low CMR, McDonald IM, Pether MJ, Steel KIM (1994): A new class of nonpeptidic cholecystokinin-B/gastrin receptor antagonists based on dibenzobicyclo[2.2.2]octane. *J Med Chem* 37:3671–3674.

22. Patek M, Drake B, Lebl M (1994): All-*cis*-Cyclopentane scaffolding for combinatorial solid-phase synthesis of small nonpeptide compounds. *Tetrahedron Lett* 35:9169–9172.

23. Carell T, Wintner EA, Sutherland AJ, Rebek J Jr, Dunayevskiy YM, Vouros P (1995): New promise in combinatorial chemistry: synthesis, characterization, and screening of small molecule libraries in solution. *Chem Biol* 2:171–183.

24. Carell T, Wintner EA, Bashir-Hashemi AJ, Rebek J Jr (1994): Novel method for preparation of libraries of small organic molecules. *Angew Chem* 106:2159–2161.

25. Carell T, Wintner EA, Rebek J Jr (1994): A solution phase screening procedure for the isolation of active compounds from a library of molecules. Angew Chem 106:2161–2164.

26. Meyers HV, Dilley GJ, Durgin TL, Powers TS, Winssinger NA, Zhu H, Pavia MR (1995): Multiple simultaneous synthesis of phenolic libraries. *Mol Diversity* 1:13–20.

27. Pavia MR, Cohen MP, Dilley GJ, Dubuc GR, Durgin TL, Forman FW, Heidger ME, Milot G, Powers TS, Sucholeiki I, Zhou S, Hangauer DG (1996): The design and synthesis of substituted biphenyl libraries. *Bioorg Med Chem* 4:659–666.

12

APPLICATION OF COMBINATORIAL CHEMISTRY TO BIOPOLYMERS

EDMUND J. MORAN

Ontogen Corp., Carlsbad, California

This chapter reviews solid-phase methodologies recently developed for the combinatorial synthesis of peptidomimetic oligomeric structures. These methodologies have provided variations of new, unnatural biopolymers: polymers with pendant side-chain functionalities similar to peptides but possessing alternative backbones of different structural and physiochemical properties relative to the polyamide backbone.

Chemical replacements for the amide bond in peptides have been investigated extensively.[1] A wide variety of isosteric and isoelectronic chemical linkages that present various combinations of the spatial conformation and hydrogen bond donor/acceptor properties of the amide bond have been proposed. The –CONH– substructure has been replaced by isosteres such as ureas,[2] sulfonyls,[3] imidazoles,[4] acyl-*gem*-diaminoalkyls,[5] and α-aminophosphonamides.[6] The goals of such isostere replacements are to increase the affinity of the target peptide to a receptor, increase its cell membrane permeability, and/or reduce its susceptibility to proteolytic degradation or other metabolic processes in mammalian systems (Figure 12.1).

The proper choice of an isosteric replacement in any given system is difficult to predict and typically involves a trial and error process. An alternative approach is the replacement of *all* amide bonds in a peptide to produce an entirely new type of biopolymer.[7] The chemical and biological properties of these unnatural biopolymers would potentially be quite different from those

Combinatorial Chemistry and Molecular Diversity in Drug Discovery, Edited by
Eric M. Gordon and James F. Kerwin, Jr.
ISBN 0-471-15518-7   Copyright © 1998 by Wiley-Liss, Inc.

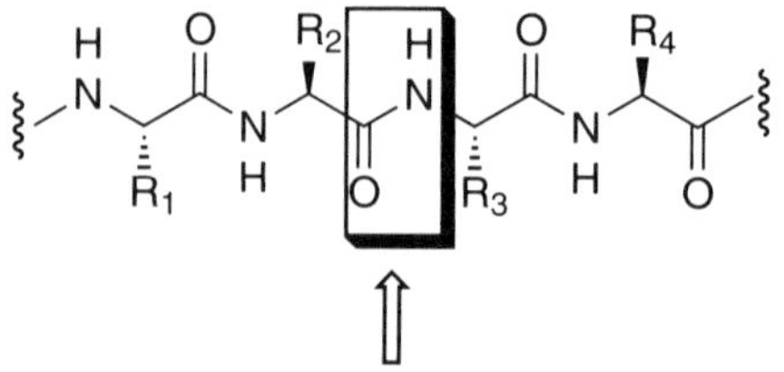

Figure 12.1. Isoteric replacement.

of peptides. Such structures may have improved pharmacodynamic properties such as increased biological half-life and bioavailability. The oligomeric nature of these structures may allow for synthesis of combinatorial libraries using the same techniques that have been developed for peptides.[8] Novel sequences discovered in biological screening of libraries would not be susceptible to rapid proteolysis and may be of use as agonists/antagonists for a variety of receptor and intracellular targets.

Schultz and coworkers have developed a method for the solid-phase synthesis of oligocarbamates, completely replacing the amide bonds in peptide-like sequences (Figure 12.2).[9,10] The monomers used for this solid-phase method are N-protected amino p-nitrophenylcarbonates. These are derived from amino alcohols, which are obtained commercially or prepared using a two-step reduction procedure on optically active α-amino acids (Scheme 12.1). The amine is protected as a carbamate (Fmoc or Nvoc) and the alcohol terminus is transformed into an activated oxycarbonyl derivative using p-nitrophenyl chloroformate. A full complement of monomers, representing the D- and L-amino acid side-chain repertoire and many unnatural side chains, have been synthesized.

The strategy for construction of oligocarbamates is derived from the C-to-N terminal synthesis technique commonly utilized in solid-phase peptide synthesis. Scheme 12.2 illustrates the synthesis of an N-acyl tetracarbamate. A solid-phase–bound amine nucleophile is treated with a solution of N-protected p-nitrocarbonate monomer to form the first carbamate bond. Coupling efficiency is evaluated using the Kaiser amine test (as employed for peptide synthesis) and is greater than 99%. The lack of 1,4 side-chain interactions in

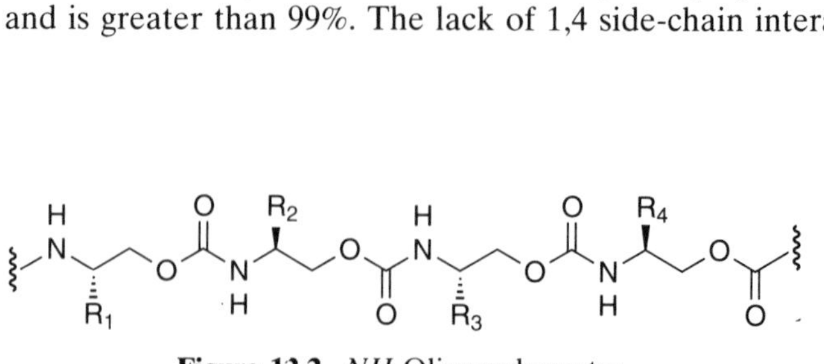

Figure 12.2. *NH*-Oligocarbamates.

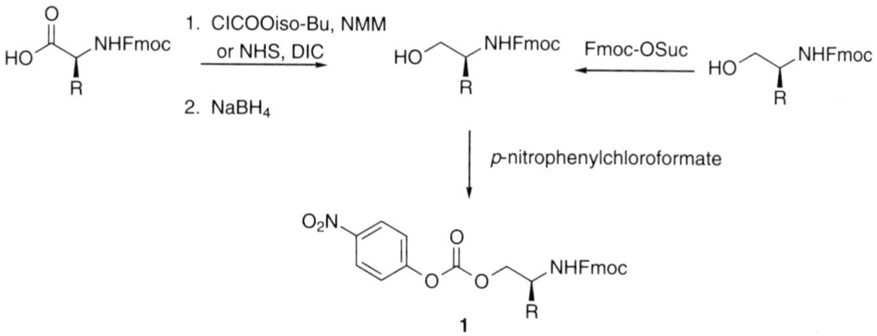

Scheme 12.1.

this polymer tends to reduce the influence of steric interference in the coupling chemistry. Following the first coupling step the amine on the monomer is deprotected, preparing the resin for a second coupling. Thus, the method requires a two-step procedure for each monomer addition to the growing biopolymer chain. After the final monomer is coupled and deprotected, the oligomer can be end-capped (via acylation, for example) and cleaved from the solid support using standard protocols.

Using this methodology sequences as long as 15-mers have been constructed on standard solid-phase resins.[11] Single diastereomers are isolated, indicating that loss of enantiomeric purity during monomer preparation or biopolymer synthesis is not observed. Split–pool library techniques have generated libraries of linear tetramers derived from a repertoire of 18–20 monomers. In addition, libraries of cycle trimers and tetramers utilizing a thiol–ether bridge (generated from thiol addition to an α-bromoacetamide terminus[12]) have been synthesized and screened against the platelet-derived glycoprotein II_bIII_a receptor.[13]

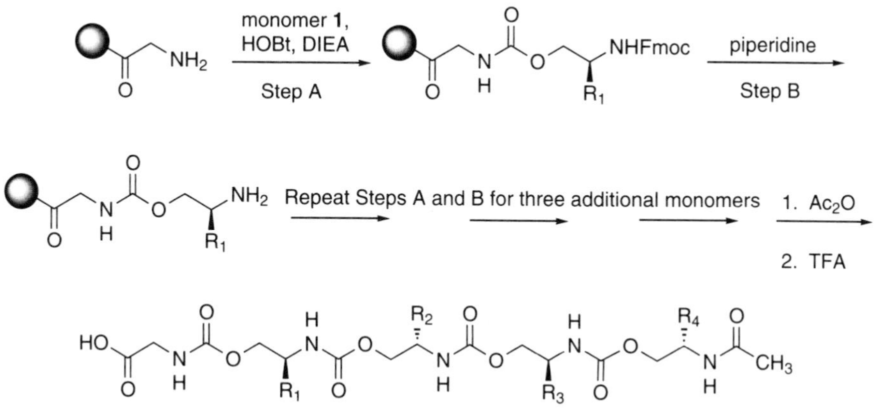

Scheme 12.2.

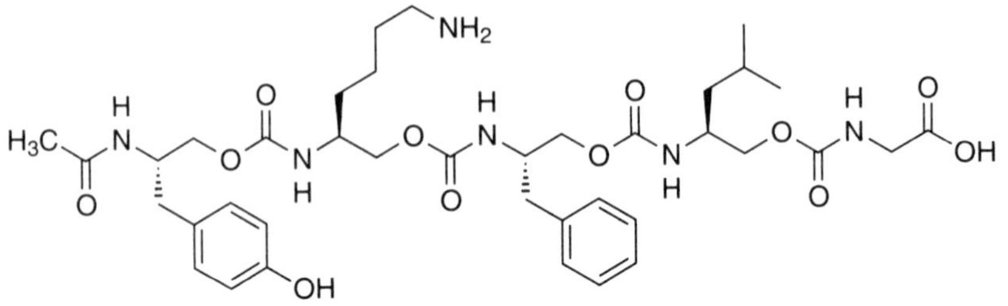

Figure 12.3. AcY^{c}K^{c}F^{c}L^{c}G.

The synthesis of a library of oligocarbamates on aminopropyl-derivatized glass has also been accomplished.[9] The use of Nvoc protecting groups (instead of Fmoc) in the preparation of monomers allows for the photodeprotection of amines with 365 nm light and thus application of the VLSIPS binary masking combinatorial library strategy.[14] A 256-member library containing all the possible single and multiple truncations of the sequence AcY^{c}F^{c}A^{c}S^{c}K^{c}I^{c}F^{c}L^c was synthesized. (The nomenclature uses single-letter amino acid designations for monomers with corresponding side chains. The superscript, c, refers to a carbamate linkage.) The library was screened with the monoclonal antibody 20D6.3, which was raised against the sequence AcY^{c}K^{c}F^{c}L^{c}G (Figure 12.3). Not only was this sequence identified but several other sequences with nanomolar affinity for the antibody were discovered.

The water/octanol partitioning coefficients for peptides are considerably higher than those for the corresponding oligocarbamate sequences (Table 12.1). Additionally, the oligocarbamates are completely resistant to proteases under in vitro conditions, which completely degrade the corresponding peptide sequences in less than 20 min.[9]

N-ALKYL OLIGOCARBAMATES

Extension of the solid-phase oligocarbamate synthesis strategy to achieve higher side-chain density has recently been accomplished by Schultz and co-workers.[15] In addition to substitution of a side chain on the backbone carbon

TABLE 12.1. Partition Coefficient of Oligocarbamates and Corresponding Peptides

Sequence	Water/Octanol Partition Coefficient
AcYKFLG-OH	90
AcY^{c}K^{c}F^{c}L^{c}G-OH	0.5
AcYIFLG-OH	10
AcY^{c}I^{c}F^{c}L^{c}G-OH	0.4

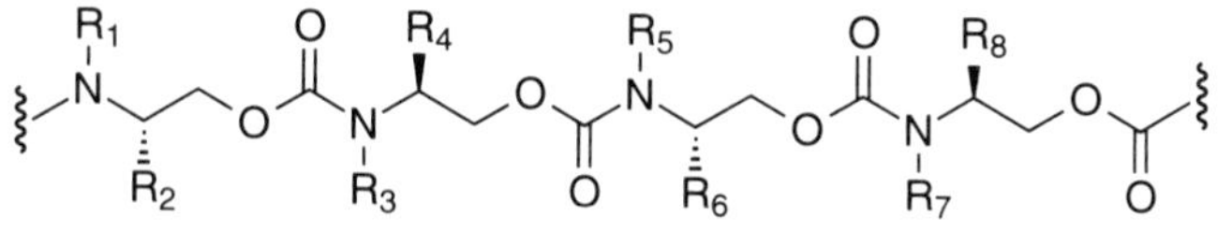

Figure 12.4. *N*-Alkyl-Oligocarbamate.

atom, the carbamate nitrogen can also be substituted with a pendant group (Figure 12.4). Thus, each monomer unit in a sequence has two sites for the introduction of diversity. This substitution removes the hydrogen bond donor properties of the biopolymer backbone but allows selection of a mix of side chains with varying properties. Side chains bearing hydrophobic, aromatic, and positively or negatively charged functionality are possible. Local folding changes are expected within the backbone as 1-*N*-alkyl-2-*C*-alkyl group interactions remove eclipsed (and perhaps gauche) conformations accessible in the *NH*-oligocarbamate sequences. Alternatively, in analogy to amide bond conformational analysis, the all-trans orientation in the –CONHR– linkage is no longer expected to be preferred relative to the cis orientation when the linkage becomes –CON(R_1)R_2–.

The synthesis of *N*-alkyl oligocarbamates proceeds with the same repertoire of monomers developed for *NH*-oligocarbamates. Following the coupling of the first monomeric unit the amino group on the growing chain is deprotected and coupled to a carboxylic acid using standard peptide bond-forming conditions (Scheme 12.3). The amide bond is then reduced on the resin, resulting in a secondary amine ready for the next round of monomer coupling. Thus, a four-step procedure per round of synthesis ensures a considerably dense packing of side-chain diversity. The commercial availability of hundreds of carboxylic acids greatly expands the building block source for oligocarbamates.

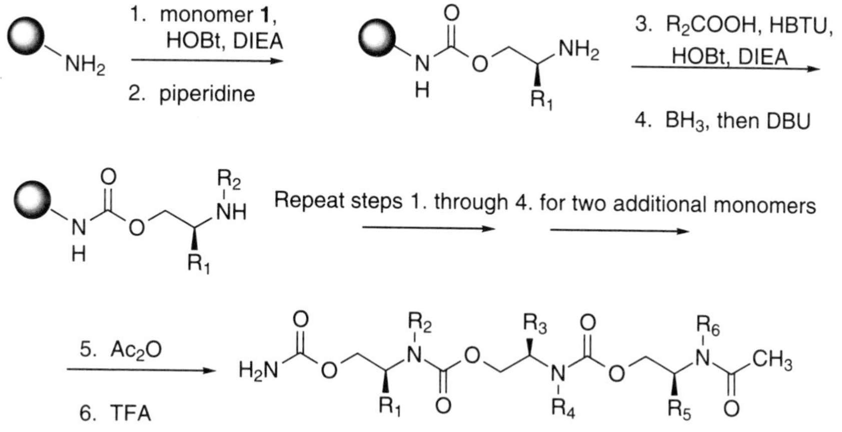

Scheme 12.3.

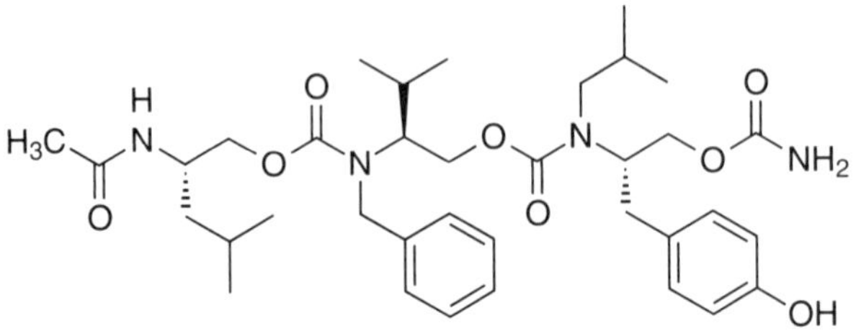

Ac-L^{c}F^{n}V^{c}L^{n}Y^c

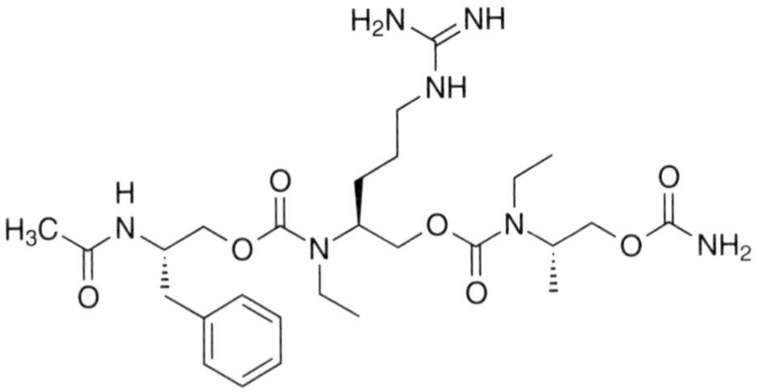

Ac-F^{c}Et^{n}R^{c}Et^{n}A^c

Figure 12.5. Ac-L^{c}F^{n}V^{c}L^{n}Y^c and Ac-F^{c}Et^{u}R^{c}Et^{n}A^c.

Yields for individual monomer couplings are greater than 96%. Synthesis of sequences such as Ac-L^{c}F^{n}V^{c}L^{n}Y^c and Ac-F^{c}Et^{n}R^{c}Et^{n}A^c have been reported (Figure 12.5). The nomenclature introduced by Schultz uses an additional superscript, n, to refer to an *N*-substitution. If the alkyl group represents an amino acid side chain the corresponding single-letter amino acid code is used to designate the group; if not, then the common alkyl group abbreviation itself is used. Investigation of the conformational and pharmacological properties of *N*-alkyl oligocarbamates is the subject of further studies.

OLIGOUREAS

Several efforts have been directed toward the preparation of unnatural oligourea biopolymers.[16] As with oligocarbamates, the diversity of monomers is derived principally from the amino acid and amino alcohol pool; the prepara-

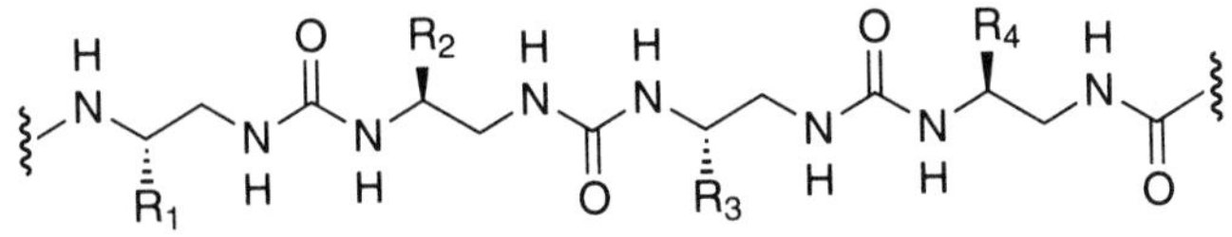

Figure 12.6. Oligourea.

tion of N-protected amino alcohols follows the general synthetic methods described above (Figure 12.6).

Burgess and coworkers have prepared diamino monomers utilizing a phthalimide/carbamate protecting group strategy (Scheme 12.4).[16a] The transformation of the hydroxyl group of an amino alcohol to a latent amine is accomplished by displacement under Mitsunobu conditions with phthalimide. This phthalimide functionality serves as the amine protecting group on the nucleophilic end of the monomer during solid-phase synthesis. Removal of the Boc group from the nonphthalimide end of the diamino monomer allows for *in situ* activation of the monomer as an isocyanate. The isocyanate carbonyl thus becomes the bridging carbonyl of the urea backbone.

Growth of the unnatural urea polymer on solid support is accomplished by addition of the activated monomer to a support bound nucleophilic amine, resulting in a new urethane linkage (Scheme 12.5). Individual coupling yields were not reported, but Kaiser tests during synthesis indicated complete reaction after 1.5 h. Following the coupling step, the next amine nucleophile is then revealed by treatment of resin with hydrazine for 15 h.

The synthesis of a tetrameric urea/monoamino acid sequence and a mixed dipeptide/dioligourea has been reported: $CH_2G^u\text{-}CH_2F^u\text{-}CH_2F^u\text{-}CH_2A^u\text{-}A\text{-}NH_2$ and $YG\text{-}CH_2G^u\text{-}CH_2F^u\text{-}L\text{-}NH_2$ (Figure 12.7), where CH_2 identifies the methylene group position relative to the side chain in the monomer unit. The superscript u denotes a urea linkage. Overall yields of purified material (RP-HPLC) for these sequences were 46 and 17%, respectively. The variability in these yields warrant further investigation of the individual coupling steps to optimize the procedure for future combinatorial library synthesis.

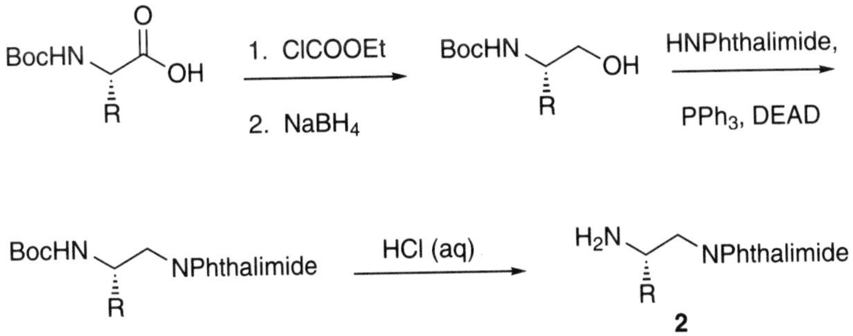

Scheme 12.4.

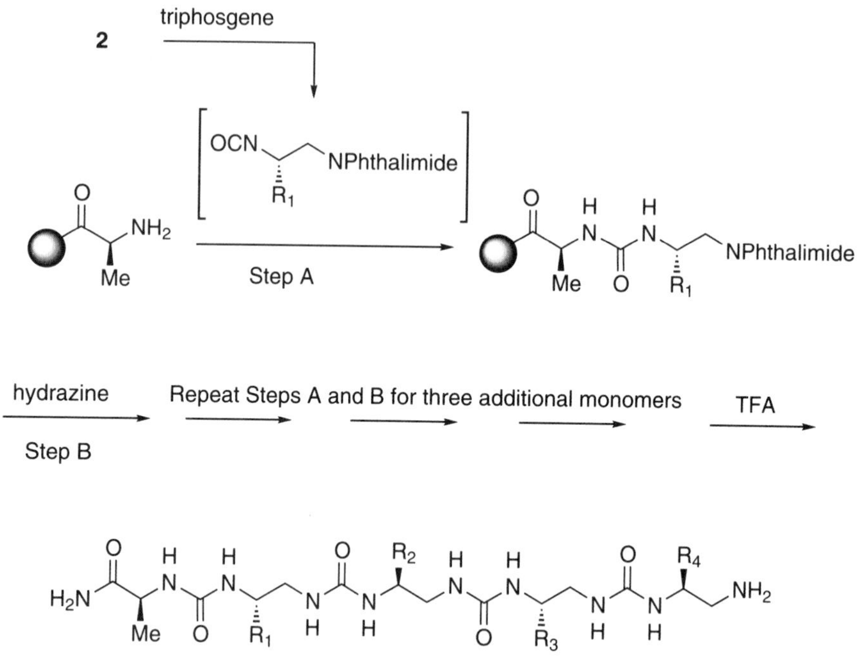

Scheme 12.5.

Schultz and coworkers have taken another approach to differential protection of the diamino monomer.[16b] In their work, an *N*-protected amino alcohol is transformed to an *N*-protected amino azide (Scheme 12.6). The carbamate-protecting group is then removed and the monomer is activated for coupling via the formation of a *p*-nitrophenyl carbamate. The carbamates are stable and can be purified on silica gel. The coupling strategy is illustrated in Scheme 12.7, which shows the synthesis of a trimer on solid support. The carbamate carbonyl is incorporated into the urea linkage and the azide then serves as the latent nucleophilic amine.

Coupling is achieved in 4 h and is monitored with Kaiser tests to ensure complete reaction. The azide functionality is reduced on resin in a relatively short time (<2 h) with a procedure that does not interfere with acid-labile side-chain protecting groups (*N*-Boc, *tert*-butyl esters). Overall, reverse-phase high-performance liquid chromatography (RP-HPLC) purified yields for alkyl urea-capped and uncapped trimeric sequences are reported between 54 and 76%. The sequences benzyl[u]-S[u]-Y[u]-A[u] and L[u]-F[u]-A[u] are shown in Figure 12.8.(The authors use an abbreviated urea nomenclature; all sequences are of the CH$_2$X orientation.) Synthesis of combinatorial libraries and the evaluation of these polymers in biological screening assays is ongoing.

The nature of a diamino monomer raises the possibility of using the same coupling strategies to build an "inverse" sequence, i.e., produce a 1,5 side-chain relationship versus a "normal" 1,6 relationship. In this case, a monomeric

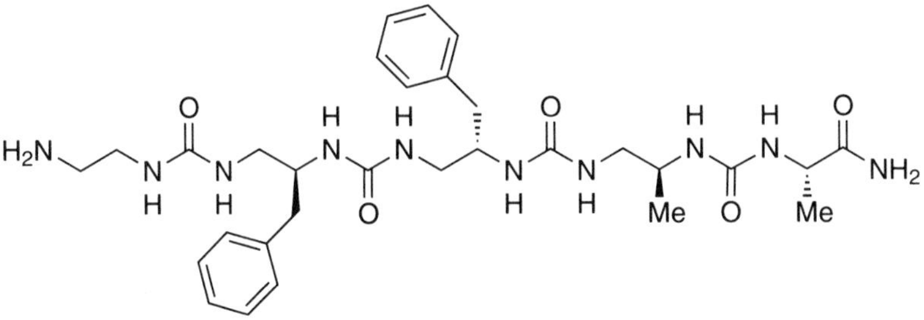

$$CH_2G^u\text{-}CH_2F^u\text{-}CH_2F^u\text{-}CH_2A^u\text{-}A\text{-}NH_2$$

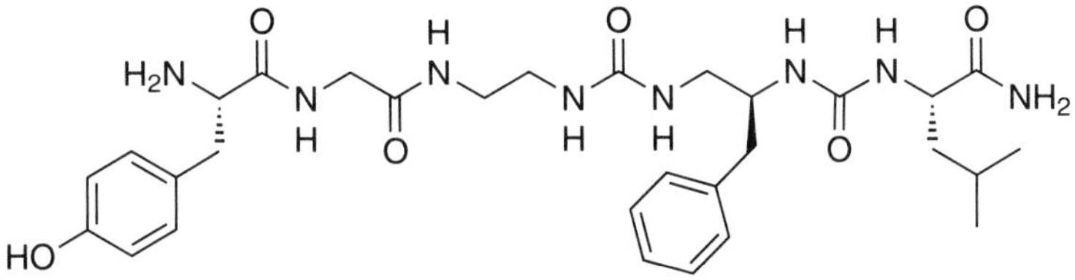

$$YG\text{-}CH_2G^u\text{-}CH_2F^u\text{-}L\text{-}NH_2$$

Figure 12.7. $CH_2G^u\text{-}CH_2F^u\text{-}CH_2F^u\text{-}CH_2A^u\text{-}A\text{-}NH_2$ and $YG\text{-}CH_2G^u\text{-}CH_2F^u\text{-}L\text{-}NH_2$.

unit in which the amine *distal* to the side-chain group is *N*-protected and the amine *proximal* to the side chain is activated would have to be prepared. This mixed strategy adds complexity to the biopolymer sequence possibilities, but also diversity. The nomenclature rule proposed by Burgess is therefore useful in differentiating the orientation of monomer units in a mixed inverse/normal oligourea sequence.

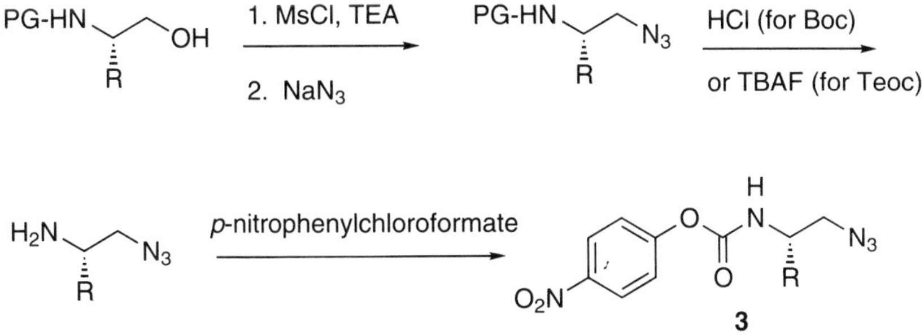

Scheme 12.6.

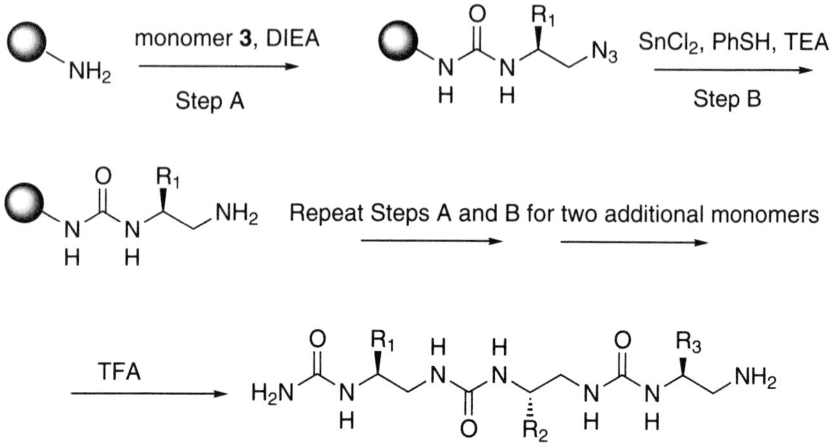

Scheme 12.7.

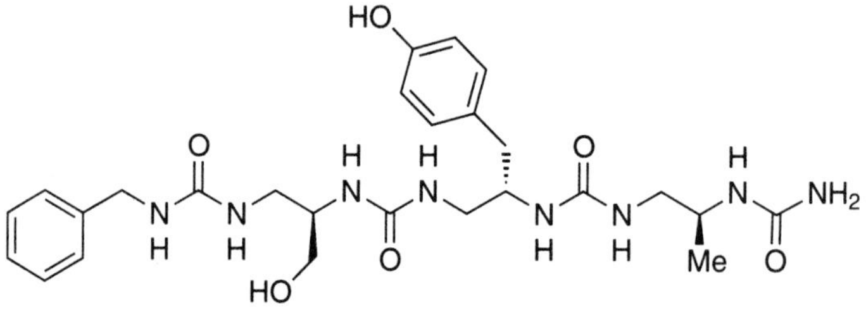

Benzyl-S^u-Y^u-A^u

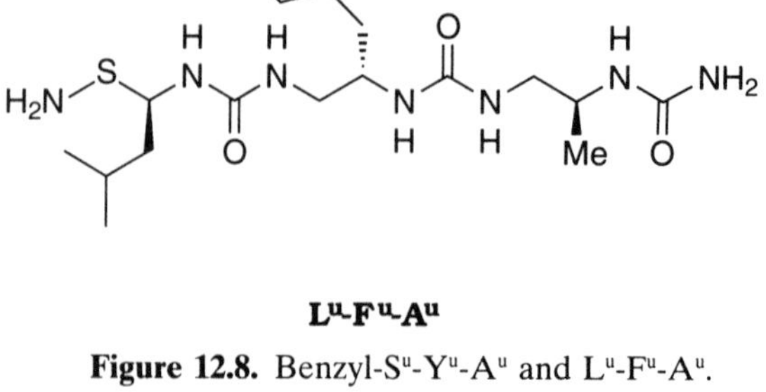

L^u-F^u-A^u

Figure 12.8. Benzyl-S^u-Y^u-A^u and L^u-F^u-A^u.

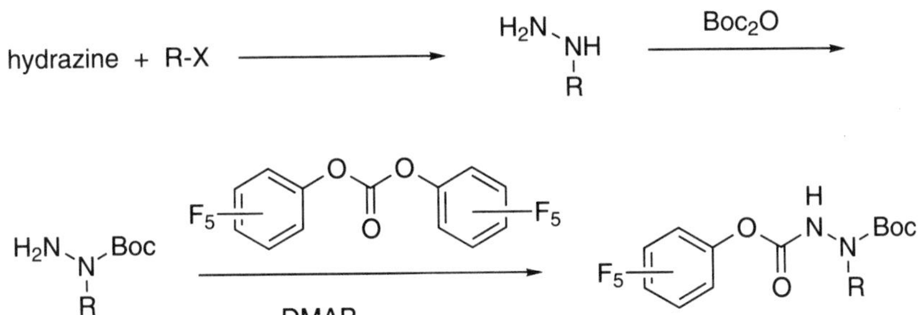

Scheme 12.8.

AZATIDES

The term azapeptide refers to a peptide in which a nitrogen atom replaces the α-carbon in one or more of the amino acids in the sequence.[17] This backbone isosteric replacement renders the polymer more resistant to enzymatic cleavage and has proved useful in the development of cysteine and serine protease inhibitors.[18] The exchange of the sp^3 carbon atom for an sp^2 nitrogen atom results in an achiral monomer and alters the expected orientation of side-chain R-groups on the peptide. This reorientation is reminiscent of the translocation of the side-chain R-groups in peptoids[19,20] (Figure 12.9).*

Janda and coworkers have reported the construction of oligomers devoid of any α-carbon amino acids.[21] These oligomers are synthesized entirely from 1-alkyl-1-*tert*-butyloxycarbonyl protected hydrazine monomers. These monomers have been called α-aza-amino acids and their preparation is shown in Scheme 12.8. Hydrazine is monoalkylated with a selection of alkyl halides and then Boc-protected. The latent carbonyl bridging unit is provided by the pentafluorophenyl carbamate, which also serves to activate the monomers for coupling. Yields for the preparation of dimers in solution range between 82 and 92%, with the sterically least-demanding alkyl substitutions providing the highest values.

The oligomeric coupling strategy follows a pattern conceptually opposite

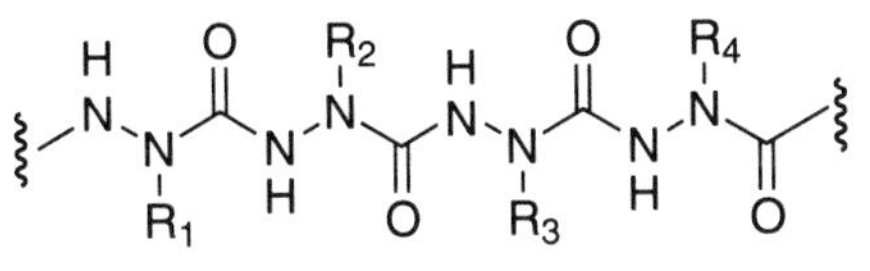

Figure 12.9. Azatide.

---

* The synthesis and utility of such oligomers have been investigated quite extensively by Zuckermann and coworkers (see Chapter 6).

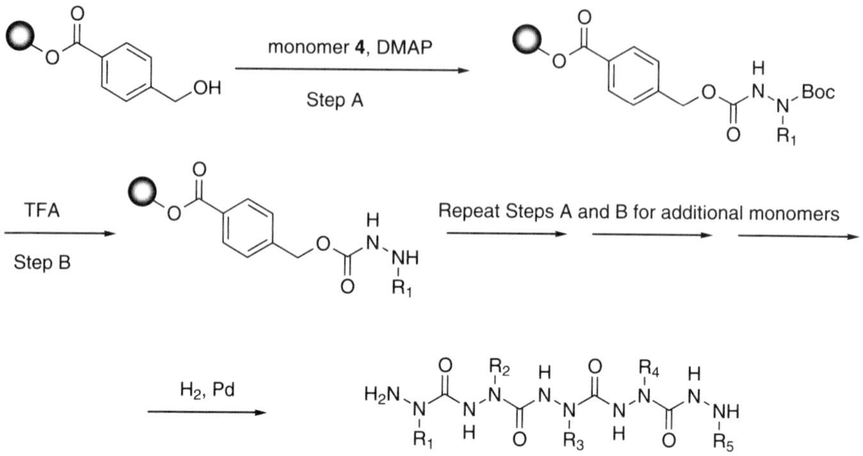

Scheme 12.9.

to solid-phase peptide synthesis (Scheme 12.9). The oligomers on support are constructed from N-to-C terminus by coupling a support bound nucleophile[22] to the electrophilic "amino" terminus of the protected α-aza-amino acid monomer. Interestingly, only N-alkyl substitution on the nucleophilic end of the monomer is tolerated. When an N-alkyl group resides on the pentafluorophenyl carbamate no coupling is observed. This is not due to steric interference from the alkyl group on the electrophile but rather is explained by the mechanism of the coupling reaction. An intermediate isocyanate must be formed *in situ* to provide a high yield coupling; N-substitution on the pentafluorophenyl carbamate precludes this. Acidic deprotection steps are used between each coupling, thus, benzyl groups are used for the protection of potentially reactive side-chain functionality. The full-length azatide is released from the benzyl alcohol linker and simultaneously deprotected via catalytic hydrogenation.

The resulting biopolymers are called azatides. They possess side chains and urea bond carbonyls in sequence register with the side chains and amide bond carbonyls in peptides. The question of peptide–azatide sequence equivalance has been investigated briefly. The sequence Tyr[a]-Gly[a]-Gly[a]-Phe[a]-Leu[a] (the superscript a denotes an azatide linkage) was unable to displace the β-endorphin antigenic peptide sequence [Leu[5]]-enkephalin from the Herz 3-E7 antibody[23] in competition ELISA experiments with up to 1 mM azatide ligand concentrations. Perhaps this result is not unexpected. Single substitutions of α-nitrogens for α-carbons in extended peptides may not disrupt peptide–receptor interactions, but the additive conformational and bond-length perturbations occurring in a pure azatide should render peptide/azatide sequences biochemically nonequivalent.

CONCLUSION

Research in unnatural biopolymers draws upon extensive work in peptidomimetic and amide isostere chemistry. Rather than following a single amide bond replacement strategy, chemists are replacing the entire backbone of peptides with a variety of chemical linkages, many quite different from the primary amide bond linkage. Considerable interest in unnatural biopolymers that mimic or replace peptides continues. Initial reports of progress in the synthesis of oligosulfones and oligosulfoxides on solid support have been reported.[11] Undoubtably, other biopolymers will be of interest, as will the combination of compatible coupling strategies to make heterocopolymers. This chapter has discussed some of the synthetic methods to produce these new biopolymers on solid support. As demonstrated with oligocarbamates, chemists can employ the powerful combinatorial synthesis methods developed for oligopeptide research to identify novel chemical structures using modern biochemical screening techniques. What will be the electronic, conformational, and solubility properties of these new polymers? Will they have an increased propensity to cross lipid bilayers and resist metabolic alteration? The work described here should provide a basis for further research into the new materials and their properties.

REFERENCES

1. (a) Spatola AF (1983): In Weinstein B, ed. *Chemistry and Biochemistry of Amino Acids, Peptides, and Proteins.* New York: Dekker, p 267; (b) Farmer PS (1980): In Ariens EJ, ed. *Drug Design,* vol 10. New York: Academic, p 119; (c) Freidinger RM (1989): Non-peptide ligands for peptide receptors. *TIPS Rev* 10:270–274, (d) Hirschmann R (1991): Medicinal chemistry in the golden-age of biology—lessons from steroid and peptide research. *Angew Chem Int Ed Engl* 30:1278–1301.

2. Lam PY, Jadhav PK, Eyermann CJ, Hodge CN, Ru Y, Bacheler LT, Meek JL, Otto MJ, Rayner MM, Wong YN, Chang C-H, Weber PC, Jackson DA, Sharpe TA, Erickson-Viittanen (1994): Rational design of potent, bioavailable, nonpeptide cyclic ureas as HIV protease inhibitors. *Science* 263:380–384.

3. Maurer R, Gaehwiler BH, Buescher HH, Hill RC, Reomer D (1982): Opiate antagonistic properties of an octapeptide somatostatin analog. *Proc Natl Acad Sci USA* 79:4815–4817.

4. (a) von Geldern TW, Hutchins CW, Kester JA, Wu-Wong J, Chiou W, Dixon DB, Opgenorth TJ (1996): Azole endothelin antagonists. A receptor model explains an unusual structure-activity profile. *J Med Chem* 39:957–967, (b) von Geldern TW, Kester JA, Bal R, Wu-Wong JR, Chiou W, Dixon DB, Opgenorth TJ (1996): Azole endothelin antagonists. Structure activity studies. *J Med Chem* 39:968–981, (c) Gordon TD, Singh J, Hansen PE, Morgan BA (1993): Synthetic approaches to the azole peptide mimetics. *Tetrahedron Lett* 34:1901–1904.

5. Chorev M, Goodman M (1993): A dozen years of retro-inverso peptidomimetics. *Acc Chem Res* 26:266–273.

6. Bartlett PA, Marlowe CK (1987): Evaluation of intrinsic binding energy from a hydrogen bonding group in an enzyme inhibitor. *Science* 235:569–571.

7. For solution phase work see (a) Smith AB, Knight SD, Sprengeler PA, Hirschmann R (1996): The design and synthesis of 2,5-linked pyrrolinones. *Bioorg Med Chem* 4:1021–1034, (b) Smith AB, Keenan TP, Holcomb RC, Sprengeler PA, Guzman MC, Wood JL, Carroll PJ, Hirschmann R (1992): Design, synthesis, and crystal-structure of a pyrrolinone-based peptidomimetic possessing the conformation of a beta-strand. *J Am Chem Soc* 114:10672; (c) Hagihara M, Anthony NJ, Stout TJ, Clardy J, Schreiber SJ (1992): Vinylogous polypeptides—an alternative peptide backbone. *J Am Chem Soc* 114:6568–6570, (d) Liskamp RM (1994): Opportunities for new chemical libraries. *Angew Chem Int Engl* 33:633–636.

8. (a) Gallop MA, Barrett RW, Dower WJ, Fodor SPA, Gordon EM (1994): Applications of combinatorial technologies to drug discovery. *J Med Chem* 37:1233–1251, (b) Janda KD (1994): Tagged versus untagged libraries. *Proc Natl Acad Sci USA* 91:10779–10785.

9. Cho CY, Moran EJ, Cherry SR, Stephans JC, Fodor SPA, Adams CL, Sundaram A, Jacobs JW, Schultz PG (1993): An unnatural biopolymer. *Science* 261:1303–1305.

10. The solution-phase synthesis of oligocarbamate linkages to fully replace the phosphodiester bond in short nucleotide sequences has been reported: (a) Stirchak EP, Summerton JE, Weller DD (1987): Uncharged stereoregular nucleic-acid analogs. *J Org Chem* 52:4202–4206, (b) Mungall WS, Kaiser JK (1977): Carbamate analogs of oligonucleotides. *J Org Chem* 42:703–706.

11. Moran EJ, Wilson TE, Cho CY, Cherry SR, Schultz PG (1995): Novel biopolymers for drug discovery. *Biopolymers (Pept Sci)* 37:213–219.

12. Barker PL, Bullens S, Bunting S, Burdick DJ, Chan KS, Deisher T, Eigenbrot C, Gadek TR, Gantzos R, Lipari MT, Muir CD, Napier MA, Pitti RM, Padua A, Quan C, Stanley M, Struble M, Tom JYK, Burnier JP (1992): Cyclic RGD peptide analogs as antiplatelet antithrombotics. *J Med Chem* 35:2040–2048.

13. Cho CY, personal communication.

14. Fodor SPA, Read JL, Pirrung MC, Stryer L, Lu AT, Solas D (1991): Light directed, spatially addressable parallel chemical synthesis. *Science* 251:767–773.

15. Paikoff SJ, Wilson TE, Cho CY, Schultz PG (1996): The solid-phase synthesis of N-alkylcarbamate oligomers. *Tetrahedron Lett* 37:5653–5656.

16. (a) Burgess K, Linthicum DS, Shin H (1995): Solid-phase syntheses of unnatural biopolymers containing repeating urea units. *Angew Chem Int Ed Engl* 34:907–909.

17. Gante J (1989): Azapeptides. *Synthesis,* 405–413.

18. Magrath J, Abeles RH (1992): Cysteine protease inhibition by azapeptide esters. *J Med Chem* 35:4279–4283.

19. Zuckermann RN, Kerr JM, Kent SBH, Moos WH (1992): Efficient method for the preparation of peptoids [oligo(N-substituted glycines)] by submonomer solid-phase synthesis. *J Am Chem Soc* 114:10646–10647.

20. The simplicity in the construction of various sequences of this type of oligomer make it attractive from a synthetic standpoint. A large variety of amine inputs can be explored, allowing for incorporation of a diverse array of unnatural side chains in addition to those of the natural amino acids.

21. Han H, Janda KD (1996): Azatides—solution and liquid-phase syntheses of a new peptidomimetic. *J Am Chem Soc* 118:2539–2544.

22. The support in this case is not a traditional solid-phase resin such as polystyrene or polyacrylamide. A linear homopolymer (MeO-PEG) is used, which is soluble in many organic solvents but can be precipitated in diethyl ether. This allows solution-phase chemistry during synthesis but solid-phase filtration of the precipitated polymer during resin washing and mechanical transfer steps. See Han H, Wolfe MM, Brenner S, Janda KD (1995): Liquid-phase combinatorial synthesis. *Proc Natl Acad Sci USA* 92:6419–6423.

23. Meo T, Gramsch C, Inan R, Hollt V, Weber E, Herz A, Reithmuller G (1983): Monoclonal-antibody to the message sequence TYR-GLY-GLY-PHE of opioid-peptides exhibits the specificity requirements of mammalian opioid receptors. *Proc Natl Acad Sci USA* 80:4084–4088.

13

OLIGOSACCHARIDE AND GLYCOCONJUGATE SOLID-PHASE SYNTHESIS TECHNOLOGIES FOR DRUG DISCOVERY

MICHAEL J. SOFIA

Transcell Technologies, Inc., Cranbury, New Jersey

Saccharides and glycoconjugates play key functions in cell biology. As integral parts of membrane-bound glycoproteins and glycolipids they modulate biological processes such as cell signaling, cellular differentiation, and cellular adhesion.[1–3] Many pathogenic organisms rely on their ability to recognize carbohydrate determinants on the surface of host cells to facilitate attachment and subsequent infection.[4,5] Certain cell-surface carbohydrates act as biological recognition elements for attachment of tumor associated cells. Also, many hormones rely on cell-surface carbohydrates to facilitate cellular binding. The saccharide portions of various natural product glycoconjugates function as key molecular recognition elements important to the biological properties of the natural product. Antineoplastic glycoconjugates, such as ciclcamycin and calicheamicin, require their oligosaccharide units for sequence-specific DNA binding.[6] Neoglycoconjugates constructed by the attachment of saccharides to biologically significant peptides provide enhanced peptide stability and increase transport across biological barriers.[7,8] In addition, because saccharides are characteristically conformationally rigid molecules rich in stereochemically defined functionality, they provide a unique platform for displaying functional elements in a defined arrangement unparalleled by any other molecular sys-

Combinatorial Chemistry and Molecular Diversity in Drug Discovery, Edited by
Eric M. Gordon and James F. Kerwin, Jr.
ISBN 0-471-15518-7   Copyright © 1998 by Wiley-Liss, Inc.

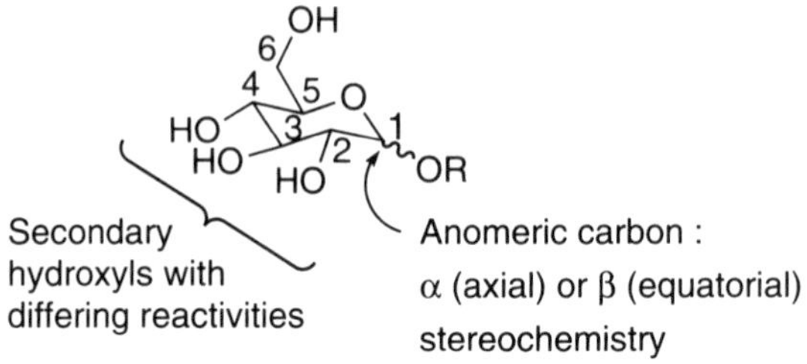

Figure 13.1. Molecular characteristics of a sugar unit.

tem.[9] The utilization of carbohydrate-based molecular scaffolds for generating molecular diversity is largely an unexplored area that holds great promise for drug discovery. However, to adequately exploit the vast potential envisioned for saccharide-based drugs, efficient and rapid methods for accessing these molecules need to be developed.

Combinatorial library technology as a new paradigm for accelerating the identification of novel lead structures by rapidly generating vast chemical diversity has gained wide acceptance in the drug discovery community.[10,11] The ability to generate complex mixtures of compounds or to rapidly synthesize large numbers of compounds in a spatial array has relied on highly efficient chemistries applied either to the solid phase or in solution.[12-14] These chemical library generation efforts have significantly expanded the capabilities of medicinal chemists.[15-21] However, the application of library technology to the synthesis of oligosaccharides and glycoconjugates has largely been ignored. One reason for the lack of focus on the generation of carbohydrate-based libraries relates to the chemical complexity of the monosaccharide building block with its multiple reactive sites and complexities of stereochemistry including the anomeric center (Figure 13.1). In addition, development has been hampered by the lack of a single generalizable glycosidic bond-forming reaction that can be applied reliably to a wide variety of substrates either in solution or on the solid phase and the absence of a well-developed solid-phase synthesis technology applicable to oligosaccharides or glycoconjugates using the various combinatorial approaches.

To ultimately take full advantage of combinatorial carbohydrate technology for drug discovery, library deconvolution strategies are necessary. For peptide and oligonucleotide libraries, microsequencing techniques and amplification technologies are used to identify active agents from a compound pool. Library encoding techniques have been developed for deconvoluting peptide and some small molecule libraries. Unfortunately, none of these techniques are available for oligosaccharides or glycoconjugates. However, existing intelligence-based deconvolution strategies and spatially arrayed reaction formats can be used with rapid and efficient methods for the construction of library constituents to solve the encoding problem. Therefore, solid-phase synthesis methods for accessing oligosaccharides and glycoconjugates would provide the technologies that are needed for the generation of carbohydrate-based chemical libraries.

SOLID-PHASE SYNTHESIS OF OLIGOSACCHARIDES

In recent years, much effort has been directed toward the development of solid-phase oligosaccharide synthesis technologies. The attractive aspects of solid-phase synthesis methods include speed and efficiency, the ability to use excess reagents to drive the reactions to completion and, therefore, obtain high chemical yields, and the ability to eliminate tedious workup and purification steps. The requirements for successful implementation of solid-phase synthesis technologies to the construction of oligosaccharides and glycoconjugates are similar to those that exist for the development of solid-phase chemical synthesis methods for other molecular systems. The protocol must include a method for attachment of a saccharide to a solid or polymeric support; it must include a method for attaching subsequent saccharides to the existing polymer-bound unit, and it must provide a method for detaching the completed saccharide from the polymeric support without destruction of the newly synthesized molecules of interest.

The solid-phase construction of oligosaccharides and glycoconjugates is further complicated by the very nature of the molecules themselves (Figure 13.1). The glycosidic bond between two saccharide units or between a saccharide unit and its aglycone conjugate can be one of two stereochemistries, α (axial) or β (equatorial); therefore, adequate control of this anomeric stereochemistry becomes an important issue. Also, in typical oligosaccharides, glycosidic linkages occur to primary and secondary alcohols. Consequently, a glycosylation technology that is sufficiently reactive to accommodate all types of substrates is needed. In addition, because of the multiple reactive sites on any monosaccharide unit, the potential exists for the construction of not only linear but branched systems. The existence of these multiple reactive sites complicates the molecular construction because of the need to control for site-selective glycosylation. Typically, in solution, control of site-selective glycosylation is accomplished by judicious choice of protecting groups. Consequently, any solid-phase oligosaccharide technology must provide a protecting group strategy as an integral part of its protocol.

Attempts to execute solid-phase synthesis of oligosaccharides using existing glycosylation chemistries began approximately 25 years ago (Figure 13.2). These attempts demonstrated that one could indeed attach a carbohydrate to a polymer support, then glycosylate the polymer-bound monosaccharide and remove the product from the support. The first reported attempt at accomplishing solid-phase oligosaccharide synthesis was provided by Fréchet and coworkers (Scheme 13.1).[22–24] They utilized a glycosyl halide as the donor in a classical alcoholysis reaction. Using the styrene–divinylbenzene copolymer as their solid support, they attached it to the anomeric carbon of the first monosaccha-

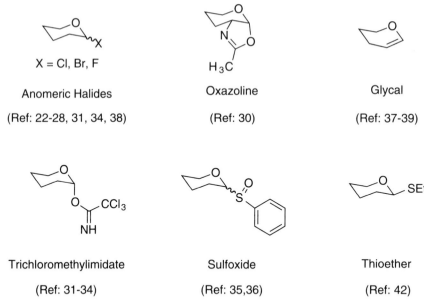

X = Cl, Br, F   H₃C

Anomeric Halides   Oxazoline   Glycal

(Ref: 22-28, 31, 34, 38)   (Ref: 30)   (Ref: 37-39)

Trichloromethylimidate   Sulfoxide   Thioether

(Ref: 31-34)   (Ref: 35,36)   (Ref: 42)

Figure 13.2. Donor technologies applied to polymer-supported oligosaccharide synthesis.

ride via a propenyl ether linker. Oxidative cleavage of this linker with ozone resulted in an oligosaccharide containing an anomeric acetaldehyde residue. Demonstrating the feasibility of this approach, a disaccharide was constructed on the solid-phase by the addition of a glycosyl bromide to a polymer-bound glycosyl acceptor; however, glycosidic bond formation was reported only to a C-6 primary hydroxyl of a pyranose nucleus. Also, this chemistry was limited by very prolonged reaction times and poor anomeric stereochemical control.

An extension of the work by Fréchet and coworkers was undertaken by Zehavi, who developed a photolabile linker to the styrene–divinylbenzene solid support.[25,26] Zehavi prepared the disaccharide isomaltose by first attaching 6-nitrovanillin through an ether linkage to chloromethylated Merrifield resin and glycosylated the linker at the reduced aldehyde functionality of the nitrovanillin linker employing a glycopyranosyl bromide donor (Scheme 13.1). Subsequent removal of the C-6 protecting group and glycosylation to form the 1→6 glycosidic linkage was followed by irradiation at 320 nm to cleave the disaccharide from the resin.

Guthrie and coworkers chose a different strategy in their early attempts to effect solid-phase oligosaccharide synthesis.[27,28] They effected the attachment of the first monosaccharide building block to a polymer support by copolymerizing 6-O-vinylbenzoylglucopyranose derivatives with styrene, thus producing soluble-polymer-bound monosaccharides. These workers also decided to use the polymer-bound saccharide unit as the glycosyl donor (the

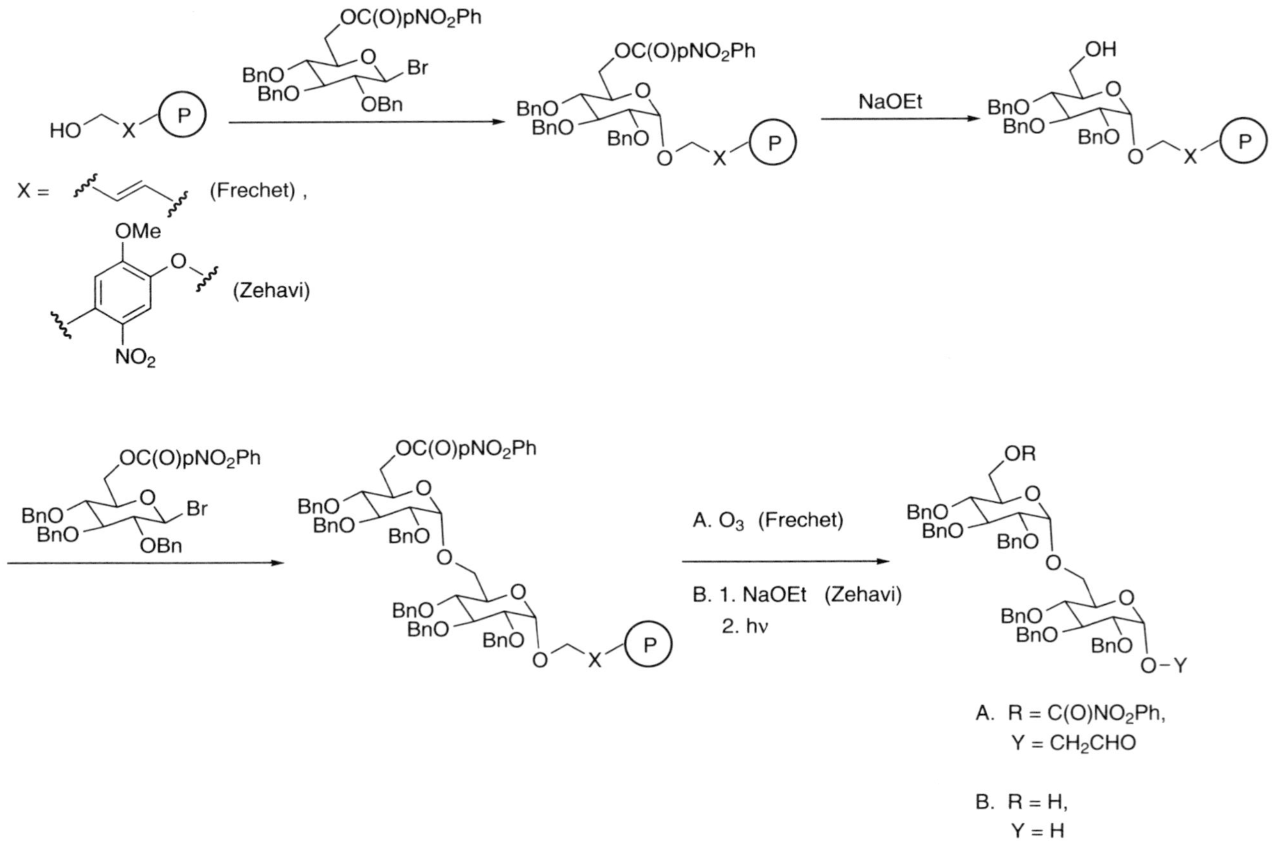

OC(O)pNO2Ph
BnO
BnO
Br
OBn
HO
X
P
NaOEt
OH
BnO
X = (Frechet) ,
OMe
O
(Zehavi)
NO2
A. O3 (Frechet)
B. 1. NaOEt (Zehavi)
2. hν
OR
O−Y
A. R = C(O)NO2Ph,
Y = CH2CHO
B. R = H,
Y = H

Scheme 13.1.

reactive species), with the acceptor as the free species in solution. This approach was a significant departure from the key observations of Merrifield in his seminal work on solid-phase peptide synthesis. Merrifield concluded that the preferred strategy for solid-phase synthesis favored the addition of the reactive species to the growing polymer-bound unit.[29] However, using the Kochetkov orthoester chemistry with an anomeric bromide donor, Guthrie was able to construct a β-(1→6)-linked disaccharide on the solid phase; α-linked saccharides were not accessible using this chemistry. Furthermore, this approach was limited by the need for long reaction times at elevated temperatures and a difficulty in precipitating the soluble polymer support.

Another early attempt at effecting solid-phase synthesis of oligosaccharides employed glycosyl oxazoline donor chemistry (Figure 13.2).[30] This approach produced a β-(1→6)-linked disaccharide, where glycosylation of the primary C-6 hydroxyl group of a pyranose system was accomplished in the presence of free secondary alcohols. Although this chemoselectivity appeared attractive, it also indicated the lack of generality of this chemistry for the glycosylation of all types of acceptors. In addition, implementation of the oxazoline donor chemistry required the use of elevated temperatures to effect coupling, indicating yet another limitation of this chemistry.

Subsequent to the early reports demonstrating the feasibility of solid-phase oligosaccharide synthesis, further implementation of those early strategies have not been forthcoming. However, new efforts directed toward solid-phase synthesis of oligosaccharides are providing new solutions to this problem. Not only the traditional solid supports, such as polystyrene resins and controlled-pore glass, but also soluble-polymeric systems based on polyethyleneglycol[31-34] have been used for polymer supported oligosaccharide synthesis. Although more traditional glycosylation chemistries continue to be investigated for polymer-supported synthesis, increasingly, the development of new and efficient glycosylation technologies is leading the way to a solution to the problem of solid-phase carbohydrate chemistry. Solid-phase technologies that have shown promise for potential practical application include those based on glycosyl sulfoxide glycosylation technology,[35,36] glycal-based glycosylation technology,[37-39] trichloromethylimidate-based glycosylation chemistry,[31-34] and enzyme-based glycosylation approaches[40,41] (Figure 13.2).

Reports by Kahne and coworkers showed glycosyl sulfoxide glycosylation chemistry to be a very versatile technology for both solution and polymer supported synthesis (Scheme 13.2).[35,36] Using either insoluble polystyrene-type Merrifield resin or soluble (polyethyleneglycol-based) polymer supports, construction of 1→6 and 1→4-linked oligosaccharides with either α- or β-glycosidic linkages proceeded rapidly and efficiently with glycosyl sulfoxide donor technology. Even disaccharides containing 2-amino sugars and trisaccharides were prepared on the solid phase. Glycosylation efficiencies on the solid phase were reported to be approximately 95% with the use of only 4 equivalents of sulfoxide donor.

Access to the reduced-end oligosaccharide was demonstrated using either

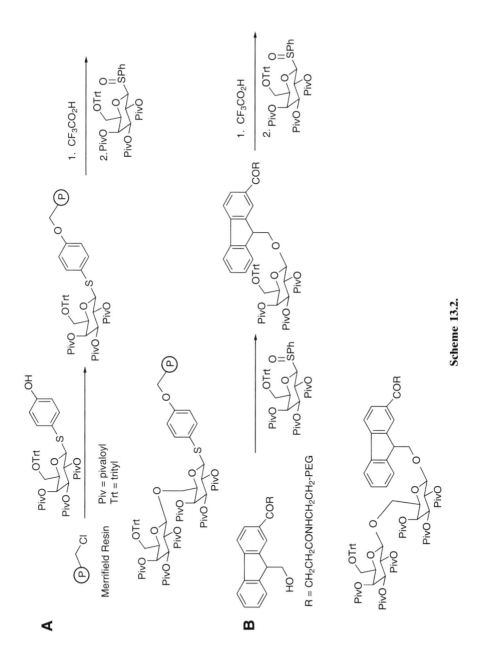

Scheme 13.2.

a thioether[35] linker or a 9-hydroxymethylfluorenyl[36] linker to the polymer support. These linkers were efficiently cleaved in the presence of mercuric trifluoroacetate (thioether linker) or an amine base (fluorenyl linker). Currently, only the glycosyl sulfoxide-based technology has demonstrated the ability to use a single glycosyl donor chemistry to construct glycosidic linkages on the solid-phase to both primary and secondary alcohols with the control of either α- or β-anomeric stereochemistry.

Solid-phase chemistry employing glycal-based glycosylation technology was shown by Danishefsky and coworkers to provide effective access to β-(1→6)-linked oligosaccharides (Scheme 13.3).[37-39] This technology was used to construct tetrasaccharides and several biologically relevant oligosaccharides. The Danishefsky solid-phase strategy attached the first glycal donor through its C-6 hydroxyl group to a polystyrene-based resin via a diphenyl- or diisopropylsilyl ether linker. The silyl ether linker was ultimately cleaved in the presence of fluoride ion to yield the free saccharide.

Glycosidic bond formation using glycal chemistry required 4–10 equivalents of acceptor and was reported as very stereoselective for the formation of β-(1→6) linkages. Although formation of glycosidic linkages to reactive secondary alcohols was reported using solid-phase glycal chemistry, glycosidic bond formation to less reactive secondary alcohols and the preparation of α-linked saccharides appeared difficult. To accomplish solid-phase glycosylation of poorly reactive secondary alcohols with the generation of α-anomeric stereochemistry, Danishefsky reverted to glycosyl fluoride donor chemistry. Also, the generation of a polymer-bound glycal donor limits the ability to drive these glycosylation reactions to completion by addition of excess reactive donor. In addition, although the capping protocol employed with this strategy prevents deletion sequences, the generation of a polymer-bound donor introduces the possibility of excessive chain termination events.

The development of glycosyl trichloromethylimidate activation for application to polymer supported synthesis was described using the soluble polyethyleneglycol (PEG)-bound solution synthesis approach. In this approach, the polymer-bound product was isolated by precipitation with ether subsequent to each glycosylation step.[31-34] This approach utilized either a base-labile succinoyl linkage (Su) to polyethyleneglycol ω-monomethyl ether (MPEG) or a reductively labile α,α'-dioxyxylyldiether linker (DOX) to the polymer and appeared to work most effectively when using silver trifluoromethanesulfonate as the promoter. In the construction of several saccharides, the MPEG polymeric support was attached to the hydroxyls at the C-1, C-2, or C-3 position of the initial monosaccharide unit providing simultaneous hydroxyl group protection. Applying this technology, Krepinsky described the construction of β-linked oligosaccharides and α-linked polymannose oligomers where glycosylation yields appeared to be in the range of 75% (Scheme 13.4). However, no protocol was provided for the general construction of α-linked saccharides. Keprinsky did demonstrate the use of glycosyl bromide donors with the MPEG-Su polymer support.

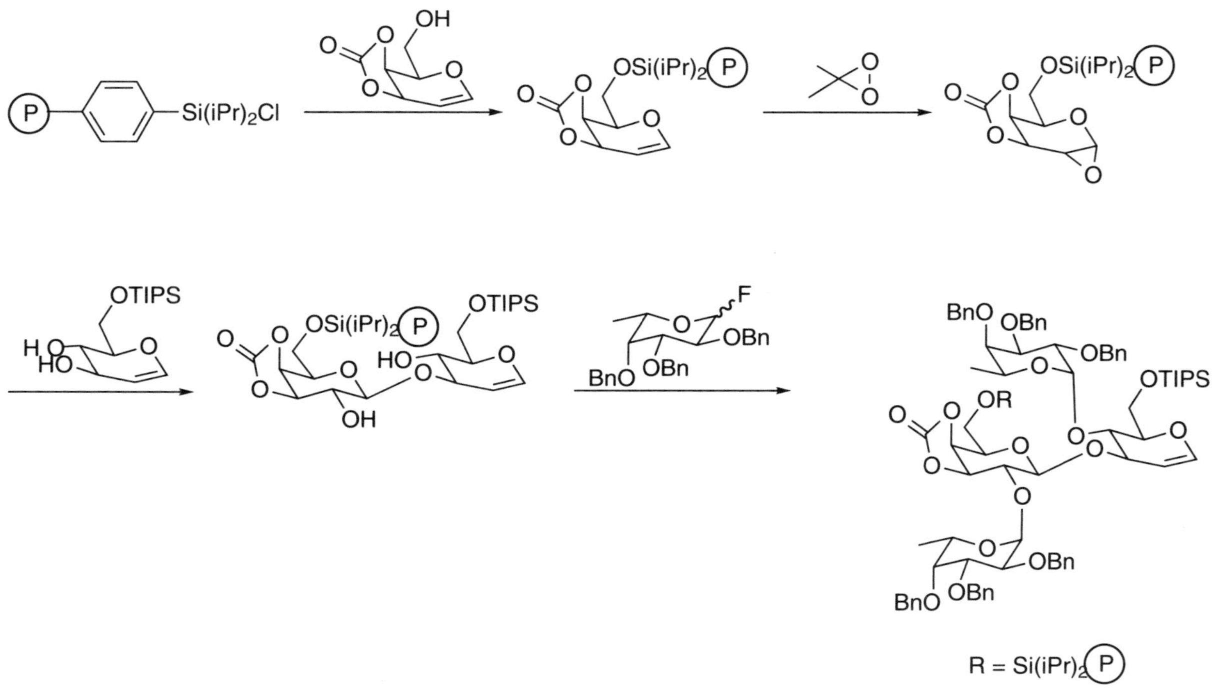

Scheme 13.3.

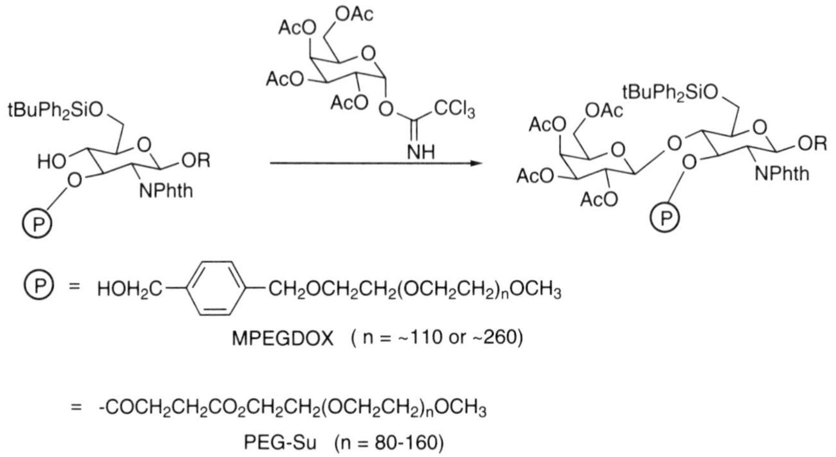

tBuPh2SiO
HO
OR
NPhth
P
AcO
OAc
AcO
AcO
CCl3
NH
OR
NPhth
P = HOH2C
CH2OCH2CH2(OCH2CH2)nOCH3
MPEGDOX (n = ~110 or ~260)
= -COCH2CH2CO2CH2CH2(OCH2CH2)nOCH3
PEG-Su (n = 80-160)

Scheme 13.4.

Polymer-supported synthesis using thioether donor technology on the soluble PEG polymer was reported by van Boom (Figure 13.2).[42] This chemistry produced a branched heptaglucoside in which the branched units were introduced via preformed disaccharides. In this synthesis, only β-(1→6) linkages were constructed and *N*-iodosuccinimide and catalytic triflic acid were selected as the promoter system. These glycosylations to the primary hydroxyl group at the 6-position of a pyranose nucleus were accomplished in the presence of a free C-4 hydroxyl, thus indicating a potential limitation of this chemistry for the construction of glycosidic linkages to less reactive secondary alcohols.

Recently, several reports by Wong and coworkers have described the use of enzyme technology to construct oligosaccharides on solid supports (Scheme 13.5).[40,41] The application of enzyme technology for the synthesis of oligosaccharides has the potential advantages of providing efficient and stereospecific glycosidic bond formation and eliminating the need for protecting group manipulations associated with the chemical-based approaches. However, application of enzyme technology to solid-phase synthesis of oligosaccharides presents several limitations. The need to obtain the required glycosyltransferases for the construction of each different type of glycosidic linkage and the requirement for the difficult to obtain sugar–phosphate–nucleotide cosubstrate pose serious practical problems. In addition, the limited ability of glycosyltransferase enzymes to accept nonnatural substrates, thus restricting the use of monosaccharide analogs, and the significant reduction in reaction rates for enzymatic reactions when applied to solid supported substrates also constrains this approach.[43]

Irrespective of the potential limitations associated with an enzyme-based solid-phase strategy, it was employed to construct several oligosaccharides (Scheme 13.5).[40,41] In fact, the only reported solid-phase synthesis of an oligo-

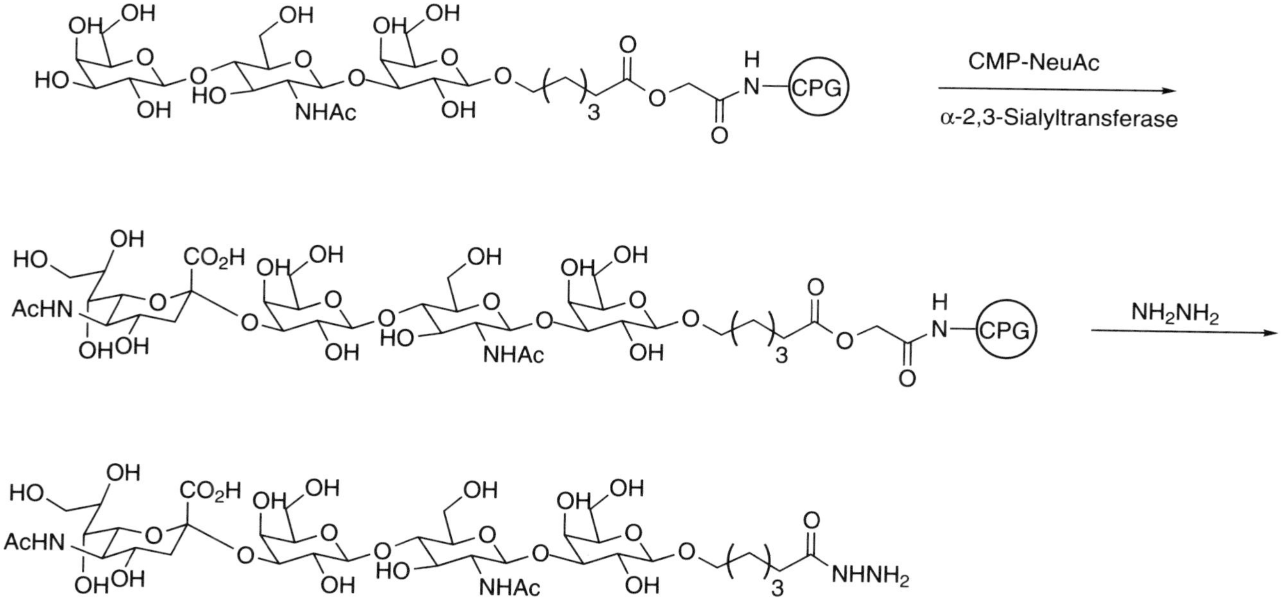

CMP-NeuAc

α-2,3-Sialyltransferase

NH₂NH₂

Scheme 13.5.

253

saccharide that incorporated the biologically significant sialic acid residue was accomplished using enzyme technology.[41] These successful solid-phase oligosaccharide syntheses applied N-iodoacetyl aminopropyl controlled-pore glass as the solid support. An intervening spacer between the support and the acceptor saccharide was used to relieve potential steric problems associated with the approach of an enzyme toward a polymer. To provide a hydrophilic buffer zone to facilitate enzymatic glycosylation, this spacer included an intervening monosaccharide residue between a 6-hydroxyheptanoate linker attached to the support and the first acceptor monosaccharide. Upon completion of the oligosaccharide synthesis, the product was recovered from the solid support by base cleavage of the ester bond constructed between the heptanoate linker and the derivatized solid support.

Certain technologies have been developed that have potential application to the area of solid-phase oligosaccharide synthesis as it relates to drug discovery. Vetter and Gallop employed 1-amino-1-deoxy saccharide derivatives coupled to a solid-support (TentaGelSNH$_2$ or dodecyldiamine-grafted polystyrene).[44] They attached these saccharides to the solid support via an eight-carbon linker in which an amide bond was first constructed between the 1-amino-1-deoxy saccharides and the terminal acid of the linker. This linker–saccharide conjugate was then attached to an amino functionalized solid support again via an amide linkage. The amide bond formation between the 1-amino saccharides and the acid moiety of the linker was accomplished without employing protecting groups on the hydroxyls of the saccharides. These solid supported saccharides were then screened for lectin binding by lectin immunostaining or flow cytometry techniques (Scheme 13.6).

Several of the solid-phase oligosaccharide synthesis technologies offer promise for eventual application in the generation of oligosaccharide molecular diversity within a combinatorial library scenario, but many issues remain unresolved. The ideal protocol for solid-phase synthesis of oligosaccharides would utilize a single glycosyl donor technology to construct each of the glycosidic linkages whether they be to a primary or secondary alcohol or whether they be of α- or β-stereochemistry at the anomeric center. The single-donor approach reduces the number of glycosyl donors and simplifies the optimization of chemistries required to synthesize an oligosaccharide library on the solid phase. Currently, only the glycosylsulfoxide-based solid-phase method has demonstrated this capability. An effective and practical solid-phase oligosaccharide synthesis technology must also provide a generalized solution to the issue of protecting group strategy. None of the existing solid-phase synthesis methods address this issue. In addition, the formation of α-linkages to sialic acid and β-linkages to mannose is very difficult and remains a serious challenge for current solid-phase methodologies. However, even with a number of existing limitations, certain solid-phase approaches provide the platforms for the development of a general, efficient and rapid method for oligosaccharide construction which can be married with schemes for library generation.

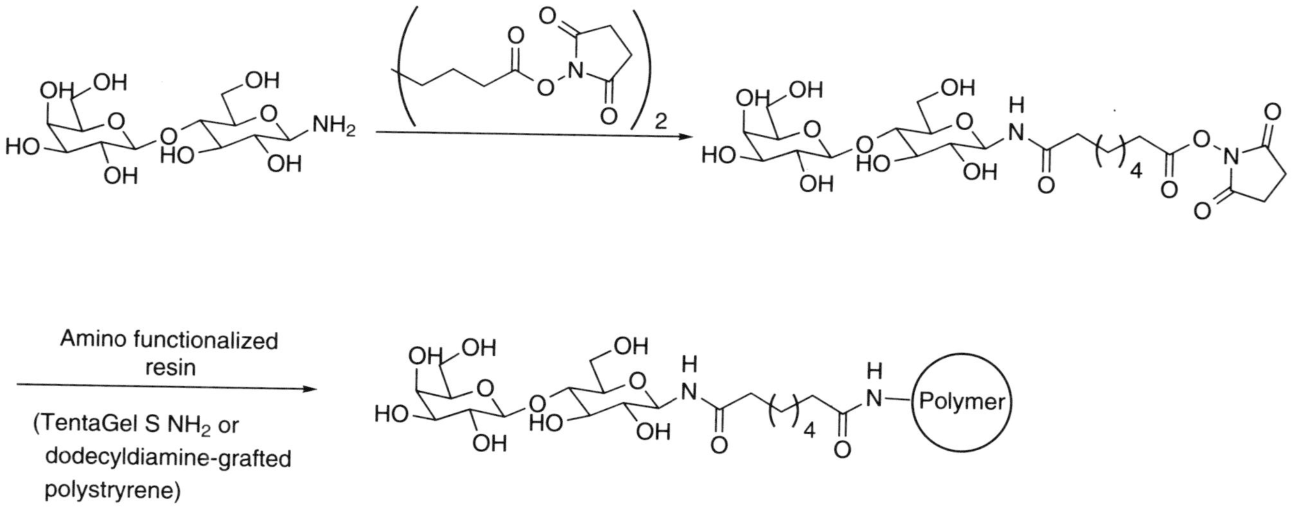
OH
OH
O
HO
OH
O
OH
O
HO
OH
NH2
O
O
O
N
O
O
2
Amino functionalized
resin
(TentaGel S NH2 or
dodecyldiamine-grafted
polystryrene)
H
N
4
Polymer

Scheme 13.6.

SOLID-PHASE SYNTHESIS OF GLYCOCONJUGATES

Solid-Phase Synthesis of Glycopeptides

Although the synthesis of several types of carbohydrate-containing conjugates have been executed on the solid phase, the development of general approaches for the construction of a particular class of glycoconjugates has focused primarily on glycopeptides because of their immense biological significance.[45,46] Two types of glycosidic linkages predominate in glycopeptide conjugates. These include N-linked glycopeptides, which are conjugated via the amide group of asparagine residues, and O-linked glycopeptides conjugated via the hydroxyl group of serine or threonine residues. In the construction of glycopeptides on the solid phase, as in the construction of oligosaccharides on the solid phase, protecting group strategy and protocols for cleavage of the product from the solid support become important issues. The peptide protecting group chemistry must be compatible with the acid-sensitive carbohydrate moieties, and any glycosylation conditions must not induce amino acid racemization or competing chemistry with the peptide.

Solid-phase O- or N-linked glycopeptide synthesis has relied on two approaches: either the convergent or the building blocks approach.[47-49] However, typically, in neither approach is a glycosylation reaction executed on the solid phase. The most successful variants of each approach take advantage of the well-established amide bond-forming reaction to build the glycopeptide. The convergent approach uses a 1-amino-1-deoxy derivatized saccharide and constructs an amide bond to an activated carboxylate group of the peptide (Figure 13.3). The bulding blocks approach requires the synthesis of desired N- or O-linked glycosylated amino acids, which are then used in standard solid-phase peptide synthesis (Figure 13.4). This method has been used extensively to build glycopeptides in which the position of the saccharide unit varies within the peptide sequence.

A convergent strategy that did investigate direct glycosidic bond formation between a fully assembled resin bound peptide and a glycosyl donor was studied by Otvös and coworkers.[50] This approach attempted direct O-glycosylation of a resin-bound serine containing peptide with an glycosyl oxazoline donor. Only 23% of the saccharide was incorporated into the peptide, with the production of several other by-products. In addition, this method required approximately a 20:1 ratio of glycosyl donor to peptide and long reaction times. No other convergent approach has attempted the preparation of O-linked glycopeptides.

Each of the more successful examples employing the general convergent approach focused on the construction of N-linked glycopeptides in which the coupling of the saccharide unit and the peptide proceeded through the formation of an amide bond between a C-1 amino functionalized saccharide and an active ester of the peptide.[51] Vetter and coworkers constructed a peptide sequence employing fluorenylmethoxycarbonyl (Fmoc) amine protection and

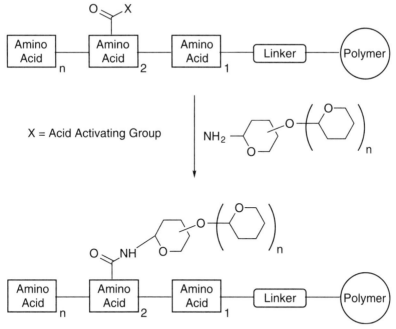

Figure 13.3. Convergent approach.

a polystyrene resin modified with a highly acid-labile linker (SASRIN) cleavable with 1% trifluoroacetic acid (Scheme 13.7).[52] Activation of a peptide carboxylate group was accomplished by preparation of the pentafluorophenyl (Pfp) ester derivative. This activation method gave coupling yields of 50–80%. The stereochemistry of the C-1 amino group of the 1-amino-1-deoxysaccharides predetermined the stereochemistry of the glycosidic linkage to the peptide. The synthetic method used to prepare the 1-amino functionalized saccharides set the anomeric stereochemistry as β unless the monosaccharide was mannose or lyxose.[53] In addition, this approach did not require the use of hydroxyl protecting groups on the saccharide unit. Utilizing this strategy, Vetter and coworkers were able to generate a small library of glycopeptides after preparing the 1-amino-1-deoxy derivatives of a series of commercially available saccharides.[52]

In another case exploring the convergent strategy, a 1-amino functionalized oligosaccharide was constructed on the solid phase and coupled to a peptide via an activated carboxylate group on the peptide. Danishefsky and coworkers used this approach with solid-phase glycal chemistry by first building a β-(1→6)-linked oligosaccharide unit, followed by amide bond formation to a free carboxylate group of an octapeptide.[54] The resulting β-glycosidic linkage to the peptide was predetermined by the method of synthesis of the 1-amino functionality on the saccharide. The oligosaccharide unit was variously pro-

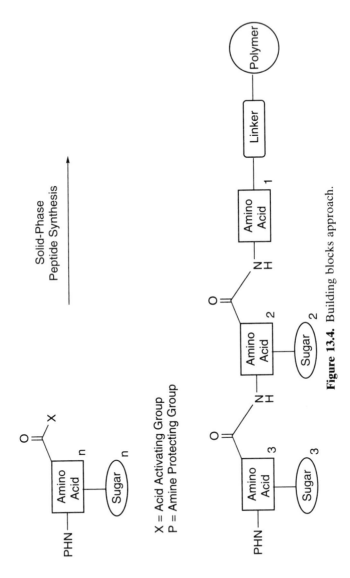

Figure 13.4. Building blocks approach.

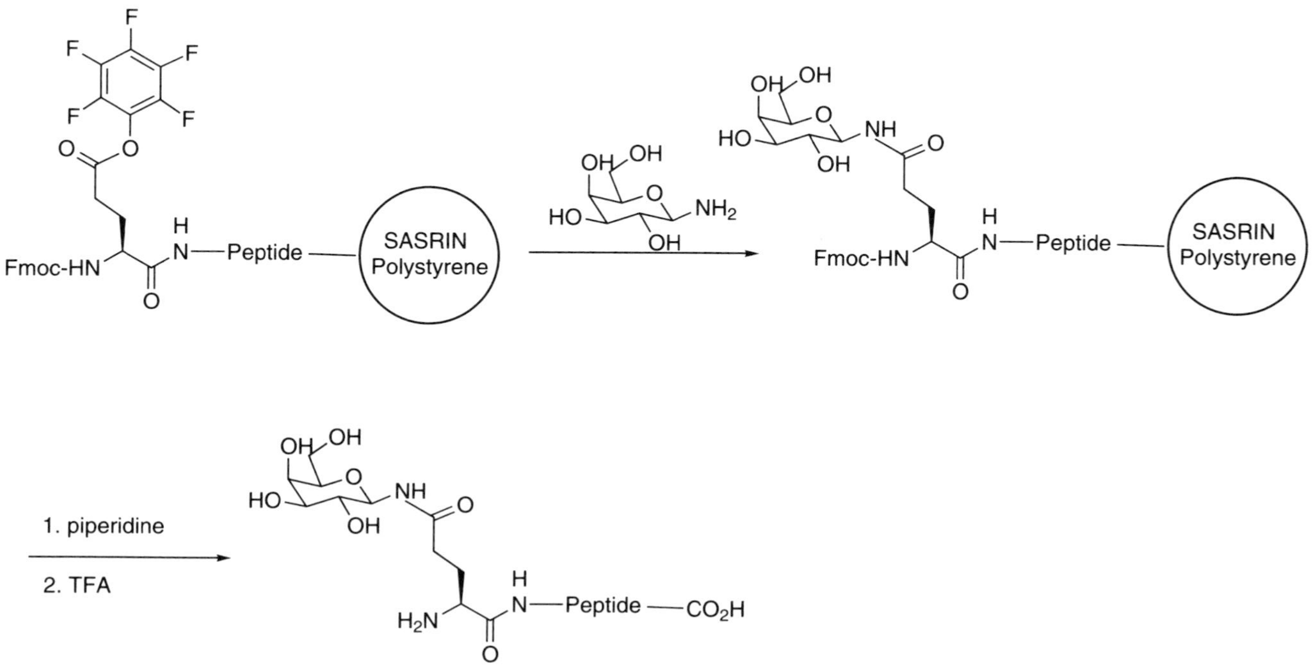
F
F
F
F
F
O
O
Fmoc-HN
H
N—Peptide
O
SASRIN
Polystyrene
OH
OH
O
HO
NH2
OH
OH
OH
O
HO
NH
O
OH
Fmoc-HN
H
N—Peptide
O
SASRIN
Polystyrene
1. piperidine
2. TFA
OH
OH
O
HO
NH
O
OH
H2N
H
N—Peptide—CO2H
O

Scheme 13.7.

tected with cyclic carbonate and benzyl ether protecting groups, and the peptide fragment was protected with benzyl ether, benzyl carbamate, and allyl ester protecting groups. Removal of the protecting groups occurred after cleavage from the solid support and was followed with a purification step. Having to execute multiple deprotection chemistries, particularly after removal from the solid support, detracts from this chemistry.

The solid-phase synthesis of glycopeptides using enzyme technology can effectively be described as a convergent approach. The use of enzymes to construct glycopeptides has focused primarily on the application of glycosyltransferases to extend the saccharide unit from a preformed glycopeptide fragment.[55,56] As for the synthesis of oligosaccharides, enzymatic technology applied to glycopeptide synthesis eliminates the need for protecting groups on the carbohydrate moieties and provides high regio- and stereoselectivity in the glycosidic bond construction. Conversely, it is limited by low reaction rates, the difficulty that enzymes have in accessing the interior of polymeric resin beads, the availability of glycosyltransferases and their sugar–phosphate–nucleotide cosubstrates, and general incompatibility with the use of monosaccharide analogs.

Several enzyme-based approaches for the synthesis of glycopeptides on polymeric supports have appeared in the literature. The work reported by Wong and Paulson on enzyme-mediated, solid-phase glycopeptide synthesis utilized an aminopropyl functionalized silica solid support to which was attached a hexaglycine spacer (Scheme 13.8).[55] The silica support was chosen because it was compatible with both aqueous and organic solvents and provided a large surface area for biomolecules. A monoglycosylated peptide unit was attached to the polyglycine spacer via a phenylalanine ester bond. The presence of the phenylalanine ester bond was critical to the overall strategy outlined by Wong and Paulson, because ultimate removal of the product from the silica support was effected by chymotrypsin cleavage of this ester bond. Glycosylation yields of 55–95% were observed with this enzymatic technology.

A similar enzymatic strategy was employed by Meldal and coworkers for the solid-phase preparation of a glycopeptide, however, they utilized a polyethylene glycol polyacrylamide copolymer (PEGA1900) as the polymer support.[56–58] The flexible nature of this polymer allowed full access to the interior of the resin for enzymatic glycosylation. In addition, they chose to employ an acid-labile 4-(α-amino-2′,4′-dimethyloxybenzyl)phenoxyacetic acid linker. Conversions of 50–95% were reported, but the glycosylation reaction times ranged from 48 to >72 h, thus limiting the practicality of this approach for rapid analog generation.

The approach used most extensively for the solid-phase construction of glycopeptides has been the building blocks approach. For problems where varying the position of the glycosylation site within a peptide chain is desired, this approach provides the most versatile strategy, since it simply requires changing the addition sequence of the amino acid derivatives in a standard solid-phase peptide synthesis. However, one criticism of the building blocks

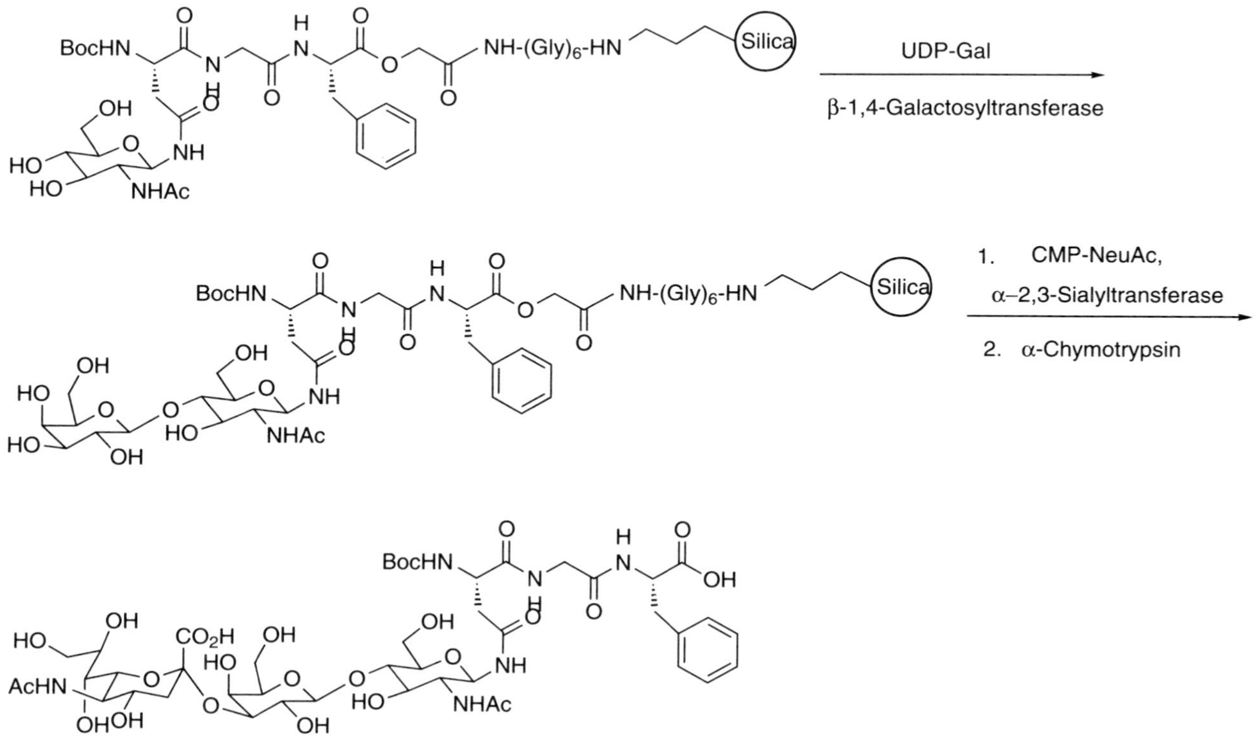

Scheme 13.8.

approach is that it requires the glycosidic linkage to survive repeated exposure to various chemistries during the construction of a full glycopeptide sequence.[49]

The laboratories of Kunz, Bock, Paulsen, and Meldal have been responsible for developing much of the chemistries and strategies for solid-phase glycopeptide synthesis using the building blocks approach.[59,60] The work has included the synthesis of both N- and O-linked glycopeptides. In addition, they have reported methods for the preparation of N-,[61,62] aliphatic O-,[63-68] and aromatic O-linked[69,70] glycosylated amino acid building blocks. Other contributions to the overall development of the building blocks approach for glycopeptide synthesis include those of Polt and coworkers. They developed general methods for the synthesis of α and β O-linked serine and threonine building blocks and used them in the solid-phase synthesis of glycosylated enkephalin analogs.[71,72]

Developing the building blocks approach required that the features of the building block be delineated as well as coupling chemistry, linker chemistry, and the nature of the solid support. The chemistry of the amine protecting group for the amino acid portion of the building block had to be compatible with the acid sensitivity of the carbohydrate moiety. A method for activating the carboxylic acid function of the building block required sufficient reactivity for solid-phase peptide synthesis, yet needed to be mild enough so as not to damage any part of the previously formed resin-bound glycopeptide. Also, a linker unit was required that allowed rapid cleavage of the product glycopeptide from the solid support without destroying any feature of the glycopeptide.

The building block construct developed by Kunz, Bock, Paulsen, and Meldal employed the Fmoc protecting group for the amine function of the amino acid, the pentafluorophenyl ester (Pfp) activation of the acid function, and base-labile (acetyl or benzoyl) protecting groups for the carbohydrate moiety (Figure 13.5).[61-70] Mild base cleavage conditions used for removal of Fmoc protecing groups made Fmoc protection ideal for use with carbohydrate systems. However, morpholine was required as the base, since the use of standard piperidine conditions resulted in substantial dehydroalanine formation with O-glycosylserine-containing peptides.

The Pfp ester derivatives of glycosylated amino acids not only provided activation of the acid for peptide coupling, but also protected the acid moiety during glycosylation of the amino acid and provided an advantage in purification of the building block. Generally, 3,4-dihydro-3-hydroxy-4-oxo-1,2,3-benzotriazine (Dhbt-OH) was used as a catalyst and monitoring agent in conjunction with the Pfp ester. Typically, either the 5-(4-aminomethyl3,5-dimethoxyphenoxy)valeric acid (PAL) or the p-(α-amino-2,4,-dimethoxybenzyl)phenoxyacetic acid (Rink linker) was used to link the glycopeptide to the solid support.[65-67,73,74] These linkers, cleavable with 95% trifluoroacetic acid, were sufficiently acid sensitive to be compatible with the acid sensitivity of the carbohydrate moiety. The solid support frequently used with this approach was the polystyrene-based poly(ethylene glycol)-grafted gel-type support (TentaGelS).[65,66,68,73,74]

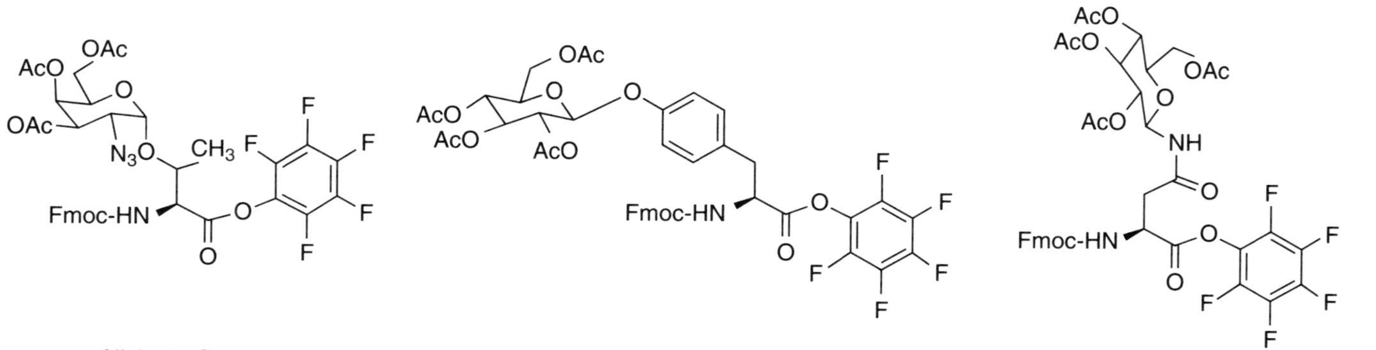

Figure 13.5. Typical glycopeptide building blocks.

Aliphatic O-linked

Aromatic O-linked

N-linked

This building blocks approach for the construction of glycopeptides was effectively applied to an automated synthesis protocol.[75,76] In addition, by applying the split-resin technique for combinatorial library generation, the building blocks strategy yielded glycopeptide libraries of modest size.[75-77]

Full access to the carbohydrate diversity of a glycopeptide is currently limited only by the availability of the necessary building blocks irrespective of which strategy one wishes to employ. The chemical-based convergent approaches require that each new oligosaccharide unit be synthesized. The enzyme-based convergent approach requires ready access to glycosyltransferases and sugar–phosphate–nucleotide cosubstrates. The synthesis of each amino acid saccharide unit is a prerequisite for application of the building blocks strategy for the solid-phase synthesis of glycopeptides. Thus, with the building blocks approach, although it is easy to vary the position of the carbohydrate-containing amino acid, it is not as easy to vary the nature of the carbohydrate itself without constructing a new building block. However, even with the current limitations of solid-phase glycopeptide construction, the technology is at a stage where significant carbohydrate molecular diversity can be introduced into peptide systems.

Polymer Supported Synthesis of Phosphate-Linked Saccharides

Solid supported synthesis of other glycoconjugate systems in which the monosaccharides were linked by phosphate units has been reported.[78-80] These efforts primarily focused on solid-phase synthesis of fragments of the capsular polysaccharides of bacterial cell walls. In these cases, the difficult to form glycosidic linkage was replaced by the well-developed phosphodiester bond construction originally developed for solid-phase oligonucleotide synthesis. Traditional solid supports, such as controlled-pore glass and the PEG-based soluble polymeric support, were used to construct these phosphate-linked systems.[79] As in the construction of glycopeptides, the preparation of these systems employed a building blocks strategy. However, since they were oligomers of the same repeating unit, typically the construction of only a single building block was necessary.

CONCLUSION

The solid-phase synthesis of oligosaccharides and glycoconjugates has progressed immensely since the seminal work of Fréchet and coworkers. The construction of complex oligosaccharides and glycopeptides is now possible using either chemical or enzymatic glycosylation technologies. Each of these approaches still requires optimization in efficiency, generalizability to multiple substrate applications and speed. Also, except for the solid-phase synthesis of glycopeptides via the 'building blocks' approach, automation has not yet been applied to these solid-phase chemistries. However, there has been sufficient technological development in all areas of carbohydrate-based solid-phase synthesis such that accessing carbohydrate molecular diversity through these technologies is now possible.

REFERENCES

1. Varki A (1993): Biological roles of oligosaccharides: all of the theories are correct. *Glycobiology* 3:97–130.

2. Rademacher TW, Parekh RB, Dwek RA (1988) Glycobiology. *Ann Rev Biochem* 57:785–838.

3. Sairam MR (1989): Role of carbohydrates in glycoprotein hormone signal transduction. *FASEB J* 3:1915–1926.

4. Gambaryan AS, Piskaren VE, Yanskov IA, Sakarov AM, Tuzika AB, Bovin NV, Nifaníev NE, Matrosovich MN (1995): Human influenza virus recognition of sialyloligosaccharides. *FEBS Lett* 366:57–60.

5. Boren T, Falk P, Roth KA, Larson G, Normark S (1993): Attachment of *Helicobacter pylori* to human gastric epithelium mediated by blood group antigens. *Science* 262:1892–1895.

6. Kahne D (1995) Strategies for the design of minor groove binders: a re-evaluation based on the emergence of site-selective carbohydrate binders. *Chem Biol* 2:7–12.

7. Fisher JF, Harnson AW, Bundy GL, Wilkinson KF, Rush BD, Ruwart MJ (1991): Peptide to glycopeptide: glycosylated oligopeptide renin inhibitors with attenuated in vivo clearance properties. *J Med Chem* 34:3140–3143.

8. Rodriguez RE, Rodriguez FD, Secristan MP, Torres JL, Valencia G, Garcia Anton JM (1989): New glycoslypeptides with high antinociceptive activity. *Neurosci Lett* 101:89–94.

9. Hirschmann R, Nicolaou KC, Pietranico S, Leahy EM, Salvino J, Arison B, Cichy MA, Spoors PG, Shakespeare WC, Sprengler PA, Hamley P, Smith III AB, Reisine T, Raynor K, Maechler L, Donaldson C, Vale W, Freidinger RM, Cascieri MR, Strader CD (1993): De novo design and synthesis of somatostatin nonpeptide peptidomimetics utilizing β-D-glucose as a novel scaffolding. *J Am Chem Soc* 115:12550–12568.

10. Gallop MA, Barrett RW, Fodor SPA, Gordon EM (1994): Applications of combinatorial technologies to drug discovery, 1: Background and peptide combinatorial libraries. *J Med Chem* 37: 1233–1251.

11. Gordon EM, Barrett RW, Dower WJ, Fodor SPA, Gallop MA (1994): Applications of combinatorial technologies to drug discovery, 2: Combinatorial organic synthesis, library screening strategies, and future directions. *J Med Chem* 37:1385–1401.

12. DeWitt SH, Schroeder MC, Stankovic CJ, Strode JE, Czarnik AW (1994): Diversomer technology: solid-phase synthesis, automation, and integration for the generation of chemical diversity. *Drug Dev Res* 33:116–124.

13. Desai MC, Zuckermann RN, Moos WH (1994): Recent advances in the generation of chemical diversity libraries. *Drug Dev Res* 33:174–188.

14. Han H, Wolff MM, Brenner S, Janda KD (1995): Liquid-phase combinatorial synthesis. *Proc Natl Acad Sci USA* 92:6419–6423.

15. Lam KS, Hruby VJ, Lebl M, Knapp RJ, Kazmierski WM, Hersch EM, Salmon SE (1993): The chemical synthesis of large random peptide libraries and their use for the discovery of ligands for macromolecular acceptors. *Bioorg Med Chem Lett* 3:419–424.

16. Pinilla C, Appel J, Blondelle S, Dooley C, Dorner B, Eichler J, Ostresh J, Houghten RA (1995): A review of the utility of soluble peptide combinatorial libraries. *Biopolymers* 37:221–240.

17. Sherman MI, Bertelsen AH, Cook AF (1993): Protein epitope targeting: oligonucleotide diversity and drug discovery. *Bioorg Med Chem Lett* 3:469–475.

18. Simon RJ Kania RS, Zuckermann RN, Huebner VD, Jewell DA, Banville S, Ng S, Wang L, Rosenberg S, Marlowe CK, Spellmeyer DC, Tan R, Frankel AD, Santi DV, Cohen FE, Bartlett PA (1992): Peptoids, a modular approach to drug discovery. *Proc Natl Acad Sci USA* 89:9367–9371.

19. Bunin BA, Plunkett MJ, Ellman JA (1994): The combinatorial synthesis and chemical and biological evaluation of a 1,4-benzodiazepine library. *Proc Natl Acad Sci USA* 91:4708–4712.

20. Campbell DA, Bermak JC, Burkoth TS, Patel DV (1995): A transition state analogue inhibitor combinatorial library. *J Am Chem Soc* 117:5381–5382.

21. Murphy MM, Schullek JR, Gordon EM, Gallop MA (1995): Combinatorial organic synthesis of highly functionalized pyrrolidines: identification of a potent angiotensin converting enzyme inhibitor from a mercaptoacyl proline library. *J Am Chem Soc* 117:7029–7030.

22. Fréchet JM, Schuerch C (1971): Solid-phase synthesis of oligosaccharides, I: Preparation of the solid support: poly[*p*-(1-propen-3-ol-1-yl)styrene]. *J Am Chem Soc* 93:492–496.

23. Fréchet JM, Schuerch C (1972): Solid-phase synthesis of oligosaccharides, III: Preparation of some derivatives of di- and trisaccharides via a simple alcoholysis reaction. *Carbohydrate Res* 22:399–412.

24. Fréchet JM, Schuerch C (1972): Solid-phase synthesis of oligosaccharides, II: Steric control by C-6 substituents in glucoside synthesis. *J Am Chem Soc* 94:604–609.

25. Zehavi U, Patchornik A (1973): Oligosaccharide synthesis on a light-sensitive solid support, I: The polymer and synthesis of isomaltose (6-*O*-α-D-glycopyranosyl-D-glucose). *J Am Chem Soc* 95:5673–5677.

26. Zehavi U, Amit B, Patchornik A (1972): Light-sensitive glycosides, I: 6-Nitroveratryl β-D-glucopyranoside and 2-nitrobenzyl β-D-glucopyranoside. *J Org Chem* 37:2281–2284.

27. Guthrie RD, Jenkins AD, Stehlicek J (1971): Synthesis of oligosaccharides on polymer supports, I: 6-*O*-(*p*-vinylbenzoyl) derivatives of glucopyranose and their copolymers with styrene. *J Chem Soc (C):* 2690–2696.

28. Guthrie RD, Jenkins AD, Roberts GAF (1973): Synthesis of oligosaccharides on polymer supports, II: Synthesis of β-D-gentiobiose derivatives on soluble support copolymers of styrene and 6-*O*-(*p*-vinylbenzoyl) or 6-*O*-(*p*-vinylphenylsulphonyl) derivatives of D-glucopyranose. *J Chem Soc Perkin* 1:2414–2417.

29. Merrifield RB (1985): Solid-phase synthesis (Nobel Lecture). *Angew Chem Int Ed Engl* 24:799–810.

30. Excoffier G, Gagnaire D, Utille JP, Vignon M (1972): Solid-phase synthesis of oligosaccharides, II: Synthesis of 2-acetamido-6-*O*-(2-acetamido-2-deoxy-β-D-glucopyranosyl)-2-deoxy-D-glucose. *Tetrahedron Lett* 5065–5068.

31. Douglas SP, Whitfield DM, Krepinsky JJ (1991): Polymer-supported solution synthesis of oligosaccharides. *J Am Chem Soc* 113:5095–5097.

32. Douglas SP, Whitfield DM, Krepinsky JJ (1995): Polymer-supported solution synthesis of oligosaccharides using a novel versatile linker for the synthesis of D-mannopentaose, a structural unit of D-mannans of pathogenic yeasts. *J Am Chem Soc* 117:2116–2117.

33. Leung O-T, Douglas SP, Whitfield DM, Pang HYS, Krepinsky JJ (1994): Synthesis of model oligosaccharides of biological significance, XIII: Synthesis of derivatives of Gal*p*NAc(β1-4)Gal*p*(β1-*O*), the common binding theme of adhesins of various bacteria: solution and polymer-supported solution approaches. An improved preparation of 2-deoxy-2-phthalimido-1,3,4,6-tetra-*O*-acetyl galactosamine and glucosamine. *New J Chem* 18:349–363.

34. Krepinsky JJ, Douglas SP, Whitfield DM (1994): Polymer-supported solution synthesis of oligosaccharides. In Lee YD, Lee RT, eds. *Methods in Enzymology,* Vol 242, *Neoglycoconjugates,* part A: *Synthesis.* New York: Academic, pp 280–293.

35. Yan L, Taylor CM, Goodnow R, Kahne D (1994): Glycosylation on the Merrifield resin using anomeric sulfoxides. *J Am Chem Soc* 116:6953–6954.

36. Wang Y, Zhang H, Voelter W (1995): A new base-labile anchoring group for polymer-supported oligosaccharide synthesis. *Chem Lett* 273–274.

37. Danishefsky SJ, McClure KF, Randolph JT, Russieri RB (1993): A strategy for the solid-phase synthesis of oligosaccharides. *Science* 260:1307–1309.

38. Randolph JT, Danishefsky SJ (1994): An interactive strategy for the assembly of complex, branched oligosaccharide domains on a solid support: a concise synthesis of the Lewis b domain in bioconjugatable form. *Angew Chem Int Ed Engl* 33:1470–1473.

39. Randolph JT, McClure KF, Danishefsky SJ (1995): Major simplifications in oligosaccharide syntheses arising from a solid-phase based method: an application to the synthesis of the Lewis b antigen. *J Am Chem Soc* 117:5712–5719.

40. Schuster M, Wang P, Paulson JC, Wong C-H (1994): Solid-phase chemical-enzymatic synthesis of glycopeptides and oligosaccharides. *J Am Chem Soc* 116:1135–1136.

41. Halcomb RL, Huang J, Wong C-H (1994): Solution- and solid-phase synthesis of inhibitors of *H. pylori* attachment and E-selectin-mediated leukocyte adhesion. *J Am Chem Soc* 116:11315–11322.

42. Verduyn R, Vanderklein PAM, Dowes M, Vandermarel GA, van Boom JA (1993): Polymer-supported solution synthesis of a heptaglucoside having phytoalexin elicitor activity. *Recl Trav Chim Pays-Bas* 112:464–466.

43. Wong C-H, Halcomb RL, Ichikawa Y, Kasimoto T (1995): Enzymes in organic synthesis: application to the problems of carbohydrate recognition, 2. *Angew Chem Int Ed Engl* 34:521–546.

44. Vetter D, Tate EM, Gallop MA (1995): Strategies for the synthesis and screening of glycoconjugates, 2: Covalent immobilization for flow cytometry. *Bioconjugate Chem* 6:319–322.

45. Kunz H (1987): Synthesis of glycopeptides, partial structures of biological recognition components. *Angew Chem Int Ed Engl* 26:294–308.

46. Paulsen H (1990): Syntheses, conformations and X-ray structure analyses of the saccharide chains from the core regions of glycoproteins. *Angew Chem Int Ed Engl* 29:823–839.

47. Meldal M (1994): Recent developments in glycopeptide and oligosaccharide synthesis. *Curr Opinion Struct Biol* 4:710–718.

48. Norberg T. In Lee YD, Lee RT, eds. *Methods in Enzymology*, Vol 242, *Neoglyco-conjugates*, part A: *Synthesis*. New York: Academic, pp 87–106.

49. Andrews DM, Seale PW (1993): Solid-phase synthesis of *O*-mannosylated peptides: Two strategies compared. *Int J Pept Res* 42:165–170.

50. Hollosi M, Kollat E, Laczko I, Medzihradszky KF, Thorin J, Otros L (1991): Solid-phase synthesis of glycopeptides: Glycosylation of resin-bound serine-peptides by 3,4,6-tri-*O*-acetyl-D-glucose oxazoline. *Tetrahedron Lett* 32:1531–1534.

51. Cohen–Anisfeld ST, Lansbury PT Jr (1993): A practical, convergent method for glycopeptide synthesis. *J Am Chem Soc* 115:10531–10537.

52. Vetter D, Tumelty D, Singh SK, Gallop MA (1995): A versatile solid-phase synthesis of N-linked glycopeptides. *Angew Chem Int Ed Engl* 34:60–63.

53. Vetter D, Gallop MA (1995): Strategies for the synthesis and screening of glycoconjugates, 1: a library of glycosylamines. *Bioconjugate Chem* 6:316–318.

54. Roberge JY, Beebe X, Danishefsky SJ (1995): A strategy for a convergent synthesis of N-linked glycopeptides on a solid support. *Science* 269:202–204.

55. Wong C-H, Schuster M, Wong P, Sears P (1993): Enzymatic synthesis of N- and O-linked glycopeptides. *J Am Chem Soc* 115:5893–5901.

56. Meldal M, Auzanneau FI, Hindsgaul O, Palcic MM (1994): A PEGA resin for use in the solid-phase chemical–enzymatic synthesis of glycopeptides. *J Chem Soc Chem Commun* 1849–1850.

57. Meldal M, Auzannear FI, Bock K (1994): PEGA, Characterization and application of a new type of resin for peptide and glycopeptide synthesis. In Epton R, ed. *Innovation and Perspective in Solid Phase Synthesis*. Kingswinford, UK: Mayflower Worldwide, pp 259–266.

58. Auzanneau FL, Meldal M, Bock K (1994): Synthesis, characterization and biocompatibility of PEGA resins. *J Pept Sci* 1:31–44.

59. Meldal M, Bock K (1994): A general approach to the synthesis of O- and N-linked glycopeptides. *Glycoconj J* 11:59–63.

60. Halcomb RL, Wong C-H (1993): Synthesis of oligosaccharide, glycoconjugates and glycolipids. *Curr Opinion Struct Biol* 3:694–700.

61. Christiansen–Brams I, Meldal M, Bock K (1993): Protected-mode synthesis of N-linked glycopeptides: single-step preparation of building blocks as peracetyl glycosylated N^{a}Fmoc asparagine OPfp esters. *J Chem Soc Perkin Trans* 1:1461–1471.

62. Meldal M, Bock K (1990): Pentafluorophenyl esters for temporary carbonyl group protection in solid phase synthesis of N-linked glycopeptides. *Tetrahedron Lett* 31:6987–6990.

63. Meldal M, Mouritsen S, Bock K (1993): Synthesis and immunological properties of glycopeptide T-cell determinants. In Garegg PJ, Lindberg AA, eds. *Carbohydrate Antigens, ACS Symposium Series 519*. New York: Academic, pp 19–33.

64. Bielfeldt T, Peters S, Meldal M, Bock K, Paulsen H (1992): A new strategy for solid-phase synthesis of *O*-glycopeptides. *Angew Chem Int Ed Engl* 31:857–859.

65. Jansson AM, Meldal M, Bock K (1992): Solid-phase synthesis and characterization of *O*-dimannosylated heptadecapeptide analogues of human insulin-like growth factor 1 (IGF-1). *J Chem Soc Perkin Trans* 1:1699–1707.

66. Rio–Anneheim S, Paulsen H, Meldal M, Bock K (1995): Synthesis of the building blocks N$^\alpha$-Fmoc-*O*-[α-D-Ac$_3$GalN$_3p$-(1–3)-α-D-Ac$_2$GalN$_3p$]-Thr-OPfp and N$^\alpha$-FMoc-*O*-[α-D-Ac$_3$GalN$_3p$-(1–6)-α-D-Ac$_2$GalN$_3p$]-Thr-OPfp and their application in the solid-phase glycopeptide synthesis of core 5 and core 7 mucin *O*-glycopeptides. *J Chem Soc Perkin Trans* 1:1071–1080.

67. Christensen MK, Melda M, Bock K (1993): Synthesis of mannose 6-phosphate-containing disaccharide threonine building blocks and their use in solid-phase glycopeptide synthesis. *J Chem Soc Perkin Trans* 1:1453–1460.

68. Paulsen H, Bielfeldt T, Peters S, Meldal M, Bock K (1994): Eine neue Strategie zur Festphasensynthese von *O*-glycopeptiden uber 2-Azidoglycopeptide. *Liebigs Ann Chem* 369–379.

69. Jensen KJ, Meldal M, Bock K (1992): In Smith JA, Rivier JE, eds. *Peptides 1991, Structural Biology, Proceedings of the American Peptide Symposium 12th.* Leiden: Pierce Chemical, pp 587–588.

70. Jensen KJ, Meldal M, Bock K (1993): Glycosylation of phenols: preparation of 1,2-*cis* and 1,2-*trans* glycosylated tyrosine derivatives to be used in solid-phase glycopeptide synthesis. *J Chem Soc Perkin Trans* 1:2119–2129.

71. Polt R, Szabol L, Treiberg J, Li YS, Hruby VJ (1992): General methods for α- or β-*O*-Ser/Thr glycosides and glycopeptides: solid-phase synthesis of *O*-glycosyl cyclic enkephalin analogues. *J Am Chem Soc* 11:10249–10258.

72. Szabo L, Li Y, Polt R (1991): *O*-Glycopeptides: a simple β-stereoselective glycosidation of serine and threonine via a favorable hydrogen bonding pattern. *Tetrahedron Lett* 32:585–588.

73. Reime KB, Meldal M, Kusumoto S, Fukase K, Bock K (1993): Small-scale solid-phase *O*-glycopeptide synthesis of linear and cyclized hexapeptides from blood-clotting factor IX containing *O*-(α-D-Xyl-1–3-α-D-Xyl-1–3-β-D-Glc)-L-Ser. *J Chem Soc Perkin Trans* 1:925–932.

74. Seitz O, Kunz H (1995): A novel allylic anchor for solid-phase synthesis: Synthesis of protected and unprotected O-glycosylated mucin-type glycopeptides. *Angew Chem Int Ed Engl* 34:803–805.

75. Meldal M, Holm CB, Bojesen G, Jakobsen MH, Holm A (1993): Multiple column peptide synthesis, part 2. *Int J Pept Protein Res* 41:250–260.

76. Peters S, Bielfeldt T, Meldal M, Bock K, Paulsen H (1992): Multiple-column solid-phase glycopeptide synthesis. *J Chem Soc Perkin Trans* 1:1163–1171.

77. Elofsson M, Roy S, Walse B, Kihlberg J (1993): Solid-phase synthesis and conformational studies of glycosylated derivatives of helper-T-cell immunogenic peptides from hen-egg lysozyme. *Carbohydrate Res* 246:89–103.

78. Veeneman GH, Brugghe HF, Vandenelst H, van Boom JH (1990): Solid-phase synthesis of a cell-wall component of *Haemophilus (Actinobacillus) pleuropneumoniae* serotype 2. *Carbohydrate Res* 195:C1–C4.

79. Kandil AA, Chan N, Chong P, Klein M (1992): Synthesis of fragments of capsular polysaccharide of *Haemophilus influenzae* type β on soluble polymeric support. *Synlett* 555–557.

80. Westerduin P, Veeneman GH, Pennings Y, Vandermarel GA, van Boom JH (1987): Preparation of a fragment of the cell wall teichoic acid of *Bacillus licheniformis* ATCC 9945 via a solid-phase approach. *Tetrahedron Lett* 28:1557–1560.

14

ENCODED COMBINATORIAL CHEMISTRY

JEFFREY W. JACOBS AND ZHI-JIE NI

Versicor, Inc., Fremont, California

One of the most powerful methods in combinatorial chemistry is split synthesis (for reviews, see Gallop et al.[1] and Gordon et al.[2]). In this process, appropriately functionalized synthesis beads are segregated into individual reaction vessels for the coupling of specific building blocks, then combined, mixed to homogeneity, and redivided for subsequent chemical steps. The size of a library prepared using this technique increases according to the mathematical product of the number of building blocks used prior to each pooling step (Figure 14.1). As an example, a four-step synthesis employing 10 building blocks at each step would afford 10,000 different compounds in only 40 (10 × 4) chemical steps. Since the bead mass is segregated for the coupling of specific building blocks, each bead bears the product of a specific reaction sequence (Figure 14.1).

The synthetic efficiency of the split-synthesis technique can be contrasted with the technical difficulties encountered when analyzing the resulting libraries. For example, the simple split-synthesis scenario outlined above results in a library consisting of 10 pools of 1000 compounds each. These compounds can be cleaved into solution and screened as soluble pools, or the ligands can remain attached to the beads and screened in immobilized form. Neither scenario is ideal for several reasons. Because of limitations on solubility, the concentrations of the individual compounds present in soluble pools must be correspondingly diminished as the pool size increases—perhaps below a

Combinatorial Chemistry and Molecular Diversity in Drug Discovery, Edited by
Eric M. Gordon and James F. Kerwin, Jr.
ISBN 0-471-15518-7   Copyright © 1998 by Wiley-Liss, Inc.

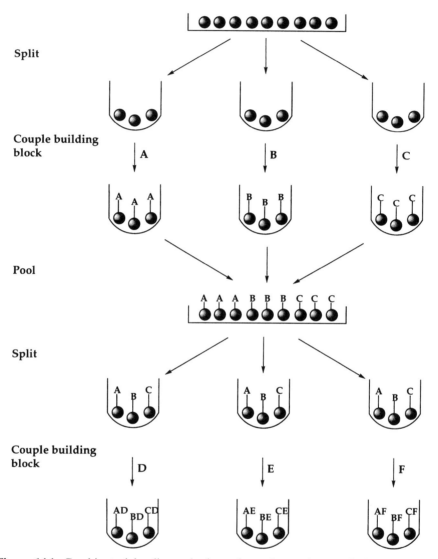

Figure 14.1. Combinatorial split synthesis performed on microscopic beads. After specific building blocks (A, B, C, etc.) are incorporated in unique reaction vessels, the bead mixture is pooled, mixed to homogeneity, and redivided for subsequent chemical steps.

desirable threshold for screening. Biological screens performed on such large mixtures of soluble compounds can be ambiguous since the observed activity could be due to a single compound or due to a collection of compounds acting either collectively or synergistically. The subsequent identification of specific biologically active members is challenging, since the number of compounds present in the pools and their often limited concentration deter their isolation and reassay. Because of this, biologically active pools are often iteratively resynthesized and reassayed as increasingly smaller subsets until activity data are obtained on homogenous compounds.[1,2]

This process of iterative resynthesis is time-consuming, requires multiple bioassays, and the deconvolution of a single pool to its individual constituents typically requires more synthetic steps than were required to prepare the parent library. When multiple pools are active, the deconvolution process becomes additively complex if each active subset is chosen for resynthesis. In addition to being inefficient, positive selection strategies such as iterative deconvolution ignore negative biological information, the knowledge of which is often important in the design of subsequent libraries.

In some instances, bead-based split-synthesis libraries can be successfully assayed with the ligands still immobilized to the beads. In this process, a reporter system is employed in the biological assay such that beads displaying active ligands can be physically distinguished from those displaying inactive compounds. Suitable reporter systems include the use of fluorescently labeled receptors, or anti-receptor antibodies similarly labeled with a reporter molecule (fluorophore, chromophore, radioisotope, etc.), that can be employed to "label" active beads. Beads thus marked are physically removed and analyzed to identify the attached ligand. This technique is limited by the capacity of the biological screen to detect immobilized ligands, as well as the sensitivity of the analytical methods employed to unambiguously identify the attached compounds.

The ability to characterize individual compounds on single beads is a challenging analytical problem, and arguably represents the greatest limitation of the split-synthesis method. A typical solid-phase synthesis bead approximately 100 μm in diameter can contain roughly 100–300 pmol of material. This is sufficient for the characterization of ligands such as peptides and oligonucleotides (via microsequencing), but is well below the detection limits of most of the analytical techniques employed in organic synthesis. The exception is mass spectroscopy (MS); however, mass detection cannot distinguish compounds of unique composition but of common mass—a typical problem when screening libraries of thousands to tens of thousands of compounds where the largest and smallest library members differ by only a few hundred atomic mass units (AMU).

While the quantities of individual compounds on single-synthesis beads are too low to permit their complete structural characterization, these amounts are more than adequate for most biological assays. For example, if 100 pmol of material were liberated in 100 μL of assay buffer (e.g., in a microtiter well), a 1 μM solution results, which is adequate for most routine high-throughput screens. Such single-bead assays avoid the complications encountered in the assay of pools since discrete compounds are screened. Thus, the quantity of material present on a single bead from a split synthesis is not assay limited, but is characterization limited. What is then required is a method for the unambiguous structural identification of individual library members obtained from single beads, independent of the bead's performance in a bioassay. If this structure elucidation problem could be solved, then the synthetic efficiency of the split synthesis method could be combined with the efficiency and analytical clarity of parallel high-throughput screening techniques.

One solution is to incorporate into the synthesis process a surrogate analyte that is in direct association with each ligand, that is unique to each ligand, and that possesses physical characteristics that allow its ready detection at concentrations well below what is present on a single-synthesis bead. This potentially general solution to the single-bead structure elucidation problem is referred to as encoding (Figure 14.2). Encoding is a technique in which easily detectable surrogate analytes, or tags, are cosynthesized with the desired ligands (Figure 14.2). Upon biological assay of the often subanalytical amounts of compound present on single beads, structural elucidation is achieved by analyzing (decoding) the associated tags.

Early encoding strategies cosynthesized DNA oligomers as the surrogate analyte because DNA can be readily amplified using the polymerase chain reaction (PCR), and sensitive sequencing methods exist for extracting its information content. Subsequent approaches utilized peptides to encode for the ligands of interest, since Edman N-terminal peptide sequencing offered sufficient sensitivity for decoding larger, high-loading beads. Although successful, both encoding techniques place a limitation on the types of chemical reactions that may be employed during library synthesis, since the encoding peptides and oligonucleotides cannot withstand many of the conditions frequently employed in routine organic syntheses (strong bases, Lewis acids, nucleophiles, etc.). More recently, highly sensitive encoding strategies have been developed that are compatible with the rigors of synthetic organic chemistry. These chemically inert, or hard, tagging strategies have been employed to encode the syntheses of many different organic compounds.

A common limitation to all of the above encoding techniques is the requirement for the cosynthesis of two unique entities: the desired ligand and its associated tag. Recent reports describe techniques that do not require the cosynthesis of tags. One approach employs radio-frequency signals to encode the synthesis, while another proposes a method that may circumvent the degeneracy problem of mass encoding. All of these technologies are reviewed in the following sections, which conclude with an analysis of the bead-handling issues and novel assay formats that will be required to more fully exploit the throughput potential of encoded combinatorial synthesis.

DNA-BASED ENCODING TECHNIQUES

One of the first reported successful ligand-encoding strategies exploited oligodeoxyribonucleic acid (DNA) as the surrogate analyte. This DNA-encoding

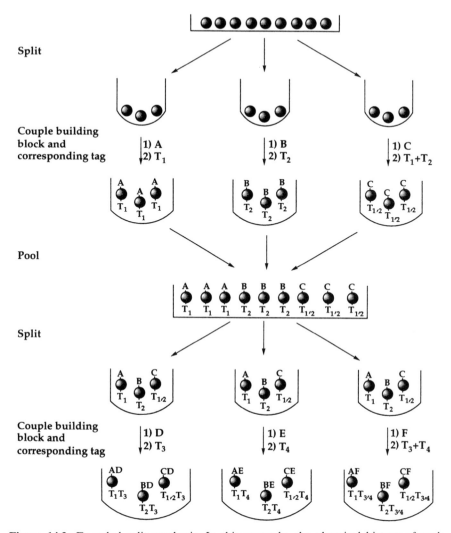

Figure 14.2. Encoded split synthesis. In this example, the chemical history of each building block is recorded through the cosynthesis of "tags" (T), with each tag or tag mixture unique to the building block and chemical step that it encodes.

concept had in fact been demonstrated in some of the first combinatorial library preparation methods ever reported—those utilizing filamentous phage particles.[3-5] In this approach, libraries of peptides are prepared biochemically from the cloning and expression of random-sequence oligonucleotides. Pools of oligonucleotides encoding the peptides of interest are inserted into an appropriate expression system, where upon translation the resulting peptides are synthesized as fusion proteins. One of the most common expression systems fuses these sequences to the gene III or gene VIII coat protein of filamentous phage particles. Each viral particle contains a unique DNA sequence that encodes only a single peptide. After screening a library in a given biological system, any viral particles displaying active peptides are isolated and the structure of the active peptides is elucidated by sequencing their encoding DNAs. A distinct disadvantage with this approach is that the molecular diversity of such systems is limited to peptides, and the amino acids that compose these peptides are restricted to the 20 encoded by genes.

Synthetic methods for combinatorial peptide synthesis are not limited to natural amino acids, nor are these techniques limited to ligands as simple as peptides. At least two laboratories recognized that powerful bead-based synthetic methods such as the split-synthesis technique might be similarly merged with a synthetic DNA encoding strategy to solve the single-bead structure elucidation problem. This approach was first articulated in the literature in a theoretical paper by Brenner and Lerner.[6]

This paper describes how an encoded library of peptides might be produced in such a way that each different peptide would be bound to a unique, single strand of DNA (the tagging molecule) in a one-to-one correspondence (Figure 14.3). Specific sequences of DNA are assigned to each building block, or monomer, such that the sequential addition of each monomer is encoded via the cosynthesis of its encoding segment of DNA. A simplified schematic of this approach is outlined in Figure 14.3, which illustrates the cosynthesis of a tripeptide and its encoding DNA sequence on a linker accommodating the chemical synthesis of both polymers. The resulting library would be screened against the receptor, enzyme, or antibody of interest and affinity selection techniques would be employed to isolate high-affinity ligands. The structure of these ligands would then be elucidated by sequencing their attached DNA tags.

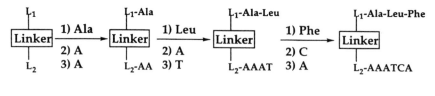

Reagent	DNA Code
Ala	AA
Gly	AC
IIe	AG
Leu	AT
Phe	CA

Figure 14.3. DNA-encoded peptides prepared in a 1:1 correspondence on a linker capable of anchoring the synthesis of both oligomers. The structure of the peptides are determined by sequencing their accompanying unique DNA sequence.

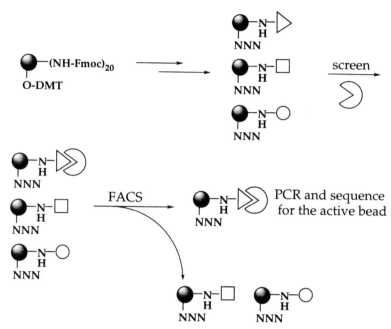

Figure 14.4. DNA-encoded peptide synthesis performed on microscopic (10-μm) beads. The peptide:DNA ratio was approximately 20:1, reflecting both the sensitivity of tag detection (PCR) as well as the desire to reserve the majority of the synthesis sites for ligand synthesis.

Gallop and coworkers at Affymax described the first functional DNA-encoding library synthesis methodology,[7] reporting the preparation and screening of a library of $\approx 10^6$ different DNA-encoded peptides which were synthesized on microscopic (10 μm) beads (Figure 14.4). To implement the chemistry, several key technical hurdles were overcome, including the development of peptide and oligonucleotide chemistries that were mutually orthogonal as well as mutually nondestructive, and the preparation of a bifunctional solid support containing synthesis sites that can anchor both peptide as well as DNA tags.

Peptides were prepared using the Fmoc/t-Boc strategy, and oligodeoxyribonucleotide synthesis proceeded using dimethoxytrityl (DMT)-protected 3'-O-methyl-N,N-diisopropyl phosphoramidites. Because peptide deprotection is performed with trifluoroacetic acid (TFA), an agent known to rapidly depurinate oligonucleotides containing either 2'-deoxyadensosine (dA) or 2'-deoxyguanosine (dG), dG was omitted from tag synthesis and 7-deaza-2'-deoxyadensosine (c^{7}dA), was substituted for dA. This was done because c^{7}dA is not sensitive to depurination, yet oligonucleotides containing this base can still be recognized under the conditions of the PCR. The phosphoramidites utilized 3'-O-methyl protection instead of the β-cyanoethyl protecting group because the former are stable to the conditions used to remove the Fmoc group from the growing peptide chain.

The solid support consisted of small (10 μm), monodisperse beads of styrene/divinylbenzene copolymer construction, functionalized with a 1,12-diaminododecane linker. The beads were differentiated by acylation with a mixture of two chemical linkers, one supporting the synthesis of peptides and the other supporting oligonucleotides. The resulting peptide/oligonucleotide ratio was approximately 20:1, reflecting both the sensitivity of the PCR as well as the desire to reserve the majority of the synthesis sites for the preparation of screenable compounds. The resulting beads supported both peptide and oligonucleotide chemistry, and oligonucleotides prepared on this support could be successfully amplified via the PCR and then sequenced using conventional techniques.

A library was synthesized consisting of approximately one million variants of the peptide sequence RQFKVVT, which is the C-terminal seven amino acid fragment of the opioid peptide dynorphin B. Using the split-synthesis method, a seven-step synthesis was performed utilizing the amino acids arginine, glutamine, phenylalanine, lysine, valine, and threonine and the unnatural amino acid D-valine at each step, affording 7^7 (823,543) unique peptides. Prior to each pooling step, a sequence of DNA specific for each amino acid was cosynthesized in the appropriate reaction vessel. At the conclusion of the synthesis, the degenerate DNA primer sequence was added to the pooled bead mass to complete the oligonucleotide synthesis.

The resulting peptide library was screened in immobilized form against an anti-dynorphin B antibody, D32.39. An aliquot of beads sufficient to statistically contain multiple copies of each ligand was first incubated with the fluorescently labeled antibody, then washed, and a fluorescence-activated cell sorter (FACS) was employed to separate antibody-bead conjugates from beads displaying low-affinity or unrecognized sequences. Active beads were isolated and their encoding DNAs amplified (PCR) and sequenced to afford a variety of peptides that bound antibody D32.39.

PEPTIDE ENCODING

Zuckermann and coworkers at Chiron recognized that peptides could be employed as tags since their information content could be extracted with high sensitivity via Edman degradation and sequencing.[8] Since the Edman degradation requires a free N-terminus, this peptide-as-code strategy could also be used to encode other peptides by acylating the N-terminus of the binding peptide strand, and leaving a free amine at the coding peptide terminus (Figure 14.5). To accommodate the parallel synthesis of both binding and coding peptides, an orthogonally protected bifunctional linker was employed that contained both acid- and base-sensitive protecting groups. This bifunctional linker resided on the cleavable Rink amide linker, such that peptide-

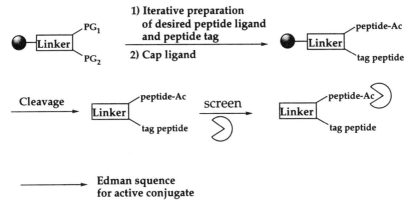

Figure 14.5. Peptide-encoded peptide libraries. Libraries of N-acylated peptides can be encoded by other peptides with free N-termini since the former are protected from Edman N-terminal sequencing. In this example, the ligand and its associated tag are synthesized in a 1:1 correspondence on a cleavable linker and released into free solution. Affinity selection techniques are employed to isolate conjugates that bind to the receptor, enzyme, or antibody target of interest.

encoded peptide conjugates would be released into solution upon treatment of the Rink linker with 95% TFA.

To test this concept, a small encoded library of 200 peptides was prepared consisting of analogs of the 10-mer Ac-RAFHTTGRII-NH$_2$, an epitope known to bind with submicromolar affinity to an anti-gp120 monoclonal antibody. Substitutions were made at three positions (X1, X2, and X3) of this peptide, as indicated: Ac-RAX$_3$HTTGX$_2$IX$_1$-NH$_2$. Each unique building block was encoded with a tripeptide sequence, each trimer unique to the amino acid it encodes. Four amino acids (Leu, Phe, Gly, and Ala) were selected as the encoding monomers, and a trimer encoding scheme was devised to encode up to 64 (4^3) unique building blocks.

Upon completion of the synthesis, the encoded library was cleaved into solution and incubated with the antibody. Affinity-selection methods were used to separate antibody-bound peptides from those sequences in free solution. The bound peptides were dissociated from the antibody and decoded via Edman sequencing, revealing three novel peptides with affinity for the antibody.

Lebl and coworkers at Selectide corporation reported a related peptide-encoding strategy (Figure 14.6).[9] In one embodiment, the free amines of the dibasic amino acid lysine were selectively protected with both the base-labile Fmoc- and acid-labile *tert*-butyloxycarbonyl (Boc-) protecting groups. This residue was immobilized to beads containing a redox-sensitive safety-catch linker (SCAL linker). Fmoc chemistry was used to prepare the desired ligands, while Boc-protected amino acids were used in tag synthesis. Thus, the resulting

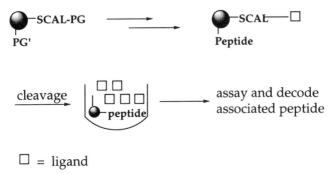

□ = ligand

Figure 14.6. Peptide encoding on a selectively cleavable linker. In this illustration of one of the Selectide approaches, ligands destined for screening are synthesized on a redox-sensitive SCAL linker, while the encoding peptide is permanently anchored to the bead. Activation and cleavage of the SCAL linker releases the ligand for assay in free solution. Active compounds can then be identified by retrieving and decoding their associated parent bead.

library displayed the ligand and its peptide-encoding tag in a one-to-one relationship via the linking lysine.

Libraries prepared in this fashion could be screened in two different ways. In one format, the ligand and its encoding peptide remain associated with the bead, and beads displaying biologically active compounds could be isolated and the encoding peptides sequenced. Alternatively, the SCAL linker could be reduced, converting it to an acid-cleavable linker. Treatment with TFA cleaved the peptide–ligand conjugates into solution for assay in that format. Alternatively, the solid support could be differentiated such that only ligands destined for biological assay are attached to the SCAL linker, while the corresponding peptide code remains permanently attached to the bead. Ligands could then be released into solution and assayed in soluble form. Structure elucidation would be performed by retrieving the parent bead and sequencing its encoding peptide, provided the assay conditions kept the peptide in association with its encoding bead.

The above peptide- and DNA-encoding techniques are not ideal because of the chemical lability of these oligomers. This places a severe restriction on the scope of the synthetic techniques that may be applied during library synthesis, and restricts the synthesis of more pharmaceutically attractive small organic molecules. What is ideally required is an encoding scheme in which the reagents used in tag synthesis are orthogonal to the conditions used in routine organic chemistry. Such chemically robust surrogate analytes are referred to colloquially as "hard" tags.

HARD TAGS

Still and coworkers reported the first encoding method utilizing such chemically stable tagging moieties.[10] The tags consisted of haloaromatic reagents linked to a carboxylic acid through an internal photochemically cleavable linker. Amide bond chemistry served to attach the tags to the beads. These haloaromatic reagents acylated the same synthesis sites used for ligand synthesis (Figure 14.7), but due to the sensitivity of tag detection this competition could be minimized. Once the haloaromatic analyte was attached to the bead it could be selectively detached into solution upon photolysis with ultraviolet light. The liberated tags could then be resolved and detected at subpicomole concentrations using electron capture capillary gas chromatography (EC GC).

A split-synthesis peptide library was prepared consisting of variants of the epitope EQKLISEEDL, in which each of the six N-terminal amino acids was substituted with any combination of seven amino acids. An efficient tagging scheme referred to as binary encoding was devised to encode the resulting library of 117,649 (7^6) peptides. The encoding strategy was binary in that unique information could be obtained from both the presence (1) or absence (0) of a given tag. For example, in the binary approach, only 20 tags would be necessary to encode for 1,048,576 different compounds (2^{20}). The library

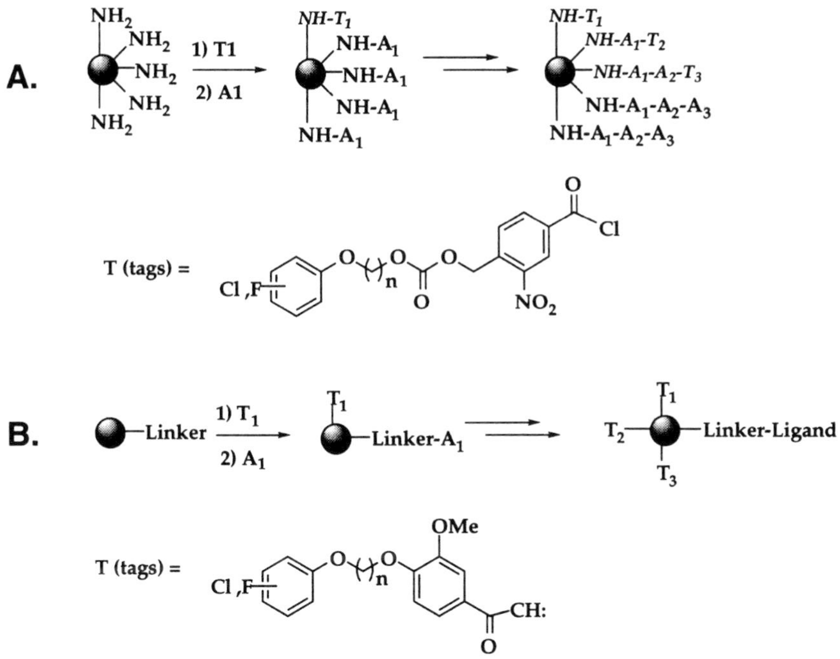

Figure 14.7. Chemically "hard" haloaromatic tags suitable for encoding applications where the beads will be exposed to rigorous synthetic conditions. The tags are released photochemically and then detected via EC GC. (A) Haloaromatic tags incorporated via amide bond chemistry at the expense of the ligand synthesis sites; (B) tags incorporated via carbene insertion. In both (A) and (B), tag concentrations are minimized to prevent the chemical derivatization of the encoded ligands or the quenching of their synthesis sites.

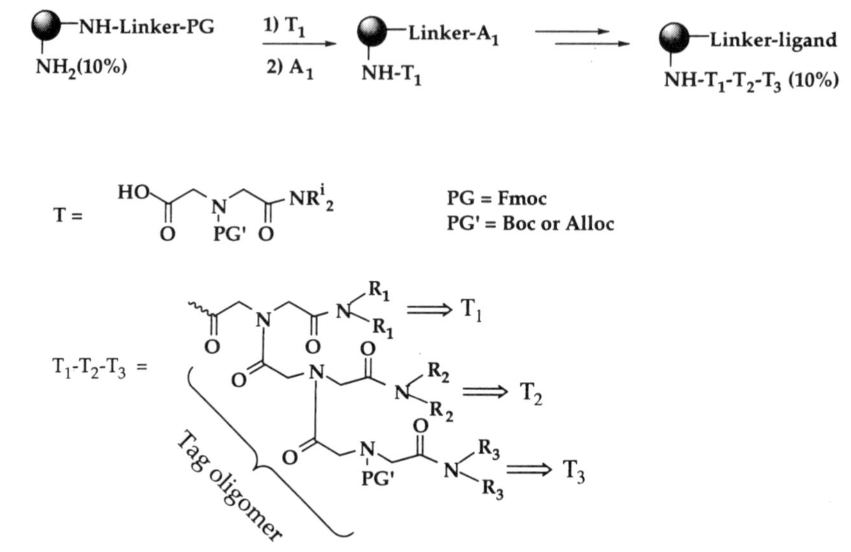

Figure 14.8. Chemically "hard" secondary amine tags incorporated into an oligomer consisting of tertiary amide bonds. The tagging polymer contains no titratable protons and is stable to most organic reagents. Hydrolysis of the oligomer in refluxing 6 N hydrochloric acid regenerates the secondary amines, which are derivatized with a reporter molecule and detected via HPLC with fluorescence detection.

described above was prepared using a 3-bit binary code, sufficient for the 7 unique tagging combinations required at each position ($2^3 - 1$; the null combination is not used). The library was screened in an immobilized format against an antibody specific for the parent sequence and stained with an alkaline–phosphatase secondary antibody. The resulting labeled beads were isolated and decoded via EC GC. Several beads containing the parent sequence were identified, as well as beads containing novel sequences.

A refinement to this approach was reported in a subsequent paper in which the carboxylic acid was replaced with a diazoketone functionality that could be converted to a reactive acylcarbene.[11] The activated tag inserted directly into the polymer matrix of the bead, and thus had no requirement for a specific functional group for its attachment. The acylcarbene can also potentially react with the compounds cosynthesized on the beads, but due to the high sensitivity of tag detection, tag concentrations were carefully minimized to limit the derivatization of the ligands thus encoded.

More recently, Ni, Gallop and coworkers described an encoding methodology that uses secondary amines as surrogate analytes.[12] These amines are incorporated into an N-(dialkylcarbamoylmethyl)glycine coding oligomer through simple amide bond chemistry (Figure 14.8), thus affording a tag oligomer. The resulting tag oligomer contains no titratable protons and is stable to most organic reagents. A further attraction of this approach was the simplicity of the tagging chemistry, which permitted the automated synthesis of encoded libraries using custom-built synthesizer hardware.

In the decoding process, acidic hydrolysis of the tagging polymer regenerates the secondary amines. While mass spectral detection of these amines was feasible, a more robust approach involved their derivatization with the reporter molecule dansyl chloride. The resulting N,N-dialkylsulfonamides were easily distinguished at subpicomole levels by reverse-phase high-pressure liquid chromatography (HPLC) with fluorescence detection. The tags could be incorporated using either Boc- or Alloc-protection strategies, and thus the tagging chemistry could be made compatible with a variety of synthesis schemes.

A binary tagging strategy was used to encode the solid-phase syntheses of several organic heterocycles including β-lactams, thiazolidinones, and pyrrolidines[19] as well as libraries of natural products. Syntheses were frequently performed on a photocleavable linker, which permitted the orthogonal liberation of library compounds into free solution for biological assay. Active beads could then be recovered and decoded as described above.

While the above hard-tagging strategies have been successfully used to encode a variety of different synthetic chemistries, a common limitation remains—the requirement for parallel synthesis (ligand and encoding tags). Since the robust preparation of a large combinatorial library is frequently a difficult synthetic challenge, it would be desirable to obviate the need for tag cosynthesis and instead delineate individual compounds by other physical means. Recent approaches that can replace or minimize the need to cosynthesize a surrogate analyte are described in the following sections.

RADIO-FREQUENCY ENCODING

Radio-frequency (rf) encoding techniques physically encapsulate an rf encodable microchip with the synthesis resin, such that the rf transponder can be scanned post-synthesis to identify its associated product (Figure 14.9). Because

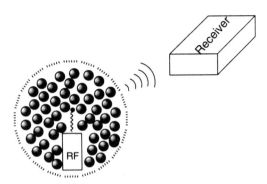

Figure 14.9. Radio-frequency encoding accomplished by the microencapsulation of bead particles with an rf transponder. The identity of a compound synthesized within a given microcapsule is determined by reading its unique rf signature.

of the size of the rf transponders (approximately $8 \times 1 \times 1$ mm), they are encapsulated with tens of milligrams of synthesis beads; hence, a small bead mass is encoded rather than a single bead. The transponders are encased in glass, and are therefore stable to solvents and other reagents.

Moran and coworkers placed a unique combination of rf transponders in each of 125 microcapsules, and used these to prepare a $5 \times 5 \times 5$ library of tripeptide derivatives.[13] The library was prepared using a modified split-synthesis strategy, and each capsule was scanned prior to redistribution to the next synthesis vessel such that its unique rf signature could be associated with that building block. Nicolaou and coworkers used a similar approach; however, the rf tags they reported had a read/write capability.[14] The devices were capable of receiving and storing information for post-synthesis retrieval, and the rf signature specific for a given building block could be input to the microcapsule during the course of the synthesis. As a proof of concept, this technique was used to encode a 24-membered peptide library.

Radio-frequency encoding successfully avoids the need to cosynthesize surrogate analytes, and also permits the larger-scale synthesis of compounds since each microcapsule can hold tens of milligrams of synthesis beads. Such quantities of resin not only allow the preparation of compounds in sufficient scale to permit their characterization using conventional analytical techniques, but also provides enough material to assay in multiple biological screens. Radio-frequency encoding is best considered a medium-throughput synthesis approach largely due to the physical size of the rf transponders, which restricts the number of these capsules that can be conveniently accommodated in standard laboratory glassware and subsequently reacted with practical volumes of reagents. Significant miniaturization will be required before libraries containing $>10^3$ members are convenient.

ISOTOPE OR MASS ENCODING

All of the reported single-bead encoding schemes require the cosynthesis of a suitable tagging moiety to record the synthetic history of each compound prepared in the library. This is inherently inefficient, since each unique compound could encode for itself if appropriate analytical techniques such as ^{1}H and ^{13}C NMR, polarimetry, etc. could be used to unambiguously assign structures to ligands present in the amounts provided by single beads. These analytical tools are obviously sample limited when performed at the single bead level. The exception is mass spectroscopy, where the detection limits for many classes of molecules are often within the constraints of single-bead chemistry. However, mass detection cannot distinguish compounds of unique composition but of common mass—a typical problem when screening libraries of thousands to tens of thousands of compounds where the largest and smallest library members differ by only a few hundred AMU.

Geysen and coworkers recently proposed a technique in which a combination of selective isotopic labeling with controlled pooling strategies might be used to solve the degeneracy problem of mass-based encoding.[15] In one embodiment of the Geysen approach, a common synthesis reagent is isotopically enriched. Different ratios of this isotopically distinct but otherwise identical reagent and its unenriched counterpart are used in the course of the synthesis. Specific ratios are employed to encode building blocks used in other steps of the synthesis that are unique but whose mass is identical to that of other reagents employed at that same synthetic step. A simplified illustration is shown in Figure 14.10. In this example, a dipeptide library is prepared on an isotopically enriched (^{14}N versus ^{15}N) cleavable linker. Upon cleavage, the unique ^{14}N/^{15}N ratio of the resulting peptide amides is used to distinguish compounds of identical mass but of unique composition (Figure 14.10). Additional "encoding" can be obtained by employing different pooling strategies— for example, the final building blocks used in a synthesis can be distinguished by simply avoiding the final pooling step.

A distinct limitation of the isotopic labeling approach is the requirement for specific isotopically enriched reagents that can be used in the synthetic scheme. A majority of synthetic schemes will not employ a degenerate reagent, and those that do will be further limited by the commercial availability of an isotopically enriched counterpart. As an alternative, the authors suggest that a tagging sequence can be prepared independent of the actual ligand synthesis chemistry. For example, a series of isotopically enriched amino acids can be cosynthesized on the bead, the ratio of which can be used to encode specific building blocks present in the desired synthetic scheme. While feasible, this approach negates the primary advantage of the mass encoding technique, which is to avoid the cosynthesis of both the desired ligand and its associated tag.

SCREENING TECHNOLOGY FOR ENCODED LIBRARIES

As described above, screening split-synthesis libraries poses several analytical challenges. While large soluble pools derived from the split-synthesis technique have been successfully screened, the ambiguities and inefficiencies associated with this approach largely outweigh any benefits. Similarly, screens requiring that the ligand remain associated with the bead cannot be adapted to many biological assay formats. A preferred method is to assay in solution phase the contents of discrete synthesis beads. This requires the resolution of several technical hurdles, including techniques for manipulating single microscopic beads, ligand release strategies that are compatible with the conditions of the biological assays, methods for keeping the released ligand in close association with its encoding bead, and in some instances the development of detection systems that can discern biological events performed under the stringencies of single-bead ligand concentrations.

Lerner and coworkers described a novel G-protein-coupled reporter assay in which peptide-associated beads were placed in contact with a lawn of

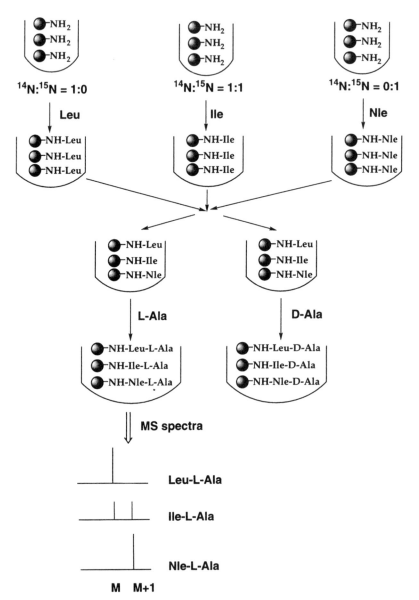

Figure 14.10. One embodiment of the Geysen group's mass encoding methodology as illustrated by the encoding of a dipeptide library. The regioisomeric amino acid building blocks L-leucine (Leu), L-isoleucine (Ile) and L-norleucine (Nle) are encoded by acylating carboxamide-releasing linkers containing isotopically unique $^{14}N/^{15}N$ ratios. After pooling, the stereoisomeric building blocks L-alanine (L-Ala) and D-alanine (D-Ala) are incorporated at the second position of the library. The identities of the latter building blocks are "encoded" by avoiding the final pooling step. Mass detection of the ligands cleaved from single beads determines the identity of the amino acid incorporated in the first step, as each dipeptide will have a unique isotopic distribution in its MS.

engineered melanophores.[16] Beads were first distributed into a crude two-dimensional matrix on a sheet of polypropylene film, and the peptides were cleaved with TFA vapor. In this gaseous environment, cleaved ligands should remain adsorbed onto their parent beads. Once applied to the cell culture, beads diffusing ligands that activated the cells could be identified by the subsequent pigment dispersion surrounding them.

Oldenburg and Yuan and coworkers have described a novel dual-culture assay designed to screen bead-based libraries for antimicrobial activity.[17] Conditions were developed for the coculture of both yeast as well as bacterial strains, both of which were engineered to express a unique fluorescent reporter molecule. After applying a lawn of beads to the cultured organisms and releasing their attached ligands, any beads expressing active compounds could be identified by a surrounding zone of clearing as determined by fluorescence microscopy. Individual compounds could then be simultaneously screened for inhibition of bacterial growth (antibacterials), yeast growth (antifungals), or the inhibition of both organisms (potential toxins). A mild ligand-release strategy was required to avoid incidental damage to the cultured organisms. Thus, libraries were prepared on beads derivatized with a photocleavable linker. Lawns of beads were created by pouring an agar suspension of the beads directly onto a culture plate, and their ligands were subsequently released via photolysis. As many as 100,000 beads were simultaneously evaluated in a petri dish 90 mm in diameter.

Schullek and Yuan and coworkers reported another high-density screening format for assaying encoded split-synthesis libraries.[18] A nanowell plate was designed that could contain >6500 wells in a footprint similar to that of a conventional 96-well microtiter plate. Each well of the plate could hold less than 1 μL of reagents. Liberation of the contents of a single bead ($\sim$100 pmol) into this volume of assay medium would provide assay concentrations as high as 100 μM. Furthermore, the nanowell approach limits or avoids the ligand diffusion that occurs in the above bead lawn formats.

The authors used this nanowell format to screen a small encoded library of carboxylic acid-based metalloprotease inhibitors against the enzyme matrilysin. The library was prepared on beads derivatized with a photocleavable linker. After a suspension of beads was applied to the nanowell plate, the beads were first mechanically driven into the wells by swiping the array surface with a straightedge tool and then photolyzed to release their compounds. Following the addition of enzyme, a fluorogenic substrate was added and time-points were taken with a CCD-coupled fluorescence imager (Figure 14.11). Nanowells emitting no or lower-intensity fluorescence were found to contain beads that released inhibitors of the enzyme.

CONCLUSIONS AND FUTURE DIRECTIONS

To summarize, the quantity of material present on a single-synthesis bead (approximately 100 pmol) is often acceptable for assay in a biological screen,

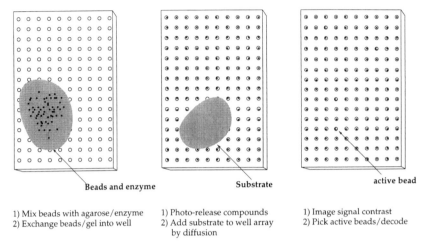

Beads and enzyme   Substrate   active bead

1) Mix beads with agarose/enzyme   1) Photo-release compounds   1) Image signal contrast
2) Exchange beads/gel into well   2) Add substrate to well array   2) Pick active beads/decode
                                       by diffusion

Figure 14.11. Assay on a nanowell plate for encoded beads.

but not sufficient to permit compound characterization using conventional analytical techniques. A variety of different encoded split-synthesis techniques have been reported that appear to solve this structural elucidation problem. In general, these encoding techniques record the synthetic history of a given bead, and in most of these approaches this information can be retrieved with high sensitivity at the single bead level. Encoding permits the direct assay of split-synthesis libraries in a variety of formats and obviates the need for iterative deconvolution and other inefficient methods for ligand identification.

Although single-synthesis beads can be screened in conventional 96-well microtiter plates, the large size of many split-synthesis libraries has necessitated the development of novel single-bead assay formats that can be performed at much higher bead densities. This miniaturization is critical because it permits not only the efficient screening of libraries of hundreds of thousands to millions of members, but also lowers the consumption of biological reagents well below what would be required to assay a similar number of compounds in conventional microtiter plates. As a more efficient alternative to the assay of single beads, pools of encoded beads could be assayed simultaneously. For example, pools of 100 beads could be screened against the enzyme or receptor of interest. The beads contained in active pools could then be segregated into individual assay wells and a second round of activity determinations could be performed to identify active single beads for decoding.

Screening mixtures of compounds has several disadvantages, as discussed previously; however, when screening for rare events where it is not necessarily imperative to generate a complete SAR from a library, such tiered assay strategies could ultimately prove more efficient. An additional requirement for this approach is the controlled tiered release of the bead-bound ligands, such that roughly equimolar portions could be released into solution for both the primary and secondary screens.

A commercially unmet need in bead synthesis technology is the development of hardware and instrumentation for performing single-bead manipulations. While 100-μm beads can be observed with relative ease in a low-power wide-field microscope, their physical manipulation into capillaries, microtiter plates, and so on, is tedious and time-consuming. Nonsettling bead suspensions can be prepared and pipetted to dispense approximate populations of beads, but this approach does not dispense consistent single-bead populations. The continued refinements of the diverse aspects of synthetic bead technology, such as bead manipulation techniques, more efficient encoding schemes, and high-density screening formats, will combine to further improve an already powerful synthetic technique.

REFERENCES

1. Gallop MA, Barrett RW, Dower WJ, Fodor SPA, Gordon EM (1994): Applications of combinatorial technologies to drug discovery, 1: background and peptide combinatorial libraries. *J Med Chem* 37:1233–1251.

2. Gordon EM, Barrett RW, Dower WJ, Fodor SPA, Gallop MA (1994): Applications of combinatorial technologies to drug discovery, 2: combinatorial organic synthesis, library screening strategies, and future directions. *J Med Chem* 37:1385–1401.

3. Scott JK, Smith GP (1990): Searching for peptide ligands with an epitope library. *Science* 249:386–390.

4. Devlin JJ, Panganiban LC, Devlin PE (1990): Random peptide libraries: a source of specific protein binding molecules. *Science* 249:404–406.

5. Cwirla SE, Peters EA, Barrett RW, Dower WJ (1990): Peptides on phage: a vast library of peptides for identifying ligands. *Proc Natl Acad Sci USA* 87:6378–6382.

6. Brenner S, Lerner RA (1992): Encoded combinatorial chemistry. *Proc Natl Acad Sci USA* 89:5381–5383.

7. Needels MC, Jones DG, Tate EH, Heinker GL, Kochersperger LM, Dower WJ, Barrett RW, Gallop MA (1993): Generation and screening of an oligonucleotide-encoded synthetic peptide library. *Proc Natl Acad Sci USA* 90:10700–10704.

8. Kerr JM, Banville SC, Zuckermann RN (1993): Encoded combinatorial peptide libraries containing nonnatural amino acids. *J Am Chem Soc* 115:2529–2531.

9. Nikolaiev V, Stierandova A, Krchnak V, Seligmann B, Lam KS, Salmon SE, Lebl M (1993): Peptide-encoding for structure determination of nonsequenceable polymers within libraries synthesized and tested on solid-phase supports. *Pept Res* 6:161–170.

10. Ohlmeyer MHJ, Swanson RN, Dillard LW, Reader JC, Asouline G, Kobayashi R, Wigler M, Still WC (1993): Complex synthetic chemical libraries indexed with molecular tags. *Proc Natl Acad Sci USA* 90:10922–10926.

11. Nestler HP, Bartlett PA, Still WC (1994): A general method for molecular tagging of encoded combinatorial chemistry libraries. *J Org Chem* 59:4723–4724.

12. Ni ZJ, Maclean D, Holmes CP, Murphy MM, Ruhland B, Jacobs JW, Gordon EM, Gallop MA (1996): A versatile approach to encoding combinatorial organic syntheses using chemically robust secondary amine tags. *J Med Chem* 39:1601–1608.

13. Moran EJ, Sarshar S, Cargill JF, Shahbaz MM, Lio A, Mjalli AMM, Armstrong RW (1995): Radio-frequency tag encoded combinatorial library method for the discovery of tripeptide-substituted cinnamic acid inhibitors of the protein tyrosine phosphatase PTP1B. *J Am Chem Soc* 117:10787–10788.

14. Nicolaou KC, Xiao XY, Parandoosh Z, Senyei A, Nova MP (1995): Radiofrequency encoded combinatorial chemistry. *Angew Chem Int Ed Engl* 34:2289–2291.

15. Geysen HM, Wagner CD, Bodnar WM, Markworth CJ, Parke GJ, Schoenen FJ, Wagner DS, Kinder DS (1996): Isotope or mass encoding of combinatorial libraries. *Chem Biol* 3:679–688.

16. Jayawickreme CK, Graminski GF, Quillan JM, Lerner MR (1994): Creation and functional screening of a multi-use peptide library. *Proc Natl Acad Sci USA* 91:1614–1618.

17. Oldenburg KR, Vo KT, Ruhland B, Schatz PJ, Yuan Z (1996): A dual culture assay for detection of antimicrobial activity. *J Biomol Screen* 1:123–130.

18. Schullek JR, Butler JH, Ni ZJ, Chen D, Yuan Z (1997): A high-density screening format for encoded combinatorial libraries: assay miniaturization and its application to enzymatic reactions. *Anal Biol* 246:20–29.

19. Maclean D, Schullek JR, Murphy MM, Ni ZJ, Gordon EM, and Gallop MA (1997): Encoded combinatorial chemistry: synthesis and screening of a library of highly functionalized pyrrolidines, Proc Nat. Acad Sci USA, 94, 2805–2810.

15

PARALLEL ORGANIC SYNTHESIS IN ARRAY FORMAT

STEVEN E. HALL

Sphinx Pharmaceuticals, A Division of Eli Lilly & Co., Durham, North Carolina

The power of combinatorial synthesis[1-3] has been documented numerous times for its utility in lead optimization and the identification of de novo lead structures.[4-6] Examples include the identification of potent nonpeptide angiotensin converting enzyme (ACE) inhibitors,[7] peptidylphosphonate inhibitors of thermolysin,[8] tripeptide endothelin antagonists,[9] peptoid ligands for $\alpha 1$-adrenergic receptors,[10] μ-opiate receptors,[10] and α-amylase,[11] and tripeptide inhibitors of serotonin reuptake.[12] Many elegant methods have been developed to aid and accelerate the deconvolution process inherent in screening large mixtures of compounds.[13] As detailed in other chapters of this book, these methods include recursive deconvolution[14] and multiple encoding approaches. In the latter process, test compounds are bound to a polymeric bead that incorporates a specific tag that can be read by analytical methods such as polymerase chain reaction (PCR) amplification,[15] mass spectrometry,[16] gas chromatography,[17-19] and radio-frequency (rf) techniques[20] to decode the structure of the compound on the bead. Methods in which the test molecule remains attached to the bead limit the assay format in that the target enzyme or receptor must be a soluble protein. Nonetheless, deconvolution methods have been successful in identifying active compounds amid a sea of candidates. Regardless of whether the compounds are tested free in solution or bound to a support, there remain a number of issues in screening large compound mixtures.

Combinatorial Chemistry and Molecular Diversity in Drug Discovery, Edited by
Eric M. Gordon and James F. Kerwin, Jr.
ISBN 0-471-15518-7   Copyright © 1998 by Wiley-Liss, Inc.

ISSUES IN SCREENING LARGE COMPOUND MIXTURES

Will molecules bound to large polymer beads bind as well or in the same way as the same ligand free in solution?

What about additivity effects when screening large mixtures?[21]

Is it really necessary to screen 10,000,000 compounds to find a de novo lead? Or would 100,000 be sufficient?

Is the time necessary to deconvolute the mixtures more or less than that required to synthesize individual compounds if one had an efficient manner of parallel synthesis?

What if there already is some information on possible pharmacophores?

These issues have prompted a number of laboratories to develop efficient approaches to parallel organic synthesis. Although chemists have been capable of preparing several compounds at a time for many years, they have only recently developed procedures for the simultaneous synthesis of hundreds of compounds using array synthesis. This chapter highlights some of the recent advances in the simultaneous array synthesis of nonpeptide, nonoligomeric compounds.

PARALLEL-ARRAY SOLID-PHASE SYNTHESIS FOR LEAD OPTIMIZATION

One of the earliest approaches, and undoubtedly the most widely discussed process for array synthesis, is the Diversomer approach from the Parke–Davis group.[22] From the outset of their work, Hobbs–Dewitt et al. were interested in a more efficient manner to develop the structure–activity relationships (SAR) around a new lead. Hobbs–Dewitt et al. described the synthesis of a series of benzodiazepines (Scheme 15.1) for a CNS program to illustrate their approach.[23]

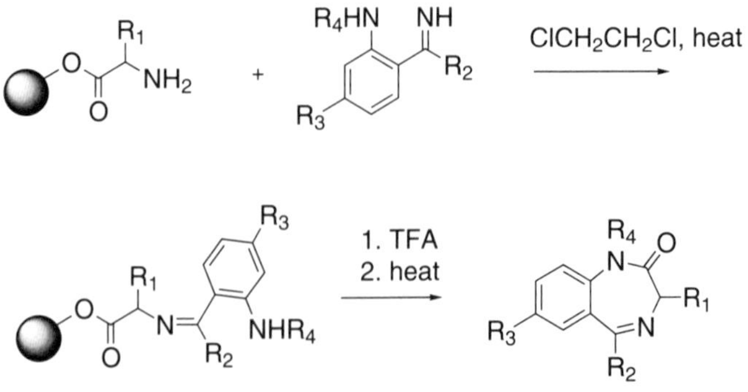

Scheme 15.1. Parke–Davis benzodiazepine synthesis.

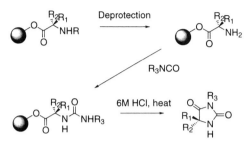

Scheme 15.2. Parke–Davis hydantoin synthesis.

A similar approach was used for the preparation of hydantoins (Scheme 15.2). Both synthetic schemes were designed such that the last reaction involved both a functional group conversion and ejection from the resin.[23] The merit of this approach is that intermediates that fail to react in the last step remain attached to the resin, thus ensuring products of increased purity.

As shown in Figure 15.1, the original apparatus allowed for 8 compounds to be prepared simultaneously.[23] More recent versions include a 40-pin synthe-

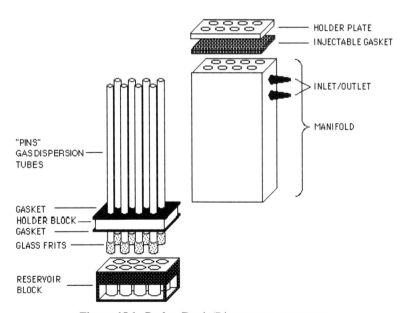

Figure 15.1. Parke–Davis Diversomer apparatus.

sizer. The 8-pin synthesizer is now commercially available from Chem-Glass under license from Parke–Davis.

PARALLEL-ARRAY SOLID-PHASE SYNTHESIS FOR LEAD GENERATION

The value of parallel organic synthesis is not limited to the directed SAR approach described by Hobbs–Dewitt. Indeed, both industrial and academic researchers have explored the utility of array synthesis to prepare large numbers of compounds. The goals in such efforts may be twofold; the synthesis of focused libraries for a particular biochemical target or the synthesis of compounds for general screening.

One of the first examples of parallel synthesis is the VLSIPS (very large-scale immobilized polymer synthesis) methodology in which large libraries are prepared on a silicon wafer.[24] This technique has been reviewed[25] and is the subject of Chapter 4.[4]

Ellman has also described the synthesis of benzodiazepines, but the goal of this work was to generate a diverse library of compounds using the benzodiazepine ring as a scaffold for side chains that would contain a variety of functional groups.[26–28] In contrast to the Parke–Davis work, the benzophenone precursor was attached to the resin and then condensed with the appropriate amino acid derivative. This chemistry was adapted to a parallel array synthesis using the Geysen pin technology.[29,30] In this manner 192 distinct benzodiazepines were prepared (Scheme 15.3). A number of these compounds were shown to bind to the cholecystokinin (CCK) receptor.

A series of cyclic β-turn mimetics have also been prepared in a parallel array format through the use of the Geysen pin technology (Scheme 15.4). Although only 11 compounds were prepared, the methodology is amenable to further scale-up.[31] The products were obtained in good purity after an eight-step sequence.

Green has recently described the synthesis of a series of benzyl amines related to lavendustin (Scheme 15.5).[32] Three scaffolds were individually coupled to the resin followed by reductive amination with five substituted benzaldehydes. At each step, the bulk resin was further subdivided so that each analog was prepared as a single component. This approach helped to ensure consistent product purity, since each individual compound was prepared using the same intermediate resin.

Ellman has recently described the solid-phase synthesis of α-amino alcohols as part of a library for aspartyl protease inhibitors (Scheme 15.6).[33] As above, development of new solid-phase synthetic transformations as well as a novel linking functionality was a prerequisite for success. Ellman prepared a resin-bound dihydropyran that allowed for the attachment of any alcohol of interest (Scheme 15.7).[34] This approach achieved both simultaneous protection of the alcohol and attachment to the resin. With this intermediate in hand, Ellman

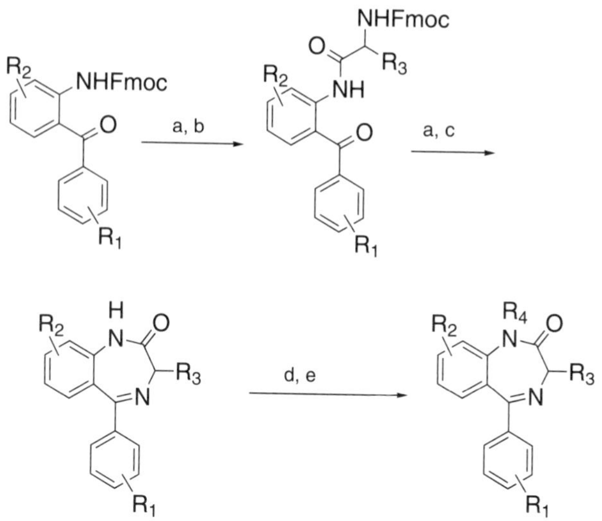

Scheme 15.3. Ellman benzodiazepine synthesis. (a) 20% piperidine/DMF; (b) N-Fmoc-aminoacid fluoride, 4-methyl-2,6-di-*tert*-butylpyridine; (c) 5% acetic acid/DMF, 60°C; (d) lithiated 5-(phenylmethyl)-2-oxazolidinone, THF, −78°C, then R_4X/DMF; (e) TFA/H_2O/Me_2S (95:5:10).

was now in a position to take advantage of the remaining masked functionality. Nucleophilic displacement of the tosylate with a variety of amines could be followed by acylation to afford ureas or reduction of the azide to provide the intermediate amine. Acylation then provided the target compounds following ejection from the resin. Adaptation of this chemistry to a multiple simultaneous synthesis format was reported to be in progress.

Researchers at Abbott[35] were faced with a similar challenge in selecting the most appropriate way to append two scaffolds of interest, a diamino diol and the related diaminoalcohol. The increased steric demands of these scaffolds rendered the dihydropyran linkers related to Ellman's work unsuitable. Successful coupling could be achieved by condensation of the diaminoalcohol with a functionalized vinyl ether to provide scaffold **3** (Figure 15.2). The related diol was condensed with a keto acid to afford ketal **4**. These products were then attached to benzhydramine resin using the pendant carboxylic acid. Synthesis of the analogs (Scheme 15.8) was performed with an Aibimed ANS422 synthesizer, which can prepare up to 48 compounds simultaneously.

Scanlan described the parallel synthesis of a small library of hydroxystilbene derivatives (Scheme 15.9).[36] Four distinct hydroxybenzaldehydes were coupled to Rink amide resin. Products were then split and condensed individually with six benzylphosphonates to provide the 24-member library.

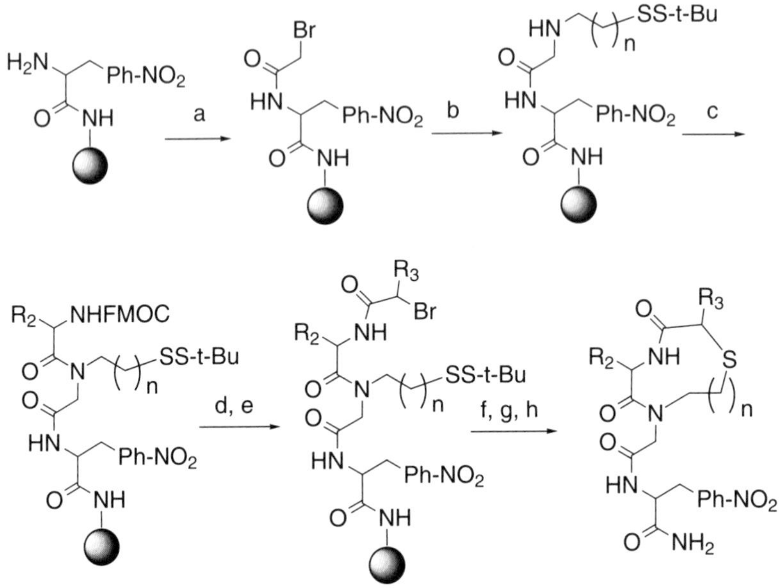

Scheme 15.4. Solid-phase synthesis of β-turn mimetics. (a) α-Bromo acid, diisopropyl-carbodiimide; (b) $H_2N(CH_2)_nSS$-t-C_4H_9; (c) N-Fmoc-α-amino acid, HATU; (d) 20% piperidine/DMF; (e) α-bromoacid anhydride; (f) tributylphosphine, H_2O; (g) tetra-methylguanidine; (h) 1:1:18 $H_2O/Me_2S/TFA$.

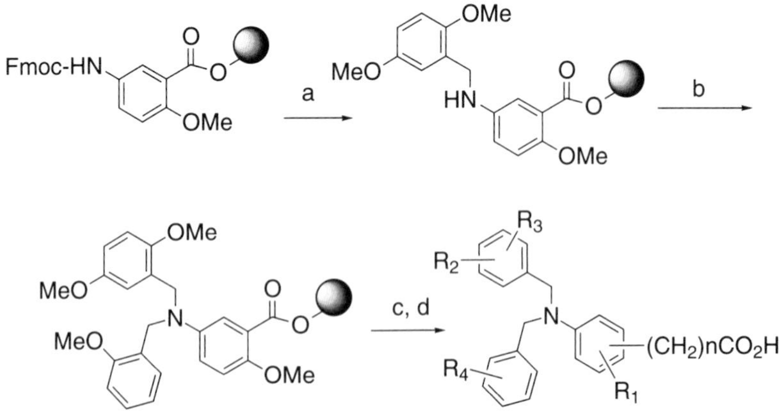

R_1-R_3 = OCH_3, H
R_4 = OCH_3, NO_2, F, H
n = 0,1

Scheme 15.5. Synthesis of lavendustin analogs.

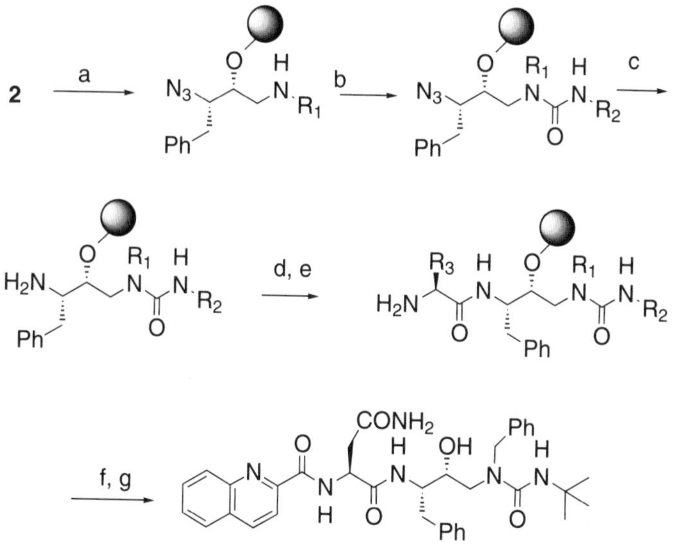

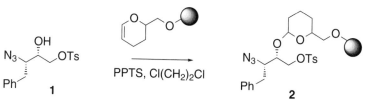

Scheme 15.6. Synthesis of aspartyl protease inhibitors. (a) R_1NH_2, NMP, 80°C; (b) R_2NCO, $ClCH_2CH_2Cl$; (c) $SnCl_2$:HSPh: Et_3N (1:4:5), THF; (d) Fmoc-amino acid, PyBOP, HOBt, i-Pr_2EtN (3 equiv), DMF; (e) 20% piperidine in DMF; (f) pentafluorophenyl ester of quinaldic acid, HOBt, Et_3N, DMF; (g) 95:5 TFA/H_2O.

PARALLEL-ARRAY SOLUTION SYNTHESIS FOR LEAD GENERATION

All of the above methods have used solid-phase synthesis techniques, but the application of parallel array synthesis is not limited to solid-phase methods. Recently, Janda et al. have described the synthesis of a small library of sulfonamides using a soluble resin (Scheme 15.10).[37] In this process, the scaffold is coupled to a low molecular polyethylene glycol(PEG) polymer that is soluble in a number of organic solvents but precipitates from solvents such as ether. The advantage of this approach is that new solid-

Scheme 15.7. Use of a resin-bound THP-protecting group.

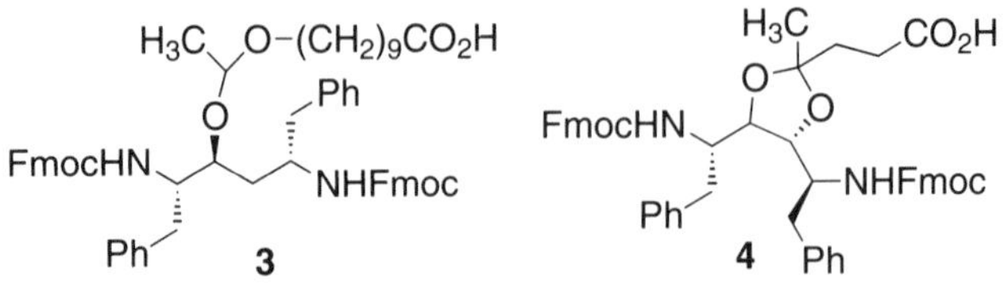

Figure 15.2. Scaffolds for the synthesis of HIV inhibitors.

phase chemistry does not need to be developed since all of the reactions are conducted in solution. One practical limitation to this approach can be the volume of solvent needed to efficiently precipitate the resin-bound intermediates.

Taking this one step further, a number of laboratories have explored solution libraries in which all components are condensed in solution. Several manifestations of this approach include the use of resin-bound reagents[38] (Scheme 15.11) as well as the use of multicomponent condensation reactions. In the latter reactions, the excess reagents are removed as a consequence of their volatility or other physical properties.

LARGE-SCALE PARALLEL-ARRAY SYNTHESIS

Many of the results described above have focused on the synthesis of a specific class of compounds for which an activity is already known. Although not

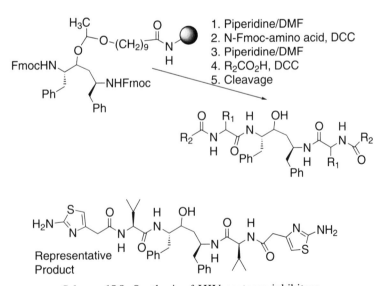

Scheme 15.8. Synthesis of HIV protease inhibitors.

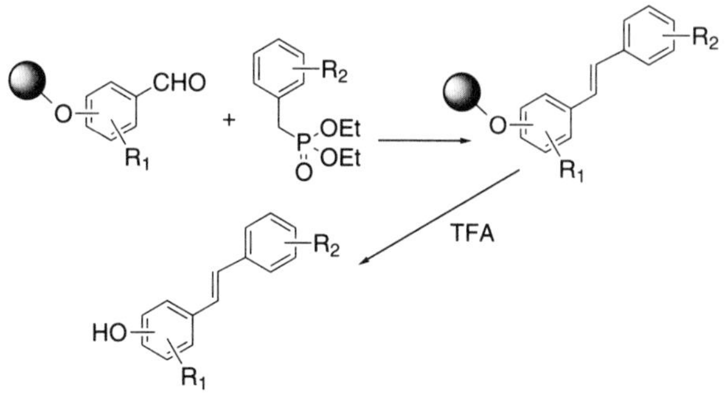

Scheme 15.9. Synthesis of substituted stilbenes.

specifically aimed at a bioilogical target, much of Ellman's work involves compounds for evaluation in broad target classes, i.e., aspartyl proteases. The use of parallel array format for the nonbiased synthesis of general chemical libraries is much less common and includes research from ArQule, Ontogen, and Sphinx. A brief description of the latter two approaches is described in the following section.

Workers at Ontogen have developed a modular reaction matrix to provide spatially differentiated chemical libraries.[20] The OntoBLOCK system was designed to prepare milligram quantities of individual compounds in a 96-well format. The OntoBLOCK reaction apparatus consists of 96 septum-sealed reaction vessels (polypropylene or Teflon) organized in the standard well pattern of a 96-well microtiter plate (Figure 15.3). The temperature of the reaction block can be controlled up to +100°C and down to −160°C. The reaction block can be sealed completely to allow for containment of volatiles during reactions and can be charged with inert gas. The reaction conditions are maintained (temperature, pressure, and inert atmosphere)

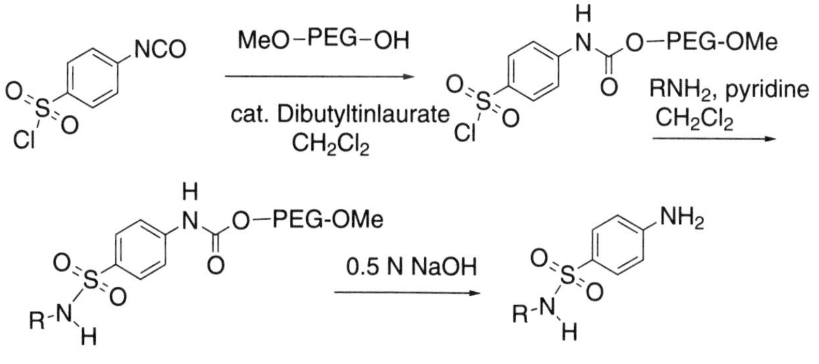

Scheme 15.10. Solution synthesis of sulfonamides.

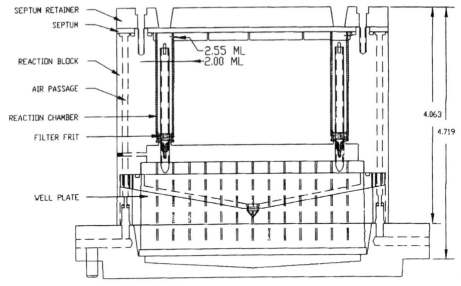

Figure 15.3. Ontogen OntoBLOCK reaction manifold.

via a universal docking station, which allows the movement of the reaction block between task specific workstations. This approach stands in contrast to the turn-key synthesis machines such as the MOS 496 from Advanced ChemTech.

The Ontogen group has used this apparatus to prepare a variety of nonpeptidic compounds, including a number of heterocyclic targets. They have adapted solution-based reactions, such as multicomponent condensation reactions, to solid-phase synthesis. In one example, diketones were condensed with aldehydes and amines to prepare highly substituted imidazoles (Scheme 15.11). The scope of the reaction was limited to the use of diaryl ketones; aliphatic ketones provided only low yields of the desired products. Either the amine or the aldehyde component could be attached to the polymeric support, usually through an ester or ether linkage. Using these conditions, yields in excess of 95% could be achieved.

Researchers at Sphinx have also pursued parallel array synthesis using the 96-well layout of microtiter plates; however, they were particularly interested in developing a low-cost, low-tech approach. In this manner, the process and equipment could be easily transferred to traditional medicinal chemistry labs. To realize this goal, the Sphinx Group maximized their use of commercially available microtiter plates and accessories. A straightforward process was developed in which a 96-well reaction vessel was constructed from a standard deep-well polypropylene plate. A small hole drilled into the bottom of each well and insertion of a frit converted each well into a filtration chamber.

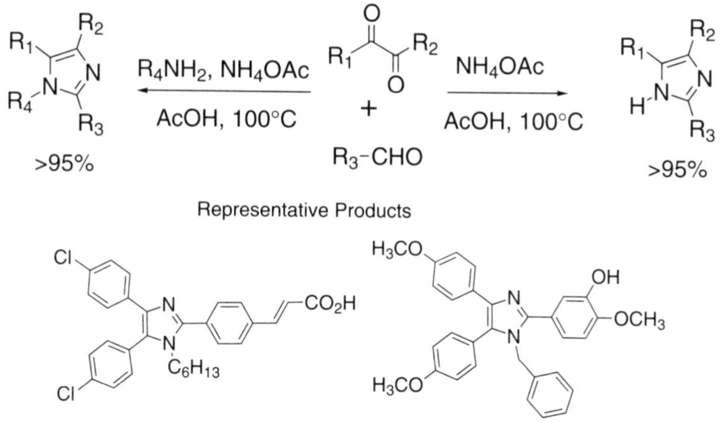

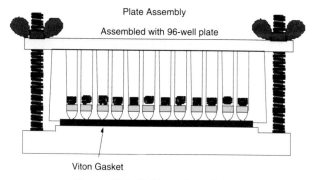

Scheme 15.11. Solid-phase synthesis of imidazoles.

Subsequent clamping of the plate with Viton gaskets to seal the wells afforded an efficient matrix for parallel array synthesis (Figure 15.4).

Meyers et al. have described the synthesis of a series of bisamide phenols, urea-amide phenols, and sulfonamide-amide phenols using this apparatus.[39] In this approach, p-amino-m-nitro-phenol was condensed with a carboxylated polystyrene resin. Analogous to much of the work already cited, the choice of resin and linker is critical to the successful synthesis of the library. An ester linkage using carboxylated polystyrene was important since a high-loading resin could be synthesized.[40] This capability allowed for the array synthesis of 10–15 mg of *each* individual product.

The first site of diversity was introduced through condensation of the amine with acid chlorides, sulfonyl chlorides, or isocyanates. Reduction of the nitro

Plate Assembly

Assembled with 96-well plate

Viton Gasket

Figure 15.4. Sphinx plate clamp.

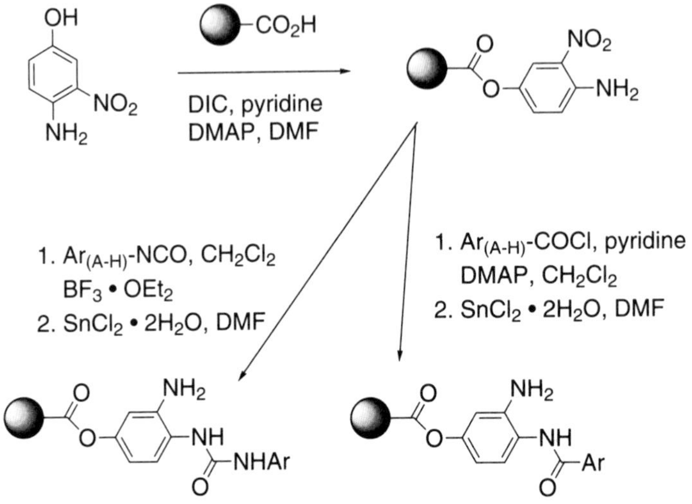

Scheme 15.12. Synthesis of intermediate nitro ureas and nitro amides.

group provided 12 unique amino resins (Scheme 15.12). These resins were loaded into a deep-well microtiter plate where each well within a column contains the same intermediate resin. Addition of the second functionalization reagent followed by hydrolysis provided 96 unique bis-functionalized phenols (Scheme 15.13). Although Meyers et al. described these products as targeted libraries for protein kinases, it should be apparent that selection of alternative side chains would allow one to prepare a diverse set of bis-functionalized phenols.

In this process, the first site of diversity is introduced in a batch synthesis mode. Although this approach has an advantage from a quality-control per-spective (i.e., all starting resin is from the same batch), disadvantages include potential logistical problems associated with handling, tracking, and loading multiple resins. We have also prepared compounds in this general class by conducting all of the chemistry in a 96-well plate. With this variant, p-amino-m-nitro-phenol is attached to the resin and the resulting product is loaded into the 96-deep well plates. Functionalization of the amine and nitro reduction are accomplished in the plate format. Using this procedure, we have prepared thousands of analogs in this structural class.

CONCLUSION

The application of combinatorial chemistry to the discovery of new human therapeutics remains an experiment in progress. Will large mixture synthesis and screening be more useful and efficient than parallel array synthesis in the

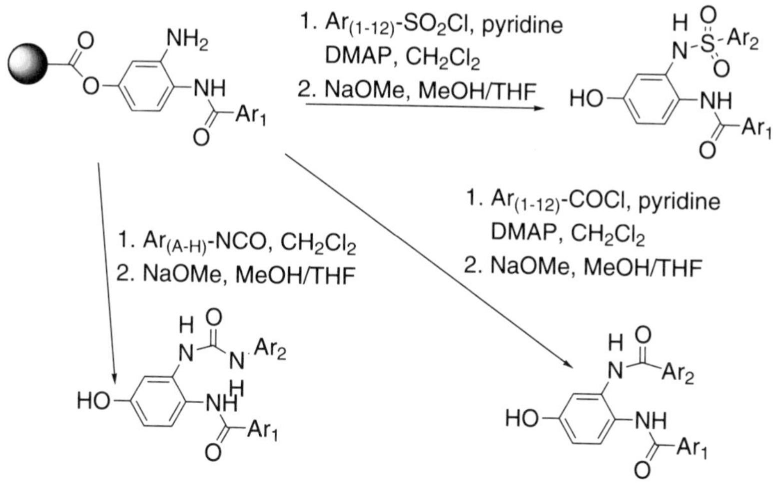

Scheme 15.13. Synthesis of bis amides and urea amides.

generation of de novo lead compounds? We will likely need to wait for several years to answer that question. However, it is already clear that parallel array synthesis is an effective vehicle to accelerate lead optimization. Anecdotal stories have emerged from several pharmaceutical organizations on the successful application of array synthesis for optimization studies.

Researchers have tended to separate combinatorial chemistry from mainstream medicinal chemistry. This division is unfounded; array synthesis should be viewed as simply an updated tool for medicinal chemistry. Put another way, medicinal chemistry of the twenty-first century has arrived, a few years early!

REFERENCES

1. Pavia MR, Sawyer TK, Moos WH (1993): The generation of molecular diversity. *Bioorg Med Chem Lett* 3:387–396.
2. Moos WH, Green GD, Pavia MR (1993): Recent advances in the generation of molecular diversity. *Annu Rep Med Chem* 28:315–324.
3. Gordon EM, Barrett RW, Dower WJ, Fodor SPA, Gallop MA (1994): Applications of combinatorial technologies to drug discovery, 2: combinatorial organic synthesis, library screening strategies, and future directions. *J Med Chem* 37:1385–1401.
4. Desai MC, Zuckermann RN, Moos WH (1994): Recent advances in the generation of chemical diversity libraries. *Drug Dev Res* 33:174–188.
5. Terrett NK, Gardner M, Gordon DW, Kobylecki RJ, Steele J (1995): Combinatorial synthesis: the design of compound libraries and their application to drug discovery. *Tetrahedron* 51:8135–8173.

6. Seligmann B, Abdul-Latif F, Al-Obeidi F, Flegelova Z, Issakova O, Kocis P, Krchnak V, Lam K, Lebl M, Ostrem J, Safar P, Sepetov N, Stierandova A, Strop P, Wildgoose P (1995): The construction and use of peptide and non-peptide combinatorial libraries to discover enzyme inhibitors. *Eur J Med Chem* 30:319–333.

7. Murphy MM, Schullek JR, Gordon EM, Gallop MA (1995): Combinatorial organic synthesis of highly functionalized pyrrolidines: identification of a potent angiotensin converting enzyme inhibitor from a mercaptoacyl proline library. *J Am Chem Soc* 117:7029–7030.

8. Campbell DA, Bermak JC, Burkoth TS, Patel DV (1995): A transition state analogue inhibitor combinatorial library. *J Am Chem Soc* 117:5381–5382.

9. Terrett NK, Bojanic D, Brown D, Bungay PJ, Gardner M, Gordon DW, Mayers CJ, Steele (1995): The combinatorial synthesis of a 30,752-compound library: discovery of SAR around the endothelin antagonist, FR-139,317. *Bioorg Med Chem Lett* 5:917–922.

10. Zuckermann RN, Martin EJ, Spellmeyer DC, Stauber GB, Shoemaker KR, Kerr JM, Figliozzi GM, Goff DA, Siani MA, Simon RJ, Banville SC, Brown EG, Wang L, Richter LS, Moos WH (1994): Discovery of nanomolar ligands for 7-transmembrane G-protein-coupled receptors from a diverse N-(substituted)glycine peptoid library. *J Med Chem* 37:2678–2684.

11. Simon RJ, Kania RS, Zuckermann RN, Huebner VD, Jewell DA, Banville S, Ng S, Wang L, Rosenberg S, Marlowe CK, Spellmeyer DC, Tan R, Frankel AD, Santi DV, Cohen FE, Bartlett PA (1992): Peptoids: a modular approach to drug discovery. *Proc Natl Acad Sci USA* 89:9367–9371.

12. Koppel G, Dodds C, Houchins B, Hunden D, Johnson D, Owens R, Chaney M, Usdin T, Hoffman B, Brownstein M (1995): Use of peptide combinatorial libraries in drug design: the identification of a potent serotonin reuptake inhibitor derived from a tripeptide cassette library. *Chem Biol* 2:483–487.

13. Janda KD (1994): Tagged versus untagged libraries: methods for the generation and screening of combinatorial chemical libraries. *Proc Natl Acad Sci USA* 91:10779–10785.

14. Erb E, Janda KD, Brenner S (1994): Recursive deconvolution of combinatorial chemical libraries. *Proc Natl Acad Sci USA* 91:11422–11426.

15. Nedels MN, Jones DG, Tate EH, Heinkel GL, Kochersperger LM, Dower WJ, Barrett RW, Gallop MA (1993): Generation and screening of an oligonucleotide-encoded synthetic peptide library. *Proc Natl Acad Sci USA* 90:10700–10704.

16. Brummel CL, Lee INW, Zhou Y, Benkovic SJ, Winograd N (1994): A mass spectrometric solution to the address problem of combinatorial libraries. *Science* 264:399–401.

17. Baldwin JJ, Burbaum JJ, Henderson I, Ohlmeyer MHJ (1995): Synthesis of a small molecule combinatorial library encoded with molecular tags. *J Am Chem Soc* 117:5588–5589.

18. Burbaum JJ, Ohlmeyer MHJ, Reader JC, Henderson I, Dillard LW, Li G, Randle TL, Sigal NH, Chelsky D, Baldwin JJ (1995): A paradigm for drug discovery employing encoded combinatorial libraries. *Proc Natl Acad Sci USA* 92:6027–6031.

19. Ohlmeyer MHJ, Swanson RN, Dillard LW, Reader JC, Asouline G, Kobayashi R, Wigler M, Still C (1993): Complex synthetic chemical libraries indexed with molecular tags. *Proc Natl Acad Sci USA* 90:10922–10926.

20. Mjalli AMM, Toyonaga BE 1992: Rapid discovery and optimization of biologically active small molecules using automated synthesis methods. In Devlin JP, ed. *High Throughput Screening: The Discovery of Bioactive Substances.* New York: Decker.

21. Freier SM, Konings DAM, Wyatt JR, Ecker DJ (1995): Deconvolution of combinatorial libraries for drug discovery: a model system. *J Med Chem* 38:344–352.

22. DeWitt SH, Schroeder MC, Stankovic CJ, Strode JE, Czarnik AW (1994): Diversomer technology: solid-phase synthesis, automation and integration for the generation of chemical diversity. *Drug Dev Res* 33:116–124.

23. DeWitt SH, Kiely JS, Stankovic CJ, Schroeder MC, Reynolds Cody DM, Pavia MR (1993): "Diversomers": an approach to nonpeptide, nonoligomeric chemical diversity. *Proc Natl Acad Sci USA* 90:6909–6913.

24. Fodor SPA, Read JL, Pirrung MC, Stryer L, Lu AT, Solas D (1991): Light-directed, spatially addressable parallel chemical synthesis. *Science* 251:767–773.

25. Jacobs JW, Fodor SPA (1994): Combinatorial chemistry applications of light-directed chemical synthesis. *Trends Biotech* 12:19–26.

26. Plunkett MJ, Ellman JA (1995): Solid-phase synthesis of structurally diverse 1,4-benzodiazepine derivatives using the Stille coupling reaction. *J Am Chem Soc* 117:3306–3307.

27. Bunin BA, Plunkett MJ, Ellman JA (1994): The combinatorial synthesis and chemical and biological evaluation of a 1,4-benzodiazepine library. *Proc Natl Acad Sci USA* 91:4708–4712.

28. Bunin BA, Ellman JA (1992): A general and expedient method for the solid-phase synthesis of 1,4-benzodiazepine derivatives. *J Am Chem Soc* 114:10997–10998.

29. Geysen HM, Meleon RH, Barteling SJ (1984): Use of peptide synthesis to probe viral antigens for epitopes to a resolution of a single amino acid. *Proc Natl Acad Sci USA* 81:3998–4002.

30. Bray AM, Maeji NJ, Geysen HM (1990): The simultaneous multiple production of solution phase peptides: assessment of the Geysen method of simultaneous peptide synthesis. *Tetrahedron Lett* 31:5811–5814.

31. Virgilio AA, Ellman JA (1994): Simultaneous solid-phase synthesis of β-turn mimetics incorporating side-chain functionality. *J Am Chem Soc* 116:11580–11581.

32. Green J (1995): Solid-phase synthesis of lavendustin A and analogues. *J Org Chem* 60:4287–4290.

33. Kick EK, Ellman JA (1995): Expedient method for the solid-phase synthesis of aspartic acid protease inhibitors directed toward the generation of libraries. *J Med Chem* 38:1427–1430.

34. Thompson LA, Ellman JA (1994): Straightforward and general method for coupling alcohols to solid supports. *Tetrahedron Lett* 35:9333–9336.

35. Wang GT, Li S, Wideburg N, Krafft GA, Kempf DJ (1995): Synthetic chemical diversity: solid-phase synthesis of libraries of C_2 symmetric inhibitors of HIV protease containing diamino diol and diamino alcohol cores. *J Med Chem* 38:2995–3002.

36. Williard R, Jammalamadaka V, Zava D, Benz CC, Hunt CA, Kushner PJ, Scanlan TS (1995): Screening and characterization of estrogenic activity from a hydroxystilbene library. *Chem Biol* 2:45–51.

37. Han H, Wolfe MM, Brenner S, Janda KD (1995): Liquid-phase combinatorial synthesis. *Proc Natl Acad Sci USA* 92:6419–6423.

38. Parlow JJ (1995): Simultaneous multistep synthesis using polymeric reagents. *Tetrahedron Lett* 36:1395–1396.

39. Meyers HV, Dilley GJ, Durgin TL, Powers TS, Winssinger NA, Zhu H, Pavia MR (1995): Multiple simultaneous synthesis of phenolic libraries. *Mol Diversity* 1:13–20.

40. Farall MJ, Frechet JMJ (1976): Bromination and lithiation: two important steps in the functionalization of polystyrene resins. *J Org Chem* 41:3877–3882.

16

SYNTHETIC ORGANIC CHEMISTRY ON SOLID SUPPORT

Stephen W. Kaldor and Miles G. Siegel

Lilly Research Laboratories, A Division of Eli Lilly & Company, Indianapolis, Indiana

Organic chemists in the pharmaceutical industry have recently come under intense pressure to generate and optimize leads at a greatly accelerated pace. Increased competition and market pressures have placed a higher premium on delivering innovative pharmaceuticals: Derivative "me-too" products are less likely to capture market share, and are destined to more rapidly suffer the brunt of generic competition. In addition, innovations in high-throughput screening over the last decade have dramatically enhanced assay capacities. Painstakingly acquired sample collections from a variety of sources (natural products, external acquisitions, archived samples from internal medicinal chemistry programs) are now assayed in their entirety in a matter of months, all too often without generating bona fide leads for further follow-up. In such circumstances, scientific teams often question the molecular diversity of existing libraries of screening samples, and challenge chemists to provide more optimal screening sets for lead generation. In the happy instance where multiple families of leads are generated for a given screen, there is increased pressure for rapid optimization work in parallel on all lead families to enhance the probability of success. Thus, both scientific and business factors have recently conspired to focus efforts on the high-throughput synthesis of pharmaceutically interesting molecules.

The basis for the vast majority of published work in the field of high-

Combinatorial Chemistry and Molecular Diversity in Drug Discovery, Edited by Eric M. Gordon and James F. Kerwin, Jr.
ISBN 0-471-15518-7   Copyright © 1998 by Wiley-Liss, Inc.

throughput organic chemistry has been solid-phase synthesis (SPS).[1,2] Solid-phase organic synthesis has been known for many years, but the majority of research in this area has been devoted to peptide and oligonucleotide synthesis. Despite the pioneering efforts of a number of academic chemists,[3-6] it is only in the last three to five years, as the need for large numbers of nonoligomeric compounds with more drug-like characteristics has grown, that there has been an explosion of interest in expanding the scope of SPS beyond amide and phosphate ester bond-forming reactions.

It is our intent in this chapter to give the reader a flavor for the strategies and tactics available to the organic chemist conducting chemistry on a solid support, with critical commentary where appropriate. We will also provide a sampling of the wide variety of reactions that have been successfully conducted on solid phase, with particular emphasis on reactions that may be of relevance to the pharmaceutical industry. A focal point of this chapter is a survey of diversifying transformations, that is, bond-forming reactions on key intermediates which serve to multiply the number of products for eventual testing. To this end, we have organized the chapter by broad reaction classes and picked representative examples in each class. We feel this format will allow practitioners in the field to quickly find relevant references. Several very good recent reviews on solid-phase synthesis have taken a complementary approach of focusing more directly on library construction.[7-19] Although the synthesis of oligomeric compounds remains a mainstay of the pharmaceutical industry, these compound classes are covered in-depth in other sections of this book. We restrict our coverage to the SPS of nonoligomeric molecules.

SOLID-PHASE SYNTHESIS

Basic Concepts

The majority of current research in the area of solid-phase synthesis stems from the early work of Merrifield on polymer-supported peptide synthesis.[20] As shown in Scheme 16.1, a solid support typically consisting of a cross-linked, insoluble polymer[2] such as polystyrene is modified to enable covalent attachment of a starting material for further chemical elaboration. This starting material, often termed a scaffold,[21] is frequently viewed as a fixed entity to which variable groups are appended during the course of a given synthesis, although this is not a requirement (path A vs. path B). It has been found particularly advantageous to employ a linker to facilitate both attachment of one's starting material to the solid support and to allow for selective detachment of intermediates and products. In many cases, the solid support–linker combination can be viewed as a large protecting group, and it is no coincidence that the majority of scaffold attachment/detachment strategies draw heavily from traditional solution phase protecting group chemistry (Table 16.1).[22]

An exception to this rule is the concept of cyclization/detachment (Scheme

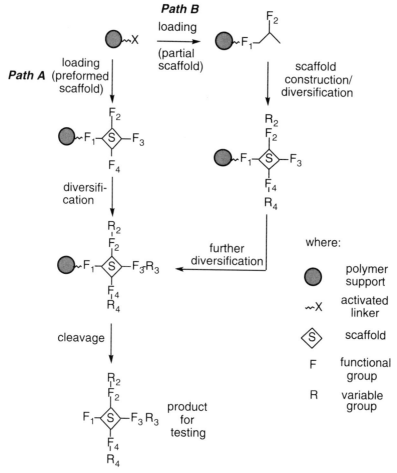

Scheme 16.1. General protocols for solid-phase synthesis.

16.2).[32-36] In this procedure, a penultimate intermediate in a synthetic sequence on solid support is positioned to cyclize and concommitantly release the solid support. This approach is elegant in that one can orchestrate one's chemistry such that only the desired cyclization product is released into solution, even if all prior reactions do not proceed to completion. In practice, this method often delivers products of very high purity, and alleviates the need for "perfect" reaction sequences consisting only of high-yielding transformations. Scheme 16.2 provides a representative example of this protocol from our laboratories in which amide **2** is induced to selectively cyclize off a solid support in the presence of a volatile base.[37] Notably, intermediates from incomplete reactions such as resin-bound acid **1** are not amenable to cleavage under the

TABLE 16.1. Representative Linkers for SPS

Scaffold Functional Group	Linker (with appended scaffold)	Cleavage Conditions	Reference
RCOOH		HF	20
RCOOH		TFA	23
RCONH$_2$		TFA	24
ROH		HCOOH	25
ROH		PPTS	26
ROH		LiOH	27
RCHO		HCl	28
HNR$_1$R$_2$		TFA	29
ArH		HF	30, 31

reaction conditions, and the desired hydantoin products **3** are obtained in high purities (average purity > 80%) after filtration and solvent evaporation.

The potential also exists for the cleavage step to serve as a diversification step as well. In a modification of the Kenner safety-catch linker, Ellman and coworkers activated an acylated sulfonamide resin with bromoacetonitrile.[38]

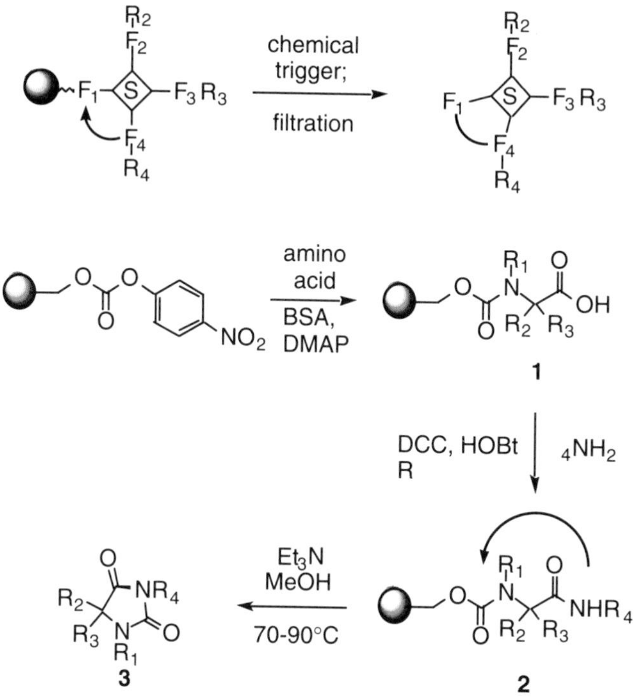

Scheme 16.2. The cyclization/detachment concept.

The cyanomethylated product is highly susceptible to nucleophilic attack by a wide range of amines, including *tert*-butyl amine and aniline. The amine can be used as the limiting reagent, since the reaction is rapid and nearly quantitative, to produce amides directly in solution free from contaminating starting material. This type of diversifying cleavage step will most likely receive more attention in the future.

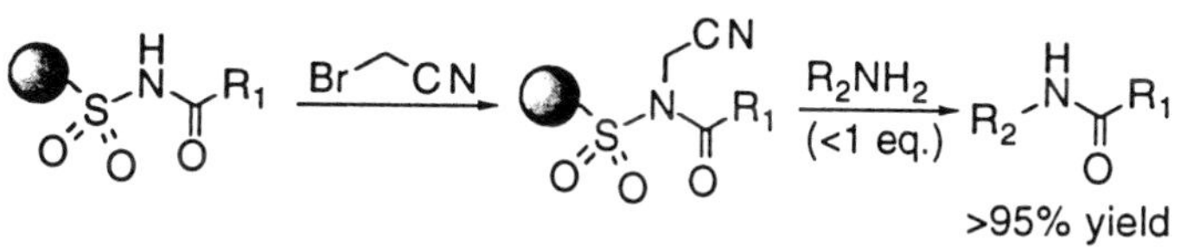

Reference 38

Functional Group Interconversions

Traditional organic chemistry is still an imperfect science. Protecting group chemistry, oxidation state manipulation, and other functional group interconversions remain important components of organic synthesis, although these operations can be viewed as cumbersome preludes to the key coupling reactions that drive the synthesis of a given target molecule. In a similar vein, such manipulations are ingrained in the art of solid-phase organic synthesis.[1] In general, SPS practitioners have drawn successfully from existing solution-phase techniques in this arena. More than 20 years ago, Leznoff performed a number of functional group interconversions on solid support (Scheme 16.3).[39] More recently, Kurth has explored chemistry for the oxidation and reduction of alcohols on solid support using standard solution-phase reagents in the context of screening library synthesis (Scheme 16.3).[25,40] Countless examples now exist of such transformations. Failures in extrapolating from solution to solid phase are typically readily explained. A recurring issue involves the use of heterogeneous catalysis: Such reactions often fail on solid supported substrates, presumably because the catalyst and substrate cannot intermingle sufficiently to react.[41] In such cases, it is frequently possible to switch to an equivalent soluble catalyst to perform the desired transformation. For example, Meyers has substituted stannous chloride for palladium on carbon in effecting nitro group reductions on solid support.[27]

Representative Diversifying Reactions

Diversifying reactions are the key bond constructions that add meaning to the phrase combinatorial chemistry (vide supra). In this section, we cover carbon–heteroatom and carbon–carbon bond-forming reactions, which have proven useful within the context of solid-phase organic synthesis. In many cases, these reactions have already been used as diversifying steps in the construction of combinatorial libraries.

Carbon–Nitrogen Bond Formation. *Nitrogen Acylation.* Due to the enormous importance of solid-phase peptide synthesis,[20] nitrogen acylation on resin is certainly the most widely performed and most highly developed solid-phase reaction. Several books are currently available dealing solely with solid-phase peptide synthesis,[42] and the solid-phase synthesis of both peptide and nonpeptide amides in the context of library construction has been reviewed recently.[11–18] The reader is directed to these reviews, as well as relevant chapters in this work, for a more detailed account of nitrogen acylation chemistry.

A recent example of a broad range of amine acylation chemistry is the synthesis of diacylated phenolic libraries by workers at Sphinx Pharmaceuticals.[27] Acylations are performed with acid chlorides to form amides, as well as isocyanates and sulfonyl chlorides to generate ureas and sulfonamides, respectively. A second acylation is performed after reduction of the aromatic nitro group.

Other examples of the application of solid-phase amine acylation to form nonpeptides and modified peptides have appeared. Workers at Merck have

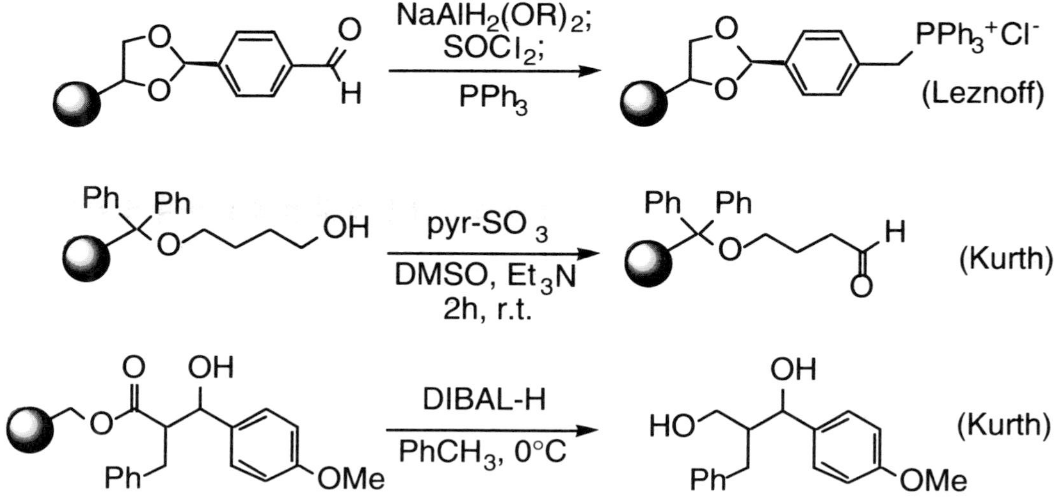

Scheme 16.3. Illustrative functional group interconversions on solid support.

utilized amine acylations to generate libraries of ureas and bisureas,[43,44] and amine acylation has been applied to the solid-phase synthesis of N-linked glycopeptides.[45]

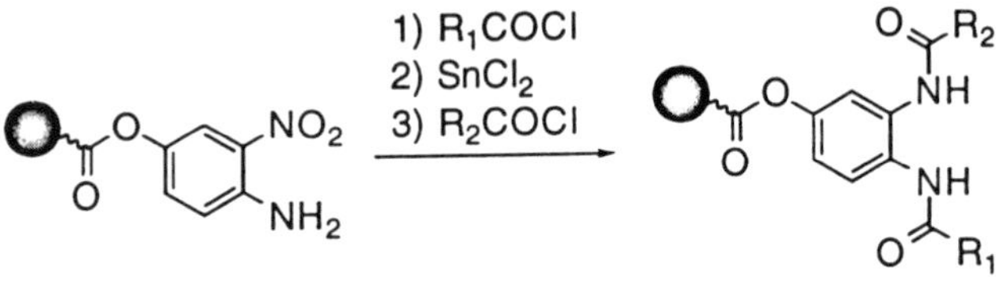

Reference 27

Nitrogen Alkylation. Both reductive amination of aldehydes and simple alkylation of nitrogen have been extensively employed on solid phase. In both types of alkylation the amine has appeared as either the solid-phase or solution-phase reactant.

Some early examples of amine alkylation are an outgrowth of solid-phase peptide synthesis and result in the formation of modified peptide structures. Coy used reductive amination of a resin-bound amino acid with a protected amino aldehyde to give pseudopeptides where one amide bond has been replaced with an amide isostere.[46]

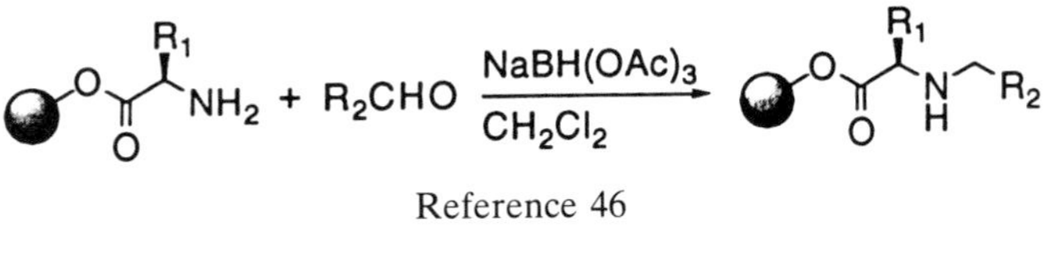

Reference 46

Similarly, workers at Pfizer have performed reductive aminations on resin-bound amino acids.[34] The N-alkylated amino acids were subsequently acylated with a second amino acid and cyclized to the corresponding piperazinediones. Ley and coworkers have extended reductive amination on solid support to ketones in good yield.[47]

Alkyl halide alkylation has also found utility in the construction of peptide-like molecules. Workers at Chiron have synthesized "peptoids" and peptide nucleic acids by alkylation of solid-phase bromoacetate with a variety of primary amines to generate *N*-alkyl glycine derivatives.[48–50] This procedure can be performed iteratively to produce oligomeric or cyclic[32] structures.

More recently, amine alkylation has been utilized to generate nonpeptide-like structures on solid phase. Ellman has synthesized ethanolamines on solid phase by displacement of a resin-bound tosylate with a primary amine.[51] These were further functionalized into a library of potential aspartyl protease inhibitors.

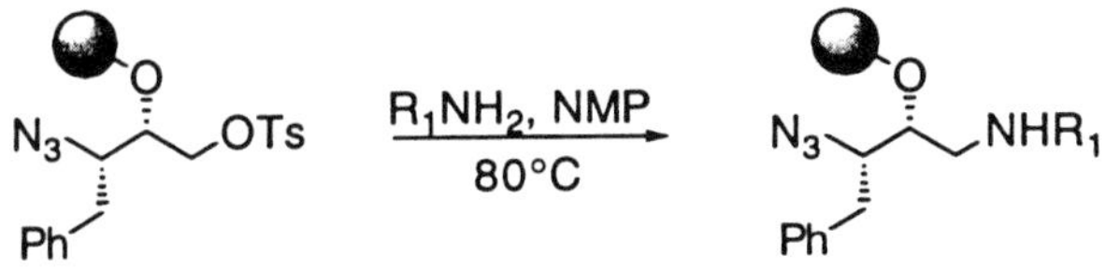

Reference 51

Green has performed a sequential reductive amination/alkylation sequence on a resin-bound aniline to generate the potent tyrosine kinase inhibitor lavendustin A and analogs, where two of the three alkyl groups on nitrogen have been introduced via a solid-phase amine alkylation.[52]

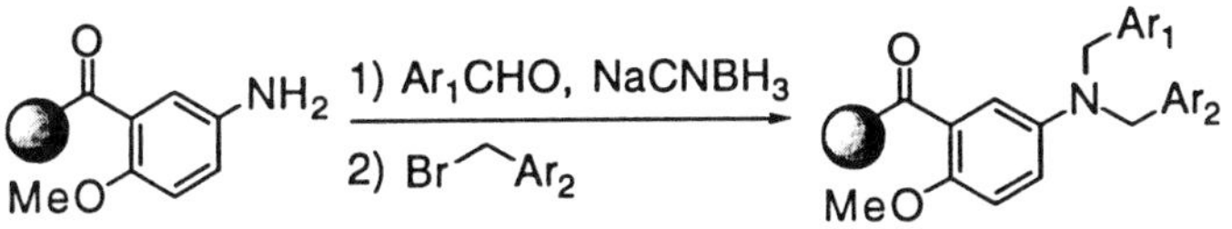

Reference 52

Aniline derivatives have been synthesized by nucleophilic aromatic substitution of aryl fluorides.[53] The reaction proceeds in good yield at room temperature with electron deficient aromatic rings. This procedure was used to make a library of N-aryl piperazines.

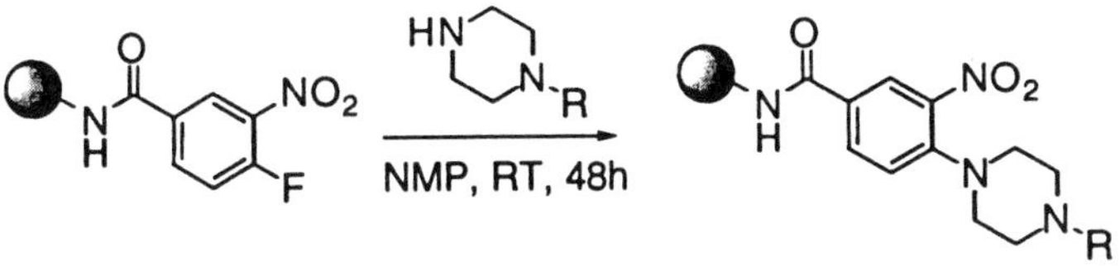

Reference 53

Several other examples of amine alkylations to generate both peptide[54,55] and nonpeptide[56,57] structures have been reported recently. In addition, amine-containing heterocycles have been synthesized by workers at Ontogen, who reacted amines with aldehydes and 1,2-diones to produce a library of highly functionalized imidazoles, where either the amine or aldehyde component is polymer-supported.[58]

Carbon–Oxygen Bond Formation. *Oxygen Acylation.* Esterifications are widely conducted as a means of attachment of carboxylic acid starting materials to a solid support. However, esterification has not been explored widely as a diversifying reaction in the construction of screening libraries. One example from the Parke–Davis group is illustrated below.[59]

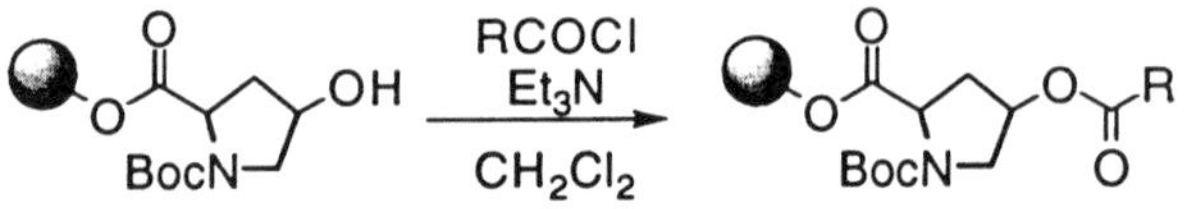

Reference 59

Oxygen Alkylation. The use of the Mitsunobu reaction for the construction of aryl alkyl ethers has been studied in-depth recently by several research groups.[60–63] For example, Rano and Chapman have determined that both polymer-bound phenols and benzyl alcohols react cleanly with TMAD/n-Bu$_3$P and a variety of electrophiles/nucleophiles to provide aryl alkyl ethers in excellent yields and purities.[61]

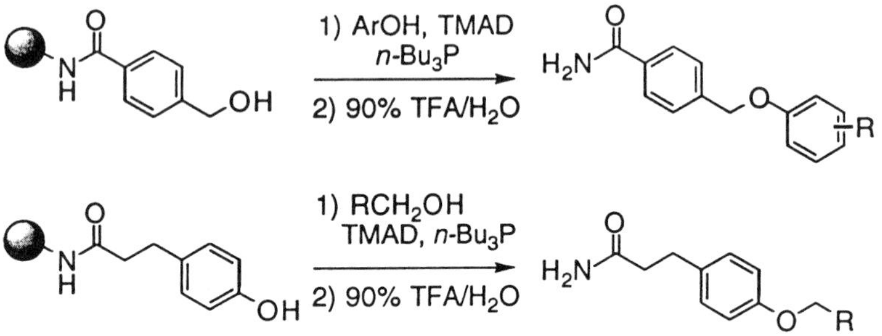

Reference 61

Workers at Pharmacopeia have performed an aldol condensation/cyclization sequence to synthesize a tagged library of dihydrobenzopyrans and dihydrobenzopyranones.[64,65]

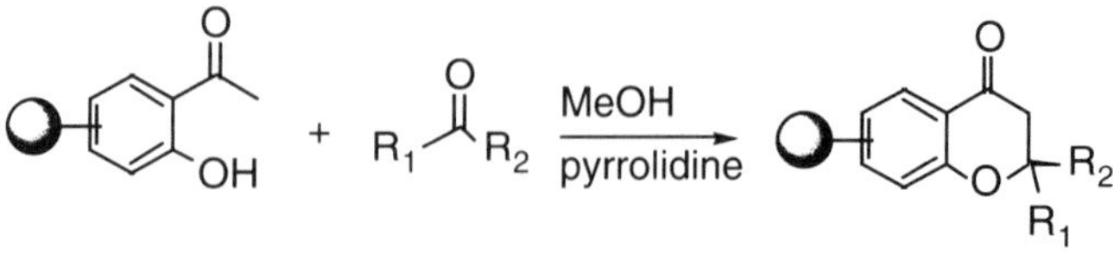

Reference 64

Although more limited in scope, more traditional oxygen alkylation reactions have been used with success on solid support.[23,66–68] Kurth has also constructed cyclic ethers in the context of a cyclization/detachment protocol.[33,69,70]

Acetal Formation. The construction of acetals has been conducted in the context of SPS primarily for the attachment of alcohol, diol, aldehyde, and ketone scaffolds to the polymer support (see Table 16.1).[26,28,59,71] However, acetal formation has been used as a diversifying step in the synthesis of oligosaccharide libraries on solid support.[72,73]

Carbon–Sulfur Bond Formation. To date, the formation of carbon–sulfur bonds on polymer support has received scant attention. The following examples serve to illustrate the potential of this area.

Olefin Additions. Kurth has demonstrated the addition of aromatic thiols to a variety of polymer-supported α,β-unsaturated ketones in the context of solid-phase split-and-mix library construction.[25] Products from a multistep sequence incorporating this chemistry were obtained in high purities as demonstrated by a variety of analytical methods. Still and coworkers have added solid-phase thiol radicals to olefins in the synthesis of a chiral stationary phase.[74]

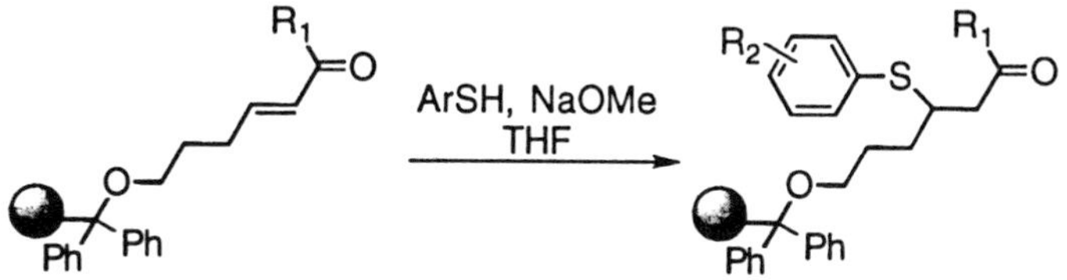

Reference 25

Heterocycle Synthesis. The reaction of a polymer-supported amino thiol with a series of aldehydes to form a diverse library of thiazolidines has recently been reported by Pátek et al.[75] Other workers at Affymax have synthesized 4-thiazolidinones and 4-metathiazanones by the reaction of polymer-supported aminoacids with aldehydes and mercaptoacids.[76,77]

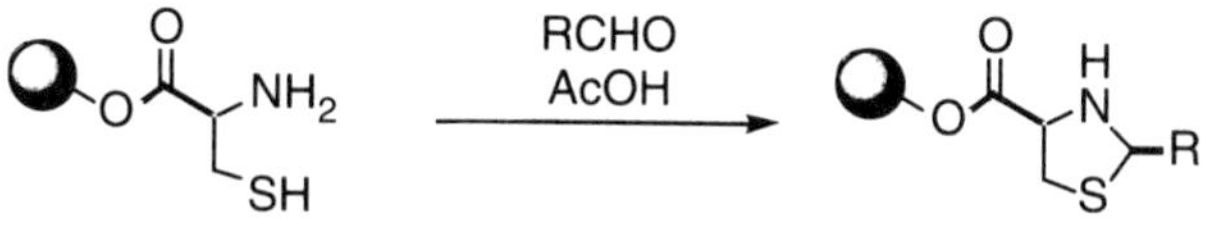

Reference 75

Macrocycle Construction. A library of cyclic β-turn mimetics incorporating a thioether backbone has recently been described by Ellman and Virgilio.[78] Ring closure was effected on solid support by intramolecular thioether formation to yield a diverse family of 9- and 10-membered cycles.

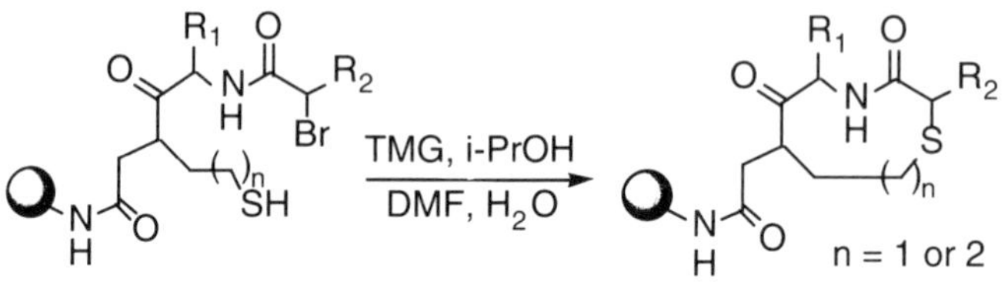

Reference 78

Carbon–Carbon Bond Formation. A variety of carbon–carbon bond-forming reactions have been performed on resin, although these are in general not as prevalent as the carbon–heteroatom bond-forming reactions described previously. Many of these reactions have been used in library construction.

Enolate Chemistry. Early work using enolates has been reviewed previously.[79] Several recent examples, however, are of interest. Workers at Pharmacopeia have condensed acetophenone derivatives with a variety of ketones in their synthesis of a benzodihydropyranone library.[65] Kurth has synthesized a small library of propanediols on resin by aldol chemistry of zinc enolates (Table 16.2).[40]

Kurth and Schore have also examined the use of resin-bound chiral enolates to achieve enantioselective enolate alkylations (Table 16.2).[80,81] In this work, propionic acid is attached to a resin-bound C_2-symmetric chiral auxiliary as the corresponding amide. Enolate alkylation of this amide with allyl iodide, followed by cleavage from the resin by iodolactonization, yields iodolactone in 87% ee, 34% yield. Neither 3,5-*cis*-iodolactone product was observed, indicating extremely high diastereoselectivities for the iodolactonization step.

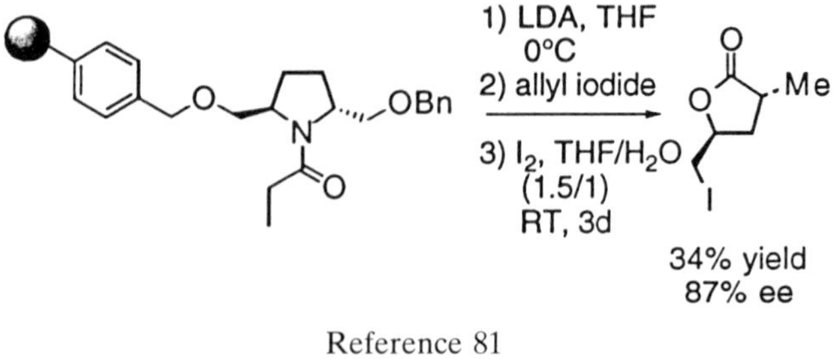

Reference 81

TABLE 16.2. Carbon–Carbon Bond Formations on Resin

Reaction	Product	Reference	Reaction	Product	Reference
Enolate: aldol		40, 59, 65	Wittig/ Horner– Emmons		3, 25, 101, 102
Enolate: alkylation (chiral)		80, 81	Pauson– Khand		89
Enolate: alkylation		82	Organo- metallic: add'n to carbonyl		3, 29, 59, 90, 103, 104
Enolate: Michael		47	1,4- addition/ trapping		92
Suzuki coupling		82, 83	Diels– Alder		93–96
Stille coupling: alkylation	R=Ar, olefin	86, 87	[3 + 2] cyclo- additions		60, 87, 90
Stille coupling: alkylation		85	Biginelli reaction		105
Heck reaction	a b	57, 84, 88	Bischler– Napieral ski/Pictet– Spengler		105, 106

Ley has reacted an enolate generated in solution with a polymer supported acrylate ester in a tandem Michael addition reaction to give bicyclo[2.2.2]octa-none systems.[47] These ketones were then further functionalized by reductive amination, then cleaved from the resin to yield amino acid, amino amide, or aminoalcohol products, depending on cleavage conditions. The products were diastereomeric mixtures.

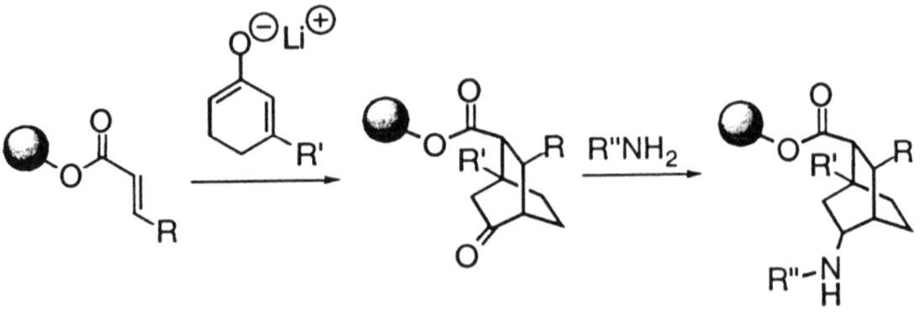

Reference 47

Ellman has alkylated a sulfonamide-derived enolate as the first step in the generation of substituted arylacetic acid derivatives (Table 16.2).[82] In this reaction, the enolate is protected from premature cleavage from the resin via the ketene by sulfonamide deprotonation. Yields for this reaction are typically very high, and dialkylation is observed only with the most reactive electrophiles (i.e., methyl iodide), and only in trace amounts (~4%).

Suzuki, Heck, and Stille Couplings. Ellman has functionalized 4-bromophenylacetic acid derivatives by performing the Suzuki coupling reaction to give biaryls. Friesen has also demonstrated the utility of the Suzuki coupling for the synthesis of a variety of biaryls (Table 16.2).[83]

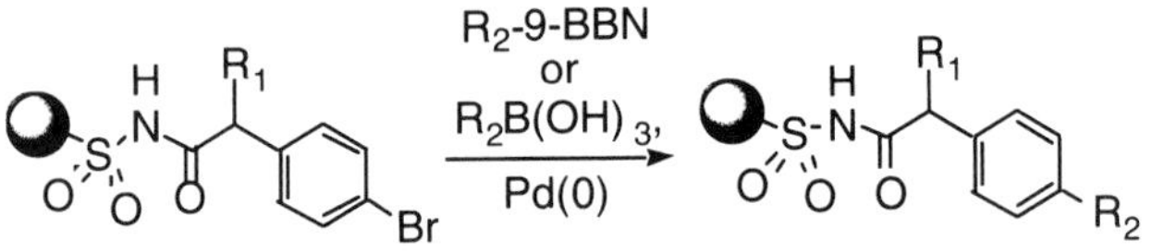

Reference 83

Other workers have explored the Heck and Stille couplings in the synthesis of biaryls and related compounds, with varying degrees of success (Table 16.2).[57,84–88] In one notable example, Hauske and coworkers utilized the Heck reaction in the solid-phase synthesis of 20- to 24-member macrocyclic peptidomimetics.[88] The macrocyclizations gave very good yields (typically 75–85%) and good purity.

Other Organometallic and Transition Metal Reactions. Schore has used the site-isolation properties inherent in many solid-phase systems to advantage by performing an otherwise low-yielding Pauson–Khand reaction with a resin-bound alkyne.[89] The reaction proceeds very cleanly and in very high yields when an excess of norbornadiene in solution is reacted with resin-bound alkyne, with a yield of 69% for the case shown in Table 16.2, compared to less than 20% in solution. It should be noted that the addition of only one equivalent of norbornadiene results in a substantial (20%) amount of product resulting from the reaction of the Pauson–Khand product with a second equivalent of resin-bound alkyne, indicating that site isolation is not complete.

Grignard reactions have been performed on resin by several workers. Ellman recently used the Grignard reaction to synthesize a series of 2-pyrrolidinemethanol derivatives in an effort to identify improved ligands for chiral catalysis.[90] Hauske describes the addition of a Grignard reagent to a resin-bound imidazolide to give the corresponding ester (Table 16.2).[29] Workers at Parke–Davis have reported a resin-bound Weinreb amide that on treatment with a Grignard reagent cleaves from the resin to give a ketone in solution.[59] Armstrong has also examined the use of polymer-supported Weinreb amides in the synthesis of ketones and aldehydes.[91]

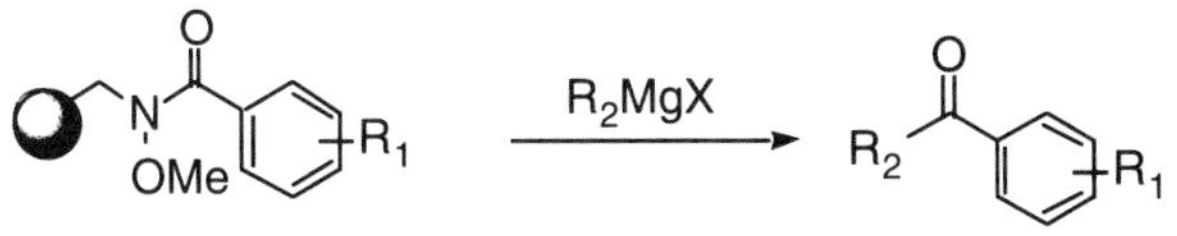

Reference 59

The Parke–Davis workers have also opened epoxides with a resin-bound phenyllithium species generated by ortholithiation of a resin-bound catechol.[59]

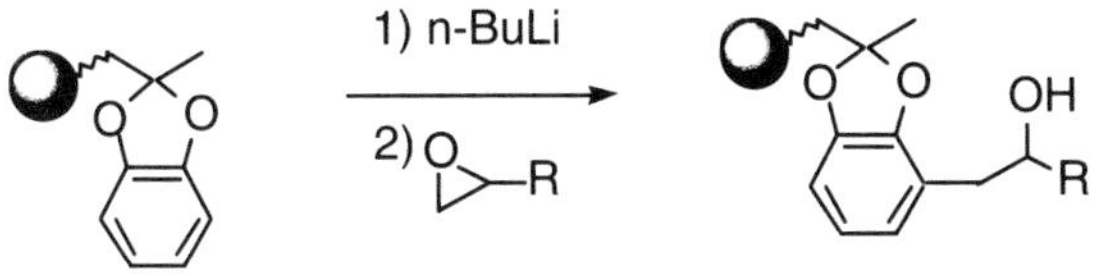

Reference 59

Ellman has used a 1,4-addition-trapping sequence to synthesize prostaglandin-like molecules on resin.[92] The reaction proceeded in approximately 50% yield (including ketone reduction and cleavage) and provided a single diastereomer.

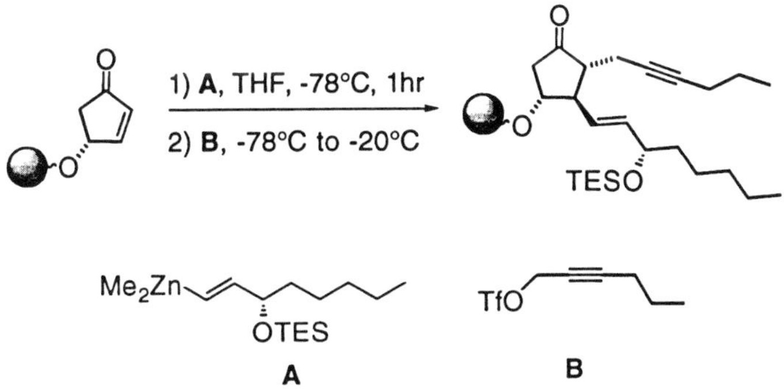

Reference 92

Pericyclic Reactions. Several pericyclic reactions have been demonstrated on solid phase, including the Diels–Alder reaction,[93-96] [3 + 2] nitrile oxide cycloaddition,[70,97] [2 + 2] cycloaddition of a resin bound imine[98] with an in situ generated ketene to yield β-lactams,[99] and [3 + 2] azomethine ylide cycloaddition.[100] In the last case, polymer-supported imines were reacted with a series of α,β-unsaturated esters, ketones, and nitriles to yield the corresponding pyrrolidines.

Reference 100

Wittig/Horner–Emmons Olefination. The solid-phase Wittig reaction has been used extensively by Leznoff in his synthesis of insect pheromones on resin.[3,79,107] In more recent work, the Horner–Emmons reaction has been used by Kurth to create a small library of β-thioketones,[25] and by Scanlan and coworkers to create a hydroxystilbene library with estrogenic activity (Table 16.2).[101] Johnson and coworkers monitored the progress of the Horner–Wadsworth–Emmons reaction by gel-phase ^{31}P NMR.[102]

Multicomponent Reactions. Multicomponent reactions (MCRs) are particularly appealing for library construction, because several points of diversity can be introduced in a single transformation. Not surprisingly then, several workers have examined MCRs on solid phase. Armstrong has performed both the Passerini reaction between an aldehyde, an isocyanate, and a polymer-supported acid,[10] and the Ugi four-component condensation between an amine, an aldehyde, a convertible isocyanide, and a polymer-supported acid.[108,109] The Ugi products are versatile intermediates that were transformed into amino acids, esters, and pyrroles before cleavage from the support. Armstrong has also exploited these intermediates in an intriguing hybrid between solution- and solid-phase synthesis termed resin capture.[110] Here the Ugi product is generated in solution using a convertible isocyanide, then reacted with a polymer-supported alcohol, which captures the Ugi product on resin as the *N*-acylaminoester. Starting impurities were removed by washing the resin, and the modified Ugi product was cleaved to give *N*-acyl aminoacids in very high purity.

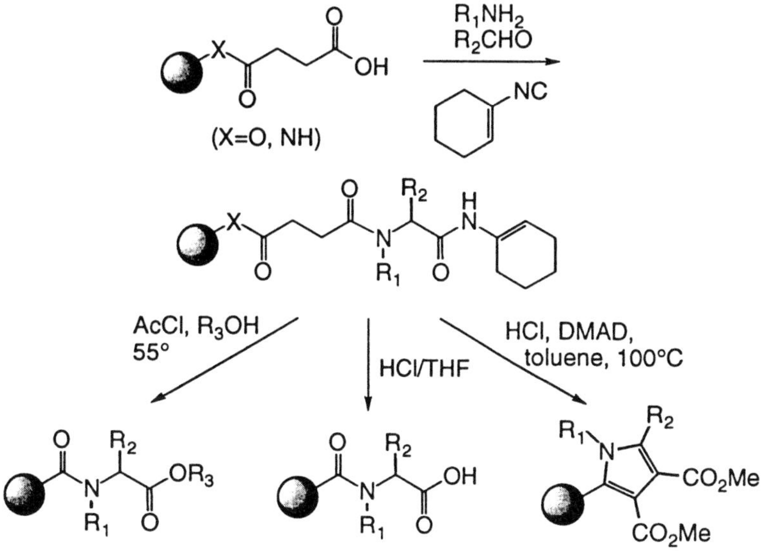

Reference 109

Several other workers have examined the Ugi reaction using a polymer-supported amine[111] and a polymer-supported isocyanide.[112] In the latter case, the condensation products were further elaborated to imidazoles.

Wipf and coworkers have carried out a resin-bound Biginelli cyclization to synthesize a series of dihydropyrimidines (Table 16.2).[113] The reaction between a β-ketoester, aromatic aldehyde, and resin-bound urea proceeds in high yield to give a product containing three points of diversity in a single step.

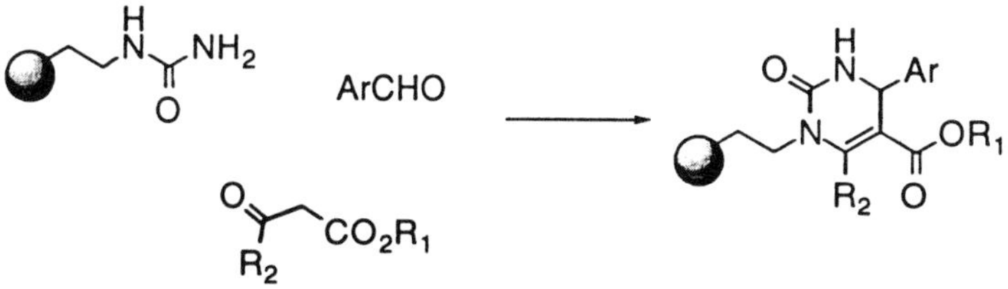

Reference 105

Gordeev and coworkers at Affymax have generated 1,4-dihydropyridines by a multicomponent condensation between an aldehyde, a β-dicarbonyl, and a polymer-supported imine (see Chapter 10).[114]

Miscellanous Reactions. Meutermans and coworkers have reported on the solid-phase Bischler–Napieralski synthesis (Table 16.2).[105] The dihydroisoquinolines synthesized in this reaction can be cleaved from the resin, or alternatively reduced to the corresponding tetrahydroisoquinoline with sodium cyanoborohydride prior to cleavage. The polymer-supported Pictet–Spengler reaction to give tetrahydroisoquinolines directly has also been reported.[106]

Advantages and Disadvantages of SPS

As the range of organic reactions that has been successfully employed on solid-phase continues to expand, the potential advantages over more traditional synthetic methodology become increasingly evident. In solid-phase synthesis, large excesses of starting materials can be employed to drive reactions to completion without fear of complicating the workup procedure; simple filtration and resin washing are typically sufficient for removal of unreacted starting materials and reaction by-products in solution. The relative site isolation of the resin-bound species can inhibit many types of intermolecular side reactions and help to stabilize reactive intermediates.[5,89] Resin-bound materials tend to be air and moisture stable and are easily manipulated. Solid-phase synthesis can frequently be utilized for multistep sequences, as has been powerfully demonstrated in the solid-phase synthesis of polypeptides and polynucleotides. Libraries of many hundreds of thousands of compounds have been generated using split-and-mix synthesis, a technique not readily translated to solution-phase chemistry; in many of these cases the solid support serves effectively as a molecular "tag" enabling the identification of individual library members.[18,115] And one of the most appealing features of solid-phase synthesis for library construction is that it is amenable to automation. Indeed, many workers in the field have devised effective automated processes for small molecule synthesis on solid phase.[19]

Nevertheless, many disadvantages associated with solid-phase chemistry have not been effectively addressed. Solid-phase reactions are more difficult to monitor by conventional techniques. Typically a large quantity of resin produces only a small amount of final product, serving to complicate analysis in many cases; increasing the loading capacity of a resin to circumvent this problem frequently leads to a concomitant decrease in the effective site isolation. Reactions are frequently significantly retarded on solid phase, and heterogeneous reagents are typically less effective. Many resins, particularly cross-linked polystyrene, display significant swelling and shrinking properties which can severely affect reaction rates and site accessibility; the optimal solvent for resin swelling may not be the optimal solvent for the desired reaction. Finally, most solid-phase synthesis necessitates two nondiversifying synthetic steps in every reaction sequence: an attachment step and a cleavage step. Typically, the cleavage step will leave in the products some common functional group or "stub," such as a carboxylic acid or alcohol, which may be an undesirable functionality.

Many workers have sought to mitigate these disadvantages, with some promising results. It is becoming increasingly straightforward, for example, to obtain both ^{1}H and ^{13}C NMR spectra of resin-bound species.[54,116,117] Resins such as Tentagel (a polyethyleneglycol modified polystyrene) are less susceptible to shrinking and swelling, and frequently decrease reaction times by increasing site accessibility,[118] although the mechanical stability and loading capacity of Tentagel is significantly decreased relative to simple polystyrene. Recent efforts to address the solid-phase stub problem have resulted in the creation of linkers that produce only a hydrogen (see Table 16.1, final example)[30,31] or methyl[119] as the common structural feature, although the generality of these approaches has not been demonstrated. Other workers have taken advantage of the stub as a diversification point and developed cleavage/diversification protocols, yielding products with diverse stubs.[38,47,59,82]

An intriguing complimentary approach to solid-phase chemistry which largely circumvents the disadvantages while maintaining many of the advantages of resin synthesis is the use of polymer-supported reagents to effect reactions between components in solution.

"INVERSE" SOLID-PHASE SYNTHESIS

The use of solid-supported reagents and catalysts in organic synthesis is an area of long-standing interest. With the possible exception of ion-exchange resins employed for acid and base catalysis, however, these intriguing materials remain largely neglected as tools for expediting the discovery of novel therapeutics. In a review covering this field, Akelah and Sherrington define a solid-supported (polymeric) reagent, as, "a reactive organic group bound to a macromolecular support and used in stoichiometric quantities to achieve the chemical modification of an added substrate" (Scheme 16.4).[120] In this section, we provide several representative examples of solid-supported reagents that point to the enormous potential of this area.[124]

By way of illustration, we draw from the field of amide bond-forming reactions. Polymer-bound active esters have been described by various research groups as convenient reagents for amide synthesis. For example, Patchornik has described the preparation of polymer-supported *o*-nitrophenyl

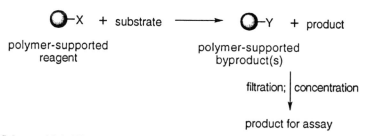

Scheme 16.4. Use of a polymer-supported reagent in organic synthesis.

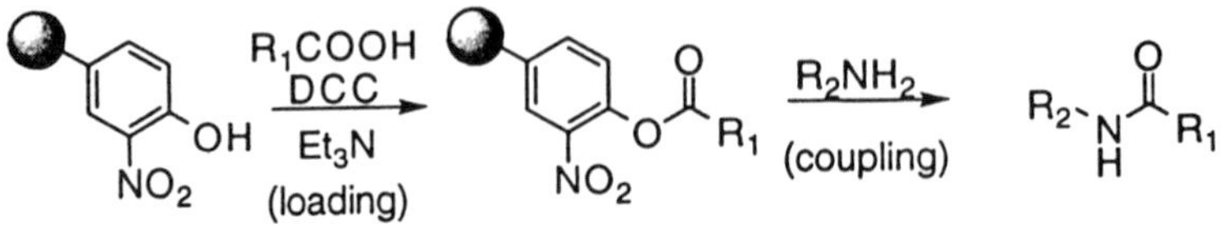

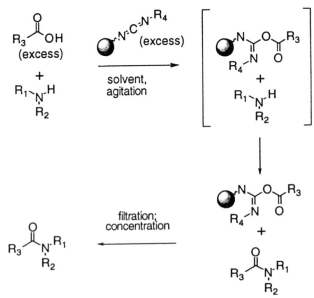

Scheme 16.5. Amide formation using a polymer-supported nitrophenyl ester.

esters and their use in automated peptide synthesis (Scheme 16.5).[41] As in Merrifield SPS, excesses of the acylating agent can be employed to drive reactions to completion, and reaction workups are greatly facilitated because the leaving group and excess active ester remain covalently appended to the polymer support. A similar reagent, polymer-supported 1-hydroxybenzotria-zole, was recently reported for the cyclization of α,ω-aminoacids to medium-ring lactams in low yield.[106]

A clever extension of this concept involves the use of polymer-supported reagents for the in situ activation of a coupling partner. This technique is particularly appealing because a separate step is not required for substrate activation, and relatively stable starting materials can be utilized (alleviating the need for frequent monitoring of their purities over prolonged periods). For example, polymer-supported carbodiimides have been successfully utilized for amide synthesis.[121] In a recent refinement of this technology, Desai and Stramiello have reported the synthesis of polymer-supported 1-ethyl-3-(3-

Scheme 16.6. Use of a polymer-supported carbodiimide for in situ acid activation and subsequent amide formation.

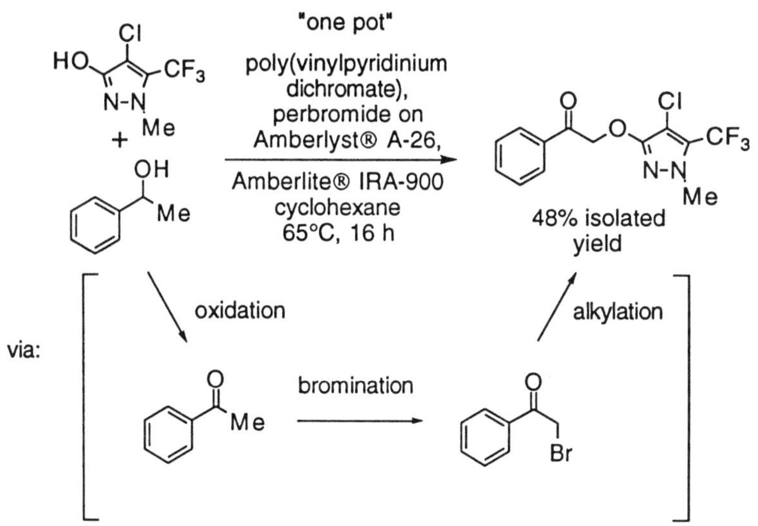

Scheme 16.7. A simultaneous multistep synthesis using polymer-supported reagents.

dimethylaminopropyl) carbodiimide (EDC) and its use in the synthesis of nonpeptide amides (Scheme 16.6).[122] An excess of acid and polymer-supported EDC are used, and simple filtration serves to remove the excess of unreacted active ester that is generated in situ.

A large family of polymer-supported reagents now exists that has been used for a variety of functional group interconversions and coupling reactions. It is important to stress that multistep sequences have been performed in high yields and purities using these reagents. In a particularly elegant application, Parlow has recently disclosed a "one-pot," three-reaction protocol in which three polymeric reagents are employed simultaneously in a single reaction vessel (Scheme 16.7).[123] The concept of site isolation (vide supra) was exploited to allow for the use of three reagents that are mutually incompatible in solution.

CONCLUSIONS

The field of solid-supported organic chemistry has seen explosive growth in the last four years, and the repertoire of reactions available to the practitioner in this area has expanded to the point where it is now a legitimate option to conduct at least portions of most of drug-discovery projects using these methods. We view standard solid-phase synthesis and inverse Merrifield synthesis as highly complementary techniques that can be applied in both lead generation and lead optimization. Although overgeneralizations are dangerous, we suggest that standard solid-phase synthesis will be particularly useful for multistep sequences, whereas inverse Merrifield synthesis using polymer-supported reagents will often prove to be the method of choice for single-step transformations, especially if preloading of one's substrate onto the polymer support is not required. We further conclude that polymer-supported reagents will see increased use in lead optimization projects involving one- to three-step reaction sequences. A likely development in the near future will be the implementation of hybrid syntheses that incorporate the best features of both methods. For example, it appears logical to conduct multistep solid-phase synthesis on solid support, and to then reveal a functional group via cleavage for final derivatization using a polymer-supported reagent. This hybrid approach would provide one potential solution to the stub problem outlined earlier.[124]

It is increasingly evident that organic chemists are building a robust and easily traversible bridge between solution- and solid-phase techniques; it will soon become impossible to ascertain by simple visual inspection of a product structure which method has been used in its synthesis.

ACKNOWLEDGMENTS

The authors thank Jon Ellman and Lorin Thompson for providing a copy of their recently completed review on the construction of small molecule libraries in advance of publication. We also thank Peter Wipf and Michael Pavia for providing us with copies of unpublished material.

REFERENCES

1. Hernkens PHH, Ottenheijm HCJ, Rees D (1996): Solid-phase organic reactions: a review of the recent literature. *Tetrahedron* 52:4527–4554.
2. Früchtel JS, Jung G (1996): Organic chemistry on solid supports. *Angew Chem Int Ed Eng* 35:17–42.
3. Leznoff CC (1978): The use of insoluble polymer supports in general organic synthesis. *Acc Chem Res* 11:327–333.
4. Letsinger RL, Kornet MJ, Mahadevan V, Jerina DM (1964): Reactions on polymer supports. *J Am Chem Soc* 86:5163–5165.
5. Jayalekshmy P, Mazur S (1976): Pseudodilution, the solid-phase immobilization of benzyne. *J Am Chem Soc* 98:6710–6711.
6. Frechet JM, Schuerch C (1971): Solid-phase synthesis of oligosaccharides, 1: preparation of the solid support: poly[p-(1-propen-3-ol-1-yl)styrene]. *J Am Chem Soc* 93:492–496.
7. Lowe G (1995): Combinatorial chemistry. *Chem Soc Rev* 309–317.
8. Gordon EM, Gallop MA, Patel DV (1996): Strategy and tactics in combinatorial organic synthesis: applications to drug discovery. *Acc Chem Res* 29:144–154.
9. Ellman JA (1996): Design, synthesis, and evaluation of small-molecule libraries. *Acc Chem Res* 29:132–143.

10. Armstrong RW, Combs AP, Tempest PA, Brown SD, Keating TA (1996): Multiple-component condensation strategies for combinatorial library synthesis. *Acc Chem Res* 29:123–131.

11. DeWitt SH, Czarnik AW (1996): Combinatorial organic synthesis using Parke–Davis's Diversomer method. *Acc Chem Res* 29:114–122.

12. Thompson LA, Ellman JA (1996): Synthesis and applications of small molecule libraries. *Chem Rev* 96:555–600.

13. Blondelle SE, Pérez–Payá E, Dooley CT, Pinilla C, Houghten RA (1995): Soluble combinatorial libraries of organic, peptidomimetic and peptide diversities. *Trends Anal Chem* 14:83–91.

14. Gordon EM, Barrett RW, Dower WJ, Fodor SPA, Gallop MA (1994): Applications of combinatorial technologies to drug discovery, 2: combinatorial synthesis, library screening strategies, and future directions. *J Med Chem* 37:1385–1401.

15. Gallop MA, Barrett RW, Dower WJ, Fodor SPA, Gordon EM (1994): Applications of combinatorial technologies to drug discovery, 1: background and peptide combinatorial libraries. *J Med Chem* 37:1233–1251.

16. Terrett NK, Gardner M, Gordon DW, Kobylecki RJ, Steele J (1995): Combinatorial synthesis: the design of compound libraries and their application to drug discovery. *Tetrahedron* 51:8135–8173.

17. Desai MC, Zuckermann RN, Moos WH (1994): Recent advances in the generation of chemical diversity libraries. *Drug Dev Res* 33:174–188.

18. Lebl M, Krchnák V, Seligmann B, Strop P, Felder S, Lam KS (1995): One-bead–one-structure combinatorial libraries. *Biopolymers (Pept Sci)* 37:177–198.

19. DeWitt SH, Schroeder MC, Stankovic CJ, Strode JE, Czarnik AW (1994): Diversomer™ technology: solid phase synthesis, automation, and integration for the generation of chemical diversity. *Drug Dev Res* 33:116–124.

20. Merrifield RB (1964): Solid-phase peptide synthesis, III: an improved synthesis of bradykinin. *Biochemistry* 3:1385–1390.

21. Pavia MR (1998): Scaffolds for small molecule synthesis. In Gordon EM, Kerwin JF, eds. *Combinatorial Chemistry and Molecular Diversity in Drug Discovery.* New York: Wiley, p 213–225.

22. Greene TW, Wuts PGM (1991): *Protective Groups in Organic Synthesis,* 2d ed. New York: Wiley.

23. Wang SS (1973): *p*-Alkoxybenzyl alcohol resin and *p*-alkoxybenzyloxycarbonylhydrazide resin for solid-phase synthesis of protected peptide fragments. *J Am Chem Soc* 95:1328.

24. Rink H (1987): Solid-phase synthesis of protected peptide fragments using a trialkoxy-diphenyl-methylester resin. *Tetrahedron Lett* 28:3787–3790.

25. Chen C, Randall LAA, Miller RB, Jones AD, Kurth MJ (1994): "Analogous" organic synthesis of small-compound libraries: validation of combinatorial chemistry in small-molecule synthesis. *J Am Chem Soc* 116:2261–2662.

26. Thompson LA, Ellman JA (1994): Straightforward and general method for coupling alcohols to solid supports. *Tetrahedron Lett* 35:9333–9336.

27. Meyers HV, Dilley GJ, Durgin TL, Powers TS, Winssinger NA, Zhu H, Pavia MR (1995): Multiple simultaneous synthesis of phenolic libraries. *Mol Diversity* 1:13–20.

28. Leznoff CC, Wong JY (1973): The use of polymer supports in organic synthesis, III: selective chemical reactions on one aldehyde group of symmetrical dialdehydes. *Can J Chem* 51:3756–3764.

29. Hauske JR, Dorff P (1995): A solid phase CBZ chloride equivalent: a new matrix specific linker. *Tetrahedron Lett* 36:1589–1592.

30. Plunkett MJ, Ellman JA (1995): A silicon-based linker for traceless solid-phase synthesis. *J Org Chem* 60:6006–6007.

31. Chenera B, Finkelstein JA, Veber DF (1995): Protodetachable arylsilane polymer linkages for use in solid-phase organic synthesis. *J Am Chem Soc* 117:11999–12000.

32. Scott BO, Siegmund AC, Marlowe CK, Pei Y, Spear KL (1996): Solid-phase organic synthesis (SPOS): a novel route to diketopiperazines and diketomorpholines. *Mol Diversity* 1:125–134.

33. Beebe X, Chiappari CL, Olmstead MM, Kurth MJ, Schore NE (1995): Polymer-supported synthesis of cyclic ethers: electrophilic cyclization of tetrahydrofuro-isoxazolines. *J Org Chem* 60:4204–4212.

34. Gordon DW, Steele J (1995): Reductive alkylation on a solid phase: synthesis of a piperazinedione combinatorial library. *Bioorg Med Chem Lett* 5:47–50.

35. DeWitt SH, Kiely JS, Stankovic CJ, Schroeder MC, Cody DMR, Pavia MR (1993): "Diversomers": an approach to nonpeptide, nonoligomeric chemical diversity. *Proc Natl Acad Sci USA* 90:6909–6913.

36. Camps F, Cartells J, Pi J (1974): Organic synthesis with functionalized polymers, IV: synthesis of 1,3-dihydro-5-phenyl-2*H*-1,4-benzodiazepin-2-ones. *An Quim* 70:848–849.

37. Dressman BA, Spangle LA, Kaldor SW (1996): Solid-phase synthesis of hydantoins using a carbamate linker and a novel cyclization/cleavage step. *Tetrahedron Lett* 37:937–940.

38. Backes BJ, Virgilio AA, Ellman JA (1996): Activation method to prepare a highly reactive acylsulfonamide "safety-catch" linker for solid-phase synthesis. *J Am Chem Soc* 118:3055–3056.

39. Wong JY, Manning C, Leznoff CC (1974): Solid-phase synthesis and photochemistry of 4,4'-stilbenedicarbaldehyde. *Angew Chem Int Ed Eng* 13:666–667.

40. Kurth MJ, Randall LAA, Chen C, Melander C, Miller RB, McAlister K, Reitz G, Kang R, Nakatsu T, Green C (1994): Library-based lead compound discovery: antioxidants by an analogous synthesis/deconvolution assay strategy. *J Org Chem* 59:5862–5864.

41. Patchornik A (1987): Synthesis using polymeric reagents. *Chemtech* 58–63.

42. Atherton E, Sheppard RC (1989): *Solid-Phase Peptide Synthesis: A Practical Approach*, Oxford, UK: IRL.

43. Hutchins SM, Chapman KT (1994): A general method for the solid-phase synthesis of ureas. *Tetrahedron Lett* 35:4055–4058.

44. Hutchins SM, Chapman KT (1995): A strategy for urea-linked diamine libraries. *Tetrahedron Lett* 36:2583–2586.

45. Vetter D, Tumelty D, Singh SK, Gallop MA (1995): A versatile solid-phase synthesis of N-linked glycopeptides. *Angew Chem Int Ed Eng* 34:60–63.

46. Coy DH, Hocart SJ, Sasaki Y (1988): Solid-phase alkylation techniques in analogue peptide bond and side-chain modification. *Tetrahedron* 44:835–841.

47. Ley SV, Mynett DM, Koot WJ (1995): Solid-phase synthesis of bicyclo[2.2.2]oc-tane derivatives via tandem Michael addition reactions and subsequent reductive amination. *SynLett* 1017–1020.

48. Zuckermann RN, Martin EJ, Spellmeyer DC, Stauber GB, Shoemaker KR, Kerr JM, Figliozzi GM, Goff DA, Siani MA, Simon RJ, Banville SC, Brown EG, Wang L, Richter LS, Moos WH (1994): Discovery of nanomolar ligands for 7-transmembrane G-protein-coupled receptors from a diverse N-(substituted)gly-cine peptoid library. *J Med Chem* 37:2678–2685.

49. Zuckermann RN, Kerr JM, Kent SBH, Moos WH (1992): Efficient method for the preparation of peptoids [Oligo(N-substituted glycines)] by submonomer solid-phase synthesis. *J Am Chem Soc* 114:10646–10647.

50. Richter LS, Zuckermann RN (1995): Synthesis of peptide nucleic acids (PNA) by submonomer solid-phase synthesis. *Bioorg Med Chem Lett* 5:1159–1162.

51. Kick EK, Ellman JA (1995): Expedient method for the solid-phase synthesis of aspartic acid protease inhibitors directed toward the generation of libraries. *J Med Chem* 38:1427–1430.

52. Green J (1995): Solid-phase synthesis of lavendustin A and analogues. *J Org Chem* 60:4287–4290.

53. Dankwardt SM, Newman SR, Krstenansky JL (1995): Solid-phase synthesis of aryl and benzylpiperazines and their application in combinatorial chemistry. *Tetrahedron Lett* 36:4923–4926.

54. Look GC, Holmes CP, Chinn JP, Gallop MA (1994): Methods for combinatorial organic synthesis: the use of fast ^{13}C NMR analysis for gel phase reaction monitor-ing. *J Org Chem* 59:7588–7590.

55. Bray AM, Chiefari DS, Valerio RM, Maeji NJ (1995): Rapid optimization of organic reactions on solid-phase using the multipin approach: synthesis of 4-aminoproline analogues by reductive amination. *Tetrahedron Lett.* 36:5081–5084.

56. Boojamra CG, Burow KM, Ellman JA (1995): An expedient and high-yielding method for the solid-phase synthesis of diverse 1,4-benzodiazepine-2,5-diones. *J Org Chem* 60:5742–5742.

57. Goff DA, Zuckermann RN (1995): Solid-phase synthesis of highly substituted peptoid 1(2*H*)-isoquinolinones. *J Org Chem* 60:5748–5749.

58. Sarshar S, Siev D, Mjalli AMM (1996): Imidazole libraries on solid support. *Tetrahedron Lett* 37:835–838.

59. Cody DR, DeWitt SHH, Hodges JC, Kiely JS, Moos WH, Pavia MR, Roth BD, Schroeder MC, Stankovic CJ (1994): Apparatus for multiple simultaneous synthesis. *US patent 5,324,483.*

60. Krchnák V, Flegelová Z, Weichsel AS, Lebl M (1995): Polymer-supported Mitsu-nobu ether formation and its use in combinatorial chemistry. *Tetrahedron Lett* 36:6193–6196.

61. Rano TA, Chapman KT (1995): Solid-phase synthesis of aryl ethers via the Mitsunobu reaction. *Tetrahedron Lett* 36:3789–3792.

62. Richter LS, Gadek TR (1994): A surprising observation about Mitsunobu reac-tions in solid-phase synthesis. *Tetrahedron Lett* 35:4705–4706.

63. Campbell DA, Bermak JC, Burkoth TS, Patel DV (1995): A transition-state analogue inhibitor combinatorial library. *J Am Chem Soc* 117:5381–5382.

64. Baldwin JJ, Burbaum JJ, Chelsky D, Dillard LW, Henderson I, Li G, Ohlmeyer MHJ, Randle TL, Reader JC (1995): Combinatorial libraries encoded with electrophoric tags. *Eur J Med Chem* Suppl 30:349s–358s.

65. Burbaum JJ, Ohlmeyer MHJ, Reader JC, Henderson I, Dillard LW, Li G, Randle TL, Sigal NH, Chelsky D, Baldwin JJ (1995): A paradigm for drug discovery employing encoded combinatorial libraries. *Proc Natl Acad Sci USA* 92:6027–6031.

66. Dankwardt SM, Phan TM, Krstenansky JL (1996): Combinatorial synthesis of small-molecule libraries using 3-amino-5-hydroxybenzoic acid. *Mol Diversity* 1:113–120.

67. Wong JY, Leznoff CC (1973): The use of polymer supports in organic synthesis, II: the syntheses of monoethers of symmetrical diols. *Can J Chem* 51:2452–2455.

68. Leznoff CC, Dixit DM (1977): The use of polymer supports in organic synthesis, XI: the preparation of monoethers of symmetrical dihydroxy aromatic compounds. *Can J Chem* 55:3351–3555.

69. Beebe X, Schore NE, Kurth MJ (1995): Polymer-supported synthesis of cyclic ethers: electrophilic cyclization of isoxazolines. *J Org Chem* 60:4196–4203.

70. Beebe X, Schore NE, Kurth MJ (1992): Polymer-supported synthesis of 2,5-disubstituted tetrahydrofurans. *J Am Chem Soc* 114:10061–10062.

71. Wang GT, Li S, Wideburg N, Krafft GA, Kempf DJ (1995): Synthetic chemical diversity: solid-phase synthesis of libraries of C_2 symmetric inhibitors of HIV protease containing diamino diol and diamino alcohol cores. *J Med Chem* 38:2995–3002.

72. Sophia MJ (1998): Oligosaccharide and glycoconjugate solid-phase synthesis technologies for drug discovery. In Gordon EM, Kerwin JF, eds. *Combinatorial Chemistry and Molecular Diversity In Drug Discovery.* New York: Wiley, Chapter 13.

73. Danishefsky SJ, McClure KF, Randolph JT, Ruggeri BR (1993): A strategy for the solid-phase synthesis of oligosaccharides. *Science* 260:1307–1309.

74. Gasparrini F, Misiti D, Villani C, Borcharct A, Burger MT, Still WC (1995): Enantioselective recognition by a new chiral stationary phase at receptorial level. *J Org Chem* 60:4314–4315.

75. Pátek M, Drake B, Lebl M (1995): Solid-phase synthesis of "small" organic molecules based on thiazolidine scaffold. *Tetrahedron Lett* 36:2227–2230.

76. Look GC, Schullek JR, Holmes CP, Chinn JP, Gordon EM, Gallop MA (1996): The identification of cyclooxygenase-1 inhibitors from 4-thiazolidinone combinatorial libraries. *Bioorg Med Chem Lett* 6:707–712.

77. Holmes CP, Chinn JP, Look GC, Gordon EM, Gallop MA (1995): Strategies for combinatorial organic synthesis: solution and polymer-supported synthesis of 4-thiazolidinones and 4-metathiazanones derived from amino acids. *J Org Chem* 60:7328–7333.

78. Virgilio AA, Ellman JA (1994): Simultaneous solid-phase synthesis of β-turn mimetics incorporating side-chain functionality. *J Am Chem Soc* 116:11580–11581.

79. Leznoff CC (1974): The use of insoluble polymer supports in organic chemical synthesis. *Chem Soc Rev* 3:65–85.

80. Moon HS, Schore NE, Kurth MJ (1994): A polymer-supported C_2-symmetric chiral auxiliary: preparation of non-racemic 3,5-disubstituted-γ-butyrolactones. *Tetrahedron Lett* 35:8915–8918.

81. Moon HS, Schore NE, Kurth MJ (1992): A polymer-supported chiral auxiliary applied to the iodolactonization reaction: preparation of γ-butyrolactones. *J Org Chem* 57:6088–6089.

82. Backes BJ, Ellman JA (1994): Carbon–carbon bond-forming methods on solid support: utilization of Kenner's "safety-catch" linker. *J Am Chem Soc* 116:11171–11172.

83. Frenette R, Friesen RW (1994): Biaryl synthesis via Suzuki coupling on a solid support. *Tetrahedron Lett* 35:9177–9180.

84. Yu KL, Deshpande MS, Vyas DM (1994): Heck reactions in solid-phase synthesis. *Tetrahedron Lett* 35:8919–8922.

85. Plunkett MJ, Ellman JA (1995): Solid-phase synthesis of structurally diverse 1,4-benzodiazepine derivatives using the Stille coupling reaction. *J Am Chem Soc* 117:3306–3307.

86. Forman FW, Sucholeiki I (1995): Solid-phase synthesis of biaryls via the Stille reaction. *J Org Chem* 60:523–528.

87. Deshpande MS (1994): Formation of carbon–carbon bond on solid support: application of the Stille reaction. *Tetrahedron Lett* 35:5613–5614.

88. Hiroshige M, Hauske JR, Zhou P (1995): Palladium-mediated macrocyclization on solid support and its applications to combinatorial synthesis. *J Am Chem Soc* 117:11590–11591.

89. Schore NE, Najdi SD (1990): Pauson–Khand cycloadditions of polymer-linked substrates. *J Am Chem Soc* 112:441–442.

90. Liu G, Ellman JA (1995): A general solid-phase synthesis strategy for the preparation of 2-pyrrolidinemethanol ligands. *J Org Chem* 60:7712–7713.

91. Dinh TQ, Armstrong RW (1996): Synthesis of ketones and aldehydes via reactions of Weinreb-type amides on solid support. *Tetrahedron Lett* 37:1161–1164.

92. Thompson LA, Ellman JA (1995): The solid-phase synthesis of prostaglandins. *209th ACS National Meeting, Anaheim, CA, 2–6 April.* Abstr no. ORGN 0262.

93. Corbridge MD, McArthur CR, Leznoff CC (1988): Asymmetric induction in the cycloaddition of 1,3-butadiene to a polymer-bound chiral acrylate. *Reactive Polymers* 8:173—188.

94. MacBeath G, Hilvert D (1994): Monitoring catalytic activity by immunoassay: implications for screening. *J Am Chem Soc* 116:6101–6106.

95. Nie B, Hasan K, Greaves MD, Rotello VM (1995): Reversible covalent attachment of C_{60} to a furan–fuctionalized resin. *Tetrahedron lett* 36:3617–3618.

96. Guhr KI, Greaves MD, Rotello VM (1994): Reversible covalent attachment of C_{60} to a polymer support. *J Am Chem Soc* 116:5997–5998.

97. Pei Y, Moos WH (1994): Post-modification of peptoid side chains: [3 + 2] cycloaddition of nitrile oxides with alkenes and alkynes on the solid phase. *Tetrahedron Lett* 35:5825–5828.

98. Look GC, Murph MM, Campbell DA, Gallop MA (1995): Trimethylorthoformate: a mild and effective dehydrating reagent for solution and solid-phase imine formation. *Tetrahedron Lett* 36:2937–2940.

99. Ruhland B, Bhandari A, Gordon EM, Gallop MA (1996): Solid-supported combinatorial synthesis of structrually diverse β-lactams. *J Am Chem Soc* 118:253–254.

100. Murphy MM, Schullek JR, Gordon EM, Gallop MA (1995): Combinatorial organic synthesis of highly functionalized pyrrolidines: identification of a potent angiotensin converting enzyme inhibitor from a mercaptoacyl proline library. *J Am Chem Soc* 117:7029–7030.

101. Williard R, Jammalamadaka V, Zava D, Benz CC, Hunt CA, Kushner PJ, Scanlan TS (1995): Screening and characterization of estrogenic activity from a hydroxystilbene library. *Chem Biol* 2:45–51.

102. Johnson CR, Zhang B (1995): Solid-phase synthesis of alkenes using the Horner–Wadsworth–Emmons reaction and monitoring by gel phase ^{31}P NMR. *Tetrahedron Lett* 36:9253–9256.

103. Fyles TM, Leznoff CC, Weatherston J (1978): Some solid-phase syntheses of the sex attractant of the spruce budworm-*trans*-11-tetradecenal. *J Chem Ecol* 4:109–116.

104. Leznoff CC and Yedidia V (1980): The solid-phase synthesis of tertiary hydroxyesters from symmetrical diacid chlorides using organomanganese reagents. *Can J Chem* 58:287–290.

105. Muetermans WDF, Alewood PF (1995): The solid-phase synthesis of dihydro- and tetrahydroisoquinolines. *Tetrahedron Lett* 36:7709–7712.

106. Kaljuste K, Undén A (1995): Solid-phase synthesis of 1,2,3,4-tetrahydro-β-carbolines: implications of combinatorial chemistry. *Tetrahedron Lett* 36:9211–9214.

107. Leznoff CC, Fyles TM, Weatherston J (1977): The use of polymer supports in organic synthesis, VIII: solid-phase syntheses of insect sex attractants. *Can J Chem* 55:1143–1153.

108. Strocker AM, Keating TA, Tempest PA, Armstrong RW (1996): Use of a convertible isocyanide for generation of Ugi reaction derivatives on solid support: synthesis of α-acylaminoesters and pyrroles. *Tetrahedron Lett* 37:1149–1152.

109. Keating TA, Armstrong RW (1995): Molecular diversity via a convertible isocyanide in the Ugi four-component condensation. *J Am Chem Soc* 117:7842–7843.

110. Keating TA, Armstrong RW (1996): Postcondensation modifications of Ugi four-component condensation products: 1-isocyanocyclohexene as a convertible isocyanide: mechanism of conversion, synthesis of diverse structures, and demonstration of resin capture. *J Am Chem Soc* 118:2574–2583.

111. Cao X, Moran EH, Siev D, Lio A, Ohashi C, Mjalli AMM (1995): Synthesis of *NH*-acyl-α-aminoamides on rink resin: inhibitors of the hematopoietic protein tyrosine phosphatase (HePTP). *Bioorg Med Chem Lett* 5:2953–2958.

112. Zhang C, Moran EJ, Woiwode TF, Short KM, Mjalli AMM (1996): Synthesis of tetrasubstituted imidazoles via α-(*N*-acyl-*N*-alkylamino)-β-ketoamides on Wang resin. *Tetrahedron Lett* 37:751–754.

113. Wipf P, Cunningham A (1995): A solid-phase protocol of the Biginelli dihydropyrimidine synthesis suitable for combinatorial chemistry. *Tetrahedron Lett* 36:3819–3822.

114. Gordeev MF, Patel DV, Gordon EM (1996): Approaches to combinatorial synthesis of heterocycles: a solid-phase synthesis of 1,4-dihydropyridines. *J Org Chem* 61:924–928.

115. Baldwin JJ, Burbaum JJ, Henderson I, Ohlmeyer MHJ (1995): Synthesis of a small molecule combinatorial library encoded with molecular tags. *J Am Chem Soc* 117:5588–5589.

116. Anderson RC, Jarema MA, Shapiro MJ, Stokes JP, Ziliox M (1995): Analytical techniques in combinatorial chemistry: MAS CH correlation in solvent-swollen resin. *J Org Chem* 60:2650–2651.

117. Fitch WL, Detre G, Holmes CP, Shoolery JN, Keifer PA (1994): High-resolution ¹H NMR in solid-phase organic synthesis. *J Org Chem* 59:7955–7956.

118. Bayer E (1991): Towards the chemical synthesis of proteins. *Angew Chem Int Ed Eng* 30:113–129.

119. Sucholeiki I (1994): Solid-phase photochemical C-S bond cleavage of thioethers: a new approach to the solid-phase production of non-peptide molecules. *Tetrahedron Lett* 35:7307–7310.

120. Akelah A, Sherrington DC (1981): Application of functionalized polymers in organic synthesis. *Chem Rev* 81:557–587.

121. Weinshenker, Shen CM, Wong JY (1988): Polymeric carbodiimide: preparation. *Org Synth Coll* 6:951–954.

122. Desai MC, Stramiello LMS (1993): Polymer-bound EDC (P-EDC): a convenient reagent for formation of an amide bond. *Tetrahedron Lett* 34:7685–7688.

123. Parlow JJ (1995): Simultaneous multistep synthesis using polymeric reagents. *Tetrahedron Lett* 36:1395–1396.

124. For a recent review covering inverse Merrifield synthesis, see: Kaldor SW, Siegel MG (1997): Combinatorial chemistry using polymer-supported reagents. *Curr Op Chem Biol* 1:101–106.

PART III

AUTOMATION, ANALYTICAL, AND COMPUTATIONAL METHODS

17

AUTOMATION OF COMBINATORIAL CHEMISTRY FOR LARGE LIBRARIES

MICHAEL NEEDELS AND JEFFREY SUGARMAN

Affymax Research Institute, Palo Alto, California

Commercial and custom research instruments abound for automated synthesis of one or several chemical compounds at a time. This chapter focuses on automated instruments for the combinatorial synthesis and subsequent handling of chemical libraries. For the purposes of this chapter, we include only library synthesis instruments where there is some combinatorial technology, such as splitting and pooling of resin. This distinction removes from this discussion multiple peptide synthesizers and other instruments on which discrete compounds are synthesized. The reader is referred to other chapters in this book (see Chapter 15) for detailed discussion of technologies where automation is used in these other areas, such as parallel synthesis.

Both solution-phase and solid-phase synthesis of combinatorial libraries have been described.[1,2] This chapter focuses on the synthesis of libraries on solid phase for two reasons: First, the use of a solid support will usually eliminate the need for difficult-to-automate procedures like extractions, filtrations, and chromatography; second, the use of a solid support facilitates the use of the split–pool technique, which offers the most efficient manner of synthesizing large libraries (10^4–10^9 compounds). Preparation of discrete compounds in these numbers is not practical. For these large libraries, individual compounds are not segregated, and the identity of compounds on individual resin particles is not known and (for practical purposes) is not needed except for compounds determined to be interesting. If a suitably sensitive analytical

Combinatorial Chemistry and Molecular Diversity in Drug Discovery, Edited by
Eric M. Gordon and James F. Kerwin, Jr.
ISBN 0-471-15518-7 Copyright © 1998 by Wiley-Liss, Inc.

technique, such as peptide sequencing,[3] is not available for compound identification on the resin, the identity information of these compounds is arrived at by deconvolution of the library,[4] or decoding of resin particles.[5,6] This chapter deals with instruments both for library synthesis and for handling of resin particles for biochemical assay and compound identification.

PURPOSES OF AUTOMATION

There are three primary motivations for automation of combinatorial chemistry processes: capability, quality, and throughput. First, automated systems perform functions that are impossible to do by hand, such as delivering exceedingly small volumes of liquids to precise locations. Second, the consistency of a process can be enhanced, with less error. Third, and perhaps most important, an appropriately designed automated system will perform functions quickly and repeatedly over long periods of time.

A hypothetical example of this last motivation follows: For a typical peptide chemistry cycle, deprotection is followed by coupling and capping. Each of these reaction steps will be followed by a number of washes sufficient to ensure that the next reaction will not be affected by the previous reagent. In this example, if three washes are used after each reaction, the total number of reagent and solvent additions per cycle is 12. An equal number of liquid removals need to occur. Then, assuming a library with 36 natural and unnatural amino acids and six reaction cycles (to form a 6-mer peptide), the total number of fluid operations is 5184 ($12 \times 2 \times 36 \times 6$). As the size of a library grows, either in the number of monomers or number of split–pool steps, the number of reagent additions and bead washing steps grows significantly. Very large libraries ($>10^6$ compounds) can theoretically be prepared by skilled chemists or technicians, but can be made efficiently only in an automated fashion.

In summary, in addition to freeing up the chemists hands, automation allows the potential of around-the-clock productivity with more consistent, error-free synthesis. Clearly, greater library synthesis productivity results in a requirement for more productive (i.e., higher throughput) assay strategies. It also puts more demands on the synthesis preparation and workup steps. With an extremely efficient automated instrument with parallel processing, the bottleneck of compound synthesis becomes building block (monomer) acquisition (by collection or synthesis) and weighing and measuring. It can be seen that the process of combinatorial chemistry for drug discovery needs to be balanced so that no component (synthesis preparation, synthesis, assay, compound identification) is sufficiently slow that it significantly compromises the potential of the other, faster components of the system.

COMBINATORIAL CHEMISTRY INSTRUMENTATION

The enablement of automated synthesis instruments has relied heavily on the development of solid-phase chemistry supports and techniques.[2,7] On solid phase, many chemistry protocols are reduced to a series of reagent additions, reaction incubations (either at room or controlled temperature), draining, and solvent washes in between reaction steps. All of these processes are straightforward enough for transfer to relatively simple synthesis instruments. Synthesis on solid phase eliminates complex separations steps such as chromatography, evaporation, crystallization, and liquid/liquid extractions, some of which would be difficult to automate.[8] These chemistry techniques could be automated into the combinatorial synthesis process as well, but with considerably more effort.

It is convenient to divide synthesis instruments into three categories. First are instruments that deliver reagents by moving a robotic pipetting arm from reagent reservoirs to the reaction vessels. Second are instruments that direct liquids spatially by opening and closing sets of valves that are spaced between reservoirs and reaction vessels. Third are instruments that combine significant amounts of both valving and robotic technology.

For instruments requiring a large number of reaction sites, the biggest advantage of a robotic system is the relatively small amount of hardware required for fluid handling. A single pipetting arm can draw from a multitude of reagent and solvent bottles and deliver the liquids to any other point in the system, limited only by the span of the arm's reach and the precision of its movements. If the system allows, the flexibility to add new fluid paths to the system can be very advantageous. In addition, robotic pipetting systems can be used to deliver very small fluid volumes without the necessity to prime delivery lines, avoiding the wastage of precious reagents. Pipetting arms can be multiplexed to deliver to 2, 4, 8, or more locations simultaneously. This multiplexing can mitigate one of the main drawbacks of many robotic systems—the long time required for processing a large number of movement and delivery cycles. Processing time in a robotic system is increased not only by the need to move (at reasonable speed) back and forth on the system, but also by the need to either rinse or replace delivery tips between different reagents.

Systems that employ valve manifolds and permanent fluid connections between reagent sources and reaction vessels offer the advantage of the possibility for massive parallel processing—sending reagents to many points simultaneously. In addition, fluid access to more three-dimensional space is possible, without the restrictions of access placed on a robot arm. These features come with the major disadvantages of these systems: the large number of system hardware components, including valves, tubes, and sensors; the interface hardware for these components; and the related maintenance and repair of this large collection of components. Indeed, the quality and durability of system components becomes a major engineering concern when, for example, any one valve in a system of 150 valves can disable the entire process.

Some examples of hybrid systems take advantages from each of the two strategies described above. One peptide library synthesis instrument[9] uses a valve manifold connected to pressurized gas and vacuum for the mixing and removal of synthesis reagents and a robot arm for reagent additions and resin manipulations. A multiple oligonucleotide synthesis instrument[10] uses a series of valve sets mounted above a 96-reaction vessel array. The valve sets allow the reduction of the robotics to a one-dimensional stage. Though this last instrument does have a considerable number of valves (upward of 100), employing the robot stage is key to a relatively simple design.

Another general distinction that can be drawn for automated processes is between open loop (no feedback) and closed loop (feedback controlled) systems. One possible form of feedback in a chemical system is analytical tests that determine the extent a reaction has gone to completion. Many early commercial peptide and oligonucleotide synthesizers operated open loop, with no reaction monitoring. Similarly, in a combinatorial instrument, monitoring of each of the multiple reactions is not required if the chemistry has been demonstrated to be sufficiently robust. A more likely need for feedback to the instrument controller arises from the requirement that the numerous reagent additions and removals and resin manipulations be performed correctly. Feedback is key to large systems where the flow resistances to multiple-reaction vessels in valved systems are neither equal nor constant. In addition, flexibility to perform an expanding portfolio of chemistries without the need to reprogram an instrument to handle fluids of different properties requires feedback, usually in the form of fluid detectors and level sensors, the use of which replaces the required calibration and periodic flow tests of open loop systems.

REQUIREMENTS AND LIMITATIONS OF AUTOMATION

Simply put, the required capabilities for an automated library synthesis instrument are the manual functions that are being replaced. These functions include opening and closing of reaction vessels, addition or removal of measured reagents or resin, providing an inert atmosphere, temperature control of reaction vessels, timing of reactions, and agitation of reacting mixtures. Some functions, such as reagent additions and removals, may be made time efficient through parallel processing of multiple reaction vessels. Other functions, such as resin manipulations, may require more time on an automated instrument (compared to a manual process) but are valuable as one element of a totally automatic system.

Ideally, the functions listed above are enabled in an instrument that has the following qualities: (1) *Inertness*—Particularly for a general organic combinatorial instrument, the array of reagents that the user might use is large, and systems that are built with glass and inert fluoropolymers afford the most general inertness. (2) *Ease of use*—The internal software and hardware of the system may be complex, but the user interface should be friendly to the operator. (3) *Feedback*—As discussed above, the system should include an appropriate level of feedback for monitoring fluid deliveries and/or reaction procession. (4) *Safeguards*—Features that enhance safe and error-free operation of the instrument increase its ultimate utility.

Furka et al.[11] described how very large numbers of compounds could be synthesized by alternating the coupling of amino acids on separated resin masses with a resin randomization step. The randomization is achieved by combining, mixing, then redistributing resin back to the reaction vessels. Indeed, if enough cycles of reaction and resin randomization are performed, each resin particle in the population will bear a unique synthesized compound. To perform this split–pool synthesis automatically, the fundamental requirement is for an automated synthesis system equipped with both multiple-reaction vessels and some capability to perform the resin manipulation steps.

The design of reaction vessels, the heart of any synthesis instrument, depends on all of the basic strategic points mentioned above, such as how reagents will be added and removed, how agitation will be performed, and what kind of temperature control will be required. General inertness, a primary concern, will be achieved with glass or fluoropolymer construction, and custom fabrications from both of these materials are widely available. The mode of separation of the resin from used reagents and wash solvents is also key to reaction vessel design. Though the separation may be performed by sedimentation, centrifugation, or magnetic means, the easiest and most common strategy is filtration. Glass, metal, or polymer frits can be easily incorporated into the reaction vessels for this purpose. Frits can serve a dual function: they are convenient for bubbling gas through a reacting mixture for agitation. Another factor to consider for reaction vessel design, discussed below, is the manner in which resin particles are randomized between synthesis cycles in the split-and-pool synthesis.

Since automated handling of dry synthesis resin is intractable, pooling and distribution of the resin is usually performed using a slurry of the resin. A solvent whose density closely matches that of the resin is ideal. The resin population, after transfer from the individual reaction vessels, is well mixed in a larger vessel and then redivided equally among the reaction vessels. Both combination and redistribution of the resin can be performed using a robotic pipetting arm[9,12] or by pressurized delivery of the resin suspension through a permanent fluid path between the reaction vessels and the mixing vessel.[13] A third strategy is to use open-topped reaction vessels which, when flooded with solvent, allow the resin to be mixed in a shared upper space. The mixed resin then portions back into the reaction vessels by draining and sedimentation.[14] With all of these techniques, repeated rinses with solvent are employed to ensure that all of the resin returns to the reaction vessels for the chemistry steps.

It is useful to discuss the limitations of instruments performing combinatorial split–pool synthesis. One of the primary limitations of any synthesis instrument is its finite volume. If the total reaction volume of an instrument is 50 mL, for example, and the resin concentration during reactions is 100 mg/mL, then the functional resin capacity of the instrument is 5 g. Following the same example with 130 μm resin having 800 particles/mg, the instrument's capacity is for four million resin particles, also the upper limit of unique molecules synthesized during a single synthesis on the instrument, assuming one compound is synthesized on each particle. Of course, using pooled monomer mixtures can result in significant increase in the library size, but with mixed compounds on each resin particle. The limit may also be increased by enlargement of the reaction vessels, though the accompanying increased consumption of reagents limits the practicality of large increases by this method.

In contrast, a 10-fold decrease in the diameter of the resin decreases the particle volume by 1000-fold, increasing the instrument's particle capacity by the same amount. Thus, limiting particle size is the key to maximizing the library size. It should be recognized, though, that the complete combinatorial system will ultimately determine the library format (particle size and diversity). If a particular assay or decoding strategy requires a quantity of synthesized compound that can be made on a resin particle with a minimum size of 100 μm, for example, it is of no use to increase diversity by decreasing particle size. Here is a case where miniaturization of the assay and improvements in analytical techniques become important goals for improving the combinatorial process.

TRANSFERRING A CHEMISTRY TO AUTOMATION

Though a combinatorial synthesis instrument may be available and functional for certain chemical protocols, new chemistries intended for library synthesis need to be scrutinized and optimized specifically with the instrument in mind. In transferring a chemistry process to automated library production, several steps need to take place to ensure success. Optimization of the solid-phase chemistry, from an automation perspective, includes making the chemistry robust to withstand slight variations in concentration, reaction times, and temperatures. It also includes using reagents that an automated instrument will be able to reliably handle.

Four examples of optimizing chemistries for instrumentation follow:

1. Solid reagents are commonly employed during manual organic synthesis, but solids are relatively difficult to handle by automated systems, so protocols with predissolved solutions should be developed, if possible.

2. A partially dissolved solution or one containing particulate matter may not be of concern when a chemist is doing manual pipetting, but it would

be inappropriate to flow such a solution through many types of valves. Filtration of the solution needs to be part of the protocol.

3. Chemistries performed in the presence of molecular sieves may be standard on the bench but prohibitively difficult on an instrument. Changing to a dehydrating solvent may be sufficient.

4. A reaction conventionally performed in a convenient acetone/dry ice bath at $-78°C$ should at least be attempted at a higher, more instrument-friendly temperature. Success at a higher temperature will result in significant savings of engineering development time and hardware expense.

While these optimizations may seem inappropriate in that the automation is making demands on the synthesis chemistry, the goal of automated library production is one that requires a multidisciplinary effort. New chemistries should be developed with the instrumentation in mind. In readying the combinatorial process, it would be inefficient for an instrument-blind chemistry to be developed and then handed off to another group for enablement on a machine.

QUALITY CONTROL

Combinatorial chemistries are inherently difficult to analyze and characterize, due to the large number of compounds that are each present in very small amounts. Typically, as the library size increases, the amount of each compound present decreases, and it becomes more difficult to obtain verification that a particular reaction has proceeded as hoped. For example, while it appears quite possible to obtain reliable mass spectrometry data on mixtures of 20 to 30 compounds (refer to Chapter 18), it is unlikely that we could verify the presence of each component in a mixture of 10^3 or 10^6 compounds, by mass spectrometry or any other analytical method. For this reason, the importance of preemptive quality control of combinatorial building blocks and reagents is critical. A high degree of analytical work should be performed during the single-compound and building block validation process described below. The level of confidence gained in these exercises allows the library synthesis to be performed with confidence even though complete analytical characterization of the library will not be undertaken. When building blocks fail the QC process, either they can be replaced with building blocks that do pass QC, or the reaction conditions can be further optimized to provide for good yields when even difficult-to-couple building blocks are employed.

The most obvious way to maintain the quality of difficult-to-analyze libraries is to "rehearse" the chemistry in a nonlibrary context. For example, every candidate building block can be employed in a generic rehearsal reaction. For the coupling efficiency from the generic reaction to be relevant to other specific reactions, the generic reaction should be nontrivial (e.g., coupling an amino acid to the secondary amine of proline). As an additional step of rehearsal, prior to library synthesis, the anticipated most difficult to synthesize compounds (e.g., due to steric or electronic properties) can be synthesized individually so that the quality of the synthesis is assessed. The quality of the products from each of these types of quality control reactions can be assessed by HPLC and MS. Library rehearsal reactions are less tedious when they are performed on an automated synthesis instrument. In fact, these reactions should be performed on the library synthesis instrument to provide a more realistic rehearsal.

Just as the quality of a library's chemistry needs to be verified, the library synthesis instrumentation should also be checked on a regular basis to enhance the likelihood that defects are found before they lead to malfunctions affecting the quality of a synthesis. The inspections can include leak testing of valves and fittings, sensor checks, and flow testing of vessels and fluid lines. Preparation of a difficult-to-synthesize compound in multiple-reaction vessels, followed by chemical analysis, is a functional test that can greatly increase the level of confidence that the instrument will be performing as expected during a combinatorial library synthesis.

POSTSYNTHESIS

Many chapters in this volume refer to applications of assay techniques for combinatorial libraries. Many of these techniques employ some sort of automation for high-throughput screening. We briefly discuss some of the instruments that can be employed in the handling of resin particles in the postsynthesis phase of the combinatorial process.

Some assay techniques employ the testing of pools of resin particles or mixtures of compounds rather than individual treatment of discrete particles. In Affymax's ESL technology[6] where 10-μm resin particles are used in DNA-encoded libraries, solid-phase assays can be performed on a bulk mixture of the resin and the positive particles selected and collected individually for decoding by fluorescence activated cell sorting (FACS). In contrast, larger resin particles (approximately 100–130 μm), from which library-synthesized compounds will be cleaved for solution assays, present different challenges. Even if pools of beads are treated together originally for compound cleavage in a solution assay scheme, there will eventually be a need to assay cleaved compounds from individual particles.[15,16] It is clear that manual pipetting of numerous resin particles, even with the assistance of a microscope, is impractical. Some examples of automated tools are described below.

One strategy uses an array of eight microcapillaries mounted on a manifold that can be switched between pressure and vacuum. The manifold is mounted on an electronic XYZ stage set. The controller directs the capillaries to repeatedly follow this series of actions: (1) Dip into a column of wells on a "source" microtiter plate and acquire resin particles by suction. The large particles are retained on the end of the capillaries, but cannot enter. The ends of the capillaries are large enough to hold only one particle at a time. (2) Move to a "destination" column on a plate, previously filled with a receiving liquid, and expel the particles by switching the capillary manifold to pressure. (3) Move to a rinse station so that any particles or fragments lodged in the capillaries can be cleaned out before the next cycle.

This process is robust enough that the instrument can operate in an open-loop fashion. Advantages of this system are that a complete plate of single particles can be dispensed in 5 min, and oversized beads can be handled. Resin handling by suction for assay formats beyond 96-well plates is also being explored. One variation is a suction plate that can be used to pick up an organized 12 × 12 array (144 particles) at a time from a large population source. The array is then transferred to microwells (1 μL) on a larger plate that may hold a total of 3456 particles, each sequestered in its own assay volume.

A second strategy uses a single, small conical reservoir to contain a source of resin particles that are fed into a rotating cup that is carefully machined to hold only one particle at a time. As above, the system, which consists of valves, a motor, and a Gilson FC 205 fraction collector, is controlled by computer. The repeated series of steps is as follows (1) A particle is fed into the measuring cup by gravity and slight suction. (2) The cup rotates to the exit port, and pressurized liquid pushes the particle into the microtiter plate below. The pressure simultaneously forces the particles in the reservoir to mix, reducing the possibility of having a permanently blocked system. (3) The fraction collector advances to the next well as the cup is rotated back to the particle-acquisition position.

Again, the process is open loop, and the time is about 5 min per plate. One advantage here is that the reservoir can be easily imaged with a microscope so that dispensing can be continued until all of the particles in the reservoir have been handled. This feature may be especially attractive for tiered-release experiments in which a pool of particles with positive assay results needs to be reassayed as single particles. Another advantage is that small particles and fragments will not be lost. Large particles that cannot fit into the cup will not be dispensed, though they can be visualized and retrieved manually. A variation on this instrument is to dispense a group of particles in the same fashion by using a rotating cup with the appropriate volume.

REFERENCES

1. Furka A (1995): History of combinatorial chemistry. *Drug Dev Res* 36:1–12.
2. Gordon EM, Barrett RW, Dower WJ, Fodor SPA, Gallop MA (1994): Applications of combinatorial chemistry to drug discovery, 2: combinatorial organic synthesis, library screening strategies and future directions. *J Med Chem* 37:1385–1401.
3. Lam KS, Wade S, Abdul–Latif F, Lebl M (1994): Application of a dual-color detection scheme in the screening of a random combinatorial peptide library. *J Immunol Methods* 180:219–223.

4. Houghten RA, Pinilla C, Blondelle SE, Appel JR, Dooley CT, Cuervo JH (1985): Generation and use of synthetic peptide combinatorial libraries for basic research and drug discovery. *Nature* 354:84–86.

5. Ohlmeyer MHJ, Swanson RN, Dillard LW, Reader JC, Asouline G, Kobayashi R, Wigler M, Still WC (1993): Complex synthetic chemical libraries indexed with molecular tags. *Proc Natl Acad Sci USA* 90:10922–10926.

6. Needels MC, Jones DG, Tate EH, Heinkel GL, Kochersperger LM, Dower WJ, Barrett RW, Gallop MA (1993): Generation and screening of an oligonucleotide-encoded synthetic peptide library. *Proc Natl Acad Sci USA* 90:10700–10704.

7. Merrifield RB (1963): Solid-phase peptide synthesis, 1: the synthesis of a tetrapeptide. *J Am Chem Soc* 85:2149–2154.

8. Lindsey JSA (1992): A retrospective on the automation of laboratory synthetic chemistry. *Chemometrics and Intelligent Laboratory Systems: Laboratory Information Management* 17:15–45.

9. Zuckermann RN, Kerr JM, Siani MA, Banville SC (1992): Design, construction, and application of a fully automated equimolar peptide mixture synthesizer. *Int J Pept Protein Res* 40:497–506.

10. Lashkari DA, Hunicke–Smith SP, Norgren RM, Davis RW, Brennan T (1995): An automated multiplex oligonucleotide synthesizer: development of high-throughout, low-cost DNA synthesis. *Proc Natl Acad Sci USA* 92:7912–7915.

11. Furka A, Sebestyen F, Asgedom M, Dibo G (1991): General method for rapid synthesis of multicomponent peptide mixtures. *Int J Pept Protein Res* 37:487–493.

12. Saneii HH, Shannon JD, Miceli RM, Fischer HD, Smith CW (1993): In *Peptides: Chem, Struct Biol, Proc Am Pept Symp, 13th,* pp 1018–1020.

13. Sugarman JH, Rava RP, Kedar H, issued US patent 5,503,805 (1996).

14. Bartak Z, Bolf J, Kalousek J, Mudra P, Pavlik M, Pokorny V, Rinnova M, Voburka Z, Zenisek K, Krchnak V, Lebl M, Salmon SE, Lam KS (1994): Design and construction of the automatic peptide library synthesizer. *Methods: A Companion to Methods in Enzymology* 6:432–437.

15. Lebl M, Krchnak V, Sepetov NF, Seligman B, Strop P, Felder S, Lam KS (1995): One-bead–one-structural combinatorial libraries. *Biopolymers* 37:177–198.

16. Salmon SE, Lam KS, Lebl M, Kandola A, Khattri PS, Wade S, Patek M, Kocis P, Krchnak V, Thorpe D, Felder S (1993): Discovery of biologically active peptides in random libraries: solution-phase testing after staged orthogonal release from resin beads. *Proc Natl Acad Sci USA* 90:11708–11712.

18

ANALYTICAL CHEMISTRY ISSUES IN COMBINATORIAL ORGANIC SYNTHESIS

WILLIAM L. FITCH, GARY C. LOOK, AND GEORGE DETRE
Affymax Research Institute, Santa Clara, California

Advancements in synthetic organic chemistry have always gone hand in hand with advancements in organic analytical chemistry. In the same way that structural chemistry of the early 1900s depended on melting-point measurement and combustion analysis, so, too, the accelerated developments of the 1950s through the present have been possible because of continued progress in sophisticated spectroscopic methods.

Combinatorial drug discovery represents a paradigm shift in how multistep medicinal chemistry synthesis is performed. For the last 100 years the synthetic dogma has not varied: (1) Use the purest available starting materials and reagents to perform a single chemical transformation. (2) Purify the reaction product(s) to homogeneity. (3) Characterize the product structure(s) rigorously in terms of purity and identity. (4) Repeat steps 1–3 for each subsequent step of the synthesis.

Exceptions were occasionally allowed when 2- to 3-step synthetic sequences might be performed without purification of intermediates. But the chemists who proposed to publish biological or physical properties of poorly characterized synthetic materials in top organic chemistry journals would find their work rejected. This dogma is codified in the characterization criteria promulgated in the first issue of each journal volume.[1–3] Similar criteria exist in each pharma-

Combinatorial Chemistry and Molecular Diversity in Drug Discovery, Edited by
Eric M. Gordon and James F. Kerwin, Jr.
ISBN 0-471-15518-7   Copyright © 1998 by Wiley-Liss, Inc.

ceutical research organization. This procedure assures research management in the companies that their structure collections are of high quality.

In contrast, the dogma of combinatorial drug discovery[4,5] can be stated succinctly as "thou shalt discover drugs." If biological testing of poorly characterized mixtures of compounds can lead to success and competitive advantage in that endeavor, then so it will be. Provided that it leads to success, it is accepted that this process will necessarily yield more fuzzy data, including false-negative results.

The popularity of the combinatorial approach partly reflects the growing understanding of how inefficient classical chemical purification and structure determination have been. A large percentage of the medicinal chemist's efforts go into eliminating a very small chance that a given structure is wrong or the biological activity due to a minor impurity. The new methods rationalize that drug discovery efforts are better off with more compounds and slightly increased chances of a false negative or temporarily misleading false positive result.

So, the absolute commandment of earlier synthetic chemistry is gone. But it is still true that the value of a combinatorial chemistry experiment is directly related to the truthfulness of its composition claim. Combinatorial synthesis must have a high degree of analytical rigor to it. This chapter highlights two key areas of concern:

1. How can we monitor chemical reactions in the solid phase?
2. What are appropriate procedures to assure the quality of combinatorial synthesis products?

In this review we do not cover those analytical methods that are exclusively relevant to peptide or oligonucleotide synthesis (e.g., Edman degradation, PCR); we do not discuss coding strategies (see Chapter 14) or other analytical methods for determining the structures of bioassay "hits." We do not explore methods for affinity separating mixtures. Finally, we do not spend much time on solid-phase "cleave and characterize" procedures, for once the solid phase is removed, all the well-known organic analytical procedures apply.

ANALYTICAL CHEMISTRY IN SOLID-PHASE SYNTHESIS

Solid-phase synthesis (SPS) is at the core of the split–pool method of combinatorial organic synthesis.[6] The wide utility of SPS of peptides and nucleotides in the new biology has led to the development of a vast array of analytical methods.[7] A brief flowering of the Merrifield approach to other fields of organic synthesis in the 1970s led to some novel analytical methods.[8,9] But nonpeptide SPS was not popular in the 1980s and has only now been revived with the advent of combinatorial chemistry. There are unmet needs in developing analytical tools for following reactions, measuring yields, and determining purity and identity of reaction products without cleavage.

Every reaction needs to be "followed" in some way. This process is critical for knowing when the reaction is complete, optimizing yield and stoichiometry, applying standard procedures to new substrates, scaling up a reaction, and so on. In a typical solution-phase reaction

$$A \quad + \quad B \quad \xrightarrow{\text{reagents}} \quad C \quad + \quad D$$

components A and B react in the presence of reagents or catalysts to give the desired product(s) C and by-products D. The chemist can follow this reaction by watching for a color phase or pH change, by following the disappearance of A or B or a reagent, or by following the appearance of C or D. The favorite tools for looking at appearance or disappearance are thin-layer chromatography (TLC), and, less frequently, gas chromatography (GC), liquid chromatography (LC), nuclear magnetic resonance spectroscopy (NMR), and mass spectrometry (MS).

Following reactions in the solid phase is much more difficult. In a typical solid-phase reaction

$$\bullet\!-\!A \quad + \quad B \quad \xrightarrow{\text{reagents}} \quad \bullet\!-\!C \quad + \quad D$$

components A and C are covalently attached to the solid support and their appearance or disappearance is not easy to follow. Materials B and reagents are typically present in large excess, so their concentration will not change during reaction. A final barrier to analysis is that the solid phase consists of 80–95% matrix plus linkers and only 5–20% reacting ligand.

There are no generally accepted properties of SPS products that are required in literature reports. Commercial resins are typically sold with quoted particle size and loading. In the pioneering days of SPS,[10-13] solid-phase intermediates and products were characterized for loading by quantitative and qualitative methods. Loading is less rigorously quoted in the recent literature.[14-16] As we shall see, there are a plethora of modern techniques for measuring loading and the structures of SPS materials. We will propose that a formal analytical stringency be applied to the description of these materials in the literature.

Physical Characterization

The physical properties of SPS beads can change dramatically with changes in ligand composition. Beads can change color or swell in solution. They can be fractured during handling. These changes may not easily be used to follow a reaction (but see Reference 17), but should certainly be part of the characterization of an SPS product. Commercial solid-phase resins are carefully meshed to assure uniform particle sizes. Particle size uniformity is critical to the correct functioning of bead distribution apparatus (see Chapter 17, and see Reference 18). FACS analysis also benefits from uniform beads. TentaGel, one of the most popular resins, is especially labile toward fragmentation during manipulation, so careful monitoring of its size distribution is necessary. Resin particle sizes can be measured with a microscope or standard laser diffraction methods.[19,20] Polystyrene spheres of known particle size distribution are available as primary reference materials.[21]

Wet Chemical Methods

Gravimetric and titrimetric analyses have been largely supplanted by modern NMR techniques in organic synthesis. In an earlier day these methods fluorished and micromethods were widely adopted.[22–24] But the difficulties in applying NMR to solids make it worth reexamining old methods. Gravimetric methods can be applied on resin[25] or after cleavage.[26] Combustion nitrogen analysis is a useful characterization of a resin.[10,13] Micro chlorine,[15] bromine,[27] iodine, and tin[28] elemental analyses have been reported recently.

Color tests are widely used to follow SPS reactions. The most popular is the Kaiser test,[29] wherein an amine is detected by the ninhydrin reaction. A negative Kaiser test is used routinely to assure the complete reaction of bead-bound amino groups. Other amine tests are purported to be more reliable or more quantitative,[30–34] but the Kaiser test is unlikely to be superseded. Another popular color test is Ellman's for free thiols.[35]

Infrared Spectroscopy

Infrared (IR) spectroscopy has long been a method of choice for polymer characterization. Most synthetic chemists will remember taking an IR spectrum of polystyrene. The earliest SPS practitioners used IR to follow their reactions qualitatively and quantitatively,[10–13] and the method is still widely used.[25,26,36,37] Subtracting the polymer absorption with difference IR was found to be a useful technique.[38] Surface IR techniques hold great promise.[39] By focusing the IR beam with a microscope, it is possible to obtain very high-quality IR spectra from a single bead.[40]

Ultraviolet Spectroscopy

An ultraviolet (UV) spectrophotometer is a mainstay of the SPS laboratory. Its main use is in measuring the dibenzofulvene–piperidine by-product from cleavage of the FMOC amine protecting group. UV measurement of this chromogen yields the FMOC number, a common measure of loading in SPS.[7] Similar qualitative and quantitative analyses use the cleavage of dimethoxytrityl (especially for oligonucleotide synthesis[41]), trityl, 4-nitrophenethyloxycarbonyl,[42] and *tert*-butyloxycarbonyl[43] protecting groups. Reflectance ultraviolet measurements have not found wide use, but fluorescence-based sorting is used in solid-bound ligand bioassay (see Reference 4) and can be used for characterizing the functional groups on beads.[44]

Mass Spectrometry

Mass spectrometry is mainly used in "cleave and characterization" studies. Early reports on the use of FAB or [252]Cf plasma desorption ionization to directly cleave bead-bound peptides for mass analysis[45,46] have not been followed up. Direct cleavage with secondary ion beams also is discouraging.[47] But promising results have been obtained by on-probe cleavage of cleavable-linked peptides and other ligands followed by time of flight mass analysis. This approach has worked both for acid-labile linkages cleaved by TFA vapor[47-49] and photo-labile linkages cleaved with a laser as an integral part of the matrix assisted laser desorption (MALDI) process.[50]

NMR Spectroscopy

NMR is the most important tool of the solution-phase organic chemist. Proton and ^{13}C NMR (and less frequently ^{19}F, ^{31}P, and 2D NMR) solve most structural questions in small molecule synthesis. But the resolution of NMR is greatly affected by the mobility of the atoms in a sample and the homogeneity of a sample matrix. Both of these properties are severely compromised with solid-bound molecules, so NMR was not available to the early SPS chemists. [13]C (and other elements such as tin[28]) NMR spectra are less effected by this immobility. Several groups, beginning in 1980, described the application of "gel-phase" ^{13}C NMR to SPS (see references cited in References 51 and 52). In this technique the mobility of the ligand is increased by swelling the beads in an appropriate solvent, creating a semisolution environment. Gel-phase ^{13}C NMR has also been used recently by several groups in the area of solid-supported small molecule synthesis.[51,53] However, routine ^{13}C NMR suffers from poor sensitivity due both to the low natural abundance of ^{13}C and the low percentage of analyte in an SPS sample. For these reasons, a ^{13}C gel-phase NMR analysis will need to be run for several hours, making it impractical for following reactions.

A solution to the sensitivity problem of gel-phase ^{13}C NMR, first suggested by Rapoport,[8] is the application of ^{13}C-labeled reactants in SPS.[52] An increase of 100-fold in sensitivity is achievable using compounds that are 100% [13]C enriched at a carbon atom. This technique generally yields very simple spectra. Figure 18.1 shows a typical case. Unlabeled glycine on TentaGel resin in benzene solvent gives resonances due to the solvent and the resin. When labeled benzaldehyde is added, the imine **1** (Scheme 18.1) is formed, which gives rise to a new resonance at 163.7 ppm (Figure 18.1a). Conversion of this imine to a thiazolidinone **2** by reaction with mercaptoacetic acid moves the

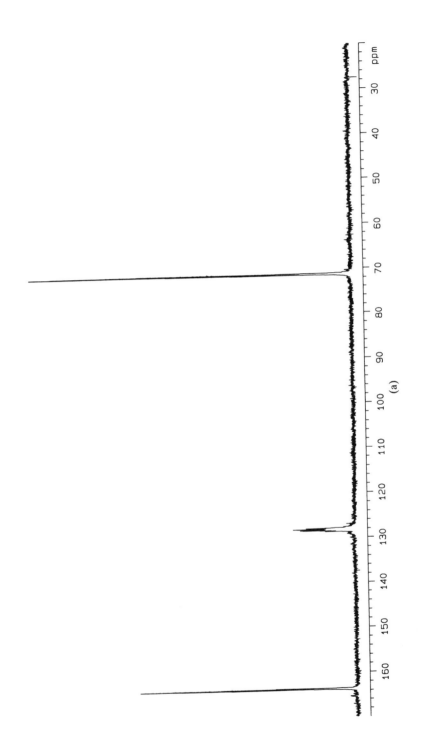

(a)

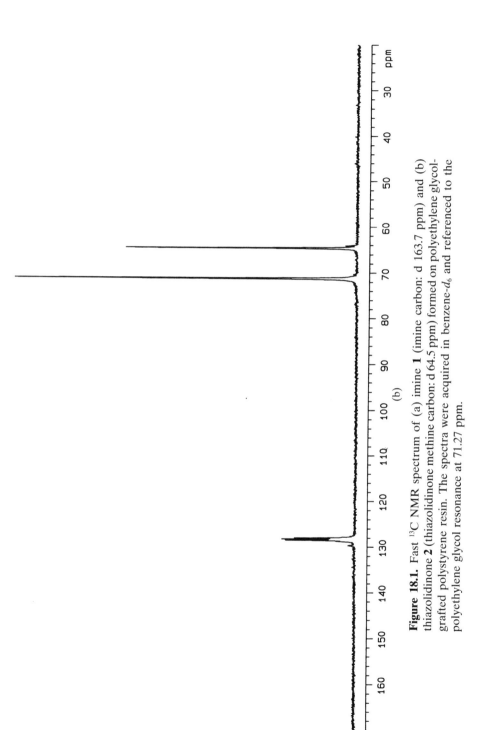

Figure 18.1. Fast ^{13}C NMR spectrum of (a) imine **1** (imine carbon: d 163.7 ppm) and (b) thiazolidinone **2** (thiazolidinone methine carbon: d 64.5 ppm) formed on polyethylene glycol-grafted polystyrene resin. The spectra were acquired in benzene-d_6 and referenced to the polyethylene glycol resonance at 71.27 ppm.

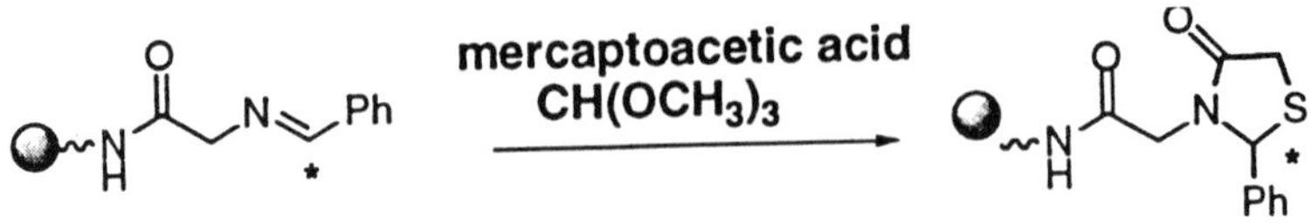

Scheme 18.1

single resonance upfield to 64.5 ppm (Figure 18.1b) in accordance with the hybridization shift. The ^{13}C NMR resonance of the methine of the resin-bound thiazolidinone correlates well with that of the solution-phase thiazolidinone. At Affymax we have found this techique to be invaluable to our chemists, who use it in a way similar to the use of TLC in solution-phase chemistry. Reactions are followed by monitoring the disappearance of one resonance and the appearance of new resonances corresponding to product(s). Generally, the data can be obtained within 30 min, where most of that time is taken up with washing and drying the beads. The spectra are obtained on the same 300- or 400-MHz NMR instruments in general use for solution-phase product or cleaved product characterization. Through the use of standard NMR tube inserts, spectra can be obtained on as little as 10 mg of resin containing only a few micromoles of compound. The quality of the ^{13}C NMR spectrum obtained is dependent on the type of solid support. The highest quality spectra are obtained from the TentaGel type, polyethylene glycol (PEG) grafted resins. The major limitation of this "fast ^{13}C gel-phase NMR" is the need for ^{13}C-labeled precursors, which are not always available.

Proton magnetic resonance is much more compromised by inhomogeneities in a sample. We have recently been able to overcome these effects to a large extent using magic-angle spinning (MAS) in the Varian Nano-NMR probe.[54] We call the technique SPMAS for solid-phase magic-angle spinning. This probe yields very high-quality ^{1}H spectra for molecules bound to the solid-phase resin TentaGel. This resin has a polystyrene (PS) core and a long-chain PEG linker, which essentially places the ligand in a free solution state. Figure 18.2a shows the SPMAS spectrum for a simple organic structure on TentaGel. The large resonance at 3.6 ppm is due to the PEG protons. All of the other resonances are readily assigned to the compound or known impurities. Figure 18.2b shows the same sample with presaturation of the PEG proton resonance. This solvent suppression experiment not only is useful for clarity of the presentation and ADC saturation-related signal/noise improvement, but has the unexpected additional benefit of leading to an enhancement of the ligand resonances relative to the PS resonances. (Note the pattern from 6 to 7.2 ppm.) This NOE enhancement is due to spin diffusion from the PEG protons to the nearby ligand resonances. The protons of the rigid PS core are more distant and are not enhanced. Of course the presaturation leads to loss of signal intensity in the 3.2- to 4.2-ppm region and the spin diffusion effect leads

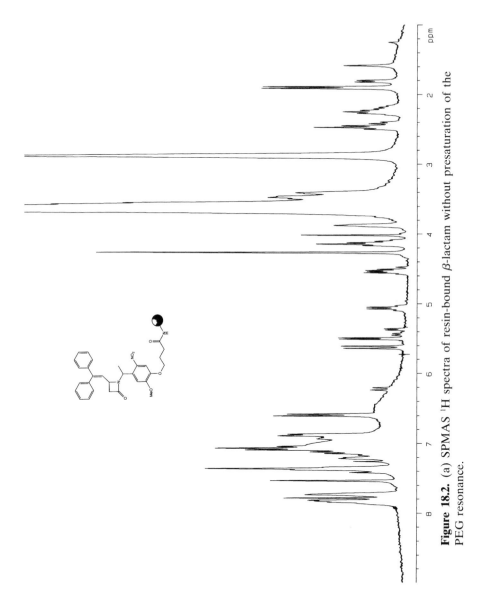

Figure 18.2. (a) SPMAS ¹H spectra of resin-bound β-lactam without presaturation of the PEG resonance.

357

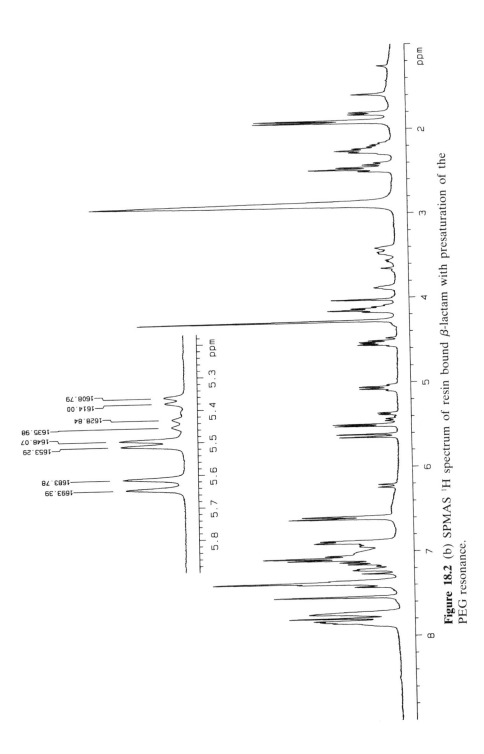

Figure 18.2 (b) SPMAS ^{1}H spectrum of resin bound β-lactam with presaturation of the PEG resonance.

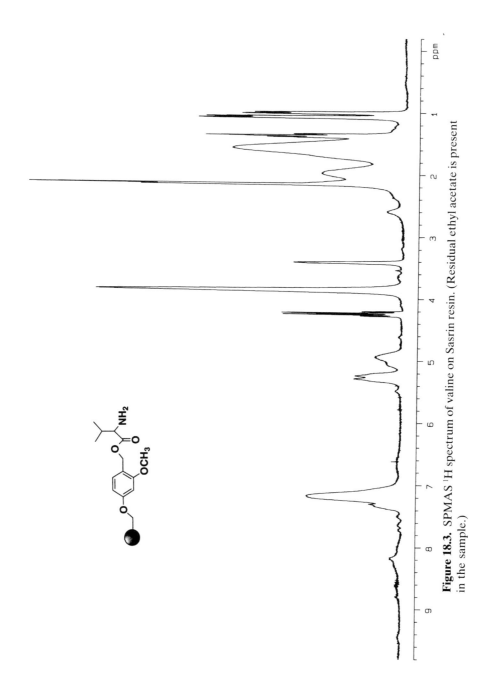

Figure 18.3. SPMAS ¹H spectrum of valine on Sasrin resin. (Residual ethyl acetate is present in the sample.)

to nonequivalent integrals for different protons. We normally run each sample both with and without presaturation to obtain maximum information.

SPMAS ^{1}H NMR of PS solid-phase resins that lack the PEG linkage yield much broader peaks. Figure 18.3 shows the spectrum for a typical case. We have attempted to take advantage of the differences in relaxation time between the PS protons of these resins and the more mobile ligand protons but have not found a satisfactory solution to this problem yet. MAS in a standard solids NMR probe[55,56] demonstrated advantages for gel-phase ^{13}C NMR and for 2D HMQC and TOCSY spectra but has disadvantages relative to the nanoprobe in quality of 1D ^{1}H proton spectra.[54]

ANALYTICAL METHODS FOR EVALUATING LIBRARIES

This section explores the available methods for analysis of libraries. Our commentary is directed to libraries made by the split-and-pool solid-phase method, but similar considerations apply to mixtures made by solution-phase methods.

The ideal library will contain all of the predicted components at known concentrations. It will not contain any by-products or inadvertent contaminants. Solid-phase libraries will consist of perfectly uniform beads with uniform and known release kinetics. Of course, these ideals cannot be met for real libraries. Indeed, with large libraries, many of these properties will not even be measurable. For these reasons we must begin by studying small, model libraries.

Libraries of one component are readily checked using the standard purity and identity criteria. We can achieve close to 100% confidence in the composition of such libraries. Similarly, for very small libraries, it is possible to imagine isolating each of the components to homogeneity (or synthesizing each individually), proving their structures, and then determining the composition of the starting mixture chromatographically using the purified components as primary reference standards for quantitative analysis.[57] But this direct approach becomes impractical as library size increases and impossible as libraries contain hundreds or thousands of members. For these larger mixtures, we will need indirect methods or statistical quality control procedures.

A traditional medicinal chemist can sometimes be cavalier about checking intermediates and following reactions carefully; the discipline of demonstrating purity and spectral identification of the isolated product will always force the resolution of any unexpected problem. Because the combinatorial medicinal chemist does not have this analytical hurdle at the end of the road, more rigor is required in the early and middle stages of the process. The first step of this process involves rehearsal of the chemistry and testing of building blocks.

Rehearsal With Single Compounds

Peptide library production did not require much rehearsal of the chemistry; 30 years of solid-phase peptide synthesis had provided a good basis for evaluating which building block combinations were difficult and which conditions were required for adequate yields. New solid-phase chemistries will need a rehearsal phase where experimental conditions and building block (BB) structural requirements are delineated. To the extent that the products cannot be evaluated, extra scrutiny must fall on the building blocks.[58] The most valuable libraries will be made from BBs with very diverse structures. This diversity also increases the burden in rehearsal. A "methyl, ethyl, butyl" library can be prepared with a high degree of confidence based solely on the results of a rehearsal of the methyl analog. However, a "methyl, methylamino, carboxymethyl" library will require more rehearsal. Commercial compounds may require structure confirmation or purity assessment before they can be used. An incorrect structure of a BB will lead to a missing structure or incorrect structure in the product. BBs also need to be tested for the compatibility in the given reaction. We have chosen 60% as the lowest acceptable yield for a BB in rehearsal. A successful rehearsal stage should allow prediction of the greatest proportion of the BB set that will behave correctly and in high yield.

A key aspect of structure determination required in the rehearsal phase is the demonstration of product optical purity.[57,59] This will not be possible after library synthesis and may be critical for deconvoluting screening data. Does the synthetic scheme create new optical centers? Are there diastereomeric mixtures in the product? Are they chromatographically separable? Does the reaction sequence lead to racemization in enantiomerically pure building blocks?

Rehearsal With Model Libraries

For new chemistries, for new automation modalities, or as an educational exercise for combinatorial chemist trainees, it will be important to extend the rehearsal phase to making small model libraries. These model libraries of 5–50 components should be chosen with analysis in mind.

For small combinatorial libraries, chromatographic and mass spectral methods are very useful to characterize the composition.[5,60–64] Gas chromatography has a resolution advantage, but liquid chromatography will be more generally useful for libraries with diverse functionalities. It may be possible to derivatize small molecule libraries to make more of the components volatile and thus take advantage of GC's higher resolution. Supercritical fluid chromatography has properties intermediate to LC and GC and may find increased applications in combinatorial chemistry. Capillary electrophoresis (CE) offers the highest resolution, but is difficult to apply to uncharged small molecules. The new technique of capillary electrochromatography (CEC) holds great promise for high-resolution, neutral molecule separations.[65,66]

Mass spectrometry can be applied with or without chromatographic separation and in single mass analysis or tandem mass analysis (MS/MS) modes. GC/MS with electron impact ionization and LC/MS with electrospray ionization are the favored methods. Complementary information can be obtained by recording both the positive and negative ion spectra.[64] The literature documents the problems in applying mass spectrometry to mixture characterization: equivalent molecular weights,[64,67] interference from isotope peaks,[64] cluster peaks (dimers, solvent adducts, alkali ion adducts) in electrospray,[68] extreme variability in compound response factor,[61,64] and interference from fragment ions.[62] MS methods will mainly be used for rehearsal libraries or for intermediate reaction products. The methods developed here will not be applicable to analysis of large complex libraries.

Production QC

In the early days of combinatorial chemistry, many studies have been model experiments with known, assayable, biologically active compounds in the target library. The presence of a known compound in the library presents the possibility of quantifying its presence by a target compound analysis or simply assuring that the activity is discoverable. The presence of a positive control is always valuable in any experiment, but could become difficult to arrange if the promise of combinatorial chemistry for random lead generation is to be met.

An in-process quality control of libraries is intermediate testing. If a finished library has too many members to test, perhaps the intermediate stage has a more testable composition.[59,64] All combinatorial syntheses will benefit from adequate application of procedures for assuring complete reaction (like the Kaiser test) or complete removal of protecting groups (FMOC). These tests work just as well on mixtures as they do on single compounds.

After the combinatorial synthesis is completed, the product should be subjected to as thorough a cleanup as possible. We find that libraries are invariably contaminated with synthetic by-products such as amines (TEA, DIEA) or triphenylmethane and plasticizers such as dibutyl phthalate. The advent of electron-impact MS long ago taught chemists to avoid stopcock grease and tygon tubing in their reaction vessels. This lesson is being rediscovered for materials used in automated synthesis instruments.

For solid-phase libraries an important general characterization is the particle size distribution. This is known for the starting commercial beads, but should be measured and reported for processed beads, which may have undergone serious degradation. This should be done in both rehearsal and production. Another general characteristic for a solid-phase library is loading. In many cases this can be measured gravimetrically or titrimetrically on the final product. In other cases it may only be practical to report the loading determined during rehearsal.

What can be done to determine the composition of a library? Analytical chemists are quite proficient at determining individual organic compounds at trace levels in complex mixtures. But these procedures involve a lengthy method development process that is critically dependent on the availability of a standard reference sample of the analyte. These are not available for the majority of the members of a combinatorial library. Early peptide libraries were characterized by amino acid analysis[69] or Edman degradation.[67] Quantitative amino acid analysis after acid hydrolysis may be useful for many combinatorial chemistries that use these versatile building blocks in amide bond formation. Comparison of observed and theoretical AAA hydrolyzate chromatograms would reveal poorly incorporated building blocks.

HPLC chromatograms of mixtures lose their information value as the number of components increases.[70] For a complex library, the main value of the chromatogram will be in documenting that the sample is indeed a complex mixture and not inadvertently contaminated with phthalates or protecting group by-products. Two-dimensional chromatographic methods such as LC/CE or LC/CEC may be important library fingerprinting tools in the future.

As stated above, compositional analysis of complex mixtures with MS methods will be difficult. But electrospray ionization of unseparated mixtures introduced by infusion or flow injection can be useful for revealing the distribution of molecular weights of the mixture components. Comparison of predicted and found average molecular weights may be one of the most generally useful measurements.[69] And MS methods may be useful for identifying target impurities in the mixtures. For example, MS/MS neutral loss and parent ion scans are valuable respectively to identify BOC and trityl incompletely deprotected components in mixtures.[62]

For solid-phase one-bead–one-component libraries, an attractive statistical process quality-control approach is to cleave the ligand from individual beads and determine the structures and concentrations. In model studies this can be done and standards can be used for quantitation by sensitive, single-component selected ion monitoring MS methods. In production libraries, with no standards available, it should still be possible to obtain molecular weight information and IR spectra from single beads. We would propose again that 80% of such spectra should be consistent with predicted members of the library.

NMR has not been used to characterize libraries to date. But the method has a long history of application to mixture analysis (e.g., petroleum). A proton NMR of a library could be an excellent fingerprint. It would reveal characteristic groups of resonances for common fragments in a library. It would reveal the presence of uncleaved protecting groups. A reasonable quality spectrum should be obtainable on 1 micromole of mixture.

CONCLUSIONS AND RECOMMENDATIONS

We have reviewed the analytical literature of SPS and combinatorial chemistry. Many useful methods of analysis have been developed over the last 30 years and the pace of discovery is accelerating rapidly. The use of these tools needs to be rigorously applied to those published solid-phase intermediates that will find general usefulness. Specifically, all single-component SPS products, such as novel linkers prepared on bead, should be well characterized for loading, particle size distribution, bead and linker structures, ligand structures), and percent composition.

Quality assurance of combinatorial library production relies heavily on a rigorous approach to rehearsal and compositional analysis of small model libraries. This rehearsal assures that all of the predicted structures in a large library should be present. But the value of compositional analysis and finger-printing of large libraries is too unclear at this point to promulgate strict standards for the compositional quality control of libraries. The available wet chemical and spectroscopic tools should be used to their fullest to build a body of knowledge as we move forward into the age of combinatorial drug discovery.

REFERENCES

1. Instructions for authors. (1995): *J Am Chem Soc* 117:9A–113A.
2. Instructions for authors. (1995): *J Org Chem* 60:7A–10A.
3. Instructions for authors. (1995): *J Med Chem* 38:7A–12A.
4. Gallop MA, Barrett RW, Dower WJ, Fodor SPA, Gordon EM (1994): Applications of combinatorial chemistry to drug discovery, 1: background and peptide combinatorial libraries. *J Med Chem* 37:1233–1251.
5. Gordon EM, Barrett RW, Dower WJ, Fodor SPA, Gallop MA (1994): Applications of combinatorial chemistry to drug discovery, 2: combinatorial organic synthesis, library screening strategies and future directions. *J Med Chem* 37:1385–1401.
6. Lebl M, Krchnak V, Sepetov NF, Seligmann B, Strop P, Felder S, Lam KS (1995): One-bead–one-structure combinatorial libraries. *Biopolymers* 37:177–198.
7. Atherton ES, Sheppard RC (1989): *Solid-Phase Peptide Synthesis: A Practical Approach.* New York: IRL, p 203.
8. Crowley JI, Rapoport H (1976): Solid-phase organic synthess: novelty or fundamental concept? *Acct Chem Res* 9:135–144.
9. Frechet JMJ (1981): Synthesis and applications of organic polymers as supports and protecting groups. *Tetrahedron* 37:663–683.
10. Frechet JM, Schuerch C (1971): Solid-phase synthesis of oligosaccharides, 1: preparation of the solid support. Poly[*p*-(1-propen-3-ol-1-yl)styrene]. *J Am Chem Soc* 93:492–496.
11. Kent SBH, Mitchell AR, Engelhard M, Merrifield RB (1979): Mechanisms and prevention of trifluoroacetylation in solid-phase peptide synthesis. *Proc Natl Acad Sci* 76:2180–2184.
12. Merrifield RB (1963): Solid-phase peptide synthesis, 1: the synthesis of a tetrapeptide. *J Am Chem Soc* 85:2149–2154.
13. Mitchell AR, Kent SBH, Engelhard M, Merrifield RB (1978): A new synthetic route to *tert*-butyloxycarbonylaminoacyl-4-(oxymethyl)phenylacetamidomethyl resin: an improved support for solid-phase peptide synthesis. *J Org Chem* 43:2845–2852.
14. Holmes CP, Jones DG (1995): Reagents for combinatorial organic synthesis: development of a new *o*-nitrobenzyl photolabile linker for solid-phase synthesis. *J Org Chem* 60:2318–2319.

15. Hiroshige M, Hauske JR, Zhou P (1995): Formation of C–C bond in solid-phase synthesis using the Heck reaction. *Tetrahedron Lett* 36:4567–4570.

16. Thompson LA, Ellman JA (1994): Straightforward and general method for coupling alcohols to solid supports. *Tetrahedron Lett* 35:9333–9336.

17. Rodionov IL, Baru MB, Ivanov VT (1992): A swellographic approach to monitoring continuous-flow solid-phase peptide synthesis. *Pept. Res.* 5:119–125.

18. Szymonifka MJ, Chapman KT (1995): Magnetically manipulable polymeric supports for solid-phase organic synthesis. *Tetrahedron Lett* 36:1597–1600.

19. McCrone WC (1991): Light microscopy. In Rossiter BW, Hamilton JF, eds. *Physical Methods of Chemistry,* vol 4. New York: Wiley, p 379.

20. Barth HG, Flippen RB (1995): Particle size analysis. *Anal Chem* 67:257–272.

21. Duke Scientific, 2463 Faber Place, PO Box 50005, Palo Alto, CA 94303.

22. Cheronis NC, Ma TS (1964): *Organic Functional Group Analysis by Micro and Semimicro Methods.* New York: Wiley Interscience.

23. Ma TS, Rittner RC (1979): *Modern Organic Elemental Analysis.* New York: Dekker.

24. Ashworth MRF (1965): *Titrametric Organic Analysis, Part I: Direct Analysis; Part II: Indirect Analysis.* New York: Wiley Interscience.

25. Beebe X, Schore NE, Kurth MJ (1992): Polymer-supported synthesis of 2,5-disubstituted tetrahydrofurans. *J Am Chem Soc* 114:10061–10062.

26. Kick EK, Ellman JA (1995): Expedient method for the synthesis of aspartic acid protease inhibitors directed toward the generation of libraries. *J Med Chem* 38:1427–1430.

27. Moon H, Schore NE, Kurth MJ (1992): A polymer-supported chiral auxiliary applied to the iodolactonization reaction: preparation of γ-butyrolactones. *J Org Chem* 57:6088–6089.

28. Forman FW, Sucholeiki I (1995): Solid-phase synthesis of biaryls via the Stille reaction. *J Org Chem* 60:523–528.

29. Kaiser E, Colescott RL, Bossing CD, Cook PI (1970): Color test for detection of free terminal amino groups in the solid-phase synthesis of peptides. *Anal Biochem* 34:595–598.

30. Chu SS, Reich SH (1995): NPIT: a new reagent for quantitatively monitoring reactions of amines in combinatorial synthesis. *Bioorg Med Chem Lett* 5:1053–1058.

31. Hodges RS, Merrifield RB (1975): Monitoring of solid-phase peptide synthesis by an automated spectrophotometric picrate method. *Anal Biochem* 65:241–272.

32. Hancock WS, Battersby JE (1976): A new micro-test for the detection of incomplete coupling reactions in solid-phase peptide synthesis using 2,4,6-trinitrobenzenesulphonic acid. *Anal Biochem* 71:260–264.

33. Krchnak V, Vagner J, Safar P, Lebl M (1988): Noninvasive continuous monitoring of solid-phase peptide synthesis by acid-base indicator. *Coll Czech Chem Commun* 53:2542–2548.

34. Gisin BF (1972): The monitoring of reactions in solid-phase peptide synthesis with picric acids. *Anal Chim Acta* 58:248–249.

35. Ellman G (1959): Tissue Sulfhydryl Groups. *Arch Biochem Biophys* 82:70–72.

36. Kurth MJ, Ahlberg Randall LA, Chen C, Melander C, Miller RB, McAlister K, Reitz G, Kang R, Nakatsu T, Green C (1994): Library-based lead compound

discovery: antioxidants by an analogous synthesis/deconvolutive assay strategy. *J Org Chem* 59:5862–5864.

37. Hauske JR, Dorff P (1995): A solid-phase CBZ chloride equivalent: a new matrix specific linker. *Tetrahedron Lett* 36:1589–1592.

38. Crowley JI, Rapoport H (1980): Unidirectional Dieckmann cyclizations on a solid and in solution. *J Org Chem* 45:3215–3227.

39. Terrett NK, Bojanic D, Brown D, Bungay PJ, Gardner M, Gordon DW, Mayers CJ, Steele J (1995): The combinatorial synthesis of a 30,752-compound library: discovery of SAR around the endothelin antagonist, FR-139,317. *Bioorg Med Chem Lett* 5:917–922.

40. Yan B, Kumaravel G, Anjaria H, Wu A, Petter RC, Jewell CF Jr, Wareing JR (1995): Infrared spectrum of a single resin bead for real-time monitoring of solid-phase reactions. *J Org Chem* 60:5736–5738.

41. Pon RT (1994): Tips for oligonucleotide synthesis. *ABRF News* 5:14–18.

42. Campbell DA, Bermak JC (1994): Solid-phase synthesis of peptidylphosphonates. *J Am Chem Soc* 116:6039–6040.

43. Kalejs, U, Gibiete I, Aukone G (1992): Qualitative control of *t*-butoxycarbonyl group removal in SPPS. In Epton R, ed. *Innovative and Perspectives in Solid-Phase Synthesis: Peptides, Polypeptides and Oligonucleotides.* New York: Oxford University Press, p 415–417.

44. Hofstraat JW, van Houwelingen GDB, van der Tol EB (1994): Determination of functional groups on particle surfaces by fluorescence and reflection spectroscopy. *Anal Chem* 66:4408–4415.

45. Aubagnac JL, Calmes M, Daunis J, El Amrani B, Jacquier R (1985): Non-destructive monitoring of solid phase peptide synthesis by mass spectrometry In *9th Am Pept Symp.* p 277–280.

46. van Veelen PA, Tjaden UR, van der Greef J (1991): Direct molecular weight determination of resin-bound oligopeptides using ^{252}Cf plasma-desorption mass spectrometry. *Rapid Commun Mass Spectrom* 5:565–568.

47. Brummel CL, Lee INW, Zhou Y, Benkovic SJ, Winograd N (1994): A mass spectrometric solution to the address problem of combinatorial libraries. *Science* 264:399–402.

48. Griffin P, Konteatis Z, Tracey K (1995): In *Proceedings of the 43rd ASMS Conference on Mass Spectrometry and Related Topics, Atlanta, Georgia,* p 490.

49. Egner BJ, Langley FJ, Bradley M (1995): Solid-phase chemistry: direct monitoring by matrix-assisted laser desorption ionization time of flight mass spectrometry—a tool for combinatorial chemisty. *J Org Chem* 60:2652–2653.

50. Wagner DS, Brown BB, Geysen HM (1995): In *Proceedings of the 43rd ASMS Conference on Mass Spectrometry and Related Topics, Atlanta, Georgia,* p 488.

51. Blossey EC, Cannon RG, Ford WT, Periyasamy M, Mohanraj S (1990): Synthesis, reactions, and ^{13}C FT NMR spectroscopy of polymer-bound steroids. *J Org Chem* 55:4664–4668.

52. Look GC, Holmes CP, Chinn JP, Gallop MA (1994): Methods in combinatorial organic synthesis: the use of fast ^{13}C NMR analysis for gel-phase reaction monitoring. *J Org Chem* 59:7588–7590.

53. Hobbs Dewitt S, Kiely JS, Stankovic CJ, Schroeder MC, Reynolds Cody DM, Pavia MR (1993): An approach to nonpeptide, nonoligomeric chemical diversity. *Proc Natl Acad Sci* 90:6909–6913.

54. Fitch WL, Detre G, Holmes CP, Shoolery JN, Keifer PA (1994): High-resolution ¹H NMR in solid-phase organic synthesis. *J Org Chem* 59:7955–7956.

55. Anderson RC, Jarema MA, Shapiro MJ, Stokes JP, Ziliox M (1995): Analytical chemistry in combinatorial chemistry: MAS CH correlation in solvent-swollen resin. *J Org Chem* 60:2650–2651.

56. Anderson RC, Stokes JP, Shapiro MJ (1995): Structure determination in combinatorial chemistry: utilization of magic-angle spinning HMQC and TOCSY NMR spectra in the structure determination of Wang-bound lysine. *Tetrahedron Lett* 36:5311–5314.

57. Bunin BA, Plunkett MJ, Ellman JA (1994): The combinatorial synthesis and chemical and biological evaluation of a 1,4-benzodiazepine library. *Proc Natl Acad Sci* 91:4708–4712.

58. Zuckermann RN, Martin EJ, Spellmeyer DC, Stauber GB, Showmaker KR, Kerr JM, Figliozzi GM, Goff DA, Siani MA, Simon RJ, Banville SC, Brown EG, Wang L, Richter LS, Moos WH (1994): Discovery of nonamolar ligands for 7-transmembrane G-protein-coupled receptors from a diverse N-(substituted)glycine peptoid library. *J Med Chem* 37:2678–2685.

59. Gordon DW, Steele J (1995): Reductive alkylation on a solid phase: synthesis of a piperazine combinatorial library. *Bioorg Med Chem Lett* 5:47–50.

60. Chen C, Ahlberg Randall LA, Miller RB, Jones AD, Kurth MJ (1994): Analogous organic synthesis of small-compound libraries: validation of combinatorial chemistry in small molecule synthesis. *J Am Chem Soc* 116:2661–2662.

61. Carell T, Wintner EA, Sutherland AJ, Rebek J Jr, Dunayevskiy YM, Vouros P (1995): New promise in combinatorial chemistry: synthesis, characterization, and screening of small-molecule libraries in solution. *Chem Biol* 2:171–183.

62. Metzger JW, Kempter C, Wiesmueller K, Jung G (1994): Electrospray mass spectrometry and tandem mass spectrometry of synthetic multicomponent peptide mixtures: determination of composition and purity. *Anal Biochem* 219:261–277.

63. Smith PW, Lai JYQ, Whittington AR, Cox B, Houston JG, Stylli CH, Banks MN, Tiller PR (1994): Synthesis and biological evaluation of a library containing potentially 1600 amides/esters: a strategy for rapid compound generation and screening. *Bioorg Med Chem Lett* 4:2821–2824.

64. Dunayevskiy Y, Vouros P, Carell T, Wintner EA, Rebek J Jr (1995): Characterization of the complexity of small-molecule libraries by electrospray ionization mass spectrometry. *Anal Chem* 67:2906–2915.

65. Boughtflower RJ, Underwood T, Paterson CJ (1995): Capillary electrochromatography: some important considerations in the preparations of packed capillaries and the choice of mobile phase buffers. *Chromatographia* 40:329–335.

66. Gordon DB, Lord GA, Jones DS (1994): Development of packed capillary column electrochromatography/mass spectrometry. *Rapid Commun Mass Spectrom* 8:544–548.

67. Stevanovic S, Wiesmueller K, Metzger J, Beck–Seckinger AG, Jung G (1993): Natural and synthetic peptide pools: characterization by sequencing and electrospray mass spectrometry. *Bioorg Med Chem Lett* 3:431–436.

68. Shah N, Fitch WF (1995): In *43rd American Society Mass Spectrometry and Allied Topics, Atlanta, Georgia*, p 909.

69. Andrews P, Boyd J, Ogorzalek Loo R, Zhao R, Zhu C-Q, Grant K, Williams S (1994): Synthesis of uniform peptide libraries and methods for physico-chemical analysis. In Crabb JW, ed. *Techniques in Protein Chemistry V*. San Diego, CA: Academic.

70. Carell T, Wintner EA, Bashir–Hashemi A, Rebek J Jr (1994): A novel procedure for the synthesis of libraries containing small organic molecules. *Angew Chem Int Ed Engl* 33:2059–2061.

19

QUANTIFYING DIVERSITY

Yvonne C. Martin, Robert D. Brown,
and Mark G. Bures

Pharmaceutical Products Division, Abbott Laboratories, Abbott Park, Illinois

As we think about molecular diversity in drug discovery it is important to remember the context in which we will use it. One objective of molecular diversity is to increase the chance that one's company will accrue a variety of valid patents. Since 2D structure diagrams of compounds form the basis of patents, 2D structural diversity in the compounds tested for biological activity is important. However, biological macromolecules bind most strongly to ligands with complementary 3D properties. As a result, to find active compounds in every screen, the compounds tested must include many combinations of such 3D properties. This dichotomy between diversity in 2D structure and homology in 3D properties is important as measures of molecular diversity are considered.

A quantitative measure of molecular diversity is necessary if one is to use the computer to select an optimal subset of molecules from a larger set. The goal is to optimize the amount of information gained per compound (synthesized and) tested. For example, one might wish to choose from a collection of existing molecules those to test for a specific biological activity; to select the optimal set of precursors for a combinatorial library or other series to follow up a lead; or to organize the thousands of molecules suggested by a computer de novo design program. A quantitative measure of diversity can also be used to compare one set of compounds with another: Is library

Combinatorial Chemistry and Molecular Diversity in Drug Discovery, Edited by
Eric M. Gordon and James F. Kerwin, Jr.
ISBN 0-471-15518-7   Copyright © 1998 by Wiley-Liss, Inc.

A or library B more diverse? Additionally, a quantitative measure of diversity can help one decide if it is more effective to make another variant of a known library or to move on to another core structure.

Workers interested in quantitative structure–activity relationships (QSAR) developed precedents for much of the current work on compound selection. In the 1970s Hansch et al.[1] pointed out that one can gain the maximum information per compound if one designs the set to span the range of physical properties of interest while maintaining no correlation between the properties. They described the approximately 100 available substituents for an aromatic ring by their electronic (Hammett σ), steric, and lipophilic (Hansch–Fujita π) values. They used hierarchical cluster analysis to group the substituents into 10 clusters and suggested that a good series would contain one member from each cluster. Such series are indeed more diverse than typical series designed without such considerations.[2] These notions were much later extended to 3D molecular field descriptors.[3] Wootton et al.[4] reported on a computer program that selects compounds so that each is a reasonable but not excessive distance from every other in multidimensional physical property space. Austel suggested that one should use the fractional factorial design method to ensure that the molecules span all physical properties of interest.[5]

In the 1980s, Willett and colleagues clustered compounds, not based on physical properties, but using the substructural descriptors.[6,7] They clustered sets of 50–200 compounds with known biological activities and asked whether the clustering would group actives with actives. The clustering was effective and they found little difference between clustering methods. Workers at Pfizer then clustered the 8500 compounds that were available in usable quantitites and chose one compound from each cluster to be added to their representatives collection for testing in new biological screens.[7]

Although these early studies are very helpful, there are still a number of issues to be resolved for larger datasets. Clearly, one cannot ignore physical properties of molecules if only to require that they be soluble enough to test. However, the properties that describe the subtle differences between analogs are not general enough to describe differences between molecules that don't share a common core. On the other hand, although the substructural descriptors and Jarvis–Patrick method performed well on small sets, there is no guarantee that they will be as effective on more diverse sets. Will any of the technologies derived from 3D database searching,[8] ligand docking to macromolecules,[9] or pharmacophore mapping[10] be useful for molecular diversity considerations?

To be of use in the problem of combinatorial libraries or selecting compounds for high-throughput screening, a computer method must be fast enough to handle hundreds of thousands of structures. It must also be accurate enough that one can demonstrate that using the method will improve the efficiency of discovering active compounds.

COMPUTER-BASED DESCRIPTORS OF MOLECULES

Physical Properties

The long tradition of QSAR[11] suggests that diversity considerations include the physical properties of the molecules. Although much progress has been made, it is still not possible to calculate for many molecules such properties as octanol–water log P or pK_a. If the calculation fails on even 10% of the compounds, these compounds are lost from consideration.

In spite of these difficulties, two groups explicitly consider physical properties in their diversity analysis.[12,13] Workers at Rhône–Poulenc–Rorer calculated 50 properties and used statistical analysis to select a subset of properties for further consideration.[12] They then partitioned their compound collection based on these properties and selected one compound from each multivariate partition for their screening set. Chiron scientists combined the physical property descriptors with others described below.[13]

We found that the 2D substructural descriptors described below encode most of the information about hydrophobicity in octanol–water and cyclohexane–water; pK_a; molecular shape, size, and flexibility; molecular connectivity indices[14]; and hydrogen-bonding character of molecules.[15] Hence, for a diversity analysis it is not necessary to include explicit information about these properties in addition to 2D substructures.

Properties That Describe 2D Structures

Connectivity of Molecules. Training in organic chemistry teaches us that the 2D structure diagram of a molecule encapsulates its reactivity and physical properties. Hence it is logical to assume that 2D substructure searching systems might provide reliable descriptors for measuring diversity. Moreover, the successes of Willett noted above suggest that some 2D descriptors might also encode enough 3D property information to be useful in predicting biological activity.[6] Existing computer programs quickly generate these substructure descriptors for every compound in a dataset. They are fast: For example, it takes about one day to generate the Daylight fingerprints[16] for 100,000 structures.

We studied a number of these substructural keys to see how much information they contain about biological activity.[17] For this purpose we investigated datasets of 1600–5000 molecules that had been tested in one of three enzyme screens. We also compared the effectiveness of various clustering strategies.

Structural Keys. The first chemical substructure searching systems introduced substructural keys to eliminate from detailed consideration those compounds that cannot possibly match the search query. The key for a structure is an array of Boolean values, each of which represents the presence or absence of a specific 2D fragment.

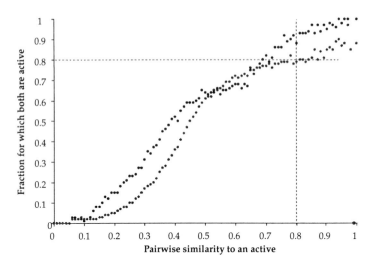

Figure 19.1. The observed probability that a compound is active given its similarity to an active compound. The biological data are derived from screening miscellaneous compounds against two different proprietary enzyme assays.

We examined 153 keys from the MACCS substructure search system. They encompass a range of types of generic and specific fragments. To quantify similarity we used the Tanimoto similarity coefficient, which is calculated as the fraction of the keys present in either of the molecules that is present in both. Figure 19-1 shows that the MACCS keys do contain information that separates active from inactive molecules. In particular, if a molecule is at least 0.85 similar to an active compound, there is an 80% chance that it will also be active. We use this 0.85 similarity number to select a clustering level: Every pair of molecules in a cluster should be at least 0.85 similar for that cluster to be biohomogenous.

Hashed Fingerprints. Hashed fingerprints are designed to describe all structures equally well and to be independent of a predefined fragment dictionary. Fingerprinting starts with a set of patterns to index. A bit string encodes all fragments in a structure that match the pattern. A pattern might, for example, be a path of length 7 bonds defined as $(atom–bond–atom)^7$, where the nature of the atoms and bonds at each point is distinguished. For example, the Daylight fingerprints encode each atom type, all augmented atoms (the central atom and the nature of the atoms and bonds attached to it), and all paths of length 2–7 nonhydrogen atoms.[16]

There are too many such fragments to permit assigning a separate bit to each fragment. Instead, a special algorithm hashes each fragment to a pattern of a few bits to set in the bit string. A given path will always set the same pattern of bits. Although the algorithm minimizes the chance that two different patterns will set exactly the same bits, it is inevitable that two different paths will set some of the same bits. Hence, a similarity measure that compares the bits set in two strings may see the same bit as set and so record a contribution to the similarity, even though that position was set by a different fragment in each of the two structures. This would distort clustering based on such finger-prints.

Recently, Patterson and coworkers described their evaluations of the relevancy of several molecular diversity descriptors to biological activity.[18] They found that using descriptors involving 2D fingerprints in similarity analyses gave generally good correlation with activity in the 20 QSAR datasets that they studied.

Molecular Connectivity Indices. Kier and Hall calculate 81 topological indices in their program MOLCONNX.[14] Some of these are related to global properties such as octanol-water log *P*, whereas others encode molecular shape and flexibility. Several groups use these and other physical properties in their analysis of diversity.[13,19]

Atom Pairs. Workers at Lederle[20] developed this descriptor to encode the types and approximate spatial distribution of features in a set of molecules. The atoms pairs are described by the atom types and the number of bonds that separate them. The atomic number, the number of nonhydrogen neighbors, and the number of π electrons distinguish the atom types. More recently this descriptor has been extended to use physicochemical properties of the atoms instead of atom types.[21]

2D Autocorrelation Analysis. This method calculates a vector based on the distances between all atoms of the structure and any property of these atoms.[22] The first step is to identify all atom pairs as described above. For each atom pair the number of bonds between them and the product of the properties is noted. Each element of the autocorrelation vector is the sum of these products for one particular distance. A separate autocorrelation vector is calculated for each property of interest, typically volume, electronegativity, hydrogen-bonding character, and hydrophobicity. Lastly, a principal components analysis reduces the number of variables. The distribution of molecules in these principal components reflects the diversity of a sample.

Atom Layers. Workers at Chiron[13] developed this descriptor to capture the spatial distribution of properties of potential precursors for a combinatorial library. Each atom layer is labeled as to its distance in number of bonds from the common backbone. Thus atom layer 1 contains the atoms that bonded directly to the backbone; layer 2, the atoms bonded to atom layer 1, and so on, up to 15 layers. Six attributes describe the receptor recognition properties of the atoms in each layer: the number of hydrogen bond donors and acceptors, the number of potential positive and negative charges, radius, and aromatic character.

Properties That Describe the 3D Structures of Molecules

It is important to consider 3D properties of molecules in a diversity analysis even though the molecules usually are entered into the computer by their 2D structures. Several immediate problems then face us: How will we generate the 3D structures of the molecules? How will we handle conformational flexibility? How will we generate the properties? How will we encode this information so that the computer can efficiently handle it? This aspect of computational molecular diversity is under intense investigation by many groups.

Generating the 3D Structures. The appearance of 3D structure searching programs has hastened the development of fast and accurate methods to generate the 3D structures of molecules from their 2D structures. Generally these convert ~90% of the structures of a database.[23] This takes less than a second per structure.

The larger issue is how to treat conformational flexibility. Some groups systematically generate "all" or "representative" conformations for 3D database searching purposes and encode the distances between key atoms in the bit string.[24] Even a rate of one per minute (1440 per day) is painfully slow for a large database.

Other 3D searching programs generate the matching conformation during searching.[25,26] Upon loading a structure, such software calculates only a distance bounds matrix between interesting atoms.

If we chose to include conformational flexibility, should we penalize higher-energy conformers? How do we even accurately calculate conformational energy? Should we consider each conformer as a distinct molecule? Should we penalize conformational flexibility since if a flexible molecule is active in a screen it might be very difficult to discover its bioactive conformation? Or should we penalize rigid compounds since they are less likely to be active in any given screen? These questions are raised, but no clear answers have been agreed on.

Descriptors Based on the Distances or Angles Between Atoms or Other Points. 3D structure searching systems provide access to screens that describe the distance between atoms, centroids of phenyl rings, points on the normal to the plane of a phenyl ring, and extension points of carbonyl oxygens. In such systems the bit map is divided into regions that correspond to particular pairs of atoms, N-O, for example. Each region is divided into bits that correspond to a particular distance range of that type of pair. Rigid screens record these distances for the input structure, and flexible screens (where the program generates the matching structure during searching) record all possible distances between particular pairs of points.

Bartlett described the diversity of different combinatorial cores by the angles between the bond vectors connecting the core to the substituent.[27]

The results were presented visually. In a related approach, a program called HookSpace evaluates diversity by looking at distances between user-defined functional groups.[28] One of the analyses involves calculating a database's coverage of a theoretical pairwise distance space. It was found that the *Available Chemicals Directory*[55] covered 85% of this space, while the Cambridge Structural Database and a benzodiazepam-derived combinatorial library covered only 34 and 13%, respectively.

Descriptors Based 3D Properties. Sheridan[29] recognized that prescreens in 3D searching will be most effective if each molecule is described in terms that reflect its potential for intermolecular interactions. The computer programs must recognize each hydrogen-bond acceptor and donor site, each potentially charged site, and often each hydrophobic site in the molecule in the 3D structure. Tautomers should be recognized and simple pK_a estimations made.

The distances or pharmacophoric triangles of these features may then be encoded in a bit string. While encoding pair distances requires a reasonable length bit string, there are 48,000 possible triangles with sides 2–15 Å, with the result that hashing is required for some applications.

Several groups[12,30–33] use as one diversity criterion the number of these pharmacophoric triangles represented in a collection of compounds. We considered distance bin widths of 1 Å and the CONCORD[34] low-energy conformation only. Workers at Rhône–Poulenc–Rorer and at Chemical Design used the variable-width distance bins devised by Chemical Design Ltd. for 3D searching based on hydrogen-bond donors and acceptors, positively charged centers, and aromatic ring centers. Considering distances up to 15 Å results in 400,000 triangles. They also considered conformational flexibility.

In our strategy for precursor selection, we transform the precursors into the molecule to be synthesized, generate the 3D structure with CONCORD, identify the interacting groups in each molecule, calculate the distances between all pairs of these features, and identify families that have all features and distances between features in common.[35] The more 3D families a dataset contains, the more diverse it is.

Similarly to the 2D autocorrelation, 3D autocorrelation of properties based on distances calculated from the 3D structures can also be used to examine the diversity of a library.[22,36] The results are presented visually, not quantitatively.

Pearlman calculates BCUT values from the first and last eigenvectors of three atom association matrices.[37] Each row and column represents one atom of the structure, and the value of the cell is the product of the properties of the two atoms. Properties related to atomic charge, atomic polarizability, and hydrogen bonding were chosen for their relevance to receptor–ligand interactions.

Cramer and colleagues have recently reported on a 3D descriptor based on CoMFA steric fields.[38] The method is useful for comparing the diversity of a set of variable groups attached to a common core, as found in small-molecule combinatorial libraries. In a validation study, they showed that this descriptor correlates well with biological activity.[18]

STRATEGIES FOR ANALYZING OR DISPLAYING DIVERSITY

Clustering

Clustering remains the classic method for diversity analysis.[39] The method may be nonhierarchical or hierarchical. Different methods differ in the exact way they form the clusters. Once the algorithm forms clusters of similar compounds, to choose a diverse set one would select one member from each cluster.

Jarvis–Patrick Clustering.[40] Nonhierarchical methods produce a single set of clusters based on user-supplied criteria. The Jarvis–Patrick method has become widely used for diversity-related tasks. The first step in the algorithm is to calculate, for every compound, its n nearest neighbors, usually 14–16. Compounds cluster together if they are on each others' list of near neighbors and if they share some number of near neighbors. The major advantage of Jarvis–Patrick clustering is that is very fast: In our experience, its major disadvantage is that it produces a large number of singletons, compounds that don't cluster with any other, under strict clustering conditions, or a few very large clusters of very diverse structures with loose conditions.

In the late 1980s, workers at DuPont vectorized the nearest-neighbors program for the CRAY-SMP, which required 42 CPU hours to calculate the near-neighbor list for 287,000 compounds.[41] Jarvis–Patrick clustering required only 1–2 min on a Vax. It was found that relaxing the requirement that compounds appear in each other's neighbor lists and that requiring 6 out of 18 near neighbors best reproduces known clusters. The clustering was used to select compounds for random screening.

Hierarchical Clustering.[39] These methods produce classifications in which small clusters of similar structures are nested within larger clusters containing more diverse structures. A dendogram shows the cluster hierarchy in which at one extreme all clusters are singletons and at the other all structures are in a single cluster. The advantage of hierarchical methods is that one can select the number of clusters on a predetermined basis, such as the required similarity of compounds to each other. Newer algorithms[42] make it possible to cluster a dataset of 100,000 compounds in reasonable time.

Diversity Selection

While clustering does group similar compounds together, it does not guarantee that the clusters will optimally explore property space: Although distinct, certain clusters may be quite similar to other clusters. For this reason, scientists at Upjohn devised a procedure to directly select compounds based on their difference from all currently selected compounds.[43,44] The procedure starts with a random compound, selects the second as that most different from the first, the next as that most different from the first two, and so on, until the required number of compounds has been selected.

The time required to select a set of compounds using the original algorithm is proportional to the square of the number of compounds considered, which makes it very time-consuming even for a database of 150,000 molecules. Holliday and Willett recently described an alternative algorithm that is substantially faster with as few as 100 structures and that also scales approximately linearly with the number of compounds in the dataset.[45] Recently, this algorithm has been used to rank the diversity of several large databases by merging all pairs of the databases studied.[46] It was found that a 160,000 tetrapeptide combinatorial library (all possible combinations of the 20 natural amino acids) was much less diverse than databases made of compounds not generated combinatorially, such as National Cancer Institute screening database and the Maybridge Chemicals database.

D-Optimal Design

Scientists at Chiron[13] describe potential monomers for a combinatorial library by considering lipophilicity, shape, chemical functionality, and receptor recognition features. They use these descriptors both to design very diverse libraries, when screening for a new lead, or to design libraries of similar structures, when optimizing a lead.

Calculated octanol/water partition coefficients describe lipophilicity; topological indices describe flexibility and branching; Daylight fingerprints describe the 2D substructure; and atom layer properties describe the receptor recognition features. They use principal components analysis or multidimensional scaling to reduce the number of topological features, 2D substructural descriptors, and atom layer properties.

Flower plots display the relative values of these properties for each precursor. A flower represents each precursor: The number of petals corresponds to the number of properties and the length of each petal is proportional to the value of that property for that compound.

The D-optimal design selects diverse subsets of structures to optimally explore property space given the number of structures to be chosen and the expected functional form of the relationship (linear, parabolic, etc.). A weakness is that if one uses linear functions, then the selection emphasizes the extremes of each property: A parabolic relationship might be missed.

The D-optimal design part of the strategy is interactive and can be forced to include certain compounds such as those already made and tested. Hence, the synthetic and computational chemists work together at the computer screen to design the set.

Partitioning Descriptor Space

Many workers have realized that an ideal set of compounds would be evenly distributed in the space of the descriptors chosen.[12,19,37] Cummins et al. use factor analysis to reduce to four dimensions 61 molecular connectivity indices plus a solvation estimate.[19] The space of these four dimensions is partitioned into hypercubes of the same volume and the number of database compounds in each cube tabulated. This is used to evaluate the diversity and overlap of different databases after removing outliers.

Pearlman uses a grid of BCUT values. Again, compounds from one or more databases can be assigned to cells in the grid, and the diversity of databases can be computed and compared by examining the populations of compounds in each cell.[37]

Genetic Algorithm Design

Workers at Merck[47] use genetic algorithms to select precursors for directed libraries such that these precursors contain features discovered by a prior structure–activity analysis. A trend vector of atom pairs encodes the desired features. The genetic algorithm selects precursors at random to design a first population of compounds and scores their fitness with the QSAR equation. As natural selection continues, certain precursors come to dominate the more fit populations. The chemists then include such precursors in their combinatorial libraries.

EXAMPLE APPLICATIONS

Clustering Large Datasets to Select Compounds for Screening

Hodes and colleagues used their own fragment-based descriptors and a "leader" clustering method to group the 232,000 compounds in the National Cancer Institute's repository.[48-50] This collection forms 117,000 clusters, including 73,757 singletons. The singletons are important because testing them provides unique SAR information.

Table 19.1 shows the performance of different molecular descriptors in

TABLE 19.1. Ability of Various Descriptors to Classify Diverse Compounds Using Ward's Clustering Method

Descriptor	Percent Actives
MACCS subset fragments	60
Unity 2D and Daylight fingerprints	56
3D feature pair distances	50
Unity 3D flexible and rigid fragments	40

separating active from inactive compounds tested for inhibition of monoamine oxidase.[17] We show the level of the hierarchy at which 290 compounds, the number of actives, are in a cluster with at least one active compound. The number reported is the percentage of these molecules that are active. Ideal clustering would thus achieve 100%. The superior performance of the 2D over 3D descriptors suggests that perhaps molecular shape and the direction of hydrogen-bonding vectors should be considered in future.

Performance of Different Selection Methods

Upjohn scientists simulated the relative performance of dissimilarity cluster selection and maximum dissimilarity selection and found that their relative performance depends on the pattern of actives[51]: If there are few actives, then the maximum dissimilarity method is superior, whereas if there are many similar clusters of active molecules, then dissimilar clustering method performs better. Since one expects that screening will identify few hits, the maximum dissimilarity method is preferred.

Taylor at Zeneca compared strategies for selecting the order of testing compounds.[52] Using many combinations of computational variables, he examined two sets of 1000 molecules with simulated biological activity (one highly active and a number weakly active of varying similarity to the highly active). He first selected a fraction of the dataset by clustering and by diversity selection, then promoted for the second round of testing those compounds that are similar to the actives discovered in the first round, (positive feedback), and so on. He asked which selection method more quickly leads to discovery of the most potent compound.

Selection by clustering and positive feedback slightly increase the rate of finding the highly active compound compared to random testing: Diversity selection retards the rate because it tests singletons early, and in these simulations singletons cannot lead to the highly active compound. However, diversity selection should identify more diverse leads. As expected, when the structure–activity information contained in the descriptors was decreased by decreasing the similarity of the weakly actives to the active, neither method was an improvement over chance.

Performance of Different Clustering Methods

Table 19.2 shows the performance of different clustering methods using the subset of MACCS descriptors on the monoamine oxidase dataset. The superior performance of Ward's clustering method is clear.[17]

Comparing the Diversity of Different Sets of Molecules

Chiron scientists compared the diversity of different compound collections estimating the number of different fragments in one or more of the com-

TABLE 19.2. Ability of Various Clustering Methods to Classify Diverse Compounds Using the Subset of MACCS Descriptors

Clustering Method	Percent Actives
Ward's	60
Group average	52
Guenoche	50
Jarvis–Patrick	38

pounds:[13] All possible nucleic acids, 200; carbohydrates, 400; peptides, 500; 7 transmembrane receptor ligands, 600; peptoid polymers, 600; benzodiazepines, 800; carbamate polymers, 1200; isoquinolines, 1200; diketopiperazines, 1300; 10 best-selling drugs, 4300; 100 best-selling drugs, 5800.[13] By this measure combinatorial libraries are clearly deficient compared to typical drug collections.

We analyze databases with three measures of diversity: the number of biohomogeneous clusters; the diversity density, that is, the mean number of compounds in biohomogeneous clusters; and the coverage of 3-point pharmacophores.[53] Table 19.3 illustrates such comparisons.

TABLE 19.3. Clustering and Pharmacophore Diversity Statistics for Various Databases

Compound Set	Number Clustered	Number of Biohomogeneous Clusters[a]	Average Cluster Size	% Pharmacophores Covered[b]
Abbott, 1995	90,361	36,125	2.5	90
Benzodiazepine virtual library[c]	125,000	31,343	4.0	78
Available Chemicals Directory, 1994[55]	81,649	23,996	3.4	85
MDDR[56]	51,745	20,102	2.6	93
Optiverse[57]	41,842	17,220	2.4	64
Abbott, 1992	33,951	13,239	2.6	87
Berdy antibiotics	10,139	3,343	3.0	84
Xanthene library[58]	52,650	3,254	16.2	38
Diverse library[54]	6,727	1,322	5.1	55
Screened for inhibition of monoamine oxidase[17]	1,645	868	1.9	42
Benzene triacid library[58]	5,832	604	9.7	21

[a] MACCS fragments and Ward's clustering.
[b] 2- to 10-Å sides in 1.0-Å bins.
[c] Trisubstituted benzodiazepine-like core library designed at Abbott.

In 1992 Abbott's collection available for testing contained compounds from several hundred projects plus a small number from outside sources. By 1995 we had added 50,000 compounds from commercial suppliers. (Not all were selected by clustering, which should give clusters with one member). The new collection has a 2.7-fold increase in the number of clusters. The average cluster size and triangle coverage are similar to those of the MDDR database of marketed and potential drugs. The chemist-selected monoamine oxidase screening set shows that one can explore many 3D pharmacophores in a small set of compounds.

Combinatorial libraries in which either a xanthene tetraacid or a benzene triacid core was functionalized with a set of amines are not very diverse. However, combinatorial libraries can approach the diversity found for the small-molecule databases, as illustrated by the library recently reported by Baldwin et al.[54]

Workers at Glaxo–Wellcome published the molecular diversity and overlap of a number of databases: CMC is the listing from *Comprehensive Medicinal Chemistry* of biologically active compounds, MDDR contains biologically active compounds in early stages of drug development, ACD contains commercially available compounds, SPECS contains specialty chemicals, and WR contains the compounds in the *Wellcome Registry*. CMC occupies 27% of the subcubes occupied by any database and 36% of those occupied by MDDR, a much larger database. The *Wellcome Registry* occupies the most subcubes and overlaps substantially with both the biological and chemical databases.

In related work, Shemetulskis and coworkers used clustering and physical property analysis to compare the diversity of the Parke-Davis database to subsets of a *Chemical Abstracts* structural database (CAS) and the Maybridge Chemicals database.[59] They combined the corporate database with the database of interest, performed substructural characterization and clustering, then analyzed the composition of the clusters. For example, they found that 78% of the compounds from the CAS database and 53% of the compounds from the Maybridge database were placed in clusters containing only CAS or Maybridge compounds, respectively. This analysis is especially useful for compound acquisition because it identifies compounds that are structurally different from those already present in a database. Using these combined databases and a histograming strategy, they also compared calculated log P, Molar refractivity, and dipole moment.

Note: Since the writing of this chapter, a more recent review of computational methods in molecular diversity has become available.[60]

REFERENCES

1. Hansch C, Unger SH, Forsythe AB (1973): Strategy in drug design: cluster analysis as an aid in the selection of substituents. *J Med Chem* 16:1212–1222.

2. Martin YC, Panas HN (1979): Mathematical considerations in series design. *J Med Chem* 22:784–791.

3. Lin CT, Pavlik PA, Martin YC (1990): Use of molecular fields to compare series of potentially bioactive molecules designed by scientists or by computer. *Tetrahedron Comp Method* 3:723–738.

4. Wootton R, Cranfield R, Sheppey GC, Goodford PJ (1975): Physicochemical-activity relationships in practice, 2: rational selection of benzenoid substituents. *J Med Chem* 18:607–613.

5. Auste V (1995): Experimental design in synthesis planning and structure–property correlations. In Van de Waterbeemd H, ed. *Chemometric Methods in Molecular Design.* Weinheim: VCH, pp 49–62.

6. Willett P (1987): *Similarity and Clustering Techniques in Chemical Information Systems.* Letchworth: Research Studies.

7. Willett P, Winterman V, Bawden D (1986): Implementation of nonhierarchic cluster analysis methods in chemical information systems: selection of compounds for biological testing and clustering of substructure search output. *J Chem Inf Comput Sci* 26:109–118.

8. Martin YC, Bures MG, Willet P (1990): Searching databases of three-dimensional structures. In Lipkowitz KB, Boyd DB, eds. *Reviews in Computational Chemistry,* vol. 1. New York: VCH, pp 213–263.

9. Kuntz ID (1992): Structure-based strategies for drug design and discovery. *Science* 257:1078–1082.

10. Martin YC, Bures MG, Danaher EA, Delazzer J, Lico I, Pavlik PA (1993): A fast new approach to pharmacophore mapping and its application to dopaminergic and benzodiazepine agonists. *J Computer-Aided Mol Design* 7:83–102.

11. Hansch C, Leo A (1995): *Exploring QSAR: Fundamentals and Applications in Chemistry and Biology.* Washington, DC: American Chemical Society.

12. Lewis R, Mclay I, Mason J (1995): Diverse property-derived sets: a novel method for selecting representative screening sets using molecular and physicochemical properties. *Chemical Design Automation News* 10(4):37–38.

13. Martin EJ, Blaney JM, Siani MA, Spellmeyer DC, Wong AK, Moos WH (1995): Measuring diversity: experimental design of combinatorial libraries for drug discovery. *J Med Chem* 38:1431–1436.

14. Hall LH, Kier LB (1991): The molecular connectivity chi indexes and kappa shape indexes in structure–property modeling. In Lipkowitz KB, Boyd DB, eds. *Reviews in Computational Chemistry,* vol 2, New York: VCH, pp 367–422.

15. Brown RD, Martin YC (1997): The information content of 2D and 3D structural descriptors relevant to ligand-receptor binding. *J Chem Inf Comput Sci,* 37:1–9.

16. Daylight Chemical Information Systems, Inc, 3951 Claremont Street, Irvine, CA 92714.

17. Brown RD, Martin YC (1996): Use of structure-activity data to compare structure-based clustering methods and descriptors for use in compound selection. *J Chem Inf Comput Sci* 36:572–584.

18. Patterson DE, Cramer RD, Ferguson AM, Clark RD, Weinberger LE (1996): Neighborhood behavior: a useful concept for validation of molecular diversity descriptors. *J Med Chem* 39:3049–3059.

19. Cummins DJ, Andrews CW, Bentley JA, Cory M (1996): Molecular diversity in chemical databases: comparison of medicinal chemistry knowledge bases and databases of commercially available compounds. *J Chem Inf Comput Sci* 36:750–763.

20. Carhart RE, Smith DH, Venkataraghavan R (1985): Atom pairs as molecular features in structure–activity studies: definition and applications. *J Chem Inf Comput Sci* 25:64–73.

21. Kearsley SK, Sallamack S, Fluder EM, Andose JD, Mosley RT, Sheridan RP (1995): Chemical similarity using physicochemical property descriptors. *J Chem Inf Comput Sci* 36:118–127.

22. Moreau G, Turpin C (1996): Use of similarity analysis to reduce large molecular libraries to smaller sets of representative molecules. *Analysis* 24:M17–M22.

23. Sadowski J, Gasteiger J, Klebe G (1994): Comparison of automatic three-dimensional model builders using 639 x-ray structures. *J Chem Inf Comput Sci* 34:1000–1008.

24. Murral NW, Davies EK (1990): Conformational freedom in 3-D databases, 1: techniques. *J Chem Inf Comput Sci* 30:312–316.

25. Hurst T (1994): Flexible 3D searching: the directed tweak technique. *J Chem Inf Comput Sci* 34:190–196.

26. Guner OS, Hughes DW, Dumont LM (1991): An integrated approach to three-dimensional information management with MACCS-3D. *J Chem Inf Comput Sci* 31:408–414.

27. Bartlett PA (1996): The caveat vector approach for structure-based design and combinatorial chemistry. *Abstracts of Papers of the American Chemical Society,* p 211.

28. Boyd SM, Beverley M, Norskov L, Hubbard RE (1995): Characterizing the geometric diversity of functional groups in chemical databases. *J Computer-Aided Mol Design* 9:417–424.

29. Sheridan RP, Nilakantan R, Rusinko A, Bauman N, Haraki K, Ventataraghavan R (1989): 3DSEARCH: a system for three-dimensional substructure searching. *J Chem Inf Comput Sci* 29:255–260.

30. Ashton MJ, Jaye MC, Mason JS (1996): New perspectives in lead generation, 2: evaluating molecular diversity. *Drug Discovery Today* 1:71–78.

31. Mason JS, Mclay IM, Lewis RA (1995): Applications of computer-aided drug design techniques to lead generation. In Dean PM, Jolles G, Newton CG, eds. *New Perspectives in Drug Design.* London: Academic, pp 225–253.

32. Pickett SD, Mason JS, McLay IM (1996): Diversity profiling and design using 3D pharmacophores: pharmacophore-derived queries (PDQ). *Journal of Chemical Information and Computer Sciences* 36:1214–1223.

33. Davies K, Briant C (1995): Combinatorial chemistry library design using pharmacophore diversity. *Network Sci* (http://www/awod.com/netsci/issues/).

34. *Concord: A Program for the Rapid Generation of High-Quality Approximate 3-Dimensional Molecular Structures.* Austin: The University of Texas; St. Louis, MO: Tripos Associates.

35. Brown RD, Bures MG, Danaher E, Lico I, Martin YC, Pavlik PA, Cesarone J, Chen R, Delazzer J, Hawe W, Liu M (1996): Structure based diversity selection

of precursors for combinatorial libraries. In *First MGMS Electronic Conference.* Molecular Graphics and Modeling Society (http://belatrix.pcl.ox.ac.uk/mgms/).

36. Sadowski J, Wagener M, Gasteiger J (1996): Assessing similarity and diversity of combinatorial libraries by spatial autocorrelation functions and neural networks. *Angew Chem Int Ed Engl* 34:23–24.

37. Pearlman RS (1996): Novel software tools for addressing chemical diversity. *Network Sci* (http://www/awod.com/netsci/issues/).

38. Cramer RD, Clark RD, Patterson DE, Ferguson AM (1996): Bioisosterism as a molecular diversity descriptor: steric fields of single topomeric conformers. *J Med Chem* 39:3060–3069.

39. Downs GM, Willett P (1994): Clustering in chemical-structure databases for compound selection. In Van de Waterbeemd H, ed. *Chemometric Methods in Molecular Design.* Weinheim: VCH, pp 111–130.

40. Jarvis RA, Patrick EA (1973): Clustering using a similarity measure based on shared nearest neighbors. *IEEE Trans Comput* C-22:1025–1034.

41. Blaney JM, Capobianco P, Eyermann CJ (1989): Clustering the Dupont compound collection; personal communication.

42. Downs GM, Willett P, Fisanick W (1994): Similarity searching and clustering of chemical structure databases using molecular property data. *J Chem Inf Comput Sci* 34:1094–1102.

43. Lajiness MS, Johnson MA, Maggoria GM (1989): Implementing drug screening programs using molecular similarity methods. In Fauchere JL, ed. *QSAR: Quantitative Structure–Activity Relationships in Drug Design.* New York: Liss, pp 173–176.

44. Johnson M, Lajiness M, Maggiora GM (1989): Molecular similarity: a basis for designing drug screening programs. In Fauchere JL, ed. *QSAR: Quantitative Structure–Activity Relationships in Drug Design.* New York: Liss, pp 167–171.

45. Holliday JK, Willett P (1995): Definitions of dissimilarity for dissimilarity based compound selection. *Quant struct-act relationship* 14:501–506.

46. Turner DB, Tyrrell SM, Willett P (1997): Rapid quantification of molecular diversity for selective database acquisition. *J Chem Inf Comput Sci* 37:18–22.

47. Sheridan RP, Kearsley SK (1995): Using a genetic algorithm to suggest combinatorial libraries. *J Chem Inf Comput Sci* 35:310–320.

48. Hodes L (1989): Clustering a large number of compounds, 1: establishing the method on an initial sample. *J Chem Inf Comput Sci* 29:66–71.

49. Whaley R, Hodes L (1991): Clustering a large number of compounds, 2: Using the connection machine. *J Chem Inf Comput Sci* 31:345–347.

50. Hodes L, Feldman A (1991): Clustering a large number of compounds, 3: the limits of classification. *J Chem Inf Comput Sci* 31:347–350.

51. Lajiness MS (1991): An evaluation of the performance of dissimilarity selection. In Silipo C, Vittoria A, eds. *QSAR: Rational Approaches to the Design of Bioactive Compounds.* Amsterdam: Elsevier Science, pp 201–204.

52. Taylor R (1995): Simulation analysis of experimental design strategies for screening random compounds as potential new drugs and agrochemicals. *J Chem Inf Comput Sci* 35:59–67.

53. Bures MG, Brown R, Martin YC (1995): Analyzing larger databases to increase the diversity of the Abbott corporate compound collection. *Abstracts of papers of the American Chemical Society,* p 210, 60-Cinf.

54. Baldwin JJ, Burbaum JJ, Henderson I, Ohlmeyer MHJ (1995): Synthesis of a small molecule combinatorial library encoded with molecular tags. *J Am Chem Soc* 117:5588–5589.

55. *Available Chemicals Directory, 1994.* San Leandro, CA: MDL Information Systems, Inc.

56. *MDDR: MACCS-II Drug Data Report, 1995.* San Leandro, CA: MDL Information Systems, Inc.

57. Ferguson AM, Patterson DE, Garr CD, Underiner TL (1996): Designing chemical libraries for lead discovery. *J Biomol Screening* 1:65.

58. Carell T, Wintner EA, Sutherland AJ, Rebek J Jr, Dunayevskiy YM, Vouros P (1995): New promise in combinatorial chemistry: synthesis, characterization, and screening of small-molecule libraries in solution. *Chem Biol* 2:171–183.

59. Shemetulskis NE, Dunbar JB, Dunbar BW, Moreland DW, Humblet C (1995): Enhancing the diversity of a corporate database using chemical database clustering and analysis. *J Computer-Aided Mol Design* 9:407–416.

60. Bures, MG, Martin YC (1998): Computational methods in molecular diversity and combinatorial chemistry. *Curr Op Chem Biol* Vol 2/3, in press.

PART IV

BIOLOGICAL DIVERSITY

20

PROTEIN SCAFFOLDS FOR PEPTIDE LIBRARIES

RONALD H. HOESS

Dupont–Merck Pharmaceutical Company, Wilmington, Delaware

The ability to synthesize molecular diversity biologically is an important new tool for drug discovery. The means for generating large peptide libraries biologically has sparked considerable interest in using these as a source for lead compounds. While peptides themselves have basically been avoided as drugs because of their poor pharmacological properties—short half-life and poor bioavailability—they nevertheless make useful starting points for non-peptidic drugs. While moving from peptide to nonpeptide is a significant hurdle, the development of rapid structural determination along with computational methods makes this an attractive approach toward drug discovery.

The advent of phage-displayed peptide libraries represented a straightforward technique for assembling and screening vast libraries of random peptides.[1-3] The initial libraries, like their chemical peptide library counterparts, had the drawback that the displayed peptides are unconstrained and extremely flexible. This presents two problems. First, since the peptide is sampling a large number of conformational states, the entropic cost upon binding the target molecule will necessarily be high. This is an important consideration if the final goal is to find a peptide that shows high selectivity and affinity. The second problem is the ability to derive meaningful structural information to guide the rational design of a nonpeptide lead. Unconstrained peptides are not amenable to either structural determination or molecular modeling and

Combinatorial Chemistry and Molecular Diversity in Drug Discovery, Edited by
Eric M. Gordon and James F. Kerwin, Jr.
ISBN 0-471-15518-7   Copyright © 1998 by Wiley-Liss, Inc.

thus their value is diminished in terms of obtaining information necessary for rational design. One possible solution to this problem would be to start with peptides that are in a constrained conformation.

A number of attempts have been made to constrain bioactive peptides on protein scaffolds. Starting with known three-dimensional structures of the scaffold protein, peptides were grafted on to the structure by inserting the peptide-encoding DNA into the gene for the scaffold. A number of scaffolds were tested, including an antibody variable domain,[4-7] lysozyme,[8,9] and alkaline phosphatase.[10] The approach has had limited success depending on where the peptide was inserted into the scaffold. It underscores how sensitive the bioactivity of a given peptide is to the conformation dictated by the scaffold as well as the peptide's surface accessibility. Nevertheless, when successful, these experiments immediately translate into the possibility of determining the bioactive conformation of the peptide. The general utility of this approach appeared limited since it required inserting a given peptide-encoding sequence by trial and error into various regions of the scaffold. What was lacking was a method by which libraries of peptides could be quickly examined in the context of a protein scaffold. The demonstration that protein domains could be displayed on the surface of phage provided the technological breakthrough to make this approach feasible.

In a sense nature has already confronted this problem and evolved a number of solutions. Antibody molecules can be thought of as scaffolds with a series of constrained peptides projecting from their surface. Each variable domain from both heavy and light chains contributes three peptide loops to form the complementarity determining region (CDR). Given the ability of the immune system to recognize an almost limitless number of shapes, these molecules would appear to be the ideal starting point for peptide leads. From the design standpoint antibodies represent a significant challenge. Often contacts with the antigen are distributed over a number of CDR loops so that binding is not wholly confined to one or even two loops. There are exceptions, of course, and these have been exploited in at least a half dozen cases where a constrained peptide derived from a CDR loop has been shown to bind specifically to its target antigen.[11] Because of the complexity of antibody structure it would be useful to start with simple alternatives. The finding that many camel antibodies are devoid of light chain variable domains, thus reducing the number of potential CDR loops that may bind a given antigen by half,[12,13] offers a potential starting point for smaller scaffolding.

Another example of small natural protein scaffolds are the cysteine-rich peptides isolated from the venoms of cone snails.[14] Many of the conotoxins are constrained into rigid molecules by the formation of multiple disulfide bonds. The intervening sequences form loops that, by analogy with CDR loops, give the toxin its specificity. These represent some of the most minimal scaffolds, ranging in size from 10 to 30 amino acids, and suggest alternative frameworks for peptide presentation.

CHOOSING A PROTEIN SCAFFOLD

A number of factors must be taken into account in selecting the proper molecular scaffold. The first consideration is choosing a protein domain for which there is structural information. This is an obvious prerequisite, since it would be difficult to choose where to insert random amino acids without having structural information as a guide. Once the domain has been selected, there must be some assessment concerning the accessibility of the randomized residues for binding to a given target. Clearly, this will depend on the target molecule as well, but generally the residues should be solvent exposed and project out from the main body of the protein.

In addition to accessibility of the random residues for binding, the scaffold itself must be able to tolerate insertions of random amino acids. This will require the scaffold to be a reasonably stable protein. If, for instance, the randomized residues are to compromise a loop, then the assumption must be that any destabilization as a result of the inserted sequence will not unduly destabilize the protein. With regard to this, it is useful, although not essential, to have some additional function associated with the scaffold that can be used as a monitor of protein folding. In this way the investigator can assess whether the insertion of random amino acids disrupts proper folding.

Since the scaffold is to be used to construct libraries of peptide sequences, ideally the protein should be able to be displayed on the surface of phage. This greatly simplifies the selection process when searching for peptides that bind a given target. Typically the vehicle of choice for display is bacteriophage M13. Since M13 is assembled by extruding the viral particle through the bacterial membrane, any displayed protein must then necessarily be secreted. While the rules for whether any given protein can be secreted are unclear, alternative phage display systems exist that do not require secretion.[15-17] The ideal scaffold should also be both small and easily overproduced, to allow for rapid structural determination by NMR or X-ray crystallography.

PROTEIN SCAFFOLD LIBRARIES DISPLAYED ON PHAGE

Loop Libraries

A number of different proteins displayed on phage have been used for the intended purpose of serving as scaffolds for peptides. A single-loop library was constructed in the B1 domain of protein G of *Streptococcus* using a reverse turn connecting two antiparallel β-strands.[18] The library was used to select phage that bound the integrin receptor II_bIII_a. Two lessons were learned from this work. Similar to results from prior peptide grafting experiments, the peptide must be made accessible for binding by extending the loop away from the main body of the scaffold. Also, the extended loop in this case required some further stabilization, since all the selected variants contained a pair of cysteine residues at the base of the loop.

In an effort to increase the potential molecular diversity, more complex scaffolds have been constructed in which multiple random peptide loops are presented. The designed minibody represents one of the early attempts at such scaffolds.[19] While these scaffolds do not follow the precept of starting with a known protein domain, the minibody design is closely modeled from the structure of the heavy-chain variable domain of immunoglobulin McPC603. The designed 61 amino acid protein consists of two β-sheets, each consisting of three antiparallel β strands (Figure 20.1A). Two loops at one end of the molecule corresponding to the canonical hypervariable loops H1 and H2 of immunoglobulins are used to present randomized amino acids. Initial reports indicated that at high concentrations of the protein needed for structural work the minibody was insoluble.[19] Reengineering of some of the framework residues has helped in reducing some of the solubility problems.[20]

The minibody has been fused to gIII of M13, and phage libraries have been constructed in which the loop residues of H1 and H2, 12 amino acids in all, have been randomized.[21] This library was then used to select for phage that would bind to human interleukin-6. Of the phage selected, one contained a sequence in the H2 loop similar to the sequence in the interleukin-6 receptor thought to be involved in binding the cytokine. When tested in vitro, the soluble form of the minibody was able to antagonize receptor binding and signal transduction. Compared to the linear peptide counterpart, the minibody was more effective at binding interleukin-6. Whether this is solely a result of constraints imposed by the minibody structure on loop H2 or whether additional residues from loop H1 or the framework contribute to the binding affinity has not been reported.

Similar to the minibody strategy is the use of Tendamistat as a protein scaffold.[22] Tendamistat is a secreted protein from *Streptomyces tendae* that binds and inhibits α-amylase. It too represents a β-sheet architecture similar to that of an immunoglobulin (Figure 20.1B). Three loops project from the upper surface of the protein. One of the loops is required for binding α-amylase and is left unchanged in the library construction since it serves as a convenient tool for testing whether any selected variant is still properly folded. The Tendamistat library was tested against a monoclonal antibody known to bind endothelin. A number of Tendamistat variants were found to bind the monoclonal antibody and to specifically compete with endothelin for binding. Mutational analysis of one of these variants using alanine substitutions indicated that all the residues required for binding were contained within one loop. An interesting point is that the residues comprising the epitope on endothelin, while discontinuous, all lie on one face of the α-helix. This suggests that extended loops can mimic residues presented by a different secondary structure.

Loop libraries need not necessarily be presented on a β-sheet scaffold. A loop library has been constructed using cytochrome b_{562}, a four-helix bundle

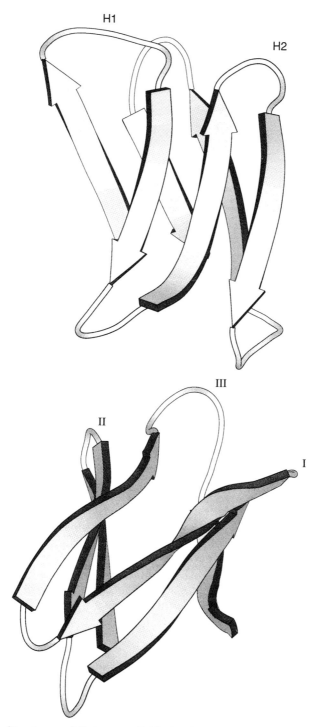

Figure 20.1. Structures of β-sheet scaffolds for loop libraries. (A) Structure of the minibody with loops used for presentation of random peptides, H1 and H2 indicated. (B) Structure of Tendamistat with randomized loops II and III indicated. Loop I is required for binding to α-amylase.

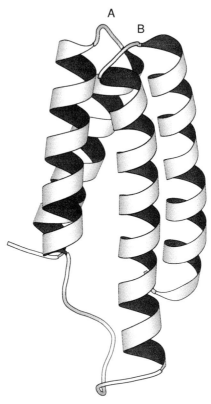

Figure 20.2. Structure of a four helix bundle scaffold, cytochrome b562, used for loop libraries. Random peptides are presented in the context of the two loops, A and B.

protein.[23] Two loops connecting the helices at one end of the protein were randomized (Figure 20.2). It has previously been shown that the protein is remarkably impervious to sequence changes and maintains its native fold when one of these loops is randomized.[24] Maintenance of a properly folded scaffold is measured by evaluating the ability to bind heme in a colorametric assay. The library was used to select for binders to *p*-nitrobenzylamine cross-linked to BSA. Interestingly, after five rounds of selection, all the phage recovered displayed cytochrome b_{562} with conserved Trp–Ser residues in loop A and a conserved Arg–Trp pair in loop B. In an effort to select for higher affinity binders, a new library was constructed in which the conserved positions were randomized at only a 20% frequency, while the remaining five residues were totally randomized. Unfortunately, the binders from this library were of no higher affinity than those selected after five rounds. When soluble mutant cytochromes (not tethered to phage) were analyzed for binding, it became clear that binding was not only to the *p*-nitrobenzylamine but also to the cross-linked BSA as well.

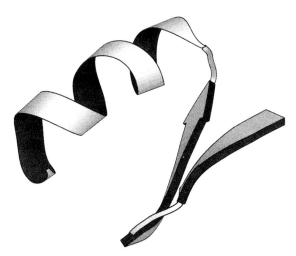

Figure 20.3. Structure of zinc-finger scaffold in which random amino acids are presented in the context of the α-helix.

Helix Libraries

In addition to loop libraries one might envision other secondary structural contexts in which random amino acids could be presented. Two recent examples have been reported in which the random amino acids are presented as part of an α-helix. The first uses a 26-amino acid zinc-finger motif displayed on phage (Figure 20.3).[25] This motif consists of two short antiparallel β-strands connected to an α-helix. From structural and phylogenetic data, residues that are conserved and required for proper folding are well defined. The remaining residues in the helix can then be randomized for construction of the library. One of the nice features of this library is that since folding of zinc fingers is dependent on metal coordination, any selected binders that require the context of a folded helix should also be dependent on the presence of zinc for binding.

A second type of helix library is represented by the display on phage of the synthetic Z domain based on the B domain of staphylococcal protein A.[26] This domain represents a three-helix bundle that has been characterized structurally by X-ray crystallography and by NMR. To construct the library, 13 residues that were determined to be solvent accessible by modeling were randomized. The residues were distributed over two helices and covered a surface of approximately 800 square angstroms. While the library was not used for binding to a particular target, a number of random members were physically analyzed. With one exception where a proline was introduced into a helix, all those analyzed showed an expected circular dichroism spectrum expected for a helical protein. Given the ease with which the protein can be produced and its remarkable stability, this is a very promising scaffold.

EXPLORING SEQUENCE SPACE

Phage-displayed peptide libraries, in which all possible sequences are thought to be represented, simply require repetitive cycles of biopanning to enrich for the rare sequence that binds the target molecule. For the protein scaffold libraries discussed thus far, the number of residues being randomized rapidly exceeds the capability of constructing complete libraries in which every possible amino acid combination is represented (10 random residues represents 1×10^{13} possibilities). Thus, in libraries where not every sequence is represented, and where the optimal sequence may not even be initially present, a somewhat different approach has to be utilized. Following each round of selection an evolutionary step must be applied. New diversity in the form of mutations is introduced into the gene pool of selected variants. This strategy allows for the gradual optimization of selected variants in a manner similar to natural selection.

Two approaches that permit an expeditious exploration of sequence space have been reported. One approach, termed sexual PCR, is similar to computer-based genetic algorithms.[27,28] Following a round of selection, DNA coding for all the selected variants is pooled and fragmented randomly using DNAse I. The resulting population of molecules is then used as both primer and template in multiple cycles of PCR. Because each fragment contains some homology with other overlapping fragments the entire gene is reassembled in this manner. Most importantly this reassembly process acts to recombine all the mutations that were acquired in the prior round of selection. Gradually, with successive rounds of selection, multiple mutations that contribute to the selected phenotype accumulate. In a rather clever step, silent mutations that do not contribute to the phenotype can be removed by mixing wild-type DNA with the selected DNA prior to the DNAse I treatment. The net effect is to recombine out those changes that do not contribute to the overall fitness, and is analogous to the classical genetic backcross.

One of the advantages that this approach has over some other combinatorial strategies is that it is very effective at recombining mutations that are widely separated from one another in the gene. This should be of particular value in scaffold libraries where the random residues are located in different parts of the protein, for example, in different loops. A potential drawback of this method concerns optimization in which the mutations fall within blocks of randomized sequences. For instance, if residues in one particular loop are selected, then how are the remaining residues in that loop optimized? Because sexual PCR is dependent on homology, similar to *in vivo* recombination, randomized blocks will tend to segregate together rather than being recombined.

An alternative approach, called recursive ensemble mutagenesis, could address this problem.[29,30] In this procedure a combinatorial library is constructed in which 5 to 6 residues are randomized. A round of selection is carried out and the selected variants are sequenced. From the selected sequences a synthetic phylogenetic tree is derived, which establishes amino acid preferences for each varied position. This information is then used in conjunction with an algorithm called CyberDope, which weights the base composition of the mutagenic oligonucleotide for the next cycle of mutagenesis and selection.[31] By doing this the search of sequence space is rapidly reduced as one moves from initially totally random sequences NNG/C to a more directed search in later rounds of selection.

ISSUES FOR THE FUTURE

Scaffold libraries represent a new technology that serves to link combinatorial libraries with structure-based peptidomimetic design. The feasibility of constructing and using these libraries has clearly been demonstrated. Nevertheless, a number of issues arise concerning the use of scaffold libraries. Since each peptide sequence is constrained in these libraries, will one scaffold library be sufficient for encompassing shape space? Another issue is whether these relatively small scaffolds will be able to effectively mimic the more complex antibody structure and evolve binders of high affinity. Finally, the most difficult area will be translating the structural information obtained from scaffold libraries into small molecule peptidomimetics. Given the rapid progress in this field, answers to these and other questions should be forthcoming shortly.

ACKNOWLEDGMENTS

I thank Hongxing Zhou for help with the graphics and Maurizio Sollazzo for the minibody coordinates.

REFERENCES

1. Cwirla SE, Peters EA, Barrett RW, Dower WJ (1990): Peptides on phage: a vast library of peptides for identifying ligands. *Proc Natl Acad Sci USA* 87:6378–6382.
2. Devlin JJ, Panganiban LC, Devlin PE (1990): Random peptide libraries: a source of specific protein binding molecules. *Science* 249:404–406.
3. Scott JK, Smith GP (1990): Searching for peptide ligands with an epitope library. *Science* 249:386–390.
4. Barbas CF III, Languino LR, Smith JW (1993): High-affinity self-reactive human antibodies by design and selection: targeting the integrin ligand binding site. *Proc Natl Acad Sci USA* 90:10003–10007.
5. Lee G, Chan W, Hurle MR, DesJarlais RL, Watson F, Sathe GM, Wetzel R (1993): Strong inhibition of fibrinogen binding to platelet receptor alpha IIb beta 3 by RGD sequences installed into a presentation scaffold. *Protein Eng* 6:745–754.
6. Sollazzo M, Billetta R, Zanetti M (1990): Expression of an exogenous peptide

epitope genetically engineered in the variable domain of an immunoglobulin: implications for antibody and peptide folding. *Protein Eng* 4:215–220.

7. Zanetti M, Filaci G, Lee RH, del Guerico P, Rossi F, Bacchetta R, Stevenson F, Barnaba V, Billetta R (1993): Expression of conformationally constrained adhesion peptide in an antibody CDR loop and inhibition of natural killer cell cytotoxic activity by an antibody antigenized with the RGD motif. *EMBO J* 12:4375–4384.

8. Yamada T, Matsushima M, Inaka K, Ohkubo T, Uyeda A, Maeda T, Titani K, Sekiguchi K, Kikuchi M (1993): Structural and functional analyses of the Arg-Gly-Asp sequence introduced into human lysozyme. *J Biol Chem* 268:10588–10592.

9. Yamada T, Uyeda A, Kidera A, Kikuchi M (1994): Functional analysis and modeling of a conformationally constrained Arg-Gly-Asp sequence inserted into human lysozyme. *Biochemistry* 33:11678–11683.

10. Langen HT, Taylor JW (1992): Alkaline phosphatase-somatostatin hybrid proteins as probes for somatostatin-14 receptors. *Proteins: Struct Funct Genet* 14:1–9.

11. Dougall WC, Peterson NC, Greene MI (1994): Antibody-structure-based design of pharmacological agents. *Trends Biotechnol* 12:372–379.

12. Hamers–Casterman C, Atarhouch T, Muyldermans S, Robinson G, Hamers C, Songa EB, Bendahman N, Hamers R (1993): Naturally occurring antibodies devoid of light chains. *Nature* 363:446–448.

13. Muyldermans S, Atarhouch T, Saldanha J, Barbosa JA, Hamers R (1994): Sequence and structure of V_H domain from naturally occurring camel heavy chain immuno-globins lacking light chains. *Protein Eng* 7:1129–1135.

14. Olivera BM, Hillyard DR, Marsh M, Yoshikami D (1995): Combinatorial peptide libraries in drug design: lessons from venomous cone snails. *Trends Biotechnol* 13:422–426.

15. Sternberg NL, Hoess RH (1995): Display of peptides and proteins on the surface of bacteriophage lambda. *Proc Natl Acad Sci USA* 92:1609–1613.

16. Maruyama IN, Maruyama HI, Brenner S (1994): Lambda foo: a lambda phage vector for the expression of foreign proteins. *Proc Natl Acad Sci USA* 91:8273–8277.

17. Dunn IS (1995): Assembly of functional bacteriophage lambda virons incorporating C-terminal peptide or protein fusions with the major tail protein. *J Mol Biol* 248:497–506.

18. O'Neil KT, DeGrado WF, Hoess RH (1994): Phage Display of Random Peptides on a Protein Scaffold. In Crabb JW (ed): "Techniques in Protein Chemistry V" New York: Academic Press, pp 517–524.

19. Pessi A, Bianchi E, Crameri A, Venturini S, Tramontano A, Sollazzo M (1993): A designed metal-binding protein with a novel fold. *Nature* 362:367–369.

20. Bianchi E, Venturini S, Pessi A, Tramontano A, Sollazzo M (1994): High level expression and rational mutagenesis of a designed protein, the minibody. From an insoluble to a soluble molecule. *J Mol Biol* 236:649–659.

21. Martin F, Toniatti C, Salvati AL, Venturini S, Ciliberto G, Cortese R, Sollazzo M (1994): The affinity-selection of a minibody polypeptide inhibitor interleukin-6. *EMBO J* 13:5303–5309.

22. McConnell SJ, Hoess RH (1995): Tendamistat as a scaffold for conformationally constrained phage peptide libraries. *J Mol Biol* 250:460–470.

23. Ku J, Schultz PG (1995): Alternate protein frameworks for molecular recognition. *Proc Natl Acad Sci USA* 92:6552–6556.

24. Brunet AP, Huang ES, Huffine ME, Loeb JE, Weltman RJ, Hecht MH (1993): The role of turns in the structure of an alpha-helical protein. *Nature* 364:355–358.

25. Bianchi E, Folgori A, Wallace A, Nicotra M, Acali S, Phalipon A, Barbato G, Bazzo R, Cortese R, Felici F, Pessi A (1995): A conformationally homogenous combinatorial peptide library. *J Mol Biol* 247:154–160.

26. Nord K, Nilsson J, Nilsson B, Uhlen M, Nygren P-A (1995): A combinatorial library of an alpha-helical bacterial receptor domain. *Protein Eng* 8:601–608.

27. Stemmer WPC (1994): Rapid evolution of a protein *in vitro* by DNA shuffling. *Nature* 370:389–391.

28. Stemmer WPC (1994): DNA shuffling by random fragmentation and reassembly: *in vitro* recombination for molecular evolution. *Proc Natl Acad Sci USA* 91:10747–10751.

29. Arkin AP, Youvan DC (1992): An algorithm for protein engineering: simulations of recursive ensemble mutagenesis. *Proc Natl Acad Sci USA* 89:7811–7815.

30. Delagrave S, Goldman ER, Youvan DC (1993): Recursive ensemble mutagenesis. *Protein Eng* 6:327–331.

31. Arkin AP, Youvan DC (1992): Optimizing nucleotide mixtures to encode specific subsets of amino acids for semi-random mutagenesis. *Bio/Technology* 10:297–300.

21

COMBINATORIAL BIOSYNTHESIS OF "UNNATURAL" NATURAL PRODUCTS

Chaitan Khosla

Departments of Chemistry and Chemical Engineering, Stanford University, Stanford, California

Throughout the history of drug discovery, natural products have been a fertile source for new lead molecules with human therapeutic, veterinary, and agrochemical uses. Commercially significant natural products include antibiotics, such as penicillins, cephalosporins, tetracyclines, streptomycin, rifamycin, and erythromycins; anticancer agents, such as doxorubicin and taxol; antifungals, such as griseofulvin, and candicidin; antiparasitic agents, such as avermectin; cholesterol-lowering agents, such as lovastatin; immunosuppressives, such as cyclosporin and FK506; veterinary products, such as tylosin, hygromycin, and monensin; and agrochemicals, such as bialophos. (For examples, see Figure 21.1). Scores of other natural products have been identified as promising leads, even though issues of toxicity and structural complexity precluded their development into marketable products. There is thus considerable interest in the application of recombinant DNA approaches to explore and exploit nature's potential for the synthesis of novel organic molecules.

A remarkable feature of natural products is that, in spite of their breathtaking structural diversity, their synthetic pathways are often mechanistically related.[1] These biosynthetic relationships govern the classification of natural products into families, such as polyketides, nonribosomal peptides, isoprenoids, and glycosides. In particular, many natural products are synthesized by

Combinatorial Chemistry and Molecular Diversity in Drug Discovery, Edited by
Eric M. Gordon and James F. Kerwin, Jr.
ISBN 0-471-15518-7   Copyright © 1998 by Wiley-Liss, Inc.

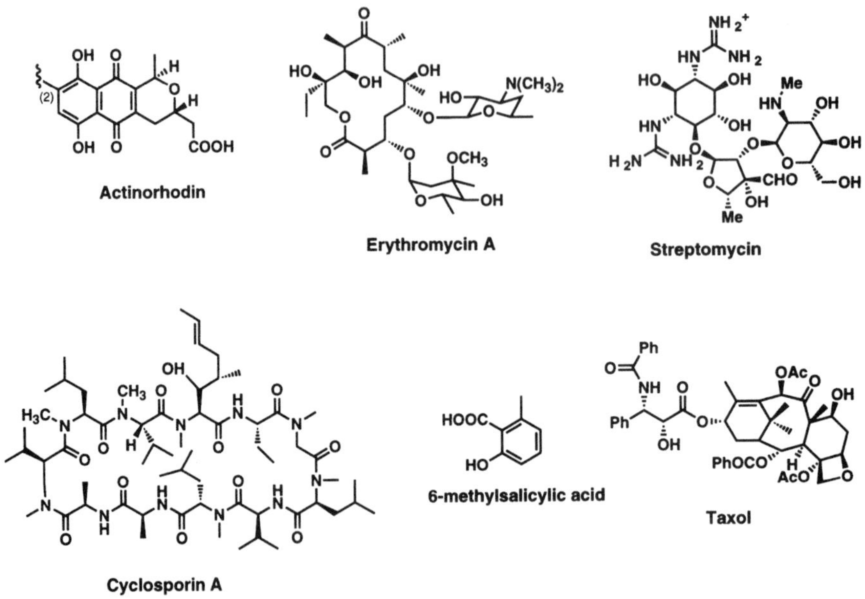

Figure 21.1. Structures of some natural products referred to in text.

the controlled assembly and modification of discrete building block units (such as organic acids, amino acids, isoprene units, or monosaccharides). This leads one to the speculation that diversity within these molecular families arose through recombination between related pathways with different specificities.

Within the past decade advances in molecular genetics have led to major advances in our understanding of the enzymatic basis for microbial natural product biosynthesis. In particular, the mechanistic relationships within families of natural products have been underscored by striking similarities between the deduced amino acid sequences of corresponding active sites. This has led to the concept of combinatorial biosynthesis, namely that homologous genes with related functions but different specificities can be shuffled via genetic engineering to generate novel natural product-like molecules.

This chapter focuses on evaluating the potential for combinatorial biosynthesis within different microbial natural product families. The reader is also directed to other recent reviews on the subject.[2] Given the rapid development of new tools and methodologies in the field, and their application to obtain a better understanding of the fundamental relationships between structure and function within enzyme families of biosynthetic relevance, one could project ahead to a time when "unnatural" natural product libraries are comparable to their natural product counterparts in terms of both size and structural diversity. This raises questions as to the richness of these engineered molecule libraries as sources of new lead compounds and even drugs. While speculation in this regard would be premature at present, several features of the libraries derived from combinatorial biosynthesis could facilitate a rapid assessment of their utility. The potential for developing novel screening and selection strategies for this unique source of biologically generated libraries of organic molecules is discussed toward the end of this chapter.

To generate combinatorial libraries of significant sizes and diversity within any given family of natural products, several conditions must be satisfied:

1. The genes of several related but distinct biosynthetic pathways must be cloned and sequenced. The existence of a high level of sequence similarity between allelic forms of different genes can be very helpful in this regard, since DNA probes can be designed.

2. The similarities and differences between related pathways must be defined. At the very least, functions must be assigned to the genes (or domains) encoding individual active sites.

3. The substrate specificity of allelic forms of such active sites must be defined.

4. The extent to which individual active sites can be added, removed, or altered without affecting other reactions in the overall biosynthetic scheme must be understood.

5. Heterologous expression systems suited for combinatorial mutagenesis of large gene clusters must be available.

6. Methods for the construction and analysis of large combinatorial libraries with minimal genetic redundancy must be developed.

The past few years have witnessed significant advances in the above directions within the context of various families of natural products. Most of this progress has been on combinatorial biosynthesis of polyketides; however, it is becoming increasingly apparent that the general concepts developed in the context of polyketides may be extended to other families as well. It should be noted that, since most genetically engineered products characterized thus far are novel chemical entities, combinatorial biosynthesis appears to represent an untapped source of molecular diversity for drug discovery (see specific references on products derived from combinatorial biosynthesis below). This can be contrasted with most current natural product screening programs, where the frequency of discovering hitherto unidentified molecules is much lower.

POLYKETIDES

The chemistry and biology of polyketide biosynthesis have been subjects of several excellent reviews within the past few years.[3] Polyketides are a large family of structurally diverse natural products possessing a wide range of biological activities, including antibiotic and other pharmacological properties. They are synthesized by multifunctional polyketide synthase enzymes (PKSs).

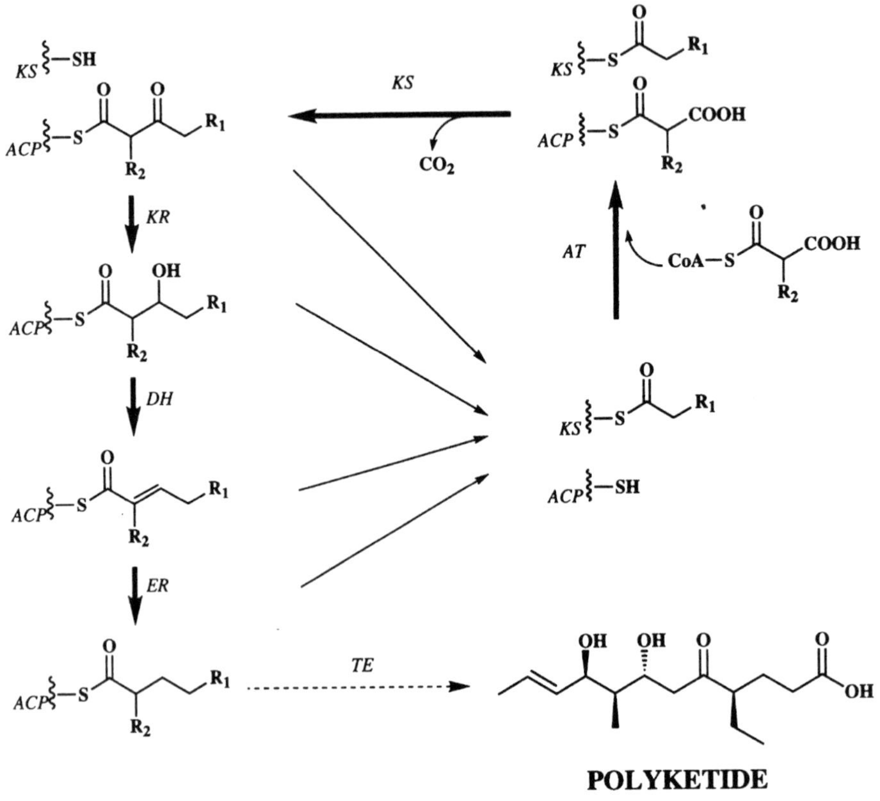

POLYKETIDE

Figure 21.2. Catalytic cycle for polyketide backbone biosynthesis. Biosynthesis of the carbon-chain backbone of a polyketide involves the following enzyme-catalyzed reactions. At the start of any condensation cycle, the growing polyketide chain is attached via a thioester linkage to a cysteine residue in the active site of the ketosynthase (KS). An acyltransferase (AT) transfers the carboxylated extender unit from CoA to the 4'-phosphopantetheine arm of the acyl carrier protein (ACP). The choice of the extender (R_2) is presumably dictated by the specificity of the AT. Following a condensation catalyzed by the KS, the β-carbonyl in the nascent chain can be subjected to all, part, or none of a series of steps catalyzed by a ketoreductase (KR), dehydratase (DH), and enoylreductase (ER). After the chain has been through a fixed number of condensation cycles, it is released from the PKS via the action of a thioesterase (TE). Wherever applicable, the stereochemistry at any asymmetric center is uniquely controlled by the PKS.

PKSs catalyze repeated condensation cycles between acyl thioesters (usually acetyl, propionyl, malonyl, or methylmalonyl). Each cycle results in the formation of a β-keto group that may undergo all, part, or none of a series of stereo-controlled reductive steps (Figure 21.2). In the case of aromatic PKSs, the β-keto groups are left largely unmodified and the resulting highly reactive polyketide backbone undergoes a series of enzyme-catalyzed regiospecific cyclizations.

Gene Cloning

Over the past decade, cloning and sequence analysis of PKS genes have yielded exciting insights into the enzymatic basis for polyketide biosynthesis. While several classification strategies have been proposed, from the viewpoint of combinatorial biosynthesis of polyketides, there are (at least) two fundamentally different types of PKSs (Figure 21.3). Complex or "modular" PKSs catalyze the biosynthesis of macrolides such as erythromycin. They are assemblies of large multifunctional proteins and carry a distinct active site for every enzyme-catalyzed step in carbon-chain assembly and modification. Active sites are clustered into modules, with each module containing a full complement of sites required for one cycle in the iterative condensation process (Figure

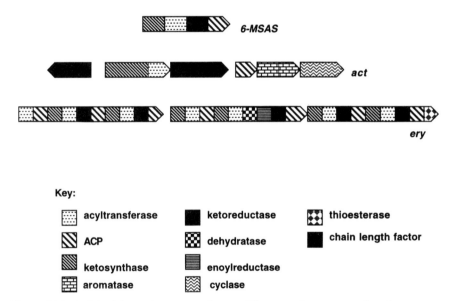

Figure 21.3. Polyketide synthase gene clusters. The gene clusters encoding three representative gene clusters are schematically shown. The *6-MSAS* gene cluster from *Penicillium patulum* encodes a multifunctional protein responsible for the synthesis of 6-methylsalicyclic acid. The *act* gene cluster from *Streptomyces coelicolor* encodes a multisubunit PKS responsible for the biosynthesis of the bicyclic precursor of actinorhodin. The *ery* gene cluster from *Saccharopolyspora erythraea* encodes three large multifunctional proteins that catalyze the biosynthesis of 6-deoxyerythronolide B. Each shaded region is indicative of a putative active site domain identified either via mutagenesis or through sequence comparison with enzymes having known functions. For further details on the functions of these active sites, see text or caption to Figure 21.1.

21.2). In contrast, the PKSs responsible for the biosynthesis of aromatic polyketides, such as 6-methylsalicyclic acid and actinorhodin, generate the entire carbon-chain backbone using a single set of iterative active sites (Figure 21.3). Iterative PKSs can exhibit considerable architectural variety. For example, bacterial aromatic PKSs, such as the actinorhodin PKS, are made up of several distinct polypeptide subunits, each containing one or two active sites, whereas fungal PKSs, such as the 6-methylsalicyclic acid synthase, consists of a single multidomain polypeptide that includes all the active sites required for the biosynthesis of 6-methylsalicyclic acid. In cases where the nascent polyketide chain undergoes subsequent cyclizations and other regiospecific modifications, some of these steps may be catalyzed by PKSs themselves.

Combinatorial Biosynthesis Technologies

To explore the combinatorial potential in PKSs, a strategy for the engineered biosynthesis of novel natural products was recently developed.[4] The strategy, which was originally developed in the context of bacterial aromatic PKSs due to their modest sizes, involves a host–vector system capable of high-level expression of subsets of biosynthetic genes. The heterologous host, *Streptomyces coelicolor* CH999, contains a chromosomal deletion of the entire *act* gene cluster, which encodes the biosynthesis of the aromatic polyketide actinorhodin. Shuttle plasmids are used to express recombinant PKSs in CH999. Such plasmids typically include a colEI replicon, an appropriately truncated SCP2* *Streptomyces* replicon, two *act*-promoters to allow for bidirectional cloning, the gene encoding the *act*II-ORF4 activator, which induces transcription from *act* promoters during the transition from growth phase to stationary phase, and appropriate marker genes. Restriction sites have been engineered into these vectors to facilitate the combinatorial construction of PKS gene clusters starting from cassettes encoding individual subunits (or domains) of naturally occurring PKSs. The primary advantages of the above strategy are that (1) all relevant biosynthetic genes are plasmid-borne and therefore amenable to facile manipulation and mutagenesis in *E. coli*; (2) the entire library of PKS gene clusters is expressed in the same bacterial host, which is genetically and physiologically well-characterized and presumably contains most, if not all, ancillary activities required for in vivo production of polyketides; (3) polyketides are produced in a secondary metabolite-like manner, thereby alleviating the toxic effects of synthesizing potentially bioactive compounds in vivo; and (4) molecules thus produced undergo fewer side reactions than if the same pathways were expressed in wild-type organisms or blocked mutants. This system, which has been successfully extended to express and manipulate modular PKSs[5] as well as fungal PKSs,[6] should be useful for combinatorial biosynthesis within other families of microbial natural products as well. In cases where large gene clusters (such as modular PKSs) are involved, an in vivo recombination strategy has been developed for high-frequency combinatorial manipulation.[5] Alternative strategies with some but not all of the above features have also been described.[4,7–10]

Bacterial Aromatic Polyketides

The potential for combinatorial biosynthesis within bacterial aromatic polyketides has been reviewed in depth elsewhere.[11] Only the salient features are discussed here. Aromatic polyketide biosynthesis begins with a primer unit loading on to the active site of the condensing enzyme, β-keto acyl synthase (KS; Figure 21.3). An extender unit (usually malonate) is then transferred to the pantetheinyl arm of the acyl carrier protein (ACP; Figure 21.3). The KS catalyzes the condensation between the ACP-bound malonate and the starter unit. Additional extender units are added sequentially until the nascent polyketide chain has grown to a desired chain length determined by the protein chain length factor (CLF; Figure 21.3), perhaps together with the KS. Thus, the KS, CLF, and ACP form a minimal set to generate a polyketide backbone, and are together called the minimal PKS. The nascent polyketide chain is then subjected to regiospecific ketoreduction by a ketoreductase (KR; Fig. 21.3) if it exists. Cyclases (CYC; Figure 21.3) and aromatases (ARO; Figure 21.3) later catalyze regiospecific ring formation events through intramolecular aldol condensations. The cyclized intermediate may then undergo additional regiospecific and/or stereospecific modifications (e.g., O-methylation, hydroxylation, and glycosylation) controlled by downstream tailoring enzymes.

Expression and analysis of numerous recombinant bacterial PKSs has led to advances in our understanding of aromatic PKS function and specificity.[4,5,9,12-19] In turn, these insights have led to the proposal of a set of design rules for combinatorially manipulating early biosynthetic steps in aromatic polyketide pathways, including chain synthesis, C-9 ketoreduction, and the formation of the first two aromatic rings.[15] It has been estimated that the number of polyketides that can be generated from combinatorial manipulation of only the first few steps in aromatic polyketide biosynthesis is on the order of a few hundred.[15] This number could potentially be expanded by including genes encoding downstream tailoring enzymes in the above libraries.[11] However, since structural diversity of intermediates increases as one combinatorially manipulates increasingly longer sequences of reactions, it is likely that the substrate specificities of tailoring enzymes will place intrinsic constraints on the fraction of intermediates produced within combinatorial PKS libraries that can be modified by tailoring enzymes. Examples of combinatorially interesting tailoring enzymes include the actinorhodin actVA hydroxylase, which can regiospecifically hydroxylate other anthraquinones,[17,20] and the tetracenomycin tcmO O-methyltransferase, which can regiospecifically methylate other carboxylated fused ring systems.[21]

Products Derived From Modular Polyketide Synthases

Although relatively few studies involving genetic manipulation of modular PKSs have been reported thus far, the one-to-one correspondence between active sites and product structure (Figure 21.3), together with the incredible chemical diversity observed among naturally occurring "complex" polyketides, suggests that the combinatorial potential within these multienzyme systems could be considerably greater than that for bacterial aromatic PKSs. For example, a wider range of primer units, including aliphatic monomers (acetate, propionate, butyrate, isovalerate, etc), aromatics (benzoate, aminohydroxybenzoate), alicyclics (cyclohexanoate), and heterocyclics (pipecolate), are found in various macrocyclic polyketides. Recent studies have shown that modular PKSs have relaxed specificity for their starter units.[22,23] The degree of β-ketoreduction following a condensation reaction can also be altered by genetic manipulation.[1,24] Likewise, the size of the polyketide product can be varied by designing mutants with the appropriate number of modules.[8,22,25] Modular PKSs also exhibit considerable variety regarding the choice of extender units in each condensation cycle, although it remains to be seen to what extent this property can be manipulated. Lastly, these enzymes are well known for generating an impressive range of asymmetric centers in their products in a highly controlled manner. It is thus possible that the combinatorial potential within modular PKS pathways could be virtually unlimited.

Fungal Polyketides

Like the actinomycetes, filamentous fungi are a rich source of polyketide natural products. The fact that fungal PKSs, such as the 6-methylsalicyclic acid synthase (6-MSAS) and the mevinolin synthase, are encoded by single multidomain proteins (Figure 21.3) suggests that they may also be well suited for combinatorial mutagenesis. Moreover, fungal PKS can be functionally expressed in bacteria.[6] As yet no evidence exists for the functional plasticity of domains within fungal PKSs. Based on a comparative analysis of fungal polyketides, however, one can speculate on the potential for designing combinatorial libraries using fungal PKSs. For example, chain lengths not observed in bacterial aromatic polyketides (e.g., tetraketides, pentaketides, and hexaketides) can be found among fungal aromatic polyketides. Likewise, the cyclization patterns of fungal aromatic polyketides are quite different from those observed in bacterial aromatic polyketides. Lastly, in contrast with modular PKSs from bacteria, branched methyl groups are introduced into fungal polyketide backbones by *S*-adenosylmethionine-dependent methyltransferases; in the case of the lovastatin PKS this activity is encoded as one domain within a monocistronic PKS.[26] It is now possible to experimentally evaluate whether these and other sources of chemical diversity in fungal polyketides are indeed amenable to combinatorial manipulation.

NONRIBOSOMAL PEPTIDES

The combinatorial biosynthesis approach developed in the context of polyketide synthases can be readily adapted for the production of novel peptides

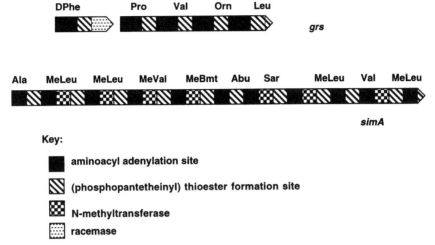

Figure 21.4. Nonribosomal peptide biosynthesis. The currently accepted model is known as the multienzyme thiotemplate mechanism. In this model, a multifunctional synthetase catalyzes a sequence of reactions leading to peptide formation. Included in these reactions are (1) activation of amino acids to their adenylylates by reaction with ATP, (2) covalent attachment of the activated amino acid to the synthetase as a AA-thioester, (3) racemization of selected AA-thioesters, (4) *N*-methylation of selected AA-thioesters using *S*-adenosylmethionine, (5) condensation to form a peptide chain growing from a phosphopantetheine cofactor, and (6) chain termination through either cyclization or peptide–thioester hydrolysis. Like modular PKSs, peptide synthetases are modular in structure; all the active sites required for the incorporation and modification of an individual amino acid into the growing peptide backbone are clustered. The grs operon and the simA gene encode for the biosynthesis of gramicidin and cyclosporin, respectively.

derived from nonribosomal thiotemplate mechanisms (Figure 21.4). Examples of such natural products include penicillins and cephalosporins, gramicidin, enniatin, and cyclosporin. DNA sequence analyses of several peptide synthetase genes have now been reported. All nonribosomal peptide synthetases examined to date demonstrate modular construction, with each module being responsible for the activation and modification of a single amino acid residue in the resulting peptide[10] (Figure 21.4). The sequence of modules specified in a synthetase gene corresponds to the sequence in the final peptide product, thus making prediction of the product sequence possible. Since the domains from a wide variety of peptide synthetases show homologous regions, it should be possible to use heterologous activation and modification domains in a combinatorial approach to expand the diversity of natural peptides. In at least one case, domain-swapping has been shown to be feasible. Replacement of the ultimate domain of the surfactin synthetase (which catalyzes the incorporation of leucine) with the second domain of a fungal ACV synthetase involved in β-lactam biosynthesis (which activates cysteine) led to the formation of a hybrid peptide.[27]

For several reasons, peptide libraries derived from combinatorial biosynthesis could be an attractive supplement to existing biologically and chemically generated peptide libraries. For example, although combinatorial synthesis of high-quality cyclic N-methylated peptides (e.g., cyclosporin) is a daunting task, a number of naturally occurring nonribosomal peptides are known to possess N-methyl groups. Likewise, in addition to the 20 "natural" amino acids, non-ribosomal peptide synthases use a number of other biologically available amino acids as substrates. Epimerase activities within peptide synthetase domains can result in the incorporation of D-amino acids into their products. Finally, in terms of cost-effectiveness, microbial fermentations can be a highly desirable alternative to the synthetic production of peptides.

AMINOCYCLITOL GLYCOSIDES

Aminocyclitol glycoside antibiotics such as streptomycin and kanamycin constitute an important class of antibiotics with diverse clinical uses. The majority of these compounds are glycosides of either streptamine or 2-deoxystreptamine, or functional modifications of these two 1,3-diaminoinositols. Additionally, glycosyl groups are often found attached to other natural products such as polyketides (erythromycin, doxorubicin, avermectin) and nonribosomal peptides (vancomycin); in virtually every case they critically contribute to the pharmaceutical properties of the compound.

Although the biosynthetic pathways of several glycosidic natural products have been elucidated, relatively little is known about the structure and function of their biosynthetic genes.[28,29] For example, the main genes involved in three branches of the streptomycin pathway (formation of streptidine, TDP-dihydrostreptose, and N-methyl-(L)-glucosamine subunits; Figure 21.5) are arranged in subclusters. Of these subclusters, the genes and enzymes of the TDP-dihydrostreptose cluster have been identified, and homologs are known to be present in various other structurally distinct 6-deoxyhexose biosynthetic pathways. Thus, in theory, the principles of combinatorial biosynthesis could be applied for the generation of novel glycosidic natural products at two levels: On one hand, new NDP-sugar units could be generated; at the same time, these NDP-sugars could be shuffled with each other (and with incoming nucleophilic groups from other natural products) through combinatorial manipulation of appropriate glycosyltransferases. However, in contrast to polyketides and peptides, whose biosynthesis proceeds via processive mechanisms, the biosynthetic pathways leading to glycosidic natural products are typically convergent. Detection of a fermentation-derived product requires activity in all converging branches of the pathway; shunt products are rarely observed. The possibility of generating nonproductive mutants by combinatorial manipulation is therefore

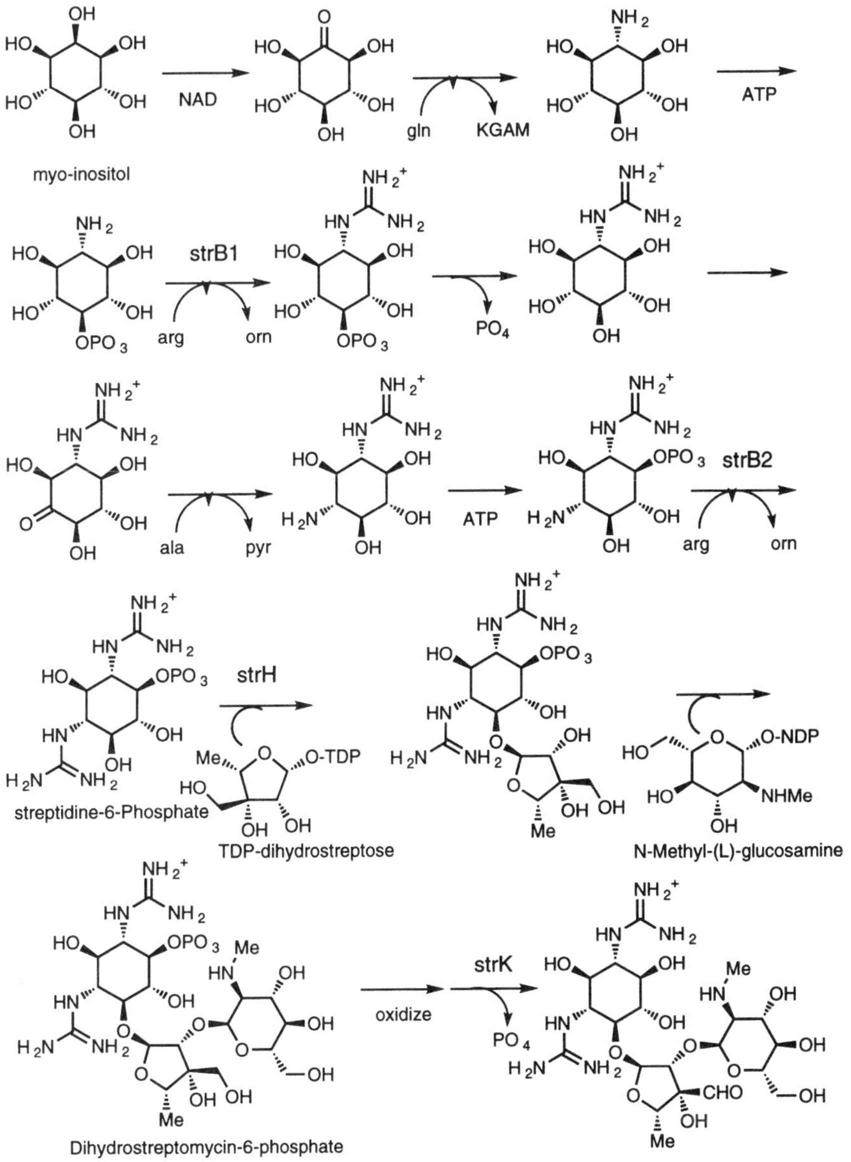

Figure 21.5. Genetics and enzymology of streptomycin biosynthesis.

considerably greater than in the case of polyketides and nonribosomal peptides.

Despite these potential difficulties, it has been shown that the glycosyltransferases involved in erythromycin[1,24] and rhodomycin[24] biosynthesis have relaxed specificities for the aglycone moiety. More significantly, the results of a recently reported study suggest that a glycosyltransferase involved in ellora-mycin biosynthesis may be able to recognize an unnatural NDP-sugar unit and attach it regiospecifically to a polyketide aglycone.[30] Together, these results suggest that at least the second strategy outlined above for the generation of novel sugar-containing molecules is promising. The importance of harnessing the potential of combinatorial biosynthesis in this class of natural products becomes even greater in light of the well-known difficulties in the chemical synthesis of stereochemically pure glycosides.

ISOPRENOIDS

Isoprenoids (terpenes or steroids) are ubiquitously distributed in plants, mammals, bacteria, and fungi. Their carbon skeletons are synthesized by highly conserved pathways involving the successive condensation of C_5 isoprene units to generate backbones containing 10 to >40 carbon atoms, which then undergo multiple regio- and stereocontrolled cyclizations giving rise to the natural product precursor (Figure 21.6). In most cases, these relatively hydrophobic cyclic intermediates undergo further oxidative modification catalyzed by downstream tailoring enzymes. Thus, unlike polyketides and peptides, where virtually unlimited diversity can arise via oligomerization of building blocks,

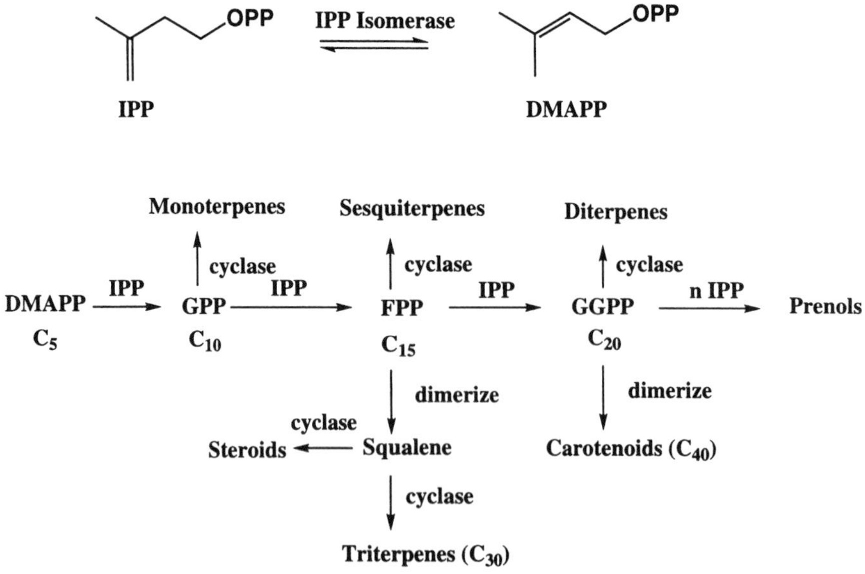

Figure 21.6. Biosynthesis of isoprenoids. The biosynthesis of the carbon-chain back-bones of terpenes, steroids, and other prenols involves the condensation of successive isoprenyl pyrophosphate units with a dimethylallyl pyrophosphate starter unit. These backbones are then cyclized in a regio- and stereocontrolled manner by diverse cyclases.

structural diversity within the isoprenoid family of natural products is predominantly the result of variations in cyclization patterns of a few carbon-chain backbones.

The genes encoding several of these remarkable cyclases have recently been cloned and overexpressed and the mechanisms of cyclization are only just beginning to be understood.[31] Many but not all of them are homologous. Nevertheless, it remains to be seen whether the regiospecificity of individual ring formation events during the biosynthesis of a polycyclic isoprenoid are controlled by genetically separable domains that can be combinatorially shuffled. If so, then it should be possible to generate large libraries of novel isoprenoids. However, even if this turns out to be unfeasible, the cloning of isoprenoid cyclases with diverse specificities from natural sources can be justified, since many of these enzymes occur in higher plants and animals. Heterologous expression of these enzymes could therefore lead to the microbial production of over 1000 core structures, since prenyl pyrophosphate substrates of terpene cyclases are naturally found in bacterial and lower eucaryotic microorganisms.

SCREENING AND SELECTION

As indicated by the wide range of biological activities that have been associated with secondary metabolites, the field of natural product discovery has a rich history of screening strategies and methods. In spite of these successes, there are three major limitations in conventional natural product screening strategies for drug discovery. The first has to do with the size of the library that can be screened. Broadly speaking, natural product screening programs can be classified into two categories: chemical screening and biological screening. While the former approach provides a more direct path toward the discovery of novel molecules, the latter has been more frequently employed since it can be adapted into a high-throughput screening format. For example, about 10,000 extracts from microbial fermentations can be generated and processed through a typical high-throughput screen annually; similar approaches have also been applied to extracts from plant sources. While this is comparable to the scale at which many other small molecule libraries are synthesized and screened, it pales in comparison to biologically derived libraries such as nucleic acids and peptides, where $>10^6$ distinct entities can be routinely generated and screened with far less time and effort requirements.

The second limitation of natural product libraries relates to the frequency at which novel molecules are currently being identified. Over the past two decades, despite the discovery of several major new classes of biologically active natural products, the frequency at which known molecules (or close analogs) have been rediscovered has increased to a point where the relevance of natural product screening programs is often questioned when compared with other combinatorial approaches toward drug discovery.

Finally, the heterogeneity of biological sources of natural products is notorious, and places unique demands on the biological and chemical skills required within drug discovery programs that screen natural products. These include the ability to clonally isolate and propagate producer microorganisms (or plants), to adapt fermentation protocols differently for the optimal production of secondary metabolites on a case-by-case basis, to screen chemically complex fermentation broths (or extracts thereof) for biological activities (usually without adequate controls), to purify bioactive molecules present in relatively trace amounts from chemically complex fermentation broths, and to elucidate the often intricate structures of natural products thus identified. Unlike many other library generation and analysis efforts that interface with high-throughput screens, the minimal effort required to operate and maintain a natural products drug discovery program is substantial and involves a strong skill base.

Combinatorial biosynthesis has considerable potential for alleviating all of the above limitations. Given knowledge of adequate numbers of catalytic degrees of freedom to manipulate, the size of libraries generated via combinatorial biosynthesis is directly limited by the transformation efficiencies of plasmid DNA into well-studied bacterial hosts such as *E. coli* and *S. coelicolor*. (The same feature also limits the sizes of other biological libraries, such as nucleic acid and peptide libraries.) As discussed above, the fraction of novel molecules within libraries derived from combinatorial biosynthesis appears to be very large; furthermore, as our understanding of the enzymatic basis for natural product biosynthesis improves, the chemical diversity within individual "unnatural" natural product libraries could be tailored via biased mutagenesis. Finally, from a technical viewpoint, the task of generating and maintaining such libraries, as well as preparing fermentation-derived samples for high-throughput screening, could be readily performed by persons ordinarily skilled in the art of molecular biology. This is primarily due to the fact that combinatorial libraries of biosynthetic gene clusters can be entirely constructed on shuttle plasmid vectors in *E. coli* and expressed using identical control elements in the same bacterial host, *S. coelicolor*. Furthermore, as controls in high-throughput screens, every producer clone can be compared with numerous clones that are genetically and physiologically identical except for their ability to produce a distinct metabolite. While the structures of metabolites derived via combinatorial biosynthesis are likely to be natural product-like, and therefore relatively complex, their elucidation is likely to be comparatively easier than their natural product counterparts, owing to the fact that they are typically present at fairly high titers and can be substantially enriched via incorporation of isotopically labeled substrates and intermediates.

In addition to the above advantages over conventional natural product screening strategies, the genetic and physiological homogeneity of producer colonies derived via combinatorial mutagenesis opens up completely new opportunities for screening and selection of these classes of compounds. For example, given the availability of a screening strategy based on the death (or survival) of a tester microorganism, it is possible to directly identify an active substance by overlaying a petri dish containing spatially segregated colonies of *S. coelicolor*. Then, again, in at least some cases, one could envision setting up a tunable positive selection system (say, for example, resistance to varying levels of neomycin) that depends on direct interaction between an "unnatural" natural product and a heterologous target protein, both of which are synthesized in *S. coelicolor* itself. Such a system would be useful not just in identifying novel lead molecules but also in their directed evolution toward increasingly potent products. Indeed, in a sense, combinatorial biosynthesis could be viewed as the method of choice for combining the selective power of molecular biology (as demonstrated in phage libraries, for instance) with the proven pharmaceutical track record of natural products.

SUMMARY

Combinatorial biosynthesis has the potential for harnessing the molecular diversity within "unnatural" natural products for drug discovery. Together with new insights into the genetics and enzymology of microbial natural product biosynthesis, recent technological developments in the field have started to facilitate the exploitation of this potential in the polyketide and, to a lesser extent, the nonribosomal peptide families of natural products. Similar strategies could also be adapted to manipulate the pathways of other secondary metabolites, such as glycosides and isoprenoids. In addition to allowing for rapid and convenient screening of these libraries, genetic engineering can also be used to optimize structurally complex lead molecules into drugs. While in vivo microbial biosynthesis has several attractive features for both drug discovery and subsequent development, the availability of active cell-free extracts from similar engineered sources could open yet another interface between combinatorial chemistry and biology through the use of multienzyme systems as catalysts for the turnover of nonnatural substrates with reasonable, but not too constrained, specificity.[20]

ACKNOWLEDGMENTS

The author thanks Gary Ashley, Mary Betlach, and Robert McDaniel for helpful comments on this manuscript. Research in the author's laboratory is supported by a NSF Young Investigator Award, a David & Lucile Packard Fellowship for Science and Engineering, and grants from the National Science Foundation (MCB-9417419) and the National Institutes of Health (CA66736-01).

REFERENCES

1. Nakanishi K, Goto T, Ito S, Natori S, Nozoe S (1974): *Natural Products Chemistry*, vol 1–3. New York: Academic.

2. Rohr J (1995): Combinatorial biosynthesis-an approach in the near future. *Angew Chem Int Ed Engl* 34:881–885.

3. Hutchinson CR, Fujii I (1995): Polyketide synthase gene manipulation—a structure-function approach in engineering novel antibiotics. *Annu Rev Microbiol.*

4. Bartel PL, Zhu CB, Lampel JS, Dosch DC, Conners NC, Strohl WR, Beale JM, Floss HG (1990): Biosynthesis of anthraquinones by interspecies cloning of actinorhodin biosynthesis genes in streptomycetes—Clarification of actinorhodin gene functions. *J Bacteriol* 172:4816–4826.

5. McDaniel R, Ebert–Khosla S, Fu H, Hopwood DA, Khosla C (1994): Engineered biosynthesis of novel polyketides—influence of a downstream enzymes on the catalytic specificity of a minimal aromatic polyketide synthase. *Proc Natl Acad Sci USA* 91:11542–11546.

6. Bedford DJ, Schweizer E, Hopwood DA, Khosla C (1995): Expression of a functional fungal polyketide synthase in the bacterium streptomyces—coelicolor A3(2). *J Bacteriol* 177:4544–4558.

7. Donadio S, Staver MJ, McAlpine JB, Swanson SJ, Katz L (1991): Modular organization of genes required for complex polyketide biosynthesis. *Science* 252:675–679.

8. Cortes J, Wiesmann KEH, Roberts GA, Brown MJB, Staunton J, Leadlay PF (1995): Repositioning of a domain in a modular polyketide synthase to promote specific chain cleavage. *Science* 268:1487–1489.

9. Shen B, Summers RG, Wendt–Pienkowski E, Hutchinson CR (1995): The streptomyces-glaucescens Tcmkl polyketide synthase and TCMN Polyketide cyclase genes govern the size and shape of aromatic polyketides. *J Am Chem Soc* 117:L6811–6821.

10. Stachelhaus T, Marahiel MA (1995): Modular structure of genes encoding multifunctional peptide synthetase required for nonribosomal peptide-synthesis. *FEMS Microbiol Lett* 125:3–14.

11. Tsoi CJ, Khosla C (1995): Combinatorial biosynthesis of unnatural natural-products—the polyketide example. *Chem Biol* 2:355–362.

12. McDaniel R, Ebert–Khosla S, Hopwood DA, Khosla C (1994): Engineered biosynthesis of novel polyketides—activi and activ genes encode aromatase and cyclase enzymes, respectively. *J Am Chem Soc* 116:10855–10859.

13. McDaniel R, Ebert–Khosla S, Hopwood DA, Khosla C (1993): Engineered biosynthesis of novel polyketides—manipulation and analysis of an aromatic polyketide synthase with unproved catalytic specificities. *J Am Chem Soc* 115:11671–11675.

14. McDaniel R, Ebert–Khosla S, Hopwood DA, Khosla C (1995): Engineered biosynthesis of nobel polyketides—analysis of Tcmn function in tetracenomycin biosynthesis. *J Am Chem Soc* 117:6805–6810.

15. McDaniel R, Ebert–Khosla S, Hopwood DA, Khosla C (1995): *Nature* 375:549–554.

16. Fu H, McDaniel R, Hopwood DA, Khosla C (1994): Rational design of aromatic polyketide natural-products by recombinant assembly of enzymatic subunits. *Biochemistry* 33:9321–9326.

17. Fu H, Ebert–Khosla S, Hopwood DA, Khosla C (1994): Engineered biosynthesis of novel polyketides—dissection of the catalytic specificity of the Act ketoreductase. *J Am Chem Soc* 116:4166–4170.

18. Fu H, Ebert-Khosla S, Hopwood DA, Khosla C (1994): Relaxed specificity of the

oxytetracycline polyketide synthase for an acetate primer in the absence of a molonamly primer. *J Am Chem Soc* 116:6443–6444.

19. Fu H, Hopwood DA, Khosla C (1994): Engineered biosynthesis of novel polyketides: evidence for temporal, but not regiospecific, control of cyclization of an aromatic polyketide precursor. *Chem Biol* 1:205–210.

20. Pieper R, Luo G, Cane DE, Khosla C (1995): Cell-free synthesis of polyketides by recombinant erthromycin polyketides synthases. *J Am Chem Soc.*

21. Fu H, Alvarez MA, Hutchinson CR, Khosla C, Bailey JE (1996): Engineered biosynthesis of novel polyketides—regiospecific methylation of an unnatural substrate by the Tcmo o-methyltransferase. Biochemistry 35:6527–6532.

22. Kao CM, Lui G, Katz L, Cane DE, Khosla C (1994): Engineered biosynthesis of a triketide lactone from an incomplete modular polyketide synthase. *J Am Chem Soc* 116:11612–11613.

23. Brown MJB, Cortes J, Cutter AL, Leadlay PF, Staunton J (1995): A mutant generated by expression of an engineered debsi protein from the erthromycin-producing polyketide synthase (PKS) in streptomyces coelicolor produces the triketide as a lacone, but the major product is the noranalog derived from acetate as starter acid. *J Chem Soc Chem Commun* 1517–1518.

24. Niemi J, Ylihonko K, Hakala J, Parssinen R, Kopio A, Mantsala P (1994): Hybrid anthracycline antibiotics—production of new anthracyclines by cloned genes from streptomyces—purpurascens in streptomyces-galilaeus. *Microbiology* 140:1351–1358.

25. Kao CM, Luo G, Katz L, Cane DE, Khosla C (1995): Manipulation of mancolide ring size by directed mutagenesis of a modular polyketide synthase. *J Am Chem Soc* 117:9105–9106.

26. Davis R, Aldrich TL, Nguyen DK, Hendrickson LE, Roach C, Vinci VA, McAda PC (1994): *Abstr of the Genetics of Industrial Microorganisms Meeting, Montreal,* p 288.

27. Stachelhaus T, Schneider A, Marahiel MA (1995): Rational design of peptide antibiotics by targeted replacement of bacterial and fungal domains. *Science* 269:69–72.

28. Piepersberg W (1994): Pathway engineering in secondary metabolite-producing actinomycetes. *Crit Rev Biotechnol* 14:251–285.

29. Liu HW, Thorson JS (1994): Pathways and mechanisms in the biogenesis of novel deoxysugars by bacteria. *Annu Rev Microbiol* 48:223–256.

30. Decker H, Haag S, Udvarnoki G, Rohr J (1995): Novel genetically-engineered tetracenomycins. *Angew Chem Int Ed Engl.*

31. Cane DE (1994): *Isoprenoid Antibiotics.* London: Butterworth–Heinemann.

PART V

SCREENING

22

STRATEGIES FOR SCREENING COMBINATORIAL LIBRARIES

BRUCE A. BEUTEL

Pharmaceutical Products Division, Abbott Laboratories, Abbott Park, Illinois

The synthesis and testing of large numbers of candidate compounds is central to a continuously successful drug discovery effort. In both drug lead discovery and lead optimization, the probability of identifying the most desirable drug candidate increases with compound number even when one is trying to avoid reliance on serendipity. This has led chemists to synthesize and archive an ever-increasing number of compounds and compound classes. The emergence of combinatorial chemistry is a natural result of the need for greater numbers of compounds.

In this quest for higher molecular diversity, there are four distinct types of synthetic methods that are commonly referred to as combinatorial chemistry. The first method is parallel synthesis in which large arrays of compounds are produced simultaneously.[1] This kind of synthesis relies on automation and parallelization to achieve the increase in synthetic productivity, because each product is synthesized in its own reaction vessel. The numbers of compounds that can be made simultaneously typically range from dozens to hundreds, and the reactions can be done in solution or on solid support.

The second method is a particular sort of solid-support-based parallel synthesis that uses chip technology to synthesize the compounds in an ordered array on the surface of a glass slide.[2] This results in an addressable library in which the spatial location of each compound can be used to identify it from the others.

Combinatorial Chemistry and Molecular Diversity in Drug Discovery, Edited by
Eric M. Gordon and James F. Kerwin, Jr.
ISBN 0-471-15518-7   Copyright © 1998 by Wiley-Liss, Inc.

The third method is mix-and-split synthesis (also called portioning–mixing, split, or pool-and-divide synthesis) in which the addition of synthons, monomers, or chemical subunits is done in separate reaction vessels as in parallel synthesis. However, after each synthetic step the products are mixed together and redivided so that the starting material for each step of the synthesis is a more complex mixture than in the previous step.[3,4] This scheme is truly combinatorial in that it leads to an exponential increase in the number of final products compared to the number of reactions that need to be completed, but it still retains some of the control that is present in parallel synthesis with the addition of each synthon being done in isolation of all others. Mix-and-split synthesis is typically done on solid support so that reactions can be forced to near completion and products can be easily purified away from reactants.

The fourth method is random incorporation synthesis in which the reactants are all mixed together and compete with each other in the same reaction vessel to form a mixture of products. This provides the highest possible ratio of products to number of required reactions but is often limited to well-characterized biopolymer (peptides and oligonucleotide) chemistries in which the reaction rates have been equalized to give roughly equimolar mixtures of products.

While chemists have been developing these four related methods of producing molecular diversity, biologists have developed methods for handling large numbers of candidates of one kind or another in various biological tests. Four relevant areas of expertise have developed during roughly the same time period as the four kinds of combinatorial chemistry. The first area is the now routine high-throughput screening, usually in 96-well microtiter plate format. This is essentially parallel screening of compounds relying on automation and parallelization to provide the payoff in numbers of tests performed, and it is analogous to the parallel chemistry that is now used to produce compounds to be tested in this fashion. The second is the ability to label proteins with different fluorescent and radioactive dyes and tags. These tags allow the tracking of soluble proteins when they bind to specific locations which are imaged by fluorescence microscopy or autoradiography.

The third area of expertise is the screening of microparticles for rare characteristics. From working with whole cells and searching for a rare mutant, using cell sorting to count or purify cells with a particular characteristic, to the use of phage display to search for novel peptide ligands, biologists have become skilled in finding the needle in the haystack in a variety of systems that have as their common characteristic the useful fact that the so-called haystack is made of insoluble particles (usually cells) that enable the search methods. The fourth area is the use of complex soluble mixtures in bioassays. From natural product screening to the affinity selection of novel oligonucleotide ligands, methods have been developed to enable the purification of a desired component from a large mixture.

When considering the chemical strategy for a combinatorial chemistry effort it is important to also consider the ramifications on the screening of the planned libraries. The creation of *useful* compounds is not necessarily the same as the creation of the highest number of compounds or the quickest route to the creation of a high number of compounds. Useful libraries must be synthesized with an understanding of how they will be screened and what chemical steps must be taken to ensure the feasibility of a robust screen. This is significantly different from the medicinal chemistry paradigm that has driven drug discovery in which chemists make large quantities of pure compounds and then a biologist tests the compound in any number of assays with little or no communication required between the synthetic chemist and the screening biologist. At the start, separate chemical and biological considerations define the ultimate aim of the research effort. The most fundamental chemical issue is the choice of chemical class desired in the synthesis and the reasonable options to synthesize such compounds. The most important biological issue is the choice of target(s) to be screened with the library.

These choices, of target and chemistry, are usually dictated by the ultimate goal of the research such as the discovery of a promising therapeutic lead compound of a particular chemical class for a particular target. Given these choices, however, many chemical and biological technological details must be addressed. The format of a combinatorial library is not confined to postsynthetic considerations but instead is directly related to the chemistry itself. For example, the choice of solid versus solution-phase chemistry, size and load of each bead if solid phase, number of compounds in each mixture, and identification of compounds from mixtures are all issues that are equally important to the chemist and the biologist.

The synergy between the chemical and biological advances has reached a point where the combination could dramatically affect the odds of success in lead discovery and optimization. The resources and needs of a given drug discovery effort, however, may make the array of choices difficult. Should compounds be made on solid support or in solution? Should compounds be tested in 96-well format? Should they be pure or in mixtures? Many interdependent decisions must be made to ensure a productive chemistry–biology interface that is critical to the success of any combinatorial chemistry program. This review is intended to discuss the strategic and technical issues that should guide these decisions so that combinatorial libraries can be efficiently screened.

SCREENING PARALLEL LIBRARIES

Compounds synthesized by parallel chemistry are technically similar to the standard archived compound libraries common in large pharmaceutical companies. These can be screened in the common high-throughput format, which is usually based on 96-well microplate and robotic liquid handling technology. This is advantageous because existing assay technology can be used so that combinatorial compounds can be seamlessly integrated with noncombinatorial libraries being screened against a given target or assay. There are also classes of compounds for which there is no realistic or efficient combinatorial alternative to parallel synthesis. Finally, testing pure individual compounds is clearly the most quantitative and robust method of screening with the highest yield of information per compound tested.

The disadvantage is that parallel libraries are, in comparison to mix-and-split or random synthesis libraries, going to produce fewer compounds so that the effect on successful lead discovery will be potentially compromised, while the demands placed on high-throughput screening could be great. Except for those cases in which robotics and miniaturization can be enhanced, these methods may not be able to handle even the modest increase in the number of compounds to be tested, even though this increase may be too small to significantly increase the discovery success rate.

Furthermore, while standard high-throughput assays cover a wide range of assay types, including enzyme, cell, and ligand binding assays, this range does not include the ability to discover new ligands for target proteins for which there is no activity assay. With an increasing rate of novel molecular targets discovered by new research in the emerging area of genomics, this reliance on standard high-throughput assays could become much more costly. Pooling compounds into solution-phase mixtures and using an affinity selection procedure with analytical identification of the best ligands (as in affinity selection/mass spectrometry) is a theoretical solution to this problem, but it is a method of screening that calls into question the decision to pursue parallel synthesis in the first place unless it was dictated by chemical feasibility.

SCREENING LIBRARIES ON SOLID SUPPORT

Libraries constructed on glass slides, chips, or beads can be screened using tagged protein targets or enzyme-linked targets and antibodies. By allowing the soluble protein to reach equilibrium in an incubation with the library, most of the protein will be bound to library compounds with the highest affinity. Examples of this kind of screening have been published for panning with libraries on chips and beads.[2,5] A variation on this has been the use of flow cytometry to quickly identify and separate library beads that have the highest signal from bound tagged protein, thereby purifying that library compound from the others.[6] This kind of screening is useful for purified molecular targets, and is therefore complementary with many high-throughput assays. Other advantages include the inherent efficiency of the miniaturization of the library and that the library can be washed to remove any bound targets and reused to screen other targets.

The primary disadvantage is not being able to screen other kinds of targets, such as those that rely on cell or enzymatic assays. In general, testing small molecules while they are tethered to a solid support also raises concerns about steric hindrance of the interaction target as well as kinetic issues that could make this kind of screening prone to both false negatives and positives.

SCREENING MIX-AND-SPLIT LIBRARIES

Libraries produced by mix-and-split synthesis on resin include a growing number of novel oligomers and nonoligomeric small molecules.[7,8] Though these libraries allow handling of thousands to millions of compounds at a time on solid support mixtures, the challenge in screening these libraries comes from the low (usually less than 1 nmol) yield of compound on each bead. A further challenge is due to the libraries being synthesized in mixtures, so that the precise identity of the compound on each bead is unknown.

There is, however, an advantage to screening these libraries that can offset these challenges. The singularly useful feature of these libraries is the fact that each bead has a unique reaction history and therefore a small solvable set of products. With proper library validation, the majority of beads will contain predominantly the single desired product of the synthetic history. Compounds can be "purified" for analysis or testing simply by removing individual beads from the mixture. This presents screening opportunities that are essentially the best of both worlds, with all the advantages of solution-phase pooling without the disadvantages of identifying compounds from intractable solution phase mixtures. Mix-and-split libraries can be screened either as individual beads with compounds attached or removed from each bead, or as pools of compounds in solution.

The screening procedures all begin with the isolation of active compounds or beads. Following isolation of the beads of interest, the compounds on the beads are identified, resynthesized by traditional methods, and retested. In this way, the focus of these procedures is primarily on the potentially active compounds in an initial screen, compared to the equality of focus that is present in a screen of purified compounds in which an equal amount of data is generated and stored for every compound. Several methods are available for screening these kinds of libraries, including gel diffusion, affinity selection, and a modified form of microtiter plate based high-throughput screening. These methods are complementary and together make mix-and-split libraries potentially useful for the widest range of targets and assays.

The screening method that best uses the advantages of mix-and-split synthesis is the gel-diffusion (or gel-permeation) method. There are examples of cell assays and enzyme assays that have been adapted to this method, which is essentially a technique for assay miniaturization.[9-11] Beads can be spread or arrayed at very high density either on top of, inside, or under an agarose gel layer and the compounds can be released either by gas-phase acid cleavage before addition to the gel or by photocleavage inside the gel. The family of techniques relies on a relatively slow diffusion of compounds in the vicinity of each bead so that the activity of each compound can be tested in near isolation from the others. This is achieved by having all assay components (cells, enzymes, substrates, etc.) present in a gel-phase matrix or media-free cell layer so that the combinatorial compounds can be active in an environment that prevents mixing of compounds but still allows the testing of pseudo-solution-phase activity.

A randomly spread layer of beads can be tested in high enough density to test thousands of compounds at once in a small (<5 mL) gel volume, effectively lowering the assay volume to a few microliters per compound tested. The low effective volume compensates for the low compound yield on each bead so that even weak hits (100 μM K_i) can be discovered if so desired. This method is quick because >10,000 compounds can be tested simultaneously without the aid of robotics, and by using partial cleavage and repetitive testing one can isolate individual beads to identify which compounds are active. Identification can be achieved by direct analytical methods such as mass spectrometry, or by the use of an encoded library.[2,12–14] Recent work shows that direct NMR detection and identification of bead-bound compounds is on the horizon.[15]

Because the gel-diffusion method is essentially a method for miniaturizing standard microtiter plate activity assays in a format that facilitates the introduction of compounds synthesized on beads, it can only work on a subset of the same kinds of assays that can be used in standard high-throughput activity screening. Enzyme assays are the easiest to adapt to the gel-diffusion method, though certain cell assays are also possible and can be developed on a case by case basis. Ligand-binding assays are likely to be the most difficult to adapt because this methodology is best used for homogeneous assays. Like microplate screening, gel diffusion has the disadvantage of not being useful for any purified targets for which there is no developed activity assay or reagents and substrates for such an assay. Mix-and-split libraries, however, can be screened by affinity selection methods that are specifically designed to identify novel ligands for any purified target.

Examples of affinity selection have been published initially for oligonucleotide libraries but more recently for small molecule libraries.[16–20] In affinity selection, compounds are initially cleaved from the resin and eluted as large solution-phase mixtures. These mixtures can contain as many as 1000–100,000 compounds in practice, depending on the precise affinity selection technology being used. Mixtures are subjected to a target-based affinity purification procedure in which the highest affinity ligands are purified from the other library members. Many chromatographic methods that can partition bound ligands from unbound compounds, including size exclusion, membrane filtration, filter binding, and solid-phase target affinity capture, can be used to purify ligands that are bound to a target protein. Iterative affinity purification can result in ever decreasing background copurification of starting library, so that even extremely rare library members can represent the majority of library remaining after selection.

The selection progress can be tracked so that binders are observed by direct physical methods, such as mass spectrometry, or the compounds can be quantitatively measured by the use of libraries synthesized with a label on each compound (radiolabel or fluorescent, for example). Identification of the bound compounds can be done directly with mass spectrometry or other spectroscopic techniques. Alternatively, compound identities can be deduced by indirect methods such as subpool testing and deconvolution.[21] Affinity selection can require large multimilligram quantities of purified target protein because the binding is driven by the protein concentration (as compared to activity assays in which the active compounds drive binding).

There is yet another option for screening mix-and-split libraries, particularly for those targets for which there is neither adequate quantity of purified target for affinity selection nor the ability to adapt an assay to the gel diffusion method. By dispensing single beads into microtiter wells and modifying a standard microplate formatted assay, virtually any microplate assay can be screened with mix-and-split libraries. A major consideration is minimizing volumes for testing the compounds on single beads to provide enough compound for testing at a reasonable stringency. If a bead contains only 100 pmol of compound, a typical 100-μL assay will have only 1 μM of each compound tested. This is considerably lower than the concentration that would normally be tested in a primary lead discovery screen. The only way of having replicate points in such a screen is to dispense replicate plates after the compounds are cleaved from the beads, thus further decreasing the amount of compound available per test. The fact that the beads are initially mixed presents a problem as well, because any hits must be identified before any other testing can be done to confirm the activity. (This is especially a problem with microplate-based assays with higher false positive rates).

For these reasons, an adapted microplate screen is considerably different from a standard high-thoughput screen with purified compounds in which the microwell location itself identifies the active compound and one can easily go back to a large stock of purified compound for further testing. Nevertheless, with proper adjustments to the logistics of a screen, the products of individual beads can be tested with replicates and individual beads can be chosen on this basis for identification. The throughput and efficiency of such a screen is certainly lower than the throughput of either gel-diffusion or affinity selection screening. This is at least partially due to the need to dispense the individual beads before testing. More generally, mix-and-split libraries are simply not an ideal match for microplate-based screening. Still, there are certain specific assays for which there may be no other choice.

SCREENING RANDOM INCORPORATION LIBRARIES

Biooligomeric libraries, such as peptides and oligonucleotides, can be produced by random incorporation synthesis. The primary advantage of these libraries is the ease with which vast numbers of unique structures can be made. In cases in which a peptide or oligonucleotide ligand is desirable (discovery of new binding sites for research purposes, for example) or in cases in which the ease of synthesis or screening is a critical factor, these libraries can be lucrative. When random synthesis is used, the library must be screened as a solution-phase mixture because each bead typically contains the entire set of library compounds. This limits the screening options to only two categories of methods. Libraries can be screened as mixtures in activity assays and active compounds can be recursively identified by synthesis and testing of ever smaller subsets.[22]

The possibility of weakly active compounds interfering with such a pooling strategy is serious. The most active pools may not contain the single most active compounds. This problem was evident in one particular published example,[23] but there is no other method for screening directly on the basis of activity with these kinds of libraries. An alternative is to screen by affinity selection, as has already been described for mix-and-split libraries. The examples of affinity selection on random synthesis oligonucleotide libraries, in particular, show the power of this method in terms of sheer numbers of compounds screened.[16-18] Iterative affinity selection can result in the enrichment of library members as rare as one part per trillion so that they are the majority of library remaining after selection. The identification of ligands can be directly achieved by cloning if oligonucleotides are being screened. If the number of compounds is lower, then the identification options that were discussed for affinity selection with mix-and-split libraries, such as mass spectrometry or subpool deductive analysis, are also possible.

Random incorporation libraries are traditionally extremely large (millions to trillions of library members are common), and the methods for screening them are powerful enough to handle these numbers even without the advantage of having one compound per bead. It should be kept in mind, however, that affinity selection can require relatively large quantities of protein targets (tens of milligrams), and biooligomer libraries are not normally considered to be the kinds of compounds that would be chosen by chemists interested in drug discovery against a variety of molecular targets. If the chemistry is altered enough to include a variety of unnatural amino acids or nucleotides (or to the point of being completely artificial oligomers) the chemistry can no longer be done by random synthesis, and mix-and-split synthesis becomes the next viable option. Once the decision to do mix-and-split synthesis is made, however, the biological advantages that enabled some of the screening and identification techniques are lost, and the unusually large number of unique species that is the hallmark of the biooligomer libraries is significantly decreased. Such modified oligomers become only one class of many that can be produced and screened by the mix-and-split methods available.

SUMMARY: COMBINATORIAL DISCOVERY STRATEGIC ISSUES

Most biological targets and assays are variations on four general types: enzyme assays, cell assays, ligand competition binding assays, and purified targets with

TABLE 22.1 Combinatorial Screening Options

	Enzyme	Cell	Ligand Binding	Pure Target Available
Parallel	Microplate	Microplate	Microplate	—
Solid array	—	—	—	Target binding
Mix and split	Gel-diffusion Microtiter (bead)	Gel-diffusion Microtiter (bead)	Microtiter Gel diffusion (bead)	Affinity select Target binding
Random incorporation	Microplate (pools)	Microplate (pools)	Microplate (pools)	Affinity select

Note. Four kinds of assays are cross-indexed with four combinatorial chemistry formats to show the available screening formats.

no available activity assay. Given any one of these and a combinatorial library produced by one of the four synthetic methods previously described (parallel, solid-array, mix-and-split, or random incorporation) there are limited technological options for screening a target with a library. The available screening options are shown in Table 22.1 and have been discussed in some detail above. Summarized in this form, it becomes clear that no single screening method can handle all kinds of biological targets. Furthermore, even with the use of a repertoire of screening techniques, only mix-and-split libraries and random incorporation libraries can handle the four kinds of targets/assays. Parallel synthesis provides compounds that are the most similar to those found in existing pharmaceutical compound archives. These libraries are, therefore, widely touted to be best for screening with existing assays, but the fact is that there is no existing high-throughput assay for many targets that one would like to screen. As noted earlier, this will become increasingly more problematic when genomics has its expected impact on target discovery.

The synthetic methods listed in Table 22.1 are also shown in ascending order of synthetic potential, in terms of the sheer numbers of compounds that can be easily produced. In theory, it is always desirable to synthesize and test larger libraries if this can be done without undue additional cost or time. It is interesting and fortunate, therefore, that those libraries that have the highest potential payoff in compound numbers (mix-and-split and random incorporation) are also the ones that have the best potential to be screened against all available and future targets. Of these two kinds of methods, the mix-and-split methods are much more widely applicable to a growing range of chemistries and compound classes. The challenge will be to make the quality of mix-and-split libraries as close as possible to the quality achieved by parallel synthesis.

The most important factor for a discovery effort is to make sure that the chemistry and biology work together. Certain combinations (those with no entry in Table 22.1) would make no sense, such as investing in solid-array library technology with a primary biological interest in cellular targets and assays. Other combinations may make more or less sense depending on the goals of a particular research effort.

As an example, a small organization interested in quickly finding any active compounds with little available resources, could be successful using mix-and-split libraries, or perhaps bio-oligomer libraries made with random incorporation synthesis. If that organization has high quantities of purified targets are available, affinity selection could be a very cost-effective route to target validation. Without the availability of large multimilligram quantities of purified target proteins, however, it would be much more advantageous to work with targets for which there are activity assays that use less, if any, purified target; therefore, affinity selection technology is probably not worth an investment compared to standard microplate and/or gel-diffusion technology.

A second example could be a large organization with many existing microplate assays and the need for optimization and structure–activity relationships (SAR) on many existing initial leads that can only be synthesized with solution-phase chemistry. In this case, parallel libraries are the best choice, require no new screening technology development, and would provide the most quantitative SAR information. As a final example, a large organization with many existing targets and the expectation of many new target discoveries in the future would benefit from the synthesis of large mix-and-split libraries with the combination of gel-diffusion, affinity-selection, and adapted microtiter assay methods all being used to address the broadest range of future targets with a large set of new diverse compounds.

Whatever the combination is that works for a particular research effort, there are fundamental differences in these kinds of technology. Not all combinatorial chemistry will provide the same results for a given task, and not all high-throughput screening is equally capable of providing the critical interface between particular targets of interest and the particular compounds in hand. If a researcher is fortunate, the method of chemical synthesis (possibly dictated by the structure of the desired compound series), and the assay technology (possibly dictated by the desired biological target), will still allow a choice of screening methods that can be used with the available resources.

REFERENCES

1. Hall S (1998): In Gordon EM, Kerwin JF Jr, eds. *Combinatorial Chemistry and Molecular Diversity in Drug Discovery.* New York: Wiley, p 291–306.

2. Jacobs JW, et al. (1998) In Gordon EM, Kerwin JF Jr, eds., Combinatorial Chemistry and Molecular Diversity in Drug Discovery. New York, Wiley, p 111–131.

3. Furka A, Sebestyen F, Asgedom M, Dibo G (1998): Cornucopia of peptides by synthesis. In *Abstr 14th Int Congr Biochem, Prague, Czechoslovakia,* Vol 5, p 47.

4. Furka A, Sebestyen F, Asgedom M, Dibo G (1991): General method for rapid synthesis of multicomponent peptide mixtures. *Int J Protein Res* 37:487–493.

5. Lam KS, Salmon SE, Hersh EM, Hruby VJ, Kazmierski WM, Knapp RJ (1991): A new type of synthetic peptide library for identifying ligand-binding activity. *Nature* 354:82–84.

6. Needels MC, Jones DG, Tate EH, Heinkel GL, Kochersperger LM, Dower WJ, Barrett RW, Gallop MA (1993): Generation and screening of an oligonucleotide-encoded synthetic peptide library. *Proc Natl Acad Sci USA* 90:10700–10704.

7. Simon RJ, Martin EJ, Miller SM, Zuckermann RN, Blaney JM, Moos WH (1994): Using peptoid libraries [poly N-substituted glycines] for drug discovery. *Tech Protein Chem* 5:533–539.

8. Baldwin JJ (1998): In Gordon EM, Kerwin JF, Jr. eds., *Combinatorial Chemistry and Molecular Diversity in Drug Discovery,* New York: Wiley, p 181–188.

9. Jayawickreme CK, Graminski GF, Quillan JM, Lerner MR (1994): Creation and functional screening of a multi-use peptide library. *Proc Natl Acad Sci USA* 91:1614–1618.

10. Quillan JM, Jayawickreme CK, Lerner MR (1995): Combinatorial diffusion assay used to identify topically active melanocyte-stimulating hormone receptor agonists. *Proc Natl Acad Sci USA* 92:2894–2898.

11. Sigal NH and Chelsky D (1998): In Gordon EM, Kerwin JF, Jr. eds., *Combinatorial Chemistry and Molecular Diversity in Drug Discovery,* New York: Wiley, p 433–443.

12. Brummel CL, Vickerman JC, Carr SA, Hemling ME, Roberts GD, Johnson W, Weinstock J, Gaitanopoulos D, Benkovic SJ, Winograd N (1996): Evaluation of mass spectometric methods applicable to the direct analysis of non-peptide bead-bound combinatorial libraries. *Anal Chem* 68:237–242.

13. Haskins NJ, Hunter DJ, Organ AJ, Rahman SS, Thom C (1995): Combinatorial chemistry: direct analysis of bead surface associated materials. *Rapid Commun Mass Spectrosc* 9:1437–1440.

14. Ni Z-J, Maclean D, Holmes CP, Gallop MA (1996): Encoded combinatorial chemistry: binary coding using chemically robust secondary amine tags. *Methods Enzymol* 267:261–272.

15. Sarkar SK, Garigipati RS, Adams JL, Keifer PA (1996): An NMR method to identify nondestructively chemical compounds bound to a single solid-phase-synthesis bead for combinatorial chemistry applications. *J Am Chem Soc* 118:2305–2306.

16. Bock LC, Griffin LC, Latham JA, Vermaas EH, Toole JJ (1992): Selection of single-stranded DNA molecules that bind and inhibit human thrombin. *Nature* 355:564–566.

17. Tuerk C, Macdougal S, Gold L (1992): RNA pseudoknots that inhibit human immunodeficiency virus type 1 reverse transcriptase. *Proc Natl Acad Sci USA* 89: 6988–6992.

18. Ellington AD, Szostak JW (1990): In vitro selection of RNA molecules that bind specific ligands. *Nature* 346:818–822.

19. Zuckermann RN, Kerr JM, Saini MA, Banville SC, Santi DV (1992): Identification of highest affinity ligands by affinity selection from equimolar peptide mixtures generated by robotic synthesis. *Proc Natl Acad Sci USA* 89:4505–4509.

20. Evans DM, Williams KP, McGuinnes B, Tarr G, Regnier F, Afeyan N, Jindal S (1996): Affinity-based screening of combinatorial libraries using automated serial-column chromatography. *Nature Biotech* 14:504–507.

21. Beutel BA (1995): Identification of groups at defined positions in a compound comprises determining differential binding to target molecules in a compound library preferably in a reiterative process. *PCT Patent WO9527072.*

22. Houghten RA, Pinilla C, Blondelle SE, Appel JR, Dooley CT, Cuervo JH (1991): Generation and use of synthetic peptide combinatorial libraries for basic research and drug discovery. *Nature* 354:84–86.

23. Dooley CT, Chung NN, Wilkes BC, Schiller PW, Bidlack JM, Pasternak GW, Houghten RA (1994): An all-D-amino acid opioid peptide with central analgesic activity from a combinatorial library. *Science* 266:2019–2022.

23

APPROACHES AND TECHNOLOGIES FOR SCREENING LARGE COMBINATORIAL LIBRARIES

Nolan H. Sigal and Daniel Chelsky

Pharmacopeia, Inc., Princeton, New Jersey

STRATEGIC ISSUES IN SCREENING

Two forces have combined to present significant challenges for the design and implementation of novel high-throughput screening strategies: the ability to create large libraries of compounds through combinatorial chemistry and the identification of thousands of potential therapeutic targets through the power of molecular biology. The challenges for assay design are expected to intensify over the next decade as the power and sophistication of combinatorial chemistry methodology increases, along with the explosion of molecular targets identified through the human genome project.

These forces may necessitate fundamental changes in how compounds are identified and developed for clinical testing. Traditionally, the first step in the drug discovery process involves the identification of a molecular target which, based on its physiology and hypothesized role in a pathophysiologic state, is considered a useful point for therapeutic intervention. Next, significant resources are committed to (1) genetic and pharmacologic validation of the target and (2) identification of lead compounds through de novo rational drug design, and high-throughput screening of sample collections and natural product broths. Progress in these activities is traditionally necessary to justify the allocation of scarce medicinal chemistry resources for the next phase of

Combinatorial Chemistry and Molecular Diversity in Drug Discovery, Edited by
Eric M. Gordon and James F. Kerwin, Jr.
ISBN 0-471-15518-7   Copyright © 1998 by Wiley-Liss, Inc.

the discovery process: lead optimization. This approach to lead identification and optimization has worked well when significant medicinal chemistry resources can be assigned to a target that has been well validated or to optimize the pharmacologic parameters of a known lead compound.

With the plethora of molecular targets available, it is unclear that scientists can use this conventional approach in a manner that will yield a steady stream of effective therapeutic agents. Combinatorial chemistry has the potential to alter this paradigm by improving the productivity of lead identification and optimization. Rather than choosing one member of a gene family that has been validated in the traditional manner, establishment of a panel of biochemical assays for genetically related targets provides the opportunity to generate selective lead molecules. The improved efficiency in lead identification provided by combinatorial chemistry can drive the process of target validation, which can affect the decision to allocate pharmacology and medicinal chemistry resources for final lead optimization. Such a strategy is feasible, however, only if the screening of the combinatorial libraries can be seamlessly interfaced into the drug discovery process.

This review discusses some of the challenges and opportunities in screening combinatorial libraries. Our focus is on the analysis of libraries produced on solid support by the mix-and-split approach. This strategy, first described by Furka et al.,[1] can generate libraries of thousands to millions of compounds in which each bead contains a single compound. While the screening of such libraries can superficially resemble high-throughput assays set up for evaluation of natural product broths and compound collections, the libraries also have special attributes that may provide some limitations in screening flexibility, but can yield some advantages as well. For example, the beads most commonly used in solid-phase synthesis contain 300 pmol of functional "handles" onto which compounds can be attached. If one wishes to run an assay in solution and removes all of the compound from the bead, one is limited to micromolar concentrations of compound in the 96-well plate format. While many scientists find this an adequate concentration for screening, since leads identified at this potency have a high probability for optimization to the nanomolar level, others may need to reconsider their screening paradigm.

Another consequence of the mix-and-split approach that has implications for traditional high-throughput screening is that the identity or structure of the "active" compound in a combinatorial library is not known until after it is found to be active in the screen. Assay of combinatorial library compounds represents a statistical sampling of all of the possible compounds in the library. Whereas traditional screens confirm their actives by running duplicate or triplicate samples, library screening involves the identification of the same structure from the library following a random sampling of its members. Thus, after sampling twice as many beads as are present in a library, one will have assayed 86% of its members at least once; there will be a 60% probability of seeing a duplicate structure and a 26% probability of a triplicate structure. Given the potential large size of combinatorial libraries, screening assays must be truly high throughput and robust, with low probability of false positives.

ON-BEAD ASSAYS

One method of rapidly analyzing bead-based combinatorial libraries is to perform on-bead assays. Typically, fluorescent- or radiolabeled receptors are prepared and incubated with the combinatorial library beads. After extensive washing, labeled beads may be selected manually by fluorescence microscopy or following exposure to film or in an automated fashion by a modified fluorescence-activated cell sorter (FACS), which sorts particles on the basis of their light-scattering or fluorescent characteristics. Such methods have been particularly useful for identifying the peptide epitope against which a monoclonal antibody interacts. One strategy that integrates spatial encoding of combinatorial libraries with on-bead fluorescent assays is the VLSPS method, described by scientists from Affymax.

While the throughput of on-bead assays is unparalleled, several pitfalls should be considered before this approach is undertaken. First, the method is largely confined to assays in which relatively pure reagents exist in soluble form; intracellular targets cannot be addressed with on-bead assays. Second, great care must be taken in establishing appropriate specificity controls, since any of the reagents used in the assay may nonspecifically interact with library compounds. Third, because of the hydrophobic environment of the bead, the library compound may exist in a conformation on the bead that is considerably different than that in solution. Furthermore, the compounds attached to the bead exist in a highly multimeric form. Therefore, the best utilization of on-bead assays may be when a multimeric ligand interaction is desired, such as in the case of antibodies or receptors. Alternatively, on-bead assays can be used to preselect a subset of compounds for more detailed, and time-consuming, analysis following elution of compound from the beads.

OFF-BEAD (SOLUTION) ASSAYS

A variety of linkers are available that can be attached to the functional handles on the bead matrix and permit detachment of compound from the bead. Among the most commonly used linkers are several that utilize ultraviolet light for compound detachment and others that are acid-labile. The former has the advantage of being "reagentless"; that is, no additional chemicals are introduced into the assay mix when the compound is released. While all linkers possess some lability with respect to certain chemical reactions that may be used in library construction, the sensitivities of photo-labile and acid-labile linkers are somewhat different. The greatest flexibility in library design can be achieved by choosing a linker that best matches the chemistry to be used in library construction.

When a library compound is removed from the bead, in theory, design and implementation of a screening assay is no different than evaluation of a compound from a sample collection. This is particularly relevant with respect to the range of molecular targets that combinatorial libraries can be assayed against. In practice, however, there are a number of differences, which potentially can speed assay throughput but also require attention. For example, in contrast to sample collection compounds, which have to be weighed out and dissolved individually, small amounts of each library compound are attached to beads and can be released directly into solution. Thus, while the need for time-consuming compound weighing and solubilization is eliminated, special methods need to be considered for moving the beads, representing small, premeasured packets of compound, into a format that makes it easy to directly integrate into the assay format.

The size of combinatorial libraries often necessitates the assay of mixtures of compounds. Deconvolution strategies require the survey of compound mixtures, but with encoded libraries, eluates from either single or multiple beads can be assayed. In deciding on a plan for assaying combinatorial libraries, one must determine whether bead manipulation or liquid manipulation will be used to generate compound mixtures. In one strategy, individual beads are placed in a container, such as a 96-well plate, the compound is released from the bead into solution, and, through commercially available robotic liquid handling systems, aliquots of the solutions are mixed together. This approach obviates the need to control the kinetics of compound release from the bead (see below), but places a considerable burden on liquid handing for all but the smallest of combinatorial libraries.

An alternative approach to generating compound mixtures is to place multiple beads in each well of the assay plate and remove part of the compound from each bead. Compound eluates are then transferred to another 96-well plate for assay, and when an active well is identified, the multiple beads from the active well are rearrayed into a new plate. The remainder of the compound is removed from the single-bead-containing wells and assayed. The active bead can then be analyzed for compound identification. Two features distinguish this approach. First, a premium is placed on accurate and efficient bead manipulation, in contrast to fluid manipulation. Second, the viability of this strategy is dependent on the ability to reproducibly control the release of compound from the bead. Typically, one-third to one-half of the compound is removed on first elution and the remainder on the second.

The choice of single-compound versus multiple-compound arraying depends on the expected frequency of actives in the library. At Pharmacopeia, random libraries, typically consisting of 50–100,000 members, are designed to maximize structural novelty and chemical diversity and serve as appropriate starting points when no structural or pharmacological information is available to guide library design. In random compound screening, when the frequency of actives is very low, the evaluation of mixtures of 10 to 30 compounds per well permits the efficient screening of such libraries. When libraries are constructed around active structures, however, the ability to assay mixtures diminishes as the frequency of active structures increases. The inhibitory effect caused by presence of a number of weakly active structures in a well obscures the capacity to distinguish such "false positives" from truly active wells.

Many of the principles discussed in the preceding paragraphs are illustrated in a recent publication from Pharmacopeia[2] describing the identification of potent inhibitors of carbonic anhydrase. In this model system, beads from two structurally distinct combinatorial libraries were arrayed individually in microtiter plate wells, and compounds were fully eluted from the beads. The high frequency of active compounds necessitated the single-bead approach, while the relatively small size of the two model libraries used made this strategy feasible. Selectivity analysis was performed on a 217-member sublibrary by dividing bead eluates between assay plates containing two isozymes of carbonic anhydrase. While this approach worked quite well in the identification of potent, selective carbonic anhydrase inhibitors from small libraries designed to contain a known pharmacophore for the enzyme, the general strategy used at Pharmacopeia for the evaluation of larger, random compound libraries follows that discussed in the preceding paragraph.

GEL-PERMEATION ASSAYS

The gel-permeation assay is based on the zone of inhibition assays used in antimicrobial screening. A population of library beads is typically distributed in a thin agarose matrix in which the biological assay is performed and subjected to photolytic release of the attached compound. As compound slowly diffuses out of the bead, a zone of high concentration is created in the immediate vicinity. In an appropriately constructed assay, active compounds are detected as zones surrounding the parent bead. The parent bead can then be removed and decoded to determine structure.

The gel-permeation approach to screening, while simple in concept, is a very powerful tool. It has the potential to address two critical needs in high-throughput screening of combinatorial libraries. First, the approach permits the rapid screening of large compound libraries with very little effort expended on assay setup or compound arraying. Active compounds can often be detected visually against a background of thousands of less active or inactive compounds. Second, there is an opportunity to expand the amount of information that can be derived from each compound prior to its resynthesis. Since the effective volume of the assay is very small, only a portion of the available compound is necessary to produce a high concentration in the area immediately surrounding the bead. By regulating photolysis, much of the compound can be left on the bead for subsequent screening, such as activity confirmation or selectivity assays.

The gel-permeation assay is adaptable to a variety of screening strategies depending on the frequency of active compounds in a library or libraries. If the library is large and the number of actives is small, screening thousands of beads at high density is very efficient. In this situation, however, the zone of observed activity may contain several beads and the active bead may not be distinguishable from its inactive neighbors. All beads in the zone can then be removed and rearrayed at lower density. Additional compound is photolyzed from the bead, and the active bead, now well separated from the other library members, is easily distinguished. When the number of test compounds on beads is not limiting and the number of assays available is large, it may not be possible to completely evaluate each library. Under these circumstances, one can rapidly survey libraries for actives. Without the need for retrieving beads or decoding tags, libraries can be screened at high density with the goal of statistically evaluating the library. Only when a library is seen to contain actives of the appropriate number and potency is that library more thoroughly investigated.

When many active compounds are found in a library, gel-permeation assays can be utilized to distinguish relative potencies of library compounds. As compound diffuses out of the bead it forms a concentration gradient, which decreases with distance from the bead. Each active compound has a minimum concentration at which it can produce a detectable signal in an assay, related to its K_i value. Highly potent compounds require a lower concentration for detection than weaker compounds. Therefore, if similar amounts of compound are released from each bead, potent compounds will have a detectable effect further from the bead than weak compounds at any given time.

We have adapted the carbonic anhydrase assay to the gel-permeation format to illustrate some of the principles discussed above. The strategy of surveying large, random libraries has been employed to search for novel structural motifs active against this enzyme. In addition, we have assayed the smaller, active libraries described previously[2] to compare active structures derived from the two formats and to assess the ability of the gel-permeation format to provide quantitative information. The latter point can be observed, for example, by comparing compounds with K_i values of 4 and 660 nM (Figure 23.1). The more active compound creates a larger zone of activity, which also lasts longer than the weaker compound. Since the amount of compound can be regulated with photolysis time, it is also possible to eliminate weaker compounds altogether by reducing photolysis so that only the most potent compounds appear (Figure 23.2).

Most of the assays to date have been performed with library beads randomly arrayed in the agarose layer of the gel-permeation assay. In the evaluation of active libraries, it may be particularly desirable to array the beads in a way that allows for their evaluation in a set of related assays. In this case, a fixed array that can be moved from one assay to another may work well. For example, a thin glass array can be constructed with many conical pores, each containing one bead and allowing for transfer of compound from the bead to

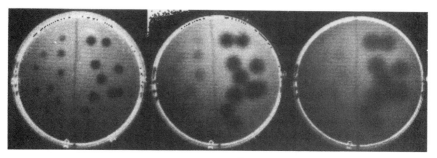

Figure 23.1. Comparison of two compounds, synthesized on beads, in the gel permeation assay. Beads containing a potent (K_1 = 4 nM, right side of plate) and a weaker (K_i = 660 nM, left side of plate) inhibitor of carbonic anhydrase were arrayed in a thin agarose layer containing the enzyme and photolyzed for 8 min. A substrate, fluorescein diacetate, was poured over the first layer and fluorescence images were collected with a CCD camera at 5, 30, and 105 min.

the assay below. The zone of activity can easily be correlated with a single bead (Figure 23.3). The glass array can then be removed and transferred to a subsequent assay for elution of additional compound.

In theory, any assay that can be performed in a 96-well mictrotiter plate can be adapted to the gel-permeation format. Enzyme assays are typically set up in two layers, one containing the beads and enzyme and the second containing the substrate, added to start the reaction after photolysis of the beads. Either the substrate or product must generate a detectable signal, preferably

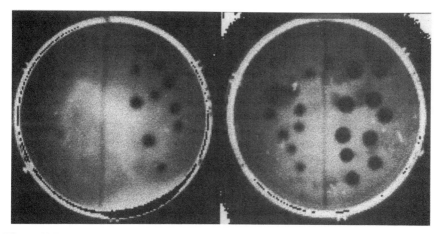

Figure 23.2. Varying compound concentration by photolysis time in the gel permeation assay. Two compounds, described in Figure 23.1, were arrayed in a carbonic anhydrase assay and photolyzed for 2.5 or 20 min. Images were collected 5 min after addition of substrate.

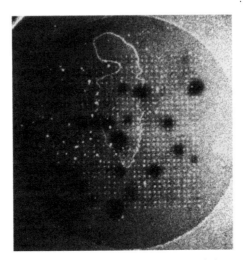

Figure 23.3. Glass array with 900 conical holes, each containing one library bead. The array was photolyzed to release compound and placed onto a gel permeation assay containing carbonic anhydrase. Inhibitors are seen as dark regions surrounding the hole containing the parent bead.

fluorescent, and the difference between them must be readily distinguished. While a range of approaches fit this concept, conversion of a nonfluorescent substrate to a highly fluorescent product is the most robust solution. An example is the conversion of fluorescein diacetate to fluorescein by an esterase such as carbonic anhydrase. A fluorescent substrate that is converted to a product with different excitation and emission characteristics can also work. By choosing appropriate bandpass filters that excite and detect only the product, a substrate such as peptidyl-4-trifluoromethyl coumarin amide can be effectively invisible until converted by a protease, such as thrombin, to the free 7-amino-4-trifluoromethyl coumarin.[3]

While enzyme assays may be the simplest in concept to adapt to the gel-permeation assay, one cell-based approach has been demonstrated using melanophore cells derived from *Xenopus laevis* epithelium.[4] These pigmented cells respond to agonists of G-protein-coupled receptors located on their surface. Depending on the G-proteins responsible for coupling the specific receptor, the pigment will either disperse, darkening the cells, or aggregate, lightening the cells. The cells can be grown on a standard tissue culture plate and overlaid with low-melt agarose. Beads can be photolyzed and applied to the surface of the agarose layer by applying a template containing an ordered array of beads or by inverting a layer of plastic film covered with a random array of beads. Compound diffuses down to the cells and can induce a response, if active. By coaddition of an agonist, screening for antagonists that block the agonist response is also possible. Figure 23.4 depicts the response of cells to a library containing compounds that stimulate an endogenous *Xenopus*

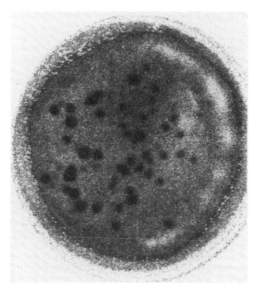

Figure 23.4. Response of *X. laevis* melanophores to compound released from library beads. Beads were arrayed on a nylon filter mesh, photolyzed, and placed over an agarose layer covering the melanophores, which are adherant to the bottom of the dish. Pigment dispersion with resultant darkening of the cells is due to an agonist effect on endogenous receptors.

receptor. These cells have also been transfected with human receptors,[5] which will respond in the same way.

One of the challenges to expanding the range of assays amenable to the gel-permeation format is the ability to detect active compounds in a large-field format. Most instruments for signal detection (i.e., fluorescence, absorbance and luminescence) are available only in cuvette or 96-well formats. Fortunately, a wide range of CCD cameras are now available that offer the ability to detect light from any relatively flat field. With the addition of appropriate optical filters, this detection can be quite selective. Selectivity is particularly important when both the substrate and product have significant but distinguishable absorbance or fluorescence characteristics. Optical filters can effectively reduce or eliminate visualization of the substrate, for example, greatly improving the ratio of product signal to background. Sensitivity can also be very important when the overall light intensity generated by the product is low. In the case of the carbonic anhydrase assay, signal is very robust with no appreciable background generated by the substrate and a product that is highly fluorescent in the visible range. A simple, inexpensive camera will suffice in this case. Most assays, however, produce much less signal but can still be detected quite easily with one of a variety of much more sensitive and low-noise CCD cameras on the market.

MICROWELL FORMATS

Microtiter plates provide a highly standardized 96-well format for screening moderate numbers of samples in relatively small volumes. There are situations, however, when a higher throughput or a smaller assay volume, to conserve limiting reagents, is required while maintaining separation between samples. For example, by reducing volume to 10 μL or less, the relatively small amount of compound available in solid-phase combinatorial libraries can be stretched over multiple assays, typically to give an activity profile on specific compounds in related screens.

Due to the large number of liquid handling and detection instruments available in the 96-well format, there is significant pressure to remain in that format, or a multiple thereof, while reducing volumes. There currently exist several small-volume wells in the 96-well format that allow for assays in the 10-μL range. In most cases, only the reduced signal resulting from the smaller reaction volume is a limitation. A variation on the 96-well theme are 384- and 864-well plates, which place four or nine wells in the space of one standard well. While these plates can be used with several automated liquid-handling devices, mechanical tolerances are often insufficient for the smallest well sizes. Detection can also be limiting, since most 96-well plate readers cannot be adapted to the denser arrays.

In response to significant interest, both liquid handling and detection of alternative assay formats are being addressed by several manufacturers. Recent models of automated liquid-handing instruments can routinely deal with 384-well plates and are moving toward tolerances that will handle smaller wells. Novel approaches, potentially allowing for submicroliter volumes, may be possible by the use of modified ink jet printers with their low delivery volumes and large number of closely packed delivery heads. In addition, several detection instruments have been introduced that are not limited to a specific format, such as the PhosphoImager and FluorImager by Molecular Dynamics or the FluoStar by SLT Instruments. CCD cameras can also be used, as described for the gel-permeation format, taking advantage of the high-resolution chips available. With resolution of up to 16 million pixels, essentially any size and pattern of assay wells is possible. Alternatively, assay wells can be eliminated altogether and replaced with droplets on a smooth surface, dependent only on surface tension, if detection and liquid handling is not limiting.

FUTURE DIRECTIONS

The opportunities afforded by the interface of large combinatorial libraries with novel high throughput screening strategies have the potential to generate unprecedented success in pharmaceutical drug discovery. Effective utilization of these technologies will require the ability to skillfully manage the vast flow of information from these activities. As the number of compounds and the number of assays through which they are evaluated increases, the resulting database should itself provide a significant resource for drug discovery. Such a database will require the correlation of pharmacophore and structural information from chemical libraries with biochemical assay results, coupled with information on families of genetically related biochemical targets. Thus, as novel gene products are identified through analysis of human genome information, the availability of a molecular structure–bioactivity database will enable one to quickly identify the most relevant combinatorial libraries for screening against the new target. Integration of this information poses a significant future challenge. Nevertheless, the opportunity to perform compound screening "in silico" and, in this manner, to rapidly identify novel lead molecules may provide the best hope of capturing the potential synergy of the human genome project and combinatorial chemistry technology for drug discovery.

REFERENCES

1. Furka A, Sebesteyen F, Asgedom M, Dibo G (1991): A general method for rapid synthesis of multicomponent peptide mixtures. *Int J Pept Protein Res* 37:487–493.
2. Burbaum J, Ohlmeyer M, Reader J, Henderson I, Dillard L, Li G, Randle T, Sigal N, Chelsky D, Baldwin J (1995): A paradigm for drug discovery employing encoded combinatorial libraries. *Proc Natl Acad Sci USA* 92:6027–6031.
3. McRae B, Kurachi K, Heimark R, Fujidawa K, Davie E, Powers J (1981): Mapping the active sites of bovine thrombin, factor IXa, plasma kallikrein, and trypsin with amino acid and peptide thioesters: development of new sensitive substrates. *Biochemistry* 20:7196.
4. Jayawickreme C, Graminski G, Quillan J, Lerner M (1994): Creation and functional screening of a multi-use peptide library. *Proc Natl Acad Sci USA* 91:1614–1618.
5. Potenza M, Graminski G, Lerner M (1992): A method for evaluating the effects of ligands upon Gs protein-coupled receptors using a recombinant melanophore-based bioassay. *Anal Biochem* 206:315–322.

PART VI

COMBINATORIAL DRUG SCREENING AND DEVELOPMENT

24

COMBINATORIAL DRUG DISCOVERY: CONCEPTS

JUDD BERMAN

Glaxo Wellcome Research Institute, Five Moore Drive, Research Triangle Park, North Carolina

RUSSELL J. HOWARD

Affymax Research Institute, Santa Clara, California

The previous chapters have discussed the exploding field of combinatorial chemistry. The rapid advances in the preparation and utilization of biological and chemical libraries is remarkable. For these advances to be translated into valuable products, significant hurdles must be overcome. This chapter highlights some of the obstacles facing drug discovery scientists and includes some speculation of what may be of value in the future. The intent is to raise questions and suggest exciting avenues of research to be pursued. It should be recognized that some of the major impact of combinatorial sciences may emerge from applications outside of the pharmaceutical industry (material sciences,[1,2] etc.), although these issues are not discussed here.

Several excellent perspectives on combinatorial drug discovery have appeared.[3-6] In this chapter we present combinatorial drug discovery from within a process-driven framework. The process of drug discovery may be viewed as a continuum. It begins with the spark of a scientific hypothesis, continues through target validation to demonstrated efficacy in the clinic, and ends with a valuable medicine being delivered to patients. Questions that are relevant from the process-driven perspective include the following:

Combinatorial Chemistry and Molecular Diversity in Drug Discovery, Edited by
Eric M. Gordon and James F. Kerwin, Jr.
ISBN 0-471-15518-7   Copyright © 1998 by Wiley-Liss, Inc.

**TABLE 24.1. Desired Attributes of a
Quality Molecule**

Chemically tractable
Novel
Synthetically accessible
Desired physicochemical properties
Bioavailable
Oral activity ($F > 30$?)
Suitable half-life
Selective
Nontoxic (at 100× therapeutic dose?)
No drug-drug interactions
Potent
In vivo (<1 mg/kg?)
In vitro (<10 μM?)

What are the rate-limiting steps in finding new medicines?

What alternative tactics might relieve these bottlenecks?

What new techniques and technologies are needed to cope with future anticipated bottlenecks?

How can combinatorial discovery approaches relieve these bottlenecks?

The scope of this chapter limits discussion of the above topics to aspects that traditionally have been considered within the bounds of preclinical discovery. What then, are the key considerations and new opportunities that combinatorial science opens up to today's drug discovery scientist? How will opportunities that result from the human genome initiative influence the application of combinatorial chemistry to drug discovery?

From a medicinal chemistry perspective, what are the desired attributes of a "quality" molecule? For a molecule to be of value there appear to be four primary considerations (Table 24.1). The molecule must be chemically tractable, bioavailable, selective, and potent. Understanding and optimizing each of these desired attributes poses unique challenges to the drug discovery scientist. Moreover, traditional approaches have often considered optimization of each of these attributes as separate and distinct stages. While combinatorial chemistry could expedite each step in a reiterative chemical program (Figure 24.1), it is unclear which stages will derive maximum benefit. Nonetheless, a

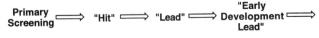

Figure 24.1. Serial screening of compound collections.

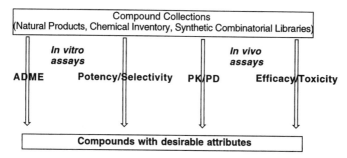

Figure 24.2. Parallel processing of compound collections.

central tenet of this article is that these types of serial approaches may be significantly improved via exploiting the opportunities to "parallel process" the chemical entities provided via combinatorial chemistry (Figure 24.2). Increases in effectiveness may result in reducing cycle time—the time it takes from when a compound is conceived until the time it is selected for evaluation in the clinic.

In essence, parallel processing may allow decision making on what types of leads could be selected until a significant amount of critical information is in hand. The major perceived benefit of this approach is that one will advance classes of compounds that more fully represent global solutions on the fitness landscape, while avoiding the commitment of resources to molecules that may be entrenched in a local minima. In other words, only molecules that appear to possess all of the attributes of a "quality" molecule deserve additional study. It is our speculation that the fullest gains from combinatorial approaches may be realized when the initial compounds being evaluated are closest to the compounds one intends to bring to market!

Specifically then, what are the major challenges that face drug discovery scientists in simultaneously optimizing four key attributes—chemical tractability, bioavailability, selectivity and potency—armed with the current advances of combinatorial chemistry?

Deciding what types of molecules to make
   What constitutes diversity?
   What structural features confer the highest likelihood of a molecule to simultaneously possessing the desired attributes?
   Conversely, what structural features should be avoided?
Deciding what assays will mirror results obtained in humans
   What throughput is required?
   What level of comfort is associated with the model?

Further to the above, what are the appropriate benchmarks that should be established as guideposts in order to measure future increases in productivity that result from combinatorial approaches?

What types of experimental approaches could harness the power of parallel processing of combinatorial samples? Different experimental tactics are currently available when one considers the nature of the molecular target. Identification of ligands for soluble targets has been greatly facilitated through the use of solid-phase and/or spatially addressable compound libraries (pins, VLSPS, tethered encoded beads, etc.). Potential pitfalls associated with the nature of solid-phase assays have been described.[6,7] Membrane-bound targets may be accessed by a variety of methodologies utilizing soluble libraries (pools/encoded/tiered release strategies, etc.). Recent investigations utilizing cell-based systems (such as melanophores[8–10]) appear quite promising. Additionally, various affinity selection methodologies have proved to be of tremendous value in identifying preferred ligands from pools of compounds[11] and these methods also hold great promise for the future. In vitro assays provide valuable information regarding the potency and selectivity (two of the desired attributes of "quality" molecules) of compounds at defined molecular targets. Soluble libraries are also additionally useful in cell-based and whole-animal studies.

One critical aspect to harnessing the full potential of combinatorial approaches relates to the types of advanced assays that can be potentially adapted to screening compound libraries. Issues relating to bioavailability, pharmacokinetics, and pharmacodynamics represent one of the major stumbling blocks and bottlenecks in drug discovery. Can combinatorial assays be developed? For example, can one utilize systems such as CACO-2 cells to screen mixtures of compounds for oral absorption? Preliminary results are encouraging (see Chapter 25). Are there methods that will allow dosing of mixtures to whole animals and subsequent identification of molecules that possess desirable properties? Again, initial results are encouraging.[12–14] With such methodologies in hand, it may be possible to dramatically increase the compound throughput of many assays. A key aspect of being able to carry out such experiments is that one is able to identify the structure of which compounds (from a much larger starting set) possess critical attributes (bioavailability, half-life) required for successful drug discovery. Advances in this area represent maximum added value.

An important consideration in realizing the full potential of combinatorial approaches in advanced assays are the analytical challenges of developing generic high-sensitivity detection methods for the identification of compounds. This is particularly true given the complexity of the biological milieu and synthetic subtleties involved in the preparation of small molecule libraries. Various mass spectrometry methods have proven to be of significant utility in the sensitive detection (<10 ng/mL) of a wide variety of structural classes of compounds. When combined with separation techniques (HPLC, CZE,

SFC), mass spectrometry affords a powerful approach to rapidly identify compounds of interest.

Many of the issues relating to chemical characterization of unique compounds from large mixtures will benefit from lessons learned in traditional natural products sciences.[15] These tactics should receive considerable and renewed attention due to the "combinatorial explosion."

In all of the above, it remains to be seen which methodologies will yield the greatest return. Nonetheless, rapid advances in combinatorial sciences appear to afford an excellent opportunity to discover more and better medicines faster!

REFERENCES

1. Xiang X-D, Sun X, Briceno G, Lou Y, Wang K-A, Chang H, Wallace–Freedman WG, Chen S-W, Schultz PG (1995): A combinatorial approach to materials discovery. *Science* 268:1738–1740.
2. Briceno G, Chang H, Sun X, Schultz PG, Xiang X-D (1995): A class of cobalt oxide magnetoresistance materials discovered with combinatorial synthesis. *Science* 270:273–275.
3. Gallop MA, Barrett RW, Dower WJ, Fodor SPA, Gordon EM (1994): Applications of combinatorial technologies to drug discovery, 1: background and peptide combinatorial libraries. *J Med Chem* 37:1233–1251.
4. Gordon EM, Barrett RW, Dower WJ, Fodor SPA, Gallop MA (1994): Applications of combinatorial technologies to drug discovery, 2: combinatorial organic synthesis, library screening strategies, and future directions. *J Med Chem* 37:1385–1401.
5. Ecker DJ, Crooke ST (1995): Combinatorial drug discovery: which methods will produce the greatest value? *Bio/Technology* 13:351–360.
6. Lyttle MH (1995): Combinatorial chemistry: a conservative perspective. *Drug Dev Res* 35:230–236.
7. Wang Z, Laursen RA (1992): Multiple peptide synthesis on polypropylene membranes for rapid screening of bioactive peptides. *Pept Res* 5:275–280.
8. Lerner MR, Potenza MN, Graminski GF, McClintock T, Jayawickreme CK, Karne, S (1993): A new tool for investigating G protein-coupled receptors. In *Ciba Foundation Symposia,* 179: *Molecular Basis of Smell and Transduction,* pp 76–87.
9. Jayawickreme CK, Graminski GF, Quillan JM, Lerner MR (1994): Creation and functional screening of a multi-use peptide library. *Proc Natl Acad Sci USA* 91:1614–1618.
10. Quillan JM, Jayawickreme CK, Lerner MR (1995): Combinatorial diffusion assay used to identify topically active melanocyte-stimulating hormone receptor antagonists. *Proc Natl Acad Sci USA* 92:2894–2898.
11. Songyang Z, Shoelson SE, Chaudhuri M, Gish G, Pawson T, Haser WG, King F, Roberts T, Ratnofsky S, et al (1993): SH2 domains recognize specific phosphopeptide sequences. *Cell* 72:767–778.
12. Potts W, Lundberg D, Peters J, Bi H, Stelman G, Sandhu P (1995): Pharmacokinetic

assessment of a mixture of compounds in the rat using simultaneous dosing and simultaneous LC/MS/MS quantitation. In *Fourth ISSX Proceeding,* vol 8, p 404.

13. Berman J, Halm K, Adkison K, Shaffer J (1997): Simultaneous pharmocokinetic screening of a mixture of compounds in the dog using APC/LC/MS/MS analysis for increased throughput. *J Med Chem* 40:827–829.

14. Olah TV, McLoughlin DA, Gilbert JD (1997): The simultaneous determination of mixtures of drug candidates by liquid chromatography atmospheric pressure chemical ionization mass spectrometry as an in vivo drug screening procedure. *Rapid Commun Mass Spectrom* 11:17–23.

15. DeSousa NJ, Ganguli BN, Reden J (1982): Strategies in the discovery of drugs from natural sources. In Jurgen–Hess H, ed. *Annual Reports in Medicinal Chemistry,* 17. New York: Academic, pp 301–310.

25

ADME/PK ASSAYS IN SCREENING FOR ORALLY ACTIVE DRUG CANDIDATES

Jacqueline A. Gibbons, Eric W. Taylor, and Rene A. Braeckman
Chiron Corporation, Emeryville, California

In targeting molecules for development as drugs, a major goal is the achievement of favorable oral bioavailability, metabolic stability, and duration of action. These endpoints can be reached only by consideration of the mechanisms by which compounds undergo absorption, distribution, metabolism, and excretion (ADME) in conjunction with the kinetics of these processes, or the pharmacokinetics (PK).

Although there is considerable diversity in the screening strategies that are currently used for chemical libraries, virtually all screening methods focus solely on pharmacologic activity. As a consequence, the compounds identified in the screens typically require chemical modifications to optimize their ADME/PK properties. This is problematic since the original activity selected in the screens is increasingly jeopardized as more extensive chemical modifications are introduced. However, such modifications of identified leads would be unnecessary if ADME/PK assays were incorporated into the screening process.

ADME/PK assays and the data that they generate could be incorporated into the pharmacologic screening process in two ways (Figure 25.1). First, these assays could be used to identify physicochemical properties that are associated with favorable ADME/PK characteristics. Such structure–activity

Combinatorial Chemistry and Molecular Diversity in Drug Discovery, Edited by
Eric M. Gordon and James F. Kerwin, Jr.
ISBN 0-471-15518-7   Copyright © 1998 by Wiley-Liss, Inc.

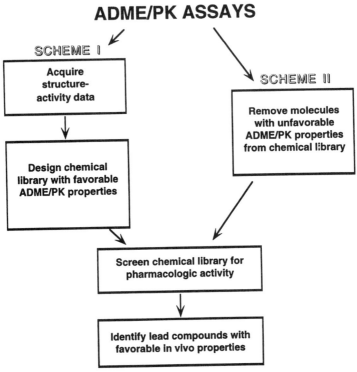

Figure 25.1. Convergence of two schemes for incorporation of ADME/PK assays into the screening process for targeting molecules with pharmacologic activity and favorable in vivo characteristics.

data could be used to design chemical libraries composed strictly of molecules known to have favorable ADME/PK properties. And second, ADME/PK assays could be used in tandem with functional assays, serving to "filter" chemical libraries by eliminating compounds with poor ADME/PK properties. Both of these methods increase the likelihood that lead compounds identified in pharmacologic assays have desirable in vivo properties and can be moved quickly into the clinic.

With the overall goal of enhancing the efficiency with which orally deliverable pharmaceutics are targeted from chemical libraries, this chapter provides an overview of ADME/PK assays that can be used to screen molecules for favorable oral absorption and metabolic stability. In addition, to facilitate the collection of structure–activity data, this chapter provides a literature review of physicochemical properties that have been proposed to affect oral bioavailability.

ADME ASSAYS

The primary factors that determine the oral bioavailability of compounds are transport across the intestinal mucosa, metabolism in the gut lumen and enterocytes, and hepatic first-pass elimination. To better understand the individual contribution and interplay among these processes, it is desirable to have appropriate model systems that allow for the examination of one process independently of the others. The following discussion therefore focuses first on absorption from the gastrointestinal (GI) tract, then turns to metabolism in the GI tract, and closes with an examination of hepatic elimination.

Assays for GI Absorption

Several in vitro test systems have been developed for measuring transport across intestinal mucosa. For example, the transport of compounds across both the mucosal and serosal layers of rat intestine has been characterized by determining the amount of drug that accumulates in the serosal lumen of sealed sacs of everted intestine.[1-3] Similarly, the in vitro transport of compounds across rat, rabbit, and monkey mucosa has been determined by mounting sections of freshly excised intestine (free of the serosal muscle layer) in specially designed side-by-side diffusion cells.[4,5] Examination of absorptive processes has been additionally accomplished using an open-ended segment or rings of everted intestinal tissue isolated from mice,[6] hamsters,[7,8] and rats.[8-13] However, of all in vitro systems that have been developed to date, the most commonly used method for determining the transport of compounds across the intestinal mucosa currently involves the use of Caco-2 cells.

Caco-2 cell lines were originally established by Jorgen Fogh approximately 20 years ago after he screened several lines derived from human colon carcinomas.[14,15] The Caco-2 cell line has been widely used for testing the GI absorption of compounds due to its ability to express the morphological features of mature enterocytes. Culture systems of Caco-2 cells have been shown to consist of a monolayer of viable, polarized, and fully differentiated villus cells, similar to those found in the small intestine.[16] In a typical transport assay, monolayers of Caco-2 cells are grown on microporous polycarbonate cell inserts and these inserts are mounted in a side-by-side diffusion apparatus.[17] Transport is then determined by measuring the rate of apical to basolateral diffusion of test compounds in the apparatus.[18] Caco-2 assays are not only simple to perform, but may be implemented at a low cost, and are therefore amenable to rapid throughput screening for GI absorption.

In some instances data generated with Caco-2 cells have been shown to be well correlated with the results of absorption assays conducted with intact intestines.[19,20] However, there have been reported cases of discrepancies. For example, the drug atenolol is transported at a much slower rate in the Caco-2 model than in isolated rat ileum,[18] and our laboratory has found several

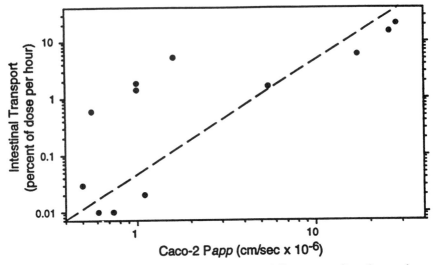

Figure 25.2. Relationship between apparent permeability across Caco-2 monolayers (Caco-2 P *app*) and transport across intact intestines for N-substituted glycine peptoids (NSG-peptoids; see Figure 25.5) and control compounds.

instances where molecules that are virtually impermeable to Caco-2 cells are rapidly transported in intact intestines (Figure 25.2). These discrepancies are probably attributed to differences in transport pathways: While the predominate means of transport across Caco-2 cell monolayers is transcellular diffusion, alternative pathways are likely to play a more prominent role in intestinal permeability in the endogenous mucosa.

Alternative pathways to transcellular diffusion are summarized in Figure 25.3. These pathways include the paracellular route, which involves diffusion

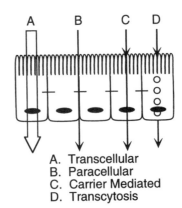

A. Transcellular
B. Paracellular
C. Carrier Mediated
D. Transcytosis

Figure 25.3. Routes of drug transport across intestinal epithelial monolayers: A, transcellular; B, paracellular; C, carrier mediated; D, transcytosis.

of a solute between adjacent intestinal cells restricted by the tight junctions.[21] Hydrophilic molecules such as atenolol and cimetidine are absorbed by this route. The tight junction acts as a sieve so that there is a dependence on the overall size of the solute. While tight junctions exist between Caco-2 cells, several investigators have speculated that the intercellular spaces are smaller in the Caco-2 cell system, resulting in a more restricted pore pathway than the in vivo epithelium. This possibility is consistent with the finding that electrical resistance in Caco-2 cell monolayers is higher than that found in normal mucosa.[16] In addition to paracellular transport, an alternative pathway that is underrepresented in the Caco-2 model is carrier-mediated transport, in which specific transporters shuttle molecules across the epithelial barrier. This pathway has been shown to play a role in the absorption of β-lactam antibiotics, ACE inhibitors, and a few renin inhibitors.[22]

Given that alternative pathways to transcellular diffusion tend to be underestimated using the Caco-2 model, many investigators prefer the use of intact intestines in in situ models of GI absorption. Currently, the most widely used in situ models are those that involve the perfusion of isolated segments of intestine. Most commonly, a rat is placed under complete anesthesia and a discrete segment of the intestine (the duodenum, jejunum, or ileum) is isolated. An inflow cannula and outflow cannula are each placed at the proximal and distal ends of the isolated segments and the luminal contents are voided by passing a wash buffer through the segment. The test compound is added to a perfusate buffer, which is passed through the intestinal segment during the course of the assay. Intestinal permeability of the test compound is then generally determined by the rate of loss of the compound from the perfusate.

Because the in situ perfusion model deviates from physiological conditions in several respects, it may be a weak predictor of absorption rates in vivo. For example, the effects of formulations, which cannot be accurately examined in the perfusion model, could have a marked impact on in vivo absorption.[23] Moreover, there may be a disruption of the mucin barrier in the perfusion models; even a slight disruption may be important, since the mucin barrier is thought to play an important role in such processes as the molecular sieving effect.[24,25] Furthermore, as noted by Hsu et al.,[26] intestinal fluids or contents are often very viscous and seem nonflowing when tested ex vivo; the high degree of viscosity of gut luminal contents in vivo contrasts strongly with the free-flowing aqueous media used in the in situ perfusion models. And, finally, binding and complexation of drugs with bile salts, food ingredients, and so on, that are present in the gut lumen in vivo are not incorporated in the perfusion models.[27] Based on these potential shortcomings of the perfusion model, alternative in situ models have been developed to examine GI absorption under more physiological conditions. A common feature of these models is that they seek to minimize the disturbance to the intestine while measuring the appearance of the test compound in the blood that drains the GI tract. One such model is illustrated in Figure 25.4 and involves bolus administration

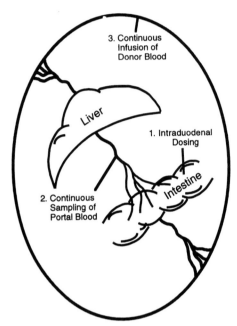

Figure 25.4. Physiologic model for intestinal absorption rate, a direct measure of transport from the GI to blood. 1. While under anesthesia with ketamine, rats are given a single bolus dose of a compound by direct injection into the duodenum. 2. Blood from the portal vein, which drains the intestines, is continuously samples via cannulation to prevent recirculation of absorbed substances. 3. Blood collected from the portal vein is replaced by transfusion of donor rat blood through a femoral vein cannula.

of the formulated test compound directly into the duodenum of an anesthetized rat.

Physicochemical Factors That Affect GI Absorption

The long-held pH partition theory holds that absorption in the GI tract can be reduced to simple diffusion across a lipid barrier.[28,29] Accordingly, the nonionized form of an acid or base, if sufficiently lipid-soluble, is predicted to be absorbed, whereas the ionized form is not. Permeability at a specific absorption site is hypothesized to be related to the fraction of the administered compound unionized at a specific absorption site. Thus, in the stomach, where the pH ranges from 1 to 3, acid and neutral molecules may be absorbed, but basic molecules are not. Weakly basic drugs ($pK_a < 5$) are essentially un-ionized throughout the intestines where the pH range from the duodenum to the colon is about 5 to 8. Strong bases (pK_a 5–11) should show pH-dependent absorption, whereas stronger bases are ionized throughout the GI tract and are predicted to be poorly absorbed. Although the pH partition theory provides a

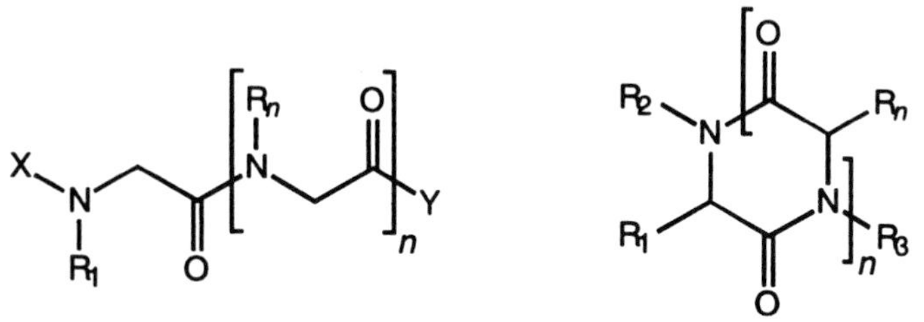

Figure 25.5. Structures of N-substituted glycine peptoids (NSG-peptoids). Each R-group consists of one of any of thousands of reactive amines, accounting for tremendous diversity of NSG-peptoids in terms of spatial features and physicochemical parameters.

basic framework for understanding GI absorption, the fact that there are many deviations from this theory demonstrates that it is an oversimplification of a more complex process. For example, certain quaternary ammonium drugs are known to be absorbed from the GI tract, suggesting that restriction of permeability to ionized forms of a molecule may not be absolute.[30]

It has further been long-held that molecular weight is a primary determinant of whether a compound is likely to pass through the intestinal wall.[31] In an effort to determine the quantitative relationship between molecular weight and GI absorption, Donovan et al.[32] studied absorption of oligomers of poly-ethylene glycol and showed that there is an inverse relationship between molecular weight and absorption, with a sharp decline in absorption above 700 daltons. However, a subsequent study with a series of ergotamine derivatives concluded that 500 daltons is the point of sharp decline.[24] In spite of these findings, several examples can be found of compounds with high molecular weights that are well absorbed. For example, cyclosporin, with a mass of 1200 daltons is reported to be 24% absorbed.[33] Moreover, when compounds were examined in our laboratory that were not confined to a series of analogs (Figure 25.5), there was no apparent relationship between size and mucosal permeability (Figure 25.6).[34] These observations suggest that it is naive to assume that molecular weight alone determines the rate and extent of GI absorption.

The octanol/buffer partition coefficient ($\log P$) was classically used to predict GI absorption of compounds.[35,36] With the general premise that $\log P$ is an index of the lipophilicity of a compound, it was believed that the most lipophilic compounds diffuse the fastest across the cellular membranes of the intestinal epithelium, the hypothesized main barrier for GI absorption. The notion that $\log P$ is related to intestinal permeability has been supported by data from a large number of investigations. For example, the relationship between $\log P$ and transcellular passage was iteratively demonstrated in in vitro studies using Caco-2 and other intestinal epithelial cells.[18,37-39] Furthermore, in seminal work by Martin,[40] the rate of disappearance from the intestine was positively correlated with the $\log P$ for more than 100 compounds, and

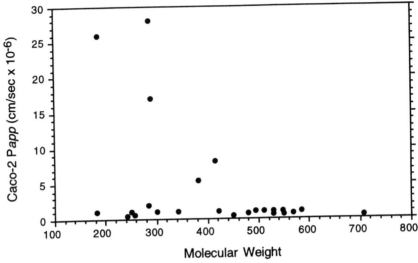

Figure 25.6. Relationship between molecular weight and apparent permeability across Caco-2 monolayers (Caco-2 P *app*) for NSG-peptoids and control compounds.

a review by Austl and Kutter[36] concluded that lipophilicity alone is the major determinant of GI absorption. However, in all of these studies of the relationship between log P and intestinal permeability, the focus was on compounds with a distribution coefficient of less than 1000. In a recent study by Wils et al.,[41] it was found that when log P values were lower than 3.5, the transepithelial permeability increased with the log P value, but for log P values ranging from 3.5 to 5.2, the intestinal permeability was inversely related to the log P value. Moreover, experiments done in our laboratory using structurally diverse compounds have shown no apparent relationship between enteral diffusivity and log P (Figure 25.6).

In describing the molecular basis of diffusion across cell membranes, Stein[31] hypothesized that desolvation is the rate-limiting step in the transport of a polar solute across a cell membrane. Thus, for a polar solute to cross a cell membrane, it is necessary for the hydrogen bonds formed between the solute and water to be broken. The energy required for breaking the hydrogen bonds between a polar solute and water may therefore be a significant barrier in the transport of solutes across lipophilic membranes. Based on this premise, several investigators have examined the relationships between the number of hydrogen bonds, hydrogen-bonding potential, and permeability to intestinal epithelia.

Using Caco-2 cells, Conradi et al.[42] showed that the number of hydrogen bonds that can be formed by a solute is a major determinant of its transepithelial permeability. In a subsequent article,[43] however, these same authors conceded that since hydrogen bonds are not energetically equivalent, permeability

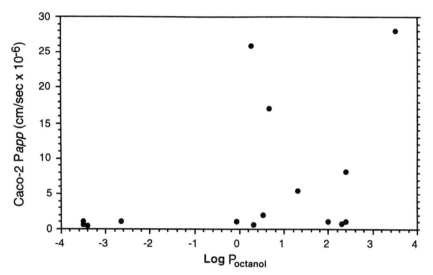

Figure 25.7. Relationship between octanol/buffer partition coefficient and apparent permeability across Caco-2 monolayers (Caco-2 P *app*) for NSG-peptoids and control compounds.

is unlikely to be strictly related to the number of hydrogen bonds. As described by Nys and Rekker,[44] hydrogen bond energies are dependent on a number of factors, including proximity effects.

Two procedures have been used to estimate hydrogen bonding potential[45]: (1) partition coefficients measured in heptane–ethylene glycol, and (2) differences measured in the partition coefficients of solutes between octanol–water and isooctane–water. A limited number of experiments with peptides have shown these estimates of hydrogen bonding potential to be well correlated with permeability across Caco-2 cells.[20,42,46] However, as shown in Figure 25.7 and Figure 25.8, data generated in our laboratory with NSG-peptoids (Figure 25.5) have not supported this relationship for other classes of compounds. Moreover, Karls[43] found that results obtained with Caco-2 cells tended to overestimate the importance of desolvation energy in intestinal transport in vivo. Because Caco-2 cells poorly emulate the leakiness of junctions between intestinal cells that exist in vivo, this overestimate was attributed to paracellular transport. Since the solute never leaves the aqueous environment as it diffuses between adjacent intestinal cells, desolvation is not required for paracellular transport. Thus, the importance of hydrogen-bonding potential to GI permeability probably pertains only to transcellular diffusion.

As outlined in this discussion, there have been several attempts to correlate GI absorption with a single physiochemical property. In general, when an analysis is limited to a series of analog compounds, trends related to a single physiochemical variable can often be gleaned from the data. However, for un-

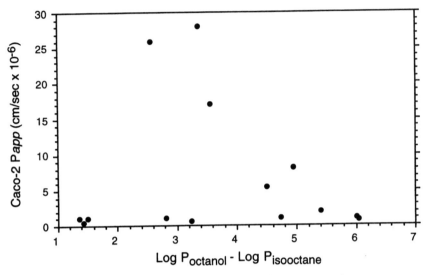

Figure 25.8. Relationship between estimated desolvation energy and apparent permeability across Caco-2 monolayers (Caco-2 P *app*) for NSG-peptoids and control compounds.

related compounds, the likelihood becomes low of identifying a single, predominant variable that is correlated with absorption. This is because physiochemical variables interact in complex ways to determine how a molecule is presented to the mucosal barrier (consider, for example, apparent molecular weight and pK_a). The most useful databases for structure–activity relationships that determine favorable GI absorption are therefore anticipated to be those that incorporate information on a host of physicochemical variables.

CATABOLISM WITHIN THE GI TRACT

Due to an abundance of enzyme systems that exist in the GI tract, the gut presents an unfavorable environment for many molecules, particularly peptidic drug candidates.[47] Enzymes that can catabolize orally administered compounds are present in the gut lumen, on the brush-border membranes that form the luminal surface of the intestinal epithelium, and in the cytosol of mucosal epithelial cells. For a molecule to be a viable candidate for development as an orally deliverable agent, its pharmacologic activity must therefore be resistant to hydrolysis by a host of peptidases, amylases, and lipases, as well as catabolism by monooxygenases. In some cases molecular entities are designed as prodrugs that are converted to active metabolites in the gut, but, in general, extensive metabolism in the GI tract is undesirable.

Unless a radiolabeled form of a test molecule is available for use (an uncommon situation for early-stage screening), routine assays that rigorously test for stability in the GI tract are fairly labor-intensive. One may, for example, incubate test compounds with homogenates of intestines, but appropriate extraction procedures and analytical methods typically must be developed on a compound-by-compound basis. Some investigators have used Caco-2 cells to screen for susceptibility to GI catabolism (reviewed in Reference 48). However, this is likely to severely underestimate rates and extents of metabolism in vivo. As a final possibility, metabolic stability in the GI can be assessed using purified enzymes. This approach can be particularly useful when discrete catabolic pathways are anticipated to be problematic, such as proteolysis of peptides. As an example, Miller et al. used enzymes representative of several known classes of relevant GI proteases to establish proteolytic stability of seemingly peptide-like N-substituted glycine peptides.[49]

HEPATIC ELIMINATION

Before reaching the general systemic circulation, all chemicals absorbed from the GI tract, except for those absorbed from the mouth and rectum, pass through the liver. In a phenomenon referred to as "hepatic first-pass elimination," a substantial amount of an administered compound can be removed from the blood in a single pass through the liver. Moreover, rapid hepatic metabolism may substantially limit the residence time of a parent compound in the body, even after intravenous administration. Evaluation of hepatic elimination is therefore critical to the determination of whether molecules are viable candidates for pharmaceutical development.

The liver is unique among organs in that chemicals in plasma come in direct contact with cells that are not separated from the plasma by vascular tissue.[50] This extensive exposure, coupled with the myriad of uptake systems expressed in hepatocytes,[51] accounts for the efficient uptake of chemicals by hepatocytes. As illustrated in Figure 25.9, on entering a hepatocyte, a molecule is generally subject to one of three fates. First, the molecule may be excreted unchanged into the hepatic sinusoidal sink that drains into the central vein, ultimately passing into the general systemic circulation. Second, the molecule may be subject to biotransformation by any of several enzyme systems that abound in the hepatocyte. And finally, the molecule may be transported into a bile canniculus, and ultimately subject to biliary excretion. These latter two possibilities are considered in the discussion that follows.

Biotransformation in the Liver

Biotransformation reactions are typically divided into two categories: phase I and phase II. In phase I reactions, molecules are largely metabolized oxidatively, although some reductive reactions do occur. Virtually all reactions catalyzed by phase I enzymes result in increased polarity of the starting mole-

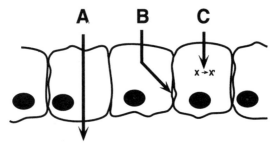

A. Direct passage to blood
B. Transport into bile
C. Biotransformation

Figure 25.9. Three potential fates of molecules after entering hepatocytes: A, direct passage to blood; B, transport into bile; C, biotransformation.

cule. The enzyme system that catalyzes most phase I reactions is found in the smooth endoplasmic reticulum and focuses on cytochromes P-450, a family of structurally related heme proteins that exists in many isoforms.[52] Although the physical differences among various isoforms are fairly subtle, pronounced differences occur in substrate specificity.[53,54] The isoforms of cytochromes P-450 that are expressed in mammalian liver vary considerably from species to species, and in many cases among members of the same species.[55–57] These differences can have a pronounced effect on not only the outcome of phase I metabolism, but the overall efficiency with which a molecule is cleared by the liver.

Phase II metabolism occurs in both the smooth endoplasmic reticulum and the cytosol. These reactions involve the conjugation of small endogenous molecules with polar functional groups and result in an increase in size and polarity of the initial molecule. The polar moieties that are conjugated may be present on the parent molecule or introduced by the action of phase I enzymes. Phase II metabolism includes the following major pathways: glucuronide formation, sulfate formation, glutathione conjugation, and acetylation.[54,58–60] As with phase I enzymes, the expression of phase II enzymes can vary substantially both among mammalian species and members of a given species.

Biliary Excretion

Biliary excretion is a process in which a compound that has entered an hepatocyte is passed into the bile canaliculi, minute canals that run between liver cells. Bile in these canals ultimately drains into the common bile duct, which feeds into the duodenum. The factors that influence the extent of biliary excretion of a compound may be categorized as either biological or physiochemical. Biological factors include species, sex, genetic factors, and protein binding.[61] The physicochemical factors that have been most widely discussed in the literature are molecular weight and polarity[50,61]; these two physicochemical factors are the focus of this section.

The importance of molecular weight to biliary excretion has been demonstrated in studies with substitutions on simple aromatic compounds (reviewed by Smith[61]). In these studies introduction of a host of markedly different atoms or groupings resulted in the same generic outcome: the extent of biliary excretion increased with the molecular size. Based on studies of the biliary excretion of diverse aromatic compounds in rats, Milburn and coworkers[62] proposed that the threshold for molecular weight is about 325 daltons; below this threshold, biliary excretion accounts for elimination of <10% of the administered dose, and above this threshold, the extent of material passed into the bile is somewhat correlated with molecular weight. However, in subsequent work by Hirom et al.[63] an examination of the biliary excretion of a large number of compounds in the rat, guinea pig, and rabbit showed that there are species-related differences with respect to the threshold molecular weight for appreciable biliary excretion. Focusing only on molecular species that appeared in the bile (which did not correspond to parent compound unless the latter was excreted in the bile unchanged), these investigators determined the following thresholds: 325 ± 50 daltons in the rat; 400 ± 50 daltons in the guinea pig; and 475 ± 50 daltons for the rabbit. It has been estimated that this threshold is between 500 and 700 daltons in humans.[61] As pointed out in a review article by Levine,[64] these figures are considered only approximate, and there is considerable latitude within an individual species. For example, the threshold for organic cations is 200 daltons and there appears to be little or no species variation.[65]

Polarity, or the presence of groups on a molecule of groups that allow the molecule to be amply water soluble at physiologic pH, appears to be an additional requirement for biliary excretion. For example, based on its high molecular weight, dieldrin (molecular weight, 381 daltons) would be expected to be efficiently eliminated in the bile. However, only 1% of this nonpolar compound appears in the bile of rats within 24 h of dosing,[61,66] suggesting that polarity is an important determinant of biliary excretion. If a parent compound is of sufficient size and polarity, it can be eliminated directly into the bile, but the majority of compounds found in bile are in the form of polar conjugates formed in phase II biotransformation reactions.

Assays for Estimating Hepatic Metabolism

One of the most important issues facing evaluation of hepatic metabolism of xenobiotics is the large potential for variability between species. Much of this variability is accounted for by differences in the expression of biotransformation enzymes in phase I and II pathways. As a consequence, identification of surrogate species or test systems for routine predictions of the metabolic fate of molecules in humans poses a daunting challenge. Some investigators have strictly adhered to the use of human tissues, but availability of such tissues is generally too limited to be of practical use in efforts aimed at large-scale screening assays. Moreover, "out-bred" human populations offer considerable variability with respect to both genotypic and phenotypic factors that affect xenobiotic metabolism.[67] In spite of these drawbacks, assays for hepatic metabolism may be useful for identifying molecules that are particularly susceptible to first-pass elimination by the liver. Removal of such molecules from chemical libraries that are screened for pharmacologic activity could serve to streamline the identification of lead compounds with desirable in vivo properties.

The most widely used method for evaluating xenobiotic metabolism involves the use of subcellular fractions derived from homogenized livers. Two subcellular fractions are primarily used: microsomal fractions, which consist of vesicles of endoplasmic reticulum (referred to as microsomes); and S9 fractions, which consist of cytosol stripped of nuclei and mitochondria.[68] Microsomes are used when the focus of analysis is on phase I reactions, particularly those catalyzed by cytochromes P-450; S9 fractions are used when incorporation of phase II metabolism is desired. Typically, a test substance is incubated directly with a subcellular fraction and the rate of metabolism is gauged by the disappearance of parent compound over time. While these test systems are extremely simple to implement, their disadvantage lies in the fact that it is virtually impossible to relate the results to concentration effects that occur in vivo. The in vitro enzyme systems are effectively bathed in substrate-rich media, and barriers that affect tissue and cellular distribution in vivo are ignored. In terms of the rates of substrate turnover and the entities of products that are generated, complex enzyme systems (such as those in subcellular fractions) tend to behave nonlinearly with respect to concentration. Thus the inability to correlate the concentrations of a substrate to which an enzyme system is exposed in vitro and in vivo may undermine the relevance of results obtained with subcellular fractions.

A second in vitro test system that is gaining popularity is the use of purified biotransforming enzymes.[69,70] This technique, which focuses on specific isoforms of cytochromes P-450 and phase II enzymes, has only recently become possible due to strides in cDNA-directed expression systems. Given that the CYP3 family of cytochromes P-450 is possibly the most important family in human drug metabolism,[53,56,57,71] screening test compounds for lability against CYP3 isoforms may present an efficient method of eliminating molecules that are susceptible to hepatic biotransformation. The main caveat to this approach mirrors that to the use of subcellular fractions: Inaccurate predictions may be generated by the use of concentrations that are irrelevant to in vivo exposure.

The use of isolated hepatocytes in evaluation of metabolism offers the potential advantage of an integrated model for hepatocellular biotransformation. Hepatocytes can be isolated from whole liver following collagenase digestion,[72] and suspension cultures of these primary hepatocytes can be used directly for metabolism studies. Primary hepatocytes retain most of their phase I and phase II metabolic capabilities for a limited period after isolation, and in some cases have been shown to be a good predictor of metabolism in vivo.[73,74] Attempts to generate monolayer cultures for long-term use in metabolism studies would appear to present an attractive alternative to suspension cultures. However, hepatocytes in monolayer culture suffer a well-documented loss of phase I metabolic capabilities over time. Another hepatocyte model that has been used involves transformed cell lines. While attempts have been made to transfect these cell lines with genes that code for major biotransforming enzymes, due to the lack of many of the metabolic pathways found in endogenous hepatocytes, these cell lines provide an inaccurate view of biotransformation reactions that occur in vivo.[75]

The use of liver slices provides another in vitro method for evaluation of hepatic metabolism. Since the structural integrity of the liver is preserved with this method,[76,77] it provides an advantage over isolated hepatocytes. While data generated with liver slices has been shown in certain cases to be correlated with hepatic metabolism in vivo,[73,76,77] little is known about the preservation of biotransforming capabilities over time.[67]

Liver perfusion is a technique that involves the surgical isolation of a liver, which is then perfused with a synthetic medium or a solution of diluted blood. (Methodological variations of this general technique are summarized in a review article by Ross.[78] In its most common application, liver perfusion is used to assess the rate of hepatic elimination of a test compound by measuring its disappearance from the perfusate. Although this in situ method is considerably more labor-intensive than alternative in vitro methods, its chief benefit lies in its preservation of hepatic architecture, which may play a large role in the quantitative aspects of liver clearance. Thus, it is assumed that the use of an intact liver provides a more accurate means than standard in vitro methods of estimating hepatic extraction in vivo.

PHARMACOKINETICS

The term "absolute oral bioavailability" is used by pharmacokineticists to refer to the percentage of an orally administered dose that reaches the general systemic circulation. This percentage reflects not only the fraction of the dose that is absorbed from the GI tract, but the portion that evades gut metabolism and hepatic first-pass elimination. Since mechanisms of absorption and elimination are obscured by this measure, it represents a far more empirical characterization than the ADME assays described in the preceding sections of this chapter. Nonetheless, in vivo determination of the oral bioavailability of a molecule provides a convenient means of summarizing its suitability for oral delivery.

In classical pharmacokinetics, oral bioavailability is determined from the plasma concentration or urinary excretion data after oral administration, with reference to data obtained at the same dose level with intravenous administration.[79,80] The most common approach is to compare the area under the time–concentration curve (AUC) for measurements of parent compound in the plasma following oral and intravenous dosing. For example, if the AUC observed with intravenous dosing is two times larger than the AUC with oral dosing, a test substance is said to be 50% orally bioavailable. Alternatively, if the test molecule is known to pass largely unchanged from the plasma to the urine, the total amount of drug excreted unchanged after oral and intravenous dosing can provide a measure of its oral bioavailability. Thus, if 4 mg of a test molecule appears cumulatively in the urine 24·h after intravenous administration, and 1 mg cumulates with oral dosing, the oral bioavailability is deduced to be 25%. Since it is experimentally simpler to measure urine than plasma, bioavailability determinations based on urinary data are more amenable to rapid throughput assays than plasma-based estimates. However, because most compounds are metabolized prior to urinary excretion, the requirement that a large fraction of the dose appear in the urine unchanged may be too stringent for screening broad classes of compounds. For certain classes of compounds (e.g., proteins), estimation of the oral bioavailability may therefore hinge on determination of the plasma pharmacokinetics.

CONCLUSION

This chapter has provided an overview of ADME/PK assays that can be used to screen molecules for favorable oral absorption and metabolic stability. These assays include tests for GI absorption, stability in the GI tract, resistance to elimination by the liver, and pharmacokinetic determinations of oral bioavailability. One potential use of these assays is to "filter" chemical libraries, removing compounds that have undesirable in vivo properties. Methods are currently being developed in our laboratory to meet this objective.

A challenge that faces the pharmaceutical industry is the transition from ADME/PK assays that are based on single compounds to assays that involve chemical mixtures. Such a transition will be key to expediting the throughput of ADME/PK assays. As illustrated in Figure 25.10, our laboratory has recently developed methods for obtaining this goal.

Our laboratory is additionally using ADME/PK to obtain structure–activity data that will assist in the design of future chemical libraries. As discussed in this chapter, much information is already published concerning physicochemical properties that influence absorption and elimination phenomena. However, these data may be of limited utility due to their unidimensional focus on single physicochemical parameters. ADME/PK events generally encompass intricate mechanisms that are driven by complex interactions of molecules with their environment. Therefore, the most useful data bases for predicting favorable

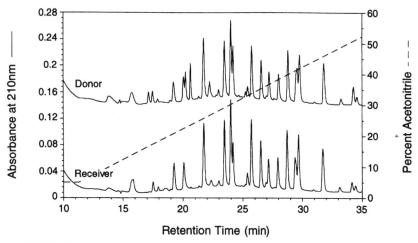

Figure 25.10. Screening chemical mixtures for favorable GI absorption. The plot shows HPLC chromatograms of samples taken from the donor and receiver chambers of a Caco-2 assay system 60 min after placement of a mixture of 24 NSG-peptoids in the donor chamber of a side-by-side diffusion apparatus. Most of the peaks were present in samples from both donor and receiver chambers, demonstrating favorable enteral permeability of the mixture overall. As demonstrated by Taylor[81] this assay can be used in conjunction with mass spectrometry to identify individual compounds that are highly permeable to Caco-2 cell monolayers.

properties of molecules in vivo are likely to be those that provide multidimensional assessments of physicochemical parameters.

REFERENCES

1. Sasaki H, Nakamura J, Konishi R, Shibasaki J (1986): Intestinal absorption characteristics of 5-fluoruracil, florafur and 6-mercaptourine in rats. *Chem Pharm Bull* 34:4265–4272.

2. Vilhardt H, Lundin S (1986): In vitro intestinal transport of vasopressin and its analogues. *Acta Physiol Scand* 126:601–607.

3. Lundin S, Pantzar N, Broeders A, Ohlin M, Westrom BR (1991): Differences in transport rate of oxytocin and vasopressin analogues across proximal and distal isolated segments of the small intestine of the rat. *Pharmaceut Res* 8:1274–1280.

4. Grass GM, Sweetana SA (1988): In vitro measurement of gastrointestinal tissue permeability using a new diffusion cell. *Pharmaceut Res* 5:372–376.

5. Grass GM, Sweetana SA (1989): A correlation of permeabilities for passively transported compounds in monkey and rat jejunum. *Pharmaceut Res* 6:857–862.

6. Kelley MJ, Chen TS (1985): Action of 5-thio-D-glucose metabolism: possible mechanism for diabetogenic effect. *J Pharmacol Exp Therapeut* 232:760–763.

7. Burston D, Mathews DM (1990): Kinetics of influx of peptides and amino acids

into hamster jejunum in vitro: physiological and theoretical applications. *Clin Sci* 79:267–272.

8. Leon del Rio A, Valazquez A, Vizcaino G, Robles–Diaz G, Gonzalez–Noriega A (1990): Association of pancreatic biotinidase activity and intestinal uptake of biotin and biocytin in hamster and rat. *Ann Nutr Metab* 34:266–272.

9. Amidon GL, Stewart BH, Pogany S (1985): Improving the intestinal mucosal cell uptake of water insoluble compounds. *J Controlled Rel* 2:13–26.

10. Osiecka I, Porter PA, Borchardt RT, Fix JA, Gardner CR (1985): In vitro drug absorption models, I: brush border membrane vesicles, isolated mucosal cells and everted intestinal rings: characterization and salicylate accumulation. *Pharmaceut Res* 2:293–298.

11. Porter PA, Osiecka I, Borchardt RT, Fix JA, Frost L, Gardner CR (1985): In vitro drug absorption models, II: salicylate, cefoxitin, α-methyldopa and theophylline uptake in cells and rings: correlation with in vivo bioavailability. *Pharmaceut Res* 2:293–298.

12. Fleisher D, Sheth N, Griffin H, McFadden M, Aspacher G (1989): Nutrient influences on rat intestinal pheytoin uptake. *Pharmaceut Res* 6:332–337.

13. Meadows KC, Dressman JB (1990): Mechanism of acyclovir uptake in rat jejunum. *Pharmaceut Res* 7:299–303.

14. Fogh J, Trempe G (1975): New human tumor cell lines. In Fogh J, ed. *Human Tumor Cells in Vitro.* New York: Plenum, pp 115–141.

15. Fogh J, Fogh JM, Orfeo T (1977): One hundred and twenty-seven cultured human tumor cell lines producing tumors in nude mice. *J Natl Cancer Inst* 59:221–225.

16. Hidalgo IJ, Raub TJ, Borchardt RT (1989): Characterization of the human colon carcinoma cell line (Caco-2) as a model system for intestinal epithelial permeability. *Gastroenterology* 96:736–749.

17. Borchardt RT, Hidalgo IJ, Hillgren KM, Hu M (1991): Pharmaceutical applications of cell culture: an overview. In Wilson G, ed. *Pharmaceutical Applications of Cell and Tissue Culture to Drug Transport.* New York: Plenum, pp 1–14.

18. Artursson P (1990): Epithelial transport of drugs in cell culture, I: a model for studying the passive diffusion of drugs over intestinal absorptive (Caco-2) cells. *J Pharmaceut Sci* 79:476–482.

19. Conradi RA, Wilkinson KF, Rush BD, Hilgers AR, Ruwart MJ, Burton PS (1993): In vitro/in vivo models for peptide oral absorption: comparison of Caco-2 cell permeability with rat intestinal absorption of renin inhibitory peptides. *Pharmaceut Res* 10:1790–1792.

20. Kim D-C, Burton PS, Borchardt RT (1993): A correlation between the permeability characteristics of a series of peptides using an in vitro cell culture model (Caco-2) and those using an in situ perfused rat ileum model of the intestinal mucosa. *Pharmaceut Res* 10:1710–1714.

21. Powell DW (1981): Barrier function of epithelia. *Am J Physiol* 241:G275–G288.

22. Kramer W, Girbig F, Gutjarh U, Kleeman HW (1990): Interaction of renin inhibitors with the intestinal uptake system for oligopeptides and beta-lactam antibiotics. *Biochim Biophys Acta* 1027:25–30.

23. Aungst BJ (1993): Novel formulation strategies for improving oral bioavailability

of drugs with poor membrane permeation or presystemic metabolism. *J Pharmaceut Sci* 82:979–987.

24. Nimmerfall F, Rosenthaler J (1980): Significance of the goblet-cell mucin layer, the outermost luminal barrier to passage through the gut wall. *Biochim Biophys Res Commun* 94:960–966.

25. Peppas NA, Hansen PJ, Buri PA (1984): A theory of molecular diffusion in the intestinal mucus. *Int J Pharmaceut* 20:107–118.

26. Hsu FH, Prueksaritant T, Lee MG, Chiou WL (1987): The phenomenon and cause of the dose-dependent oral absorption of chlorothiazide in rats: extrapolation to human data based on the body surface area concept. *J Pharmacokin Biopharmaceut* 15:369–386.

27. Chiou WL (1995): The validation of the intestinal permeability approach to predict oral fraction of dose absorbed in humans and rats. *Biopharm Drug Disp* 16:71–75.

28. Shore PA, Brodie BB, Hogben CAM (1957): Gastric secretion of drugs: a pH partition hypothesis. *J Pharmacol Exp Therapeut* 119:361–372.

29. Shanker LS (1960): On the mechanism of absorption from the gastrointestinal tract. *J Med Pharm Chem* 2:343–352.

30. Gibaldi M (1991): Gastrointestinal absorption: physicochemical considerations. In *Biopharmaceutics and Clinical Pharmacokinetics.* Philadelphia: Lea & Febiger, pp 40–60.

31. Stein WD (1967): *The movement of molecules across cell membranes.* New York: Academic.

32. Donovan MD, Flynn GL, Amidon GL (1990): Absorption of polyethylene glycols 600 through 2000: the molecular weight dependence of gastrointestinal and nasal absorption. *Pharmaceut Res* 7:863–868.

33. Hedayati S, Bernareggi A, Rowland M (1991): Absorption kinetics of cyclosporin in the rat. *J Pharm Pharmacol* 43:122–124.

34. Taylor EW, Gibbons JA, Shoemaker KR, Martin EJ, Braeckman RA (1994): Prediction of GI absorption for a series of peptoids: comparison of solvent partitioning, and in vitro and in situ measurements. *Pharmaceut Res* (Abstr.)

35. Leo A, Hansch C, Elkins D (1977): Partition coefficients and their uses. *Chem Rev* 71:526–616.

36. Austl V, Kutter E (1983): *Quantitative structure–activity relationships of drugs.* New York: Academic.

37. Artursson P, Karlsson J (1991): Correlation between oral drug absorption in humans and apparent drug permeability coefficients in human intestinal epithelial (Caco-2) cells. *Biochem Biophys Res Commun* 175:880–885.

38. Hilgers AR, Conradi RA, Burton PS (1990): Caco-2 cell monolayers as a model for drug transport across the intestinal mucosa. *Pharmaceut Res* 7:902–910.

39. Wils P, Legrain S, Frenois E, Scherman D (1993): HT29-18-C1 intestinal cells: A new model for studying the epithelial transport of drugs. *Biochim Biophys Acta* 1177:134–138.

40. Martin YC (1981): A practitioner's perspective of the role of quantitative structure–activity analysis in medicinal chemistry. *J Med Chem* 24:229–237.

41. Wils P, Warnery A, Phung-Ba V, Legrain S, Scherman D (1994): High lipophilicity

decreases drug transport across intestinal epithelial cells. *J Pharmacol Exp Therapeut* 269:654–658.

42. Conradi RA, Hilgers AR, Ho NFH, Burton PS (1991): The influence of peptide structure on transport across epithelial cells, I. *Pharmaceut Res* 8:1453–1460.

43. Karls MS, Rush BD, Wilkinson KF, Vidmar TJ, Burton PS, Ruwart MJ (1991): Desolvation energy: a major determinant of absorption, but not clearance, of peptides in rats. *Pharmaceut Res* 8:1477–1481.

44. Nys GG, Rekker RF (1974): The concept of hydrophobic fragental constants (*f*-values), II: extension of its applicability to the calculation of lipophilicities of aromatic and heteraromatic structures. *Eur J Med Chem* 9:361–375.

45. Burton PS, Conradi RA, Hilgers AR, Ho NFH, Maggiora LL (1992): The relationship between peptide structure and transport across epithelial cell monolayers. *J Controlled Rel* 9:87–98.

46. Conradi RA, Hilgers AR, Ho NFH, Burton PS (1992): The influence of peptide structure on transport across Caco-2 cells, II: peptide bond modification which results in improved permeability. *Pharmaceut Res* 9:435–439.

47. Amidon GL, Lee HJ (1994): Absorption of peptide and peptidomimetic drugs. *Annu Rev Pharmacol Toxicol* 34:321–341.

48. Audus KL, Bartel RL, Hidalgo IJ, Borchardt RT (1990): The use of cultured epithelial and endothelial cells for drug transport and metabolism studies. *Pharmaceut Res* 7:435–451.

49. Miller SM, Simon RJ, Ng S, Zuckermann RN, Kerr JM, Moos WM (1994): Proteolytic studies of homologous peptide and N-substituted glycine peptoid oligomers. *Bioorg Med Chem Lett* 4:2657–2662.

50. Klassen CD, Watkins JB (1984): Mechanisms of bile formation, hepatic uptake, and biliary excretion. *Pharm Rev* 36:1–67.

51. Goresky CA (1980): Uptake in the liver: the nature of the process. *Int Rev Physiol* 21:65–101.

52. Guengerich FP (1987): *CRC Mammalian Cytochromes P-450.* Boca Raton, FL: CRC Press.

53. Wrighton SA, Stevens JC (1992): The human hepatic cytochromes P450 involved in drug metabolism. *CRC Crit Rev Toxicol* 22:1–21.

54. Mulder GJ (1992): Glucuronidation and its role in regulation of the biological activity of drugs. *Annu Rev Pharmacol Toxicol* 32:25–49.

55. Nelson DR, Strobel HW (1987): Evolution of cytochrome P-450 proteins. *Mol Biol Evol* 4:572–593.

56. Guengerich FP (1992): Characterization of human cytochrome P450 enzymes. *FASEB J* 6:745–748.

57. Gonzalez FJ (1992): Human cytochromes P450: problems and prospects. *Trends Pharmacol Sci* 13:346–352.

58. Tephly TR, Burchell B (1990): UDP-glucuronosyltransferases: a family of detoxifying enzymes. *Trends Pharmacol Sci* 11:276–279.

59. Pickett CB, Lu AYH (1989): Glutathione *S*-transferases: gene structure, regulation and biological function. *Annu Rev Biochem* 58:743–764.

60. Evans DAP (1989): *N*-Acetyltransferase. *Pharmacol Ther* 42:157–234.

61. Smith RL (1973): *The Excretory Function of Bile: The Elimination of Drugs and Toxic Substances in Bile.* London: Chapman & Hall.

62. Milburn P, Smith RL, Williams RT (1967): Biliary excretion of foreign compounds: biphenyl, stilboesterol and phenolphthalein in the rat: molecular weight, polarity and metabolism factors in biliary excretion. *Biochem J* 105:1275–1281.

63. Hirom PC, Milburn P, Smith RL, Williams RT (1972): Species variations in the threshold of molecular-weight factor for biliary excretion of organic ions. *Biochem J* 129:1071–1077.

64. Levine WG (1978): Biliary excretion of drugs and other xenobiotics. *Annu Rev Pharmacol Toxicol* 18:81–96.

65. Hughes R, Milburn P, Williams R (1973): Molecular weight as a factor in the excretion of monoquaternary ammonium cations in the bile of the rat, rabbit and guinea pig. *Biochem J* 136:967–978.

66. Williams RT, Milburn P, Smith RL (1965): The influence of enterohepatic circulation on toxicity of drugs. *Ann NY Acad Sci* 123:110–124.

67. Wrighton SA, Vandenbranden M, Stevens JC, Shipley LA, Ring BJ (1993): In vitro methods for assessing human hepatic drug metabolism: their use in drug development. *Drug Metab Rev* 25:453–484.

68. DeDuve C, Wattiaux R, Buadhuim P (1962): Distribution of enzymes between subcellular fractions in animal tissues. *Adv Enzmol* 24:291–358.

69. Rettie AE, Korzekwa KR, Kunze KL, Lawrence RF, Eddy AC, Aoyama T, Gelboin HV, Gonzalez FJ, Trager WF (1992): Hydroxylation of warfarin by human cDNA-expressed cytochrome P-450: a role for P4502C9 in the etiology of (S)-warfarin-drug interactions. *Chem Res Toxicol* 5:54–59.

70. Flammang AM, Gelboin HV, Aoyama T, Gonzalez FJ, McCoy GD (1992): Nicotine metabolism by cDNA-expressed human cytochromes P-450. *Biochem Arch* 8:1–8.

71. Watkins PB (1990): Role of cytochromes P450 in drug metabolism and hepatotoxicity. *Semin Liver Dis* 10:235–250.

72. Berry MN, Halls HJ, Grivell MB (1992): Techniques for pharmacological and toxicological studies with isolated hepatocyte suspensions. *Life Sci* 51:1–16.

73. Berthou F, Ratanasavanh D, Riche C, Picart D, Voirin T, Guillouzo A (1989): Comparison of caffeine metabolism by slices, microsomes and hepatocytes cultures from adult human liver. *Xenobiotica* 19:401–417.

74. Wortelboer HM, deKruif CA, deBoer WI, vanIersel AA, Falke HE, Blauboer BJ (1987): Induction and activity of several isozymes of cytochrome P-450 in primary cultures of rat hepatocytes in comparison with in vivo data. *Mol Toxicol* 1:373–381.

75. Wrighton SA, Brian WR, Sari M-A, Iwasaki M, Guengerich FP, Raucy JL, Molowa DT, Vandenbranden M (1990): Studies on the expression and metabolic capabilities of human liver cytochrome P450IIIA5 (HLp3). *Mol Pharm* 38:207–213.

76. Sipes IG, Fisher RL, Smith PF, Stine ER, Gandolfi AJ, Brendel K (1987): A dynamic liver culture system: a tool for studying chemical biotransformation and toxicity. *Arch Toxicol Suppl* 11:20–29.

77. Brendel K, Fisher RL, Krumdieck CL, Gandolfi AJ (1990): Precision-cut rat liver slices in dynamic organ-culture for structure toxicity studies. *J Am Coll Toxicol* 9:621–627.

78. Ross BD (1972): Liver perfusion. In Ross BD, ed. *Perfusion Techniques in Biochemistry*. Oxford, UK: Clarendon, pp 90–130.

79. Rowland M, Tozer TN (1989): The extravascular dose. In *Clinical Pharmacokinetics Concepts and Applications*. Philadelphia: Lea & Febiger, pp 33–48.

80. Gibaldi M (1991): Introduction to pharmacokinetics. In *Biopharmaceutics and Clinical Pharmacokinetics*. Philadelphia: Lea and Febiger, pp 1–13.

81. Taylor EW, Gibbons JA, Braeckman RA (1997): Intestinal absorption screening of mixtures from combinatorial libraries in the Caco-2 model. *Pharm Res* 14:572–577.

26

COMBINATORIAL TECHNOLOGIES: PROSPECTS AND FUTURE ISSUES

JAMES F. KERWIN, JR.

Pharmaceutical Products Division, Abbott Laboratories, Abbott Park, Illinois

Much has been written about combinatorial chemistry and molecule diversity, and most of it is true! Recent letters and debates[1,2] highlight the rapid emergence of combinatorial chemistry and the natural growth pains to which a new technology is subject. Before we address the question of what is combinatorial chemistry and where is it headed, it is appropriate to reflect on the driving forces and circumstances behind this emerging technology. As a starting point we may envision the economic environment from which molecular diversity and combinatorial chemistry emerged.

The costs of discovering and developing a drug have become astronomical over the past half a century, due not only to the forces of inflation and cost effectiveness, but also to factors such as regulation, market sizes, and competitive alternative therapies.[3] The ability to routinely enter the marketplace with a second or third "me-too" drug and expect profitability has become an antiquated strategy. While combinatorial chemistry approaches may address inefficiencies in finding new entities, the approach is not likely to have a major impact on new drug development costs, which are contained mostly in the clinical development of new entities. However, combinatorial chemistry and its allied technologies will have a major impact on the ability to find the "best fastest." In an industry where being second may not allow profitability, that is an important criterion—the ability to remain competitive! Since combinatorial technologies will decrease the costs of discovering new drugs, but will

Combinatorial Chemistry and Molecular Diversity in Drug Discovery, Edited by
Eric M. Gordon and James F. Kerwin, Jr.
ISBN 0-471-15518-7   Copyright © 1998 by Wiley-Liss, Inc.

not significantly affect the overall development cost of bringing a new product forward, the economic drivers for adopting combinatorial technology are based on timeliness—efficiency in discovery that will translate into competitiveness in delivery.

Emergence of techniques that simultaneously allowed the miniaturization of assay methods coupled with an appreciation of the broader applicability of solid-phase synthesis made the appearance of combinatorial chemistry within the pharmaceutical industry a natural event. Why combinatorial approaches have not been as well developed or as fully exploited in other industries may be due primarily to the lack of rapid and high-volume selection methods as much to the fact that fast followers in other industries may continue to be adequately competitive for their marketplace. In such circumstances there may presently be less financial incentive to adopt and capitalize a new technology.

Historically, the pharmaceutical industry has relied on the screening of natural products from which to derive new leads or drugs. Even today several breakthrough drugs owe their origins to natural product leads. Pharmaceutical industry experience showed that success in developing novel drugs was linked to the quantity, quality, and rate of evaluation of compounds in high-throughput screens. A benefit is derived from screening larger numbers of compounds, and combinatorial technologies indeed improves the number of compounds available. Our knowledge of biology and pharmacology is ever expanding, and the number of potential high-throughput screens is increasing, especially driven by advances in genetic and genomic information. If the number of potential targets/screens is increasing and an ever-increasing number of individual, though related, chemical entities is evaluated at these targets, the result of these increased numbers will be an ever-mushrooming reservoir of data.

The challenge of combinatorial chemistry will then be to continue to find ways to identify and focus interest more rapidly on those compounds of highest commercial potential, which may or may not be the most potent compounds in a high-throughput screen, which may or may not be the most metabolically stable, which may or may not be the most well absorbed, but which do represent those aggregate properties which make the best possible drug for the given therapeutic use. Hence, more than one criterion has to be met for success. The drug hunter will need to derive new methods not only for measuring these properties rapidly and with small quantities of compounds, but also for integrating the data in a manner that will allow identification of such desired compounds.

The application of combinatorial technologies to an array of desired drug properties or stressors should provide the most robust candidates as commercializable entities. What is required is the ability to measure these desired properties using sets of compounds, ideally produced in or from an organized array format. Progress in this direction has been noted with some adapting metabolic, toxicological or pharmacokinetic screens while others utilize theoretical or empirical calculations to drive libraries toward more optimal drug candidates. At a larger level, the emergence of combinatorial drug development will place a greater demand on two critical areas. First, the volume and types of data generated will require new methods of information management and intelligent assessment of the data. Second, since priorities must be set on what is done, an emphasis will be placed on defining higher payoff targets early.

With the completion of the human genome project and a multitude of efforts aimed at identifying new molecular targets for the pharmaceutical industry, it is obvious that combinatorial chemistry will be increasingly utilized.[4] To assess other relevant properties for drug candidates, new paradigms and methods will have to be developed. Some of these have been discussed in preceding chapters, for example, pharmacokinetic parameters,[5] metabolic stability,[6] and enzymatic stability. More work will be done to extend the throughput of these assays and add additional criteria, such as mutagenicity, tetratogenicity, toxicity, and hemodynamics. The truest test of combinatorial drug development will be its ability to predict best of class compounds based on large aggregate sets of data.

And how will this data be processed? While even large sets of simple property systems can be handled now, the ability to accurately identify novel compound possibilities in the future where n-variated sets are used may depend on using new means of processing and weighting data sets and means of visualization of data that permit end users to grasp multivariate properties of sets of compounds quickly. Similar concepts have permeated the measurement of chemical diversity.[7] These information processing technologies may allow the intelligent processing and parsing of compounds of interest, perhaps without direct human intervention.

The continuing emphasis on project/molecular target selection will be further compounded by these emerging combinatorial technologies at many different levels. With a relatively level playing field and the ability to contract appropriate specialists at will, a competitive edge based on in-house expertise in an area will become more difficult to maintain and will erode over time. Instead, the competitive edge will go to those with proprietary technologies, proprietary knowledge bases, and staffs that execute drug discovery programs well (know how). Thus, there will be an increased emphasis to define targets that make business sense even before data supporting commercialization is available.

To this end advances in genetics and genomics may answer the question in part. With solid information that a molecular target is involved in a disease etiology, selection of that molecular target has lower risk than targets that have less well-defined commercialization points. This is not to imply that feasibilities need not be weighted into the target choice, but rather that, in an era where later entries in a new class of therapeutics may not garner as much profitability, the choice of molecular targets will be more deliberate upon program initiation. A further outcome of the use of genomic information will be the ability to predict which populations will respond to drug therapies and even those that may be at risk for adverse events, the concept of pharmaco-genomics.[8]

Taken in parallel these trends suggest a selection of possible drugs based on genomic-derived targets and a selection of patient populations based on the genomic profile and the drug together, which lead to a more personalized form of pharmacotherapy. Naturally, a prerequisite for this scenario is the development of appropriate diagnostics to predict who should be treated and, perhaps more importantly, when they should be treated.

At the same time that combinatorial chemistry offers to expand the opportunities to find new therapeutic entities, it poses interesting and far-reaching problems for the management and analysis of information (See Appendix). With combinatorial chemistry, existing chemical information can grow at exponential rates and the current means of evaluating, storing, and managing this information may not be adequate or even appropriate. More and more data can be generated both in terms of chemical information and its related biological data. Yet what remains and needs to emerge from the endless datapoints is a synthesis of large data sets into new knowledge. More succinctly, this is the difference between information management (deposition and manipulation of data) and knowledge management (distillation of critical knowledge from data). Because of the increasing volume of information the combinatorial chemist will require new tools with which to manage, communicate, and develop knowledge. It is not surprising that one of the most pressing issues within pharmaceutical companies practicing combinatorial chemistry today is data and information management related to the libraries generated and those designed.

The ability to successfully manage information has already had an impact in the negative sense within patent application offices. The ability of patent examiners to handle the already expanded biotechnology arena has already slipped and the increased information load and lack of adequate systems to handle combinatorial arrays is likely to slow patent granting processes further.

Returning to the question of how to define combinatorial chemistry, this author would echo the sentiments already espoused; "Combinatorial chemistry will end up being what combinatorial chemists do."[2] No one still doubts that combinatorial chemistry is having an impact, but where is it headed? The areas for further improvement are easily identified. Besides the need for information management systems, a continued need for improved automation is evident. In fact, several companies are competing to provide automation solutions both in terms of chemical synthesis and high-throughput screening. There is also increased interest in integrating these systems and linking them to those that perform analytical services, entity registration, storage, and dispensing. The increase in automation will allow higher screening rates, but many scientists are finding that there are optimal numbers of entities (3–10) per assay well or tube that are dependent on the current screens and types of libraries. With this empirical result and the potential for greater numbers of compounds being produced, new assay methodologies are being discovered and evaluated that could accelerate the process of lead identification.

Synthetically, problems related to product purity and identity are being addressed by synthetic techniques[9] and coupled analytical systems.[10] The breadth of chemical reactions employed continues to grow, as does the knowledge and use of solid-phase organic synthesis techniques. However, the larger problem is the transformation of medicinal chemists and the education of the combinatorial chemists. First and foremost, practicing combinatorial chemists have had to transform their thinking from looking at one product to looking at sets of products; from evaluating one reaction to evaluating a manifold of reactions; from determining individual integrity to determining the integrity of a set of compounds. Chemists no longer look at individual data points, instead, they examine large sets of data and pick out the outliers. Rather than assuring the chemical identity up front in a process, combinatorial chemists are interested in enabling methodologies that will allow retrospective identification of chemical identity or identities! Combinatorial chemists no longer focus on the discrete, intimate step—Did this reaction produce what was desired?—but, rather, they ask, Did this methodology derive a new lead or greater insight?

Of all the downstream impacts of combinatorial chemistry, the most significant will be the cultural shift in the thinking of medicinal chemists, and how the medicinal/combinatorial chemistry laboratory will be maintained in the future. Increased needs and future emphasis on automation, analytical techniques, and information technology have been highlighted, but how will this translate into a new educational paradigm for future chemists? Tomorrow's chemists will need to be just as strong synthetically, but must have higher "technology quotients" and a broader perspective of science in general. The combinatorial technology landscape now encompasses small companies with competitive strengths in combinatorial chemistry, encoding methodologies, analytical techniques, information management, screening methodologies, therapeutically focused research, automation, and combinations of these. Several of these companies, such as Affymax, Sphinx, and Selectide, have been acquired already and others are tied to a few larger pharmaceutical companies.

Nearly all large pharmaceutical houses have integrated some form of combinatorial chemistry into their research portfolio. While some further acquisitions of the remaining specialized "combi companies" may occur, the trend will be in finding integrated research platforms with collaborators that provide a sound technology basis and the opportunity to reach commercial viability. Licensing, cross licensing, and utilization of critical core technologies will continue, since no company is likely to corner all the different techniques and technologies that exist or that can be discovered. At large pharmaceutical companies the more pressing issue is how to disperse or disseminate combinatorial technology among research staff. Issues of centralization versus decentralization and specialized equipment versus standard equipment are common. Much like the advent of NMR spectroscopy and HPLC, combinatorial techniques will become part of the routine arsenal of the medicinal chemist. The question of how to best organize research units and educate staff members so as to facilitate this transformation is an ongoing experiment, the results of which will not be evident for some years.

Combinatorial technologies have hit an important nerve. They fill the gap for an efficient and reproducible means of generating potential commercial products in record times. While combinatorial chemistry has continued to expand the breadth and reliability of the technology, the biological frontier is being pushed toward selecting for more and more relevant drug properties. Advances in genetics, genomics, and molecular and cellular biology assure an increasing number of potential targets. Advances in information processing and management are still needed to unleash the fullest potential of this methodology. The number of core technologies within this field continues to grow and will likely continue at least until the successful strategies and technologies are differentiated from the rest. Combinatorial chemistry is a technology that should be widely used, though not necessarily by all the same time. Mechanisms to educate staffs and train the chemists of the future will be needed if this technology is to continue to grow and develop.

REFERENCES

1. *Chem Eng News* 10 Feb 1997, p 10; 17 Mar 1997, p 6.

2. Burgess K, Czarnik A (1997): *Chem Eng News* 14 Apr 1997, p 4.

3. Casadio Tarabusi C, Vickery G (1998): Globalization in the pharmaceutical industry, Part I *Int J Health Serv* 28:67–105.

4. Guyer MS, Collins FS (1995): How is the Human Genome Project doing, and what have we learned so far? *Proc Natl Acad Sci USA* 92:10841–10848.

5. Berman J, Halm K, Adkinson K, Shafer J (1997): Simultaneous pharmacokinetic screening of a mixture of compounds in the dog using API LC/MS/MS analysis for increased throughput. *J Med Chem* 40:827–329.

6. Gibbons JA, Taylor EW, Braeckman RA (1997): In Gordon EM, Kerwin JF, eds. *Combinatorial Chemistry and Molecular Diversity in Drug Discovery.* New York: Wiley–Liss, p 453–474.

7. Martin EJ, Blaney JM, Siani MA, Spellmeyer DC, Wong AK, Moos WH (1995): Measuring diversity: experimental design of combinatorial libraries for drug discovery. *J Med Chem* 38:1431–1436; Hassan M, Bielawski JP, Hempel JC, Waldman M (1996): Optimization and visualization of molecular diversity of combinatorial libraries. *Mol Diversity* 2:64–74.

8. Drews J (1995): Intent and coincidence in pharmaceutical discovery: the impact of biotechnology. *Arzneimittelforschung* 456:934–939.

9. Kaldor SW, Siegel MG, Fritz JE, Dressman BA, Hahn PJ (1996): *Tetrahedron Lett* 27:7193; Kaldor SW, Fritz JE, Tang J, McKinney ER (1996): Discovery of

antirhinoviral leads by screening a combinatorial library of ureas prepared using covalent scavengers. *Bioorg Med Chem Lett* 6:3041.

10. Holt RM, Newman MJ, Pullen FS, Richards DS, Swanson AG (1997): High-performance liquid chromatography/NMR spectrometry/mass spectrometry: further advances in hyphenated technology. *J Mass Spectrom* 32:64–70.

27

APPENDIX: COMBINATORIAL CHEMISTRY INFORMATION MANAGEMENT

DAVID WEININGER

Daylight Chemical Information Systems, Inc., Santa Fe, New Mexico

This chapter provides an overview of the special problems associated with managing information derived from combinatorial chemistry experiments, and describes approaches that can be used to address these problems.

WHY CHEMICAL INFORMATION SYSTEMS WORK

Modern chemical information systems are very effective for management of data about small molecules. Such systems are used for databases ranging in size from chemical catalogs (10^2–10^4 structures) to corporate in-house databases (10^3–10^6 structures) to comprehensive bibliographic databases of the chemical literature (10^6–10^7 structures). There have been significant improvements in database methodologies used for chemical databases in the past decade. However, the effectiveness of current chemical database systems is not primary due to database methodology per se. The primary reason that such systems have become more effective in the last decade is that the problem has become easier by at least 10-fold due to the rapid increase in the capability of our tools (digital computers). In the last 10 years, computational capacity, speed, and price/performance of digital computers have each improved by a factor

Combinatorial Chemistry and Molecular Diversity in Drug Discovery, Edited by
Eric M. Gordon and James F. Kerwin, Jr.
ISBN 0-471-15518-7   Copyright © 1998 by Wiley-Liss, Inc.

of $10\times$ to $30\times$. In the same period, the size of the problem (the amount of chemical information known) has increased by less than two-fold.

For chemical information, the amount of available processing capacity increases faster than the amount of available chemical information. One can easily observe the effect of this phenomenon over a decade. Consider management of the structures themselves, a core functionality of a chemical information system. Small chemical structures contain an average of 10–20 bytes of pure information (assume 16B for this argument). The inefficiencies inherent in encoding structures are about 2–$20\times$, depending on the methodology used; for example, the average SMILES might use 32 bytes, a conventional data structure 320B (assume 100B/structure here). Ten years ago, approximately 1.0×10^7 structures were available (i.e., *CAS Registry*), roughly 1.0 GB of information to process. At that time, only the largest, special-purpose computer clusters were capable of effectively processing that quantity of information. Today we have approximately 1.8×10^7 discrete small molecule structures, requiring processing about 1.8 GB. Although this is not a trivial amount of information, it is certainly in the scope of workgroup servers that are available locally to most chemists involved in drug design. In the absence of confounding factors, we could safely assume that 10 years hence systems capable of processing all known structures could be available on local, personal, desktop computers.

THE CHALLENGE OF COMBINATORIAL
SYNTHESIS INFORMATION

The advent of combinatorial synthesis challenges the principle that the amount of available processing capacity increases faster than the amount of available chemical information, at least with respect to sheer numbers of synthesized molecules as in the above argument. For instance, all natural hexapeptides have been synthesized in a single combinatorial experiment (as 20 mixtures of 20^5 hexapeptides each); that is, 20^6 different molecular structures were synthesized. This single experiment represents 64 million molecules, over three times the total number of discrete molecular structures in the chemical literature.

Obviously, doing the all-hexapeptide experiment does not quadruple the total amount of chemical information that is known. Just as obviously, storing all possible hexapeptides would in fact quadruple the amount of storage required, if the only available storage method is encoding individual structures. Chemical information systems based on storage of individual molecular structures are not an effective way of storing information about very large mixtures.

Conventional chemical information systems are suitable for processing conventional synthesis results. Combinatorial (nonconventional) synthesis requires development of nonconventional methods for effective information processing.

INFORMATION COST ANALYSIS

Another way of looking at the same effect is information cost analysis. In this analysis one compares the relative cost of producing data vs. processing the same data. Consider the incremental cost of producing and storing a single discrete chemical structure by conventional methods. Assume that the cost of conventional chemical synthesis is represented by (no less than) 2 weeks of a synthetic chemist's salary plus overhead: $5000 ten years ago and $10,000 today. Assume also that the incremental cost of processing the information thus produced is proportional to the cost of the hardware used for high-performance mass storage. Ten years ago the most cost-effective, random-access, high-performance, mass storage device was a 380-MB Winchester disk drive with an average price of $1800, that is, $4.7/MB. Assuming 100B/structure, this translates to an incremental cost of 4.7×10^{-4} per structure. Today the most cost-effective storage medium is still a Winchester disk, but it now costs $2250 for a 9.0-GB disk, that $0.25/MB, so the incremental cost is 2.5×10^{-5} per structure (19× less expensive). The relative, incremental cost of producing a new discrete substance versus storing its structure was therefore 1.1×10^7 ten years ago and 4.0×10^8 today, that is, 38× more favorable. The conclusion from this is that chemical structure storage is very cost effective (the cost of synthesis is millions of times more expensive than storage) and that it is getting more so (38× in 10 years).

One can do a similar analysis for other forms of storage, for example, random access memory (RAM). RAM is more expensive than disk (currently $0.004/structure) and RAM prices have decreased less dramatically than disk prices in the last 10 years (from 300 to 40 $/MB, or 7×). However, one arrives at similar conclusions: Structure storage in RAM is cost effective (synthesis is 2.5×10^6 times more expensive than storage) and getting more so (14× in 10 years).

Now consider the all-hexapeptide library synthesis. With current, semiautomated methods, the incremental cost of synthesizing another combinatorial library of this size is probably no more than 5× the cost of a single molecule synthesis (10 man-weeks of $50,000), that is, 7.8×10^{-4} per structure. In this case, the cost of storage of individual structures has not changed (2.5×10^{-5} per structure on disk, 4×10^{-3} in RAM), but there are 6.4×10^7 of them! If stored as individual structures, the relative, incremental cost of producing a new substance in a very large mixture versus storing its structure is only 31× on disk and 0.20× in RAM (i.e., it would cost 5× more to store the 6.4×10^7 hexapeptides in RAM than it costs to synthesize them). This result might seem less surprising if one considers that 6.4 GB of structural data is generated from a single experiment.

The all-hexapeptide library is an extreme example, but it does demonstrate that the data produced by combinatorial synthesis experiments can exceed the capacity of our current data systems. To reiterate: Combinatorial (nonconventional) synthesis requires development of nonconventional methods for effective information processing.

COMBINATORIAL SYNTHESIS DATA MANAGEMENT TASKS

Although the design and synthesis of combinatorial libraries is different from that of traditional drug molecules, most data processing needs have not fundamentally changed:

- **Combinatorial library registration:** Combinatorial mixtures need to be registered in the same way as other proprietary substances, for similar reasons.
- **Combinatorial library display:** One needs to be able to view combinatorial libraries graphically.
- **Combinatorial library specification:** One needs to be able to define mixtures conveniently, accurately, and interactively.
- **Documentation:** Accurate documentation is critical for corporate combinatorial synthesis (more important than in traditional synthesis).
- **Data retrieval and exploratory data analysis:** Data retrieval and analysis operations used with pure substances have counterparts with combinatorial libraries. There are also a few new data operations that are specific to mixtures.

Each of the above tasks are described in more detail in the following sections.

Combinatorial Library Registration

Motivations for structure-based registration also apply to combinatorial mixtures.

Is This (Set of) Molecule(s) Identical to Another (Set of) Molecule(s)? A novelty check is needed for the most basic information processing operations, i.e., Is this available? and Have we made this before?

The most commonly used approach to establishing identity of a set of molecules is to provide means of unambiguously describing all components as molecular structures. As discussed above, a naive approach (i.e., listing all component molecular structures) cannot be expected to work well for large combinatorially synthesized mixtures. A natural shortcut is to express components combinatorially, for example, to create an expression for a set of molecules from lists of molecular components and combinatorial operators. The CHORTLES nomenclature referred to below is an example.

Although the component list approach is closest to existing small molecule methods, other approaches are possible. One approach that is equally valid is to describe a set of combinatorially synthesized molecules by unambiguously describing the combinatorial reaction scheme used to create that set. The reaction scheme approach has some significant advantages over the component list approach in that it is more accurate: few (if any) large combinatorial libraries contain all intended components. With a well-designed reaction scheme nomenclature, it is possible to enumerate individual (intended) components and do all other operations possible with a component list approach. Further, reaction scheme nomenclatures make it simple to add information about the reactions per se. The main disadvantages of such nomenclatures are that they are less efficient and less intuitive than those based on component lists.

Combinatorial nomenclature systems are extremely powerful and efficient for describing the sets of molecules produced by combinatorial synthesis. For instance, the CHORTLES nomenclature is able to describe the all-hexapeptide library in 492 bytes, which is 2,300,000 times more efficient than noncombinatorial CHUCKLES (1.15 GB) and 40,000,000 times more efficient than atom-level SMILES (20 GB). Although noncombinatorial chemistries fall outside the scope of this book, it must be noted that not all useful libraries (or mixtures) are combinatorial in nature. For instance, it is possible to create very large pseudo-random mixtures. Combinatorial nomenclatures are not suitable for describing such mixtures.

Provide a Unique Identifier for This (Set of) Molecule(s). Which (Set of) Molecule(s) Is Defined for This Identifier? Unique IDs are indispensable in chemical research for practical, legal, and information processing purposes. Most data processing operations concerning chemical substances are done without reference to the chemical structure (or structures) per se. Nearly universal in practice, a unique registration number is generated for each entity, whether it be a pure substance, a small mixture, a pool, or a library. For the purposes of combinatorial synthesis data analysis, it is important to assign a unique identifier (typically a registration number) to each collection of molecules for which information might be known. For instance, it is useful to register combinatorial libraries even though the components might never be mixed beyond the pool level (or at all). Such registration numbers provide an identifier to which information can be attached about the design, synthesis, and relationships with other libraries, pools, subpools, and individual structures.

There is usually no natural one-to-one mapping of registration numbers and structures (or sets of structures). For optimal accuracy and utility, information should be attached to an identifier that specifically identifies the entity that the information is known to be about. In practice, data are usually obtained about a vial identified by a registration number. (Except for results of calculations, data are rarely known about structures themselves.) A common and convenient approach is to associate data with a registration number (which describes a sample or batch) and associate registration numbers with a structure (or set of structures). With this approach, one can accurately describe what is known and correct errors as they are discovered.

Store and Retrieve Identifier-Based Data. As described above, most data (synthetic, physical, biological) are associated with an unambiguous identifier, such as a registration number. For instance, if one sends a vial containing a particular pool or substance for GCMS analysis or bioassay, the result is reported for that vial (typically labeled with a registration, sample, or batch number), not for the (presumed) molecular structure(s) per se. Ultimately, an analysis will be done that requires both measured results and chemical structure. A chemical information system must therefore be able to efficiently store and retrieve identifier-based data as well as structure-based data.

With combinatorial chemistry data, it is especially important to maintain the relationships between identifiers and structures. There are (at least) three approaches, all of which are used by one commercial system or another: (1) One can attempt to establish a 1:1 mapping between structural identity and registration number, such as by assigning a new registration number to each entity with a new structural identity. This is essentially a computerized "dictionary" with discrete entries. (2) One can store structural information as data about entities that are authoritatively identified by registration number. This is a nonstructurally oriented database where structure is a kind of data. (3) One can connect data to both structural and nonstructural identifiers as appropriate and store relationships between identifiers as data. This is the "thesaurus" approach.

Although a full comparative analysis of these approaches is beyond the scope of this chapter, a few general remarks are offered. Method 1 is not suitable for data systems managing combinatorial synthesis data, because of the inherent error in structural identification. In practice, methods 2 and 3 are both suitable for use with combinatorial systems, but differ in their emphasis. Method 2 is a more suitable approach when structural information is of lower importance and when relationships are well known and fixed; it is suitable for an implementation using an RDB (relational database). Method 3 is more suitable when structural information is of equal importance and when relationships are unknown and dynamic.

Correctly Handle Asymmetries (Induced Chirality). Chirality is an important part of combinatorial chemistry and it is important that data systems designed for combinatorial chemistry information handle chirality correctly. The simplest chiralities are those that involve chiral monomers, scaffolds, or substituents. This class is roughly equivalent to libraries in which chiral reagents are used in the combinatorial synthesis. Data handling is no different from that in a conventional database of pure substances: One must provide a mechanism that is adequate to enter, store, and retrieve known chiralities and not lose track of them. For instance, a data system that can distinguish D-Ala-L-His-O from L-Ala-D-His-O should have no trouble distinguishing the mixtures [D-Ala;L-Ala]-L-His-O and L-Ala-[D-His;L-His]-O.

Some forms of asymmetry that do not normally occur in pure-substance databases become important in combinatorial chemistry databases. A common example of this is simple induced chirality, that is, nonchiral reactants forming chiral products via a reaction such as addition across an archiral double-bond. More complex induced chiralities are also possible, such as asymmetries created by the interaction of more than one reaction in the combinatorial synthesis scheme. A straight-forward method of dealing with such problems is to keep track of protochirality, that is, to distinguish between substituent positions on otherwise achiral atoms. A similar situation occurs with reactions producing relative stereochemistry, such as addition across an alicyclic double bond (typically requiring representation of relative stereochemistry).

Asymmetries produced by combinatorial synthesis are generally well known and have established methods of processing. However, the chirality representation in database systems designed for pure substance management (e.g., CIP) may be inadequate to represent combinatorial synthesis products.

Combinatorial Library Display

One of the core functions of a chemical information system is to maintain the chemical identity of each entity of interest. At some point, this information must be made available to the end-user in an understandable manner. Unfortunately, the chemical identity of a combinatorial library may include thousands (or millions) of different structures. Displaying such information in a way that is understandable is not a trivial problem.

2-D Structure Diagrams. In data systems dealing with single structures, the preferred method is a 2D structure diagram. Chemists are used to such diagrams; no other method is even remotely as acceptable. Combinatorial data systems need to provide the same sort of information in a graphical manner, for the same reasons. For single molecules, sets of molecules, mixtures, and reagent sequences, the most useful displays of structure are variations of the 2D structure diagram.

What Are the Components of This Library (Set of Molecules, Mixture)? The most important display of chemical identity is the one that displays component structures. Showing all component structures is reasonable up through sets of perhaps 100 structures. Beyond that, one is faced with multipage display of postage stamp-size diagrams, which become confusing due to high volume and visual overload.

How Was This Library (Set of Molecules, Mixture) Made? There is an intimate relationship between the components in a combinatorial library and the sequence of reagents used in its synthesis. The ability to display both is important. Linear synthesis schemes are relatively easy to display (or at least no harder than a component display). Fortunately, most combinatorial synthesis schemes in use today are strictly linear. There is strong indication that more complex combinatorial synthesis schemes will become popular in the near future, for example, mixed variable-length, multiway branched, and even cyclic schemes. Providing effective methods for display of complex combinatorial synthesis schemes is a outstanding problem. Until it is solved, an effective work-around is to break complex schemes into linear fragments, register intermediate libraries, and display the linear schemes annotated with their relationship in the synthesis. (Although formally correct, this work-around does not provide an easy-to-understand single view of such a library.)

Combinatorial Display. The only practical way to display combinatorial libraries containing more than a few hundred components is combinatorially. For commercial use, a display method is required to be suitable for interactive use and be understandable when printed with a monochrome laser printer.

An intuitive approach is to produce a Markush-like diagram: a 2D display of the common portion of structure (i.e., scaffold or backbone) with points of variability indicated plus lists of possible substituents at each position of variability (or monomer units, in the case of oligomers). Substituent positions with common variability can share substituent diagram lists. Advantages of this approach are that it produces a diagram with the minimum visual density for a given library; it is suitable for complex libraries; it is suitable for printing with a monochrome printer; and in many cases can be designed to reflect the underlying combinatorial chemistry. Disadvantages are that no component molecule is shown in its entirety and that the scaffold diagram for libraries synthesized with ring closure reactions often includes more than one ring atom as a point of variability, compromising the intuitiveness of the 2D diagram. Overall, this is probably the single most flexible and practical type of combinatorial display.

Variations on a Theme. A number of improvements in combinatorial data displays are possible if one relaxes the environmental constraints to allow color or user interaction. A display based on example and variations is particularly effective on color media. Variable positions are divided into two groups; intrinsic (in-ring or multivalent variations) and extrinsic (ex-ring and monovalent variations), with the provision that at least one variable position is assigned to the extrinsic group. Furthermore, extrinsic choices are sorted by complexity (number of atoms). For each combination of intrinsic variability, a complete example molecule is drawn with the template in one color, intrinsic substituents in a second color, and extrinsic substituents in other colors. This list of examples is accompanied by color-coded lists of intrinsic substituents. The main advantage of this approach is that the resultant diagram is very intuitive to a chemist: A covering set of simplest cases is presented (as complete molecules) and all ring systems are drawn in their entirety (except in rare pathological cases). The main disadvantages are that it is more verbose than the Markush-type diagram (thus less suitable for very complex libraries) and it is not suitable for monochrome media such as a laser printer.

Interactive displays allow even more flexibility. The example and variations method described above is particularly effective when adapted for interactive use: Extrinsic constituents lists are implemented as graphical menus, which, when a substituent is selected, causes the example to be replaced with the appropriate substituent. This has the further advantage of providing the chemist with a 2D diagram of any desired component and hiding much of the combinatorial complexity. However, it is suitable for interactive display only.

An interesting alternative implementation of the interactive example and variations display uses menus of reagents rather than extrinsic substituents. Selecting a reagent causes the example to reflect the appropriate constituent molecule. Assuming that one has an adequate underlying data system, this display is not much harder to implement than one based on substituents. The advantage of an interactive reagent-based display is that it is also suitable for library design, where the menu items are obtained from databases of available reagents (rather than a description of a fixed library).

Combinatorial Library Specification

Given that combinatorial libraries need to be registered by structural content, one must establish a practical method of specifying the structures in a particular library. This is a significantly more complex task than the corresponding operation in a pure substance database. To start, there are typically 10 to $50\times$ more structural entities to be specified (the scaffold plus all substituents added in all steps). One must enter not only the partial structures themselves but also their points of attachment and relationship in the combinatorial synthesis.

Combinatorial Structure Entry. Is it not practical to specify all the structures in most combinatorial libraries, due to the prohibitively large number of components. The only reasonable way of specifying structures in a combinatorial library is combinatorially. This means using the combinatorial nature of the library to simplify structural entry: One must specify a structure including positions of variability and lists of alternative substituents for each position of variability. The main disadvantage of this approach is that data entry becomes a more complex operation, since both structures and relationships must be specified. The advantage is that it drastically reduces the amount of structural information that needs to be specified. This advantage in complexity reduction is so profound that there appears to be little choice but to accept a more complex method of structure entry.

Oligomeric Versus Parent-Substituent Libraries. Combinatorial synthesis evolved from two fundamentally different synthetic approaches. One approach is oligomeric synthesis, that is, producing molecules organized (and synthesized) as linear units. The archetype of this approach is peptide synthesis (in fact, peptide synthesis methodology provided the spur to the development of combinatorial chemistry). The other approach was developed from more conventional methods used for synthesis of pure substances, often producing products defined in terms of parent–substituent relationships. The archetype of this approach is probably the classic benzodiazepine synthesis.

At first blush it would seem that these two types of libraries require different entry methods: one based on linear chains of structural units with variability, the other on a fixed scaffold with variable substituents. However, from a data entry point of view they are entirely equivalent. In terms of the combinatorial structure entry presented above, an oligomer can be treated as a molecule, with the scaffold being the backbone (where "substituents" embody the variability of monomeric units), or equivalently be treated as a molecule with the scaffold reduced to nothing (the monomeric unit at each position being a substituent).

The equivalence of various combinatorial representations is a two-edged sword. It is fortunate that a universal representation is available, because there are, in fact, a multiplicity of combinatorial library types (not just the two extreme archetypes): It would be impractical to develop a whole new representation each time a new combinatorial library synthesis scheme was invented. On the other hand, there are multiple equivalent representations for almost all combinatorial libraries. This creates a problem when dealing with combinatorial structures that does not exist when dealing with pure substances. In particular, does one define a "right" way that structures should be specified in a combinatorial library among the multiple equivalent ways available? If not, how can one recognize when two libraries are equivalent (and other such problems)? If so, how does one decide which way is the "right" way? One of the stickiest problems in system design is how to build a system for combinatorial library entry that protects the user from having to deal with such issues.

Building Blocks: An Intermediate Level of Abstraction. It is possible to build combinatorial structure entry systems that are based entirely on atom-level structure specifications, that is, the user specifies all structural features at the atomic level just as with specification of pure substances. In general, such systems are very flexible, require a minimal amount of premastering, and can be made to look similar to conventional structure specification.

There are a variety of reasons why it may be desirable to operate at a higher level of structural abstraction than the atomic level. Defining structural building blocks allows one to assign properties to the building block, such as name, synthetic reagents, and conditions. On input, one can then refer to the building block by name without having to be concerned with atomic level details. Further, one might expect that a building block level specification could be further processed in multiple ways; for example, it could be converted to a component list at reagent sequence. Most importantly, data stored about such building blocks can be reused. This is important because some of the most important information about combinatorial chemistry concerns the building blocks themselves (synthetic methods). Another presumed advantage of working at a higher level of abstraction is that it is possible to increase processing efficiency.

The concept of working with building blocks is neither new nor particularly complicated. Conventional oligomeric nomenclature uses such a concept; for example, peptides are normally specified as residue-size building blocks Ala, His, Tyr, and so on. To be completely general, it appears that building blocks must define at least the partial molecular graph with fully defined internal and external symmetry.

The building block approach has some inherent disadvantages. Once building blocks are defined, they must be used and to do this one needs to identify them. In the case of peptides, referencing building blocks by name is convenient since it is easy to learn the vocabulary of 20 amino acid abbreviations. Unfortunately, names are not very useful in general. For example, there are about 4000 commercially available primary amines that might be used as peptoid building blocks. Memorizing 4000 names or abbreviations is unreasonable; one needs to provide helper tools, such as a searchable database of building blocks. Since the most natural way of searching such a database is by atom-level structure, it is questionable whether one gains anything from the extra level of abstraction (at least with respect to structural entry). In any case, an intermediate level of abstraction requires extra machinery and maintenance.

Several different systems based on the building block concept have been developed and have been in production use for one or more years. Both the advantages (efficiency, purity of representation) and disadvantages (higher complexity and representational overhead) are becoming obvious. At this point, it is uncertain whether the advantages outweigh the disadvantages.

User Interface. A user interface must be provided that allows accurate and convenient specification of combinatorial structures. Whether based at the atom- or building block level, a number of common requirements can be identified for such a user interface.

The user interface must present an editable library in a way that is understandable and verifiable. This almost certainly means that it needs to incorporate a graphical component that displays 2D structure diagrams. The interactive example and variations method described above appears to provide a good starting framework.

The resultant interface should be convenient. The ability to do rapid combinatorial specification is an advantage but it is not as important as the above requirement of being understandable and verifiable. One must recognize that combinatorial structure entry is a much bigger job than single structure entry. If it takes a user 2 min to enter a single structure and verify it, one might expect that 100 min would be required to enter a simple combinatorial library containing 50 reagents. One hopes to do better, of course, but if one builds a data entry system based on a design goal of rapid entry, something else is probably going to break. Since the average user doesn't have an attention span of 100 min at a single sitting, the library entry process should be broken up into smaller problems. A natural division is to separate the process into specification of the combinatorial framework and specification of each combinatorial step.

A possibility for conceptual simplification is to divide the specification problem functionally. Most users are familiar with at least one molecular editor. One might imagine building an user interface for combinatorial library specification at the atom level that does not provide atom-level specification at all, but rather relies on molecular specifications from a user-selected editor. The advantage of such an interface is that it would be considerably simpler, being limited to specification of the combinatorial relationships in the library. A further advantage would be that the user would not have to learn to use a new molecular editor.

Finally, one must consider the generality of the user interface. Despite the fact that the development of an effective user interface for combinatorial structure specification seems difficult enough today, it is certain that more complex combinatorial designs will be of importance in the near future. Including a concept such as a "combinatorial outline" is a method for future-proofing such user interfaces. Most current combinatorial libraries are built based on a single combinatorial synthesis scheme, but that is not an inherent limitation in library design. People have already started to combine combinatorial pools to form libraries and to use such products as reagents for further combinatorial synthesis (synthesis paths form a tree). One can be sure that if people do it, they're going to want to enter it into a data system.

Documentation

One must record not only what substances are in a given combinatorial library, but also exactly how it was made. Although documentation is important for any corporate synthesis, it is critical for combinatorial library synthesis. The object of making libraries is to eventually resynthesize and retest components; the ability to reproduce the work is essential.

In many corporate settings, data management is divided between a chemical information system (responsible for managing only chemical structure data) and a general-purpose database system (responsible for managing all other data about samples). Communication between such systems is generally restricted to lists of common identifiers. Although this data model works well for pure subtances, it tends to break down when used with combinatorial library data. The reason is that a number of essential datatypes about combinatorial library structures are fundamentally chemical in nature. There are at least two approaches to dealing with this problem: Incorporate more sophisticated data handling in the chemical information system or increase the sophistication of communication between the data systems.

How (When, Where, Why, by Whom) Was it Made? At the simplest level, one must be able to go back to the how, when, where, and why a library was made a day, month, or year later. By itself, this requirement could be achieved by managing such data as if it were normal (non-chemical) sample-based data. However, there is an intimate relationship between what structures are in a library and the reagent sequence used to create the library, both of which are best described in terms of chemical structure and for which chemical information processing is useful.

What was it intended to be? Versus What is it? In general, only very reliable chemistries are used for combinatorial library synthesis. Even so, the fraction of design structures that appear in actual libraries is probably no better than 90% on the average. This problem is compounded by the fact that structures that are not in the design "volunteers" often produce hits that need to be tracked down. The distinction between, "What was it intended to be?" (library design data) and, "What is it?" (structure elucidation data) presents much more of a problem with combinatorial than single-molecule synthesis data. As library components are elucidated, combinatorial information systems need to provide a mechanism to record such information.

What Is the Relationship of This Entity to Others? Combinatorial synthesis introduces new relationships that must be maintained by a data system. For example, combinatorial mixture deconvolution introduces ancestor, sibling, and descendent relationships. These relationships have a dual nature: They represent both independent knowledge ("Pool 12040-23 was synthesized as part of library 12040-1, which was suggested by hits in library 12021-1 . . .") and structural relationships ("Pool 10240-23 is a subset of library 12040-1," "library 12021-1 and pool 12040-23 share a nonempty subset," etc.) Given the combinatorial nature of the structural representation, responsibility for managing such data falls to the chemical information system.

Data Retrieval and Exploratory Data Analysis

Data retrieval and analysis operations used with pure substances have counterparts with mixtures. There are also a few new data operations that are specific to mixtures.

"Normal" Structure Searches. Chemical information systems typically provide a number of structural search functions, such as finding substructures, superstructures, similar structures, equivalent tautomers, enantiomers, and structural clusters. All these search functions are useful when applied to databases of combinatorial libraries. Conventional structure searching is based on atom-level connectivity. Special techniques are required to provide high-performance atom-level searches for databases containing large libraries (e.g., the all-hexapeptide library). Systems based on building block representation also provide structural searches at the building block level. Building block level structural searches are normally very fast but not very flexible.

Associated Data Searches. As with conventional pure-substance databases, combinatorial information systems need to provide data manipulation functions based on data associated with libraries, pools, and components; that is, one must find, sort, and select by associated data.

Searches Specific to Sets of Molecules. Combinatorial synthesis products can be viewed as sets of molecules (whether they actually exist as mixtures or not). Sets of molecules have relationships that individual molecules do not. Since these relationships are important in practice, information systems dealing with such data should provide searches based on such relationships. Examples include finding subsets, supersets, and similar sets of molecules. Additional searches need to be provided for properties derived from the structural identity of sets of molecules; that is, one must find, sort, and select by number of components.

Searches Specific to Combinatorial Syntheses. As mentioned previously, there are a few new relationships which are specific to combinatorial synthesis syntheses per se. It is particularly important to be able to manipulate such information for exploratory data analysis. For example, find ancestors, parents, siblings, children, descendants of a given library, pool, or mixture.

Similarly, it may be important to manipulate data based on properties which are specific to combinatorial synthesis per se, e.g., find, sort and select by number of variable positions.

Outstanding Issues

Data management of combinatorial information is still very much in development. This is partly due to the fact that it is a relatively new problem (we're just figuring out what to do with it) and partly because the whole field of combinatorial synthesis is in flux (it's a moving target). A few of the major outstanding issues in combinatorial chemistry data management are discussed below.

What was it intended to be? Versus What is it? Currently, no general solution is effective at storing knowledge about this dichotomy. Most existing systems register combinatorial libraries (pools, mixtures) with the design set (i.e., What was it intended to be?) This is important, but it is not everything. We also need to be able to store What is it? information as it becomes known (e.g., this design molecule wasn't made, this other nondesign molecule was). In principle most extant data systems can store such information as associated data, but since it is not incorporated into the structural representation, this information is not reflected in search results. The prospect of managing long exception lists (perhaps 10% of the size of the library) is not a good one: We lose most of the efficiency we gain from combinatorial representation. Ideally, we would not only see the results of such structural exceptions reflected in the representation of a particular set of molecules, but also have the results propagated to dependent pools and cross-referenced by the generating reaction.

Impure Combinatorial Libraries and Noncombinatorial Libraries. The information-processing techniques described above are effective when working with "pure" combinatorial libraries, that is, a set of molecules synthesized by a single combinatorial design. They are much less effective when used with "impure" combinatorial libraries (such as collections of different combinatorial products that can't be expressed by a single combinatorial design) and are not at all effective with noncombinatorial libraries (for example, random collections of molecules, the first 1000 bottles in the stockroom). The bottom line is that some libraries exist today that are not encodable by any extant system except as fully expanded lists of individual structures.

One can imagine extending combinatorial representation to allow representation of some impure combinatorial designs, but at some point our shortcuts become more trouble than they're worth. It's hard to imagine methods for management of random libraries that are much more efficient than dealing with the individual structures. If we're lucky, computer capacity will increase fast enough to make this a moot point. Unfortunately, that's also hard to imagine.

Integration of Chemical Information and Reaction Intelligence. The above discussion has repeatedly mentioned the intimate relationship between the structural identity of a combinatorial library and the reaction sequences used to generate the library. Some early combinatorial data systems were implemented in which this relationship was even more than intimate: It was the identity relationship. In such systems, the encoding of monomeric units was synonymous with that of the reagent used to synthesize the corresponding library component. Such combinatorial representations of libraries were designed to be used to record what was made, to order reagents, and to program the synthetic robot directly. All this, programmed only once! Although such systems worked well as prototypes with simple oligomeric library designs, they were ultimately abandoned due to problems arising from the false assertion that the generating reagent and the generated (sub)structure could be treated as synonyms. (Such a system can't handle even simple ambiguities, such as multiple routes to the same structure.)

For all the faults of the early systems, they achieved a level of information integration that is unmatched by current combinatorial information systems.

Granted, the current systems operate in a much more complex and nonprototypical environment. But as we've extended the data systems to handle more kinds of combinatorial designs, we've lost something in simplicity of integration and have increased the required amount of special-purpose human interaction. An outstanding challenge in combinatorial information system design is to build a system that reverses this trend and integrates library design, reagent ordering and tracking, synthetic robot programming, and assay result data merging.

Similar Mixtures.   One of the most useful concepts in exploratory data analysis for discrete molecules is that of structural similarity. Similarity is useful for searching and characterizing databases (e.g., clustering). A variety of well-tested similarity metrics are available for discrete molecules. The equivalent concepts of similarity have not been developed for mixtures or sets of molecules. As we obtain significant amounts of data about sets of molecules it is likely that a similarity metric will become valuable; it is currently one of the outstanding challenges of this field.

INDEX